THE QUEST FOR THE ORIGINAL

UNDERDRAWING AND TECHNOLOGY IN PAINTING

SYMPOSIUM XVI
Bruges, September 21st, 22nd and 23rd, 2006

THE QUEST FOR THE ORIGINAL

Edited by
Hélène VEROUGSTRAETE and Colombe JANSSENS DE BISTHOVEN

with the collaboration of
Jacqueline COUVERT and Anne DUBOIS

UITGEVERIJ PEETERS

LEUVEN – PARIS – WALPOLE, MA

2009

The Symposium XVI for the study of underdrawing and technology in painting
was organized by

- the *Laboratoire d'étude des œuvres d'art par les méthodes scientifiques* (UCL)
- the *Centrum Vlaamse Miniaturisten, Illuminare* (KULeuven)
- the *Royal Institute for the Cultural Heritage* (KIKIRPA)

with the assistance of
the *Dienst Stedelijke Musea Brugge* and Hilde Lobelle-Caluwé

The assistance of the *Fonds National de la Recherche Scientifique* is gratefully acknowledged

Members of the *Laboratoire d'étude des œuvres d'art par les méthodes scientifiques*:

Hélène Verougstraete (dir.)
Jacqueline Couvert
Elisabeth Bruyns
Colombe Janssens de Bisthoven

A CIP record for this book is available from the Library of Congress.

FRONT COVER: Hiëronymus Francken II, *La boutique de Jan Snellinck*, bois, 94 × 124,7 cm, Bruxelles, Musées royaux des Beaux-Arts de Belgique, inv. 2628 détail (photo: Élisabeth Bruyns).

BACK COVER: Comparative images of blue (azurite) at different wavelengths (© Images courtesy of the Straus Center for Conservation, Harvard University Art Museums; composite by Molly Faries).

ISBN 978-90-429-2172-6
D. 2009/0602/21

CONTENTS

AVANT-PROPOS

Le volume intitulé *The Quest for the Original* de la série des *Actes des colloques pour l'étude du dessin sous-jacent et de la technologie dans la peinture* est, comme les quinze volumes précédents, consacré en majeure partie à la peinture flamande et dans une moindre mesure à la peinture étrangère et à d'autres secteurs des arts (la miniature, le vitrail, la céramique, les cadres...). Quelques auteurs se sont penchés sur des questions méthodologiques. Exceptionnellement, nous avons inclus certains articles qui ne correspondent pas à des présentations faites lors du colloque.

Pour réaliser cette publication, le travail de relecture des textes a été partagé avec nos collègues de la Katholieke Universiteit Leuven et de l'Institut royal du patrimoine artistique. Nous les remercions pour leur collaboration, en particulier Jan Van der Stock, son assistante Annelies Vogels (KULeuven) et Christina Currie (KIKIRPA). Ainsi se concrétise le projet de partager l'organisation des colloques futurs avec ces deux institutions belges. Le prochain colloque sera pris en charge par Jan Van der Stock à Louvain en octobre 2009. Le thème en sera Van der Weyden. Le Musée communal de Louvain consacrera à cet artiste une importante exposition dans des bâtiments entièrement rénovés, inaugurés pour l'occasion.

Au cours du colloque XVI (Bruges, 21, 22 et 23 septembre 2006), Jan Van der Stock (KULeuven) a dirigé une session consacrée à l'examen technique des manuscrits. Molly Faries (professeur émérite de l'Indiana University, Bloomington et de la Rijksuniversiteit Groningen) a géré une session consacrée aux techniques et équipements de l'infrarouge qui ont connu au cours de ces années des améliorations décisives. Molly Faries est une des rares scientifiques à avoir expérimenté divers équipements et comparé les résultats. Elle a fait une mise au point précieuse que nous sommes heureux de publier dans le présent volume. Qu'elle soit remerciée pour sa contribution! Elle fut un 'pilier' de ces réunions depuis la première heure. Parmi les contributeurs à ce volume, il y a d'autres 'piliers' fidèles que nous remercions. Leurs noms apparaissent souvent en tant qu'auteurs d'articles dans ces actes. Les citer comporterait le risque d'en oublier, d'autant que certains qui étaient présents à la première réunion en 1975 ont pris leur retraite parfois depuis longtemps déjà! Un des buts poursuivis dans ces colloques a été de réserver une plage horaire à des jeunes scientifiques prometteurs. Ce fut le cas cette fois encore, mais cela a obligé parfois à prendre des décisions difficiles.

Pour ce volume, Anne Dubois a réalisé une bibliographie de l'infrarouge. Elle publie ces bibliographies depuis 1991. Nous la remercions pour sa persévérante collaboration.

Les participants à la réunion à Bruges se souviendront certainement de l'équipe d'étudiants de l'UCL qui nous ont aidés. Ils ont été efficaces et le colloque n'aurait pas été le même sans eux. Je songe au défi de présenter tout un colloque en *power point*, sans faille, dans un enchaînement parfait. Quand ils sont partis le premier soir vers leur hôtel, à la veille de la réunion, avec les ordinateurs sous les bras, les disques CD et DVD, clés USB et autres fantaisies compatibles et incompatibles parmi lesquelles bon nombre de 'last minute', ils avaient le sort du colloque entre leurs mains. Pendant trois jours, ils ont pris de nombreuses initiatives, toutes charmantes et bienvenues. Merci à ces étudiants, Guillaume Galoppin, Geoffrey Espel, Cécile Heureux, Sarah Kokot et Sophie Simon, d'être venus jeter l'ancre à Bruges ces jours-là, et d'avoir permis à ce navire de colloque de voguer sur des eaux tranquilles. Il aurait pu sombrer! Nos amis brugeois, il faut le dire, sont pour beaucoup dans le bon déroulement des évènements. La ville de Bruges, ses fonctionnaires de divers services, les conservateurs des musées, les archivistes, les restaurateurs… nous accueillent avec gentillesse depuis si longtemps! Hilde Lobelle est même 'sortie de sa tanière de pensionnée' pour donner encore une fois ce coup de main efficace auquel elle était rodée depuis plusieurs colloques. Peeters a imprimé, une fois de plus à titre gracieux, le programme de la réunion. Etant donné l'enthousiasme que ces réunions génèrent, la très petite équipe du Laboratoire de l'UCL s'est toujours sentie encouragée et soudée pour réaliser le long travail de préparation, pour gérer des centaines de mails et dominer une mise en ordre après, y compris le long travail de la publication des actes. Pour les actes qui voient ici le jour, Colombe Janssens de Bisthoven a — comme pour le volume précédent — dominé avec brio le *lay-out*. Élisabeth Bruyns, plongée dans la rédaction de son doctorat sur l'encadrement au 17e siècle, nous a soutenues par ses encouragements. Elle nous livre pour ce volume un article qui témoigne de l'excellente progression de son travail. Louise Decq, ingénieur diplômée de la KULeuven prise de passion pour l'examen des tableaux, a travaillé deux ans avec nous. Elle nous a aidées par sa présence stimulante. Sous des apparences trompeuses de discrétion, Jacqueline Couvert est la talentueuse chef d'orchestre de toute l'organisation.

Hélène VEROUGSTRAETE

HANS MEMLING'S *PORTRAIT OF A WOMAN*, 1480. HIDDEN INFORMATION?

Hilde Lobelle-Caluwé

Abstract: In 1480 Hans Memling (Seligenstadt before 1440 - Bruges 1494) painted the so-called Sibylla Sambetha, the only extant feminine portrait by his hand. Clothed in the fashion of Memling's time, she is no doubt a member of the higher local bourgeoisie. Nothing allows to identify her. Two texts referring to the prophecy of Christ's coming by the Persian Sibylla Sambetha are late additions (16th or beginning 17th century). One text is painted in a cartouche upper left in the painting. The other is painted on a banderole on the lower rail of the frame. Is the banderole by Memling's hand? Is there some hidden information under text or banderole? Inscriptions on 15th century Netherlandish panel painting are not unusual especially in Gothic writing. Several, executed in a pictorial technique, can be attributed to painters. But the hypothesis of a co-operation between painters and professional scribes should be considered. On several frames of Memling's paintings in the St.John's Hospital texts are written in purely humanistic letterings. These are among the first in the Southern Netherlands referring to professional humanistic calligraphy.

The portrait of the so-called Sibylla Sambetha, now in the collection of the Saint-John's Hospital in Bruges, is considered as the only Memling portrait of an individual woman to have survived (fig. 1).[1] The background seems plain dark, but part of the surface surrounding the sitter is green, a mixture of azurite and malachite. At the bottom a small brown stroke suggests a parapet, painted on the spur of the moment. The shadow of the fingers points to an original creation. The artist painted this small edge probably at the last moment to give an illusion of depth, in a colour possibly imitating stone, in harmony with the colours of the frame. The lady puts both hands on the edge in a natural way. Some fingertips are even painted on the frame.

The young woman is clothed in the fashion that prevailed in the third quarter of the 15th century, in accordance with the date 1480 on the original frame. Costume and jewelry suggest that she is a member of the high bourgeoisie.[2] The frame is decorated with a purplish/blue, grey marbling, painted with thin brush strokes and a back edge painted in black, probably an overpainting.

On account of the inscriptions on the cartouche top left and on the banderole on the lower rail of the frame this portrait is also known as the 'Sibylla Sambetha'. The text on the banderole refers to the prophecy by the Persian Sibyl. Although belonging to the pagan Greek-Roman antiquity, Sybils predicted, inspired by God, Christ's

[1] Oak, 38.5 × 26.5 cm (inside the frame), 46.5 × 35.2 cm (with original frame); inv. 0.SJ174.I. T.-H. BORCHERT, *De portretten van Memling*, Ghent-Amsterdam, 2005, 17, p. 167; D. DE VOS, *Hans Memling, The Complete Works*, Bruges, 1994, 36, p. 168-169.

[2] For a detailed description of costume and jewelry: H. LOBELLE-CALUWÉ, "Vrouwenportret of de Sibylla Sambetha (1480)", in *Catalogus 800 jaar Sint-Janshospitaal*, Bruges, 1976, vol. II, S4, p. 512 and H. LOBELLE-CALUWÉ, *Memlingmuseum, Bruges*, Bruges, 1985, p. 80-81.

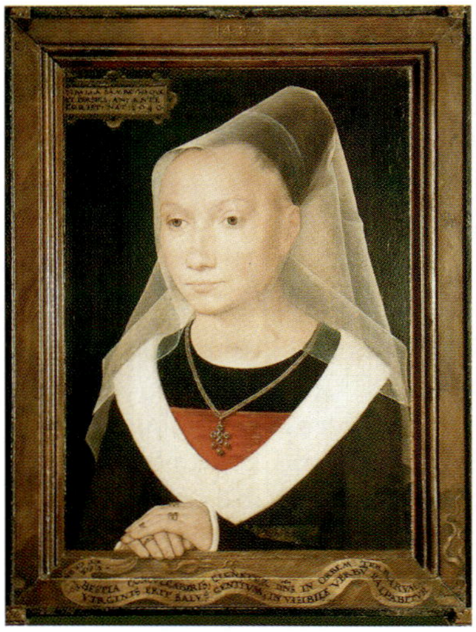

Fig. 1. Hans Memling, *Portrait of the so-called Sibylla Sambetha*, 1480, Bruges, St. John's Hospital.

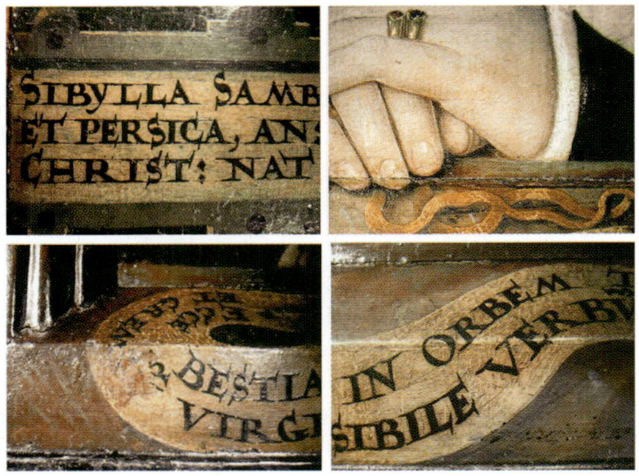

Fig. 2. Hans Memling, *Portrait of the so-called Sibylla Sambetha*, 1480. a. Text on the cartouche. b. Banderole touching the shadow of the fingers. c. Lacuna in the banderole uncovering original painting of the frame and remains of the word GREMIV(M). d. Text on the banderole.

coming.[3] At the end of the 5th century, the name 'Sambetha' appears and was given to the first Sibyl, "a Persian Sibylla". During the Middle Ages and the Renaissance, twelve Sibyls were often represented with clothing of exotic appearance, usually in the company of the twelve prophets from the Christian tradition. This iconography, very popular in Bruges during the 16th century, inspired painters as Albert Cornelis, the Master of the Holy Blood and Ambrosius Benson.

The inscriptions

The text on the cartouche is SIBYLLA SAMBETHA QVAE / ET PERSICA, AN(NO) ANTE / CHRIST(VM): NAT(VM): 2040. On the banderole is written: ECCE, BESTIA, CONCVLCABERIS. GIGNETVR D(OMI)N(V)S IN ORBEM TERRARIUM / ET GREMIV(M) VIRGINIS ERIT SALVS GENTIVM. INVISIBILE VERBV(M) PALPABITVR. Both texts are in Roman capitals.[4] The choice of the words and the syntax are classic Latin. The language is clearly inspired by the Bible. The texts are probably borrowed from one of the comments on Saint John's Apocalypse, more precisely by the description of the woman protecting her new born child against the seven-headed red dragon.[5]

[3] P. HUYSE, "Bildnis einer Dame oder Sibylla Sambetha", in *Orientalia Lovaniensia Periodica 25* (1994), p. 49-59.

[4] The abbreviations are between brackets. The date is 2040 and not Z04O. "*Bestia*" between commas as it is considered as a title. The comma before a new sentence is changed in a point as in medieval manuscripts and early prints a comma was often used instead of a point. We thank professor Gilbert Tournoy, *Seminarium Philologiae Humanisticae*, KULeuven for this information.

[5] According to G. Tournoy, the word *conculcare* appears in the texts of Cicero and other classical authors but is more frequently used in the bible and by the Church Fathers.

Both texts were submitted to technical analysis.[6] The renaissance cartouche, a *trompe l'œil* imitation of metal, is roughly painted in grey on top of the original background with some highlights on the ornaments. The light seems to come from above left, but not systematically. The infrared reflectography reveals no underdrawing. The left part of the cartouche frame seems hidden by the original frame. The black letters of the text are roughly painted on a whitish background. The *textura* lines are painted in black; after execution of the inscription, they were partially hidden with white paint. The roman capitals are all different in size and do not comply with the proper classical system. Almost all the serifs and the lines composing the letters have the same thickness. The *textura* lines are not respected. Identical letters and serifs (S, T) are not uniform and some letters are even joined together. We may conclude that the cartouche and the letters are the work of a painter who did not execute them with precision.

The banderole swings on the bottom rail of the frame from left to right and up aside. Memling painted banderoles in other paintings. Could this one be original, overpainted and hiding original information? Other banderoles on Memlings' works are imitations of *rotuli*, parchment strips with scrolling up ends, quite common in paintings and manuscripts of the time. Here the artist imitates a banner with split ends.

Brown/orange spots, painted with rough brush strokes, can be observed all over the original frame, except in an area around the original date. The technique reminds of the shadows at the end of the banner. Nowhere is the banderole in contact with the fingers. At the right, the banderole just touches their shadow. On the left a small strip is overpainted. Again the whitish background of the banderole is painted with rough brush strokes in different directions. The contours are sloppy, quickly executed. The lines of the *textura* are marked with a pointed instrument in the soft paint layer. Several scratches suggest the place of the letters. The black lines are fragmentary, broken, and even crossing each other. The letters are poorly executed, like on the cartouche. Cartouche, banderole and texts are from the same hand. The more recent spots all over the frame are probably done by the same painter in order to integrate his additions.

Years ago, looking with Hélène Verougstraete at the bottom rail of the frame, we had the impression we discovered some original paint on the left side of the banderole. Could there be some hidden information? We were thinking about initials, or a coat of arms, often represented in Memling's œuvre (fig. 2). But this was not confirmed. The stereomicroscope revealed a lacuna under the overpainting, and some original painting with a few black strokes, remains of the letters IV from the word GREMIV(M), which are fragments of the Sibylla text. The infrared reflectogram assembly does not reveal any underdrawing or other details. The X-radiograph confirms the impression that cartouche and banderole are additions; so far no hidden information could be detected. The identity of the young woman remains an enigma.

[6] With Robert van Nevel we made a technical analysis with the stereomicroscope. The infrared reflectogram assembly is made by Maryan Ainsworth in 1993 and the X-radiograph by Guido Van De Voorde (KIKIRPA Brussels).

A date for the cartouche, the banderole and their inscriptions

Gilbert Tournoy dates the cartouche and the banderole back to the second half of the 16th century or to the beginning of the 17th century, insisting it is difficult to date a text written in capitals. Dirk De Vos attributes the additions on the *Portrait of a woman* to the Bruges painter Pieter Claeissens II (Brugge 1530/36-1612)[7] (fig. 3). He compares the inscriptions with texts on Claeissens'*Allegory of Peace in the Netherlands* (Bruges, Groeningemuseum), painted after 1577. At the bottom of Claeissens'-painting a strip with different inscriptions is divided in six sections with gothic dutch letters, latin roman capitals, french roman capitals and french gothic script.[8] D. De Vos compared the latin texts (3-4) with the inscriptions on Memling's woman's portrait (fig. 4). As the texts on banderole and cartouche seem to be painted by the same hand, we have compared the cartouche inscription with both sections in the strip under Claeissens' painting. The *textura* lines in the *Allegory* are scratched with a pointed instrument like on the banderole of Memling's woman's portrait. However, taking into account the restorations and some overpaints, Claeissens' letters are much more sophisticated. They are carefully written inside the *textura* lines, except for the ascenders and descanters. The tracing of serifs is regular. The size of the letters is uniform. The lines composing some letters vary in thickness. The author of these texts was familiar with the humanistic calligraphy which he carefully applied. It is difficult to believe that Claeissens II painted the cartouche and banderole on Memling's portrait.

Several inscriptions are seen all over the surface of Claeissens' painting. While studying their different types and styles, a fascinating question raised. Did Claeissens II painted the inscriptions himself? Or did he collaborate with a professional scribe? It is well known that painters often painted inscriptions themselves. The irregular word *populus* and other written identifications on the Claeissens'*Allegory* are obviously executed by a painter.

Inscriptions on Memling's paintings

There is no discussion about the authenticity of the date 1480 on the upper rail of the frame of Memling's *Portrait of a young woman*. The letters, different in size, are composed with two brush strokes. On the left a light color and on the right a dark shadow. This surely is the technique of a painter creating the illusion of metal. Several other texts are executed in this technique, like the initials of Jan Floreins on the frame of his triptych, the date and initials on the frame of the *Reins triptych*.[9] On the *Diptych with Maarten van Nieuwenhove*, Memling painted in the left window the motto "IL YA

[7] D. DE VOS, 1994, p. 168 and this attribution seems very convincing to T.-H. BORCHERT, 2005, 17, p. 167.

[8] From left to right: 1-2: Gothic script, Dutch; 3-4: Roman capital: Latin; 5: Roman capital: French; 6: Gothic script: French.

[9] D. DE VOS, 1994, *Triptych of Jan Floreins*, 1479, Bruges, Saint-John's Hospital, inv. 0.SJ173.I, 32, p. 158-161 and *Triptych of Adriaan Reins*, 1480, Bruges, Saint-John's Hospital, inv. 0.SJ177.I, 37, 1994, p. 170-172. For the inscriptions on the frames of the *Triptych of Jan Floreins,* see illustration: D. DE VOS, 1994, p. 371.

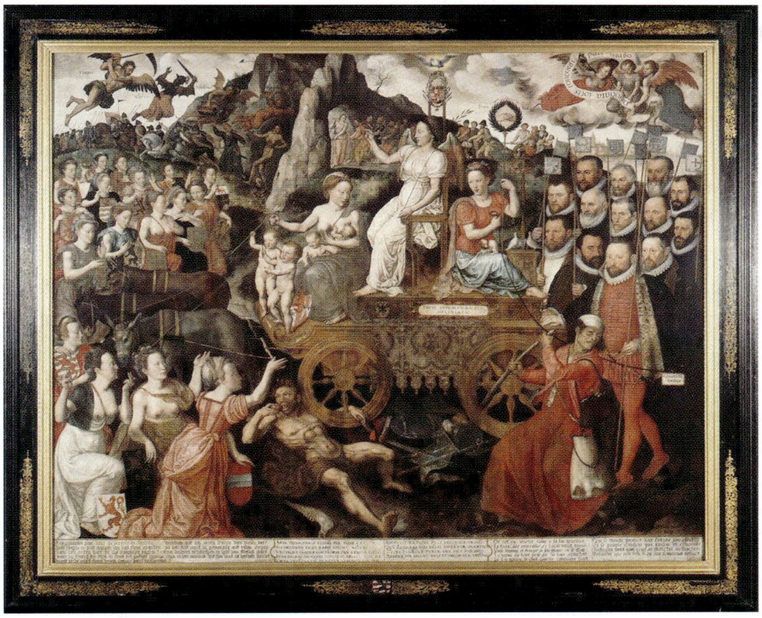

Fig. 3. Pieter II Claeissens, *Allegory of Peace in the Netherlands*, painted after 1577, Bruges, Groeninge Museum.

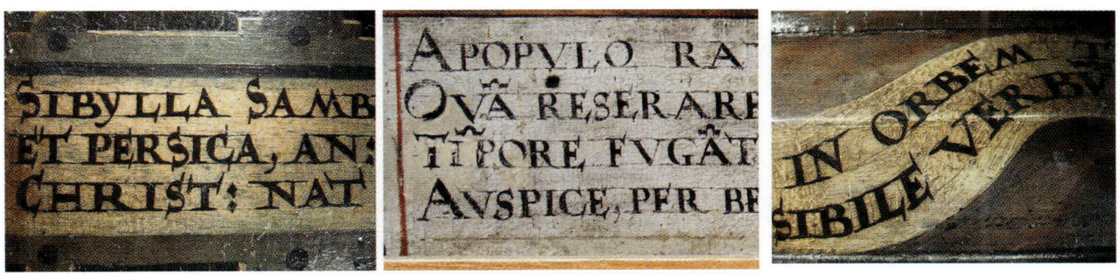

Fig. 4. a and c. Cartouche and text on banderole in Hans Memling's *Portrait of the so-called Sibylla Sambetha*. b. Detail with text in Pieter II Claeissens, *Allegory of Peace in the Netherlands*.

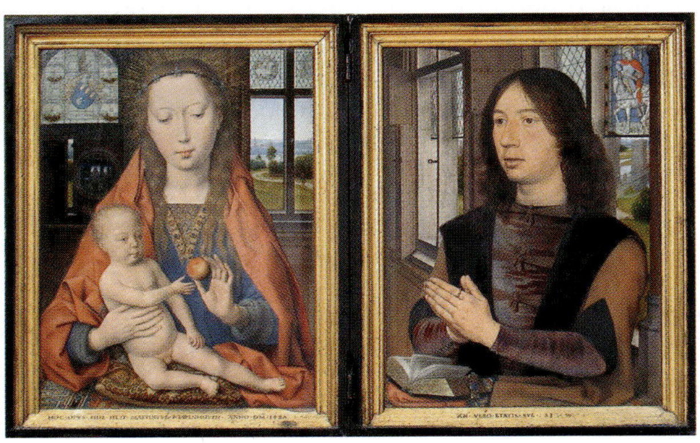

Fig. 5. Hans Memling, *Diptych with the Madonna and Maarten van Nieuwenhove*, 1487, Bruges, St. John's Hospital.

CAVSE", as part of the coat of arms of the sitter[10] (fig. 5). The not really skilled humanistic capitals are painted in gray blue in the same style as the whole window. Recently it has been suggested that Memling himself revised this part of the painting. These texts are by Memling's hand, executed with techniques familiar to a painter.

We can thus compare this 'handwriting' with other inscriptions on Memling's paintings especially on the frames of works in the Saint John's Hospital. The inscriptions on the bottom rails of the frames of the *Triptych of Saint John the Evangelist and Saint John the Baptist,*[11] *Floreins triptych* (both from 1479) and the *Diptych with Maarten van Nieuwenhove* (1487) look very similar. Let us concentrate on the *Diptych with Maarten van Nieuwenhove.* Since the end of the sixties, Memling developed his own style, defined by a spontaneus underdrawing and a simplified, time saving pictorial technique. Between Memling's handwriting, illustrated above with "IL YA CAVSE" and the inscriptions on the bottom rails of these frames, there is a world of difference (fig. 6). The black letters can be compared with the precision and spacing in manuscript writing. No question here of imitating stone or metal. Here we can refer to calligraphy which required labor and time, in contradiction with Memling's mature style. More important is the fact that these texts are among the first purely humanistic letterings in panel painting in the Southern Netherlands.

Collaboration between professional scribes and painters?

We like to discuss the hypothesis that Memling collaborated with a professional scribe for the inscriptions on the mentioned frames. Memling indeed had connections with book craftsmen. Together with the illuminator Willem Vrelant, Memling was a member of the confraternity of our Lady of the Snow. After 1478, Memling painted two wings for Vrelant, an addition to the retable decorating the altar of the guild of the librarians in the Eekhoute Abbey.[12]

We have long been intrigued by the little dragons at the end of both inscriptions on the lower rails of the frame of the Van Nieuwenhove Diptych. We discovered a similar dragon as line ending in the manuscript of Gillion de Trazegnies, illuminated by Lieven van Lathem for Lodewijk van Gruuthuse[13] (fig. 7). The text is attributed to David Aubert, active between ca 1467-1485. Nicolas Spierinc, active between 1453 and 1499, was another scribe working with Lieven van Lathem. We are not to pretending that one of them painted these texts, but the little dragon on Memling's frames is executed in a

[10] J. HAND, C. METZGER and R. SPRONK, *Prayers and Portraits, Unfolding the Netherlandish Diptych,* Washington- New Haven- London, 2006, 26, p. 178-185, p. 292-293, p. 319. Inv. 0.SJ178.I.

[11] D. DE VOS, 1994, p. 150-157; inscriptions on this frame see transcription: H. LOBELLE, 1976, p. 506.

[12] D. DE VOS, 1994, p. 21, 36.

[13] With Professor A. De Schryver (†) we had the permission to study the manuscript during the Gruuthuse exhibition in Bruges 1992. See also M.P.J. MARTENS ed., *Lodewijk van Gruuthuse, mecenas en Europees diplomaat,* 4, *Gillion de Trazegnies, copy attributed to David Aubert, illuminator Lieven Van Lathem, 1464, Chatsworth, Collection Duke of Devonshire and the Chatsworth Settlement Trustees, ms. 7535,* Bruges, 1992, p. 164-169. Such "Dragon like" line endings are not exceptional in manuscript texts. C. JAKOBI-MIRWALD, *Buchmalerei, Ihre Terminologie in der Kunstgeschichte,* Berlin, 1997, p. 51, mentions these as *drolerie-Motiven.*

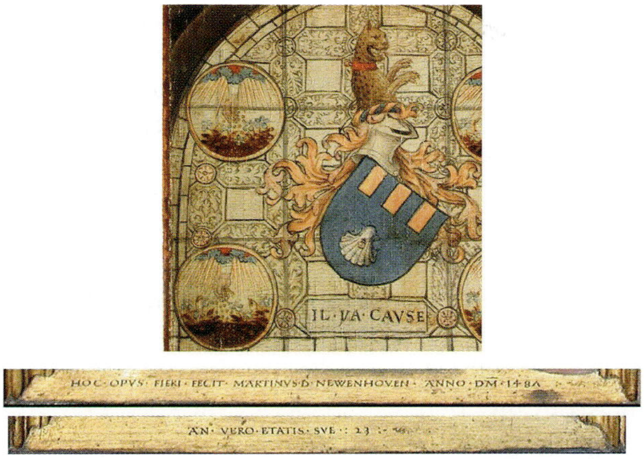

Fig. 6. Hans Memling, *Diptych with the Madonna and Maarten van Nieuwenhove*. a. Coat of arms in the window left of the head of the Virgin. b and c. Texts on the bottom rails of both original frames.

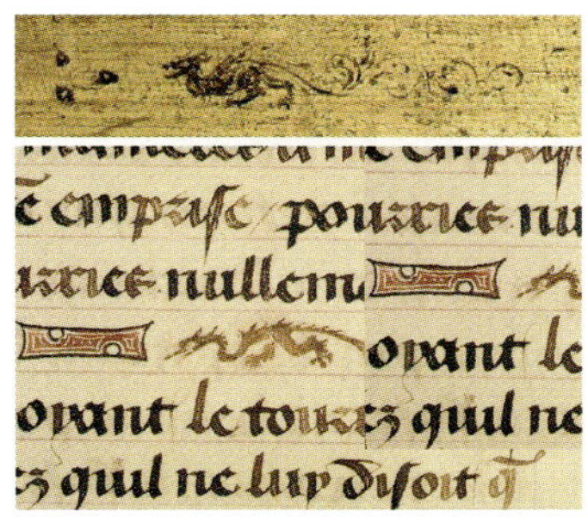

Fig. 7. Hans Memling, *Diptych with the Madonna and Maarten van Nieuwenhove*. a. Detail with the little dragon after the inscription on the frame of the Maarten. b. Little dragon as line ending on f° XXVII v° of the manuscript of *Gillion de Trazegnies*, 1464, Chatsworth (Collection Duke of Devonshire and the Chatsworth Settlement Trustees, ms. 7535).

calligraphic style, possibly with a pen. Memling, and may be other painters, had artistic relationships with scribes. Another clue lies in the relationship between Memling and Jan Crabbe, abbot (1457-1488) of the Ten Duinen abbey in Koksijde which had a 'refuge' in Bruges.[14] Jan Crabbe visited Italy in 1457/8. He was interested in antique literature and Italian humanism. Crabbe surely contributed to introduce humanism in Bruges, already at the end of 1460. In 1473/4 he commissioned a group of manuscripts, copies from Italian humanistic writers as Petrarca and Boccaccio and antique authors as Vergilius, Sallustius, Cicero and Caesar, mostly written in a humanistic or roman letter. This was a *primeur* in Bruges!

Noël Geirnaert illustrated convincingly that these books were made in a co-operation with Bruges parchment-rulers, illuminators and bookbinders. But the texts are written by professional Italian copyists who had established a workshop in Bruges. One

[14] N. GEIRNAERT, "Classical texts in Bruges around 1473: cooperation of Italian scribes, Bruges parchment-rulers, illuminators and book-binders for Johannes Crabbe", in *Transactions of the Cambridge Bibliographical Society*, vol. X, part 2, p. 173-181.

of them, Georgius Hermonymus, working between 1450 and 1510, was native from Florence. The hypothesis that one of them may be the author of the frame inscriptions is credible.[15] Memling also painted the Crabbe triptych for the chapel of the Bruges refuge of Ten Duinen Abbey in the Snaggaardstraat. On the banderole of the angel of the Annunciation decorating the outside of the wings, we see an earlier text in splendid humanistic or roman capitals. This probably is also the intervention of a professional copyist.

Memling worked for many foreign patrons, established in Bruges or visiting the town. Especially the painter's connection with Italians attracts attention. More than probably Memling had come to Bruges after the death of Rogier Van der Weyden in 1464.[16] In 1465 Memling is registered in the 'Burghers Books' of the City. However his name is not mentioned in the registers of the Bruges painter's corporation. He did not purchase a free master status. This may indicate a rather exceptional position. It seems logical that Memling made profit of Rogier's contacts with Italian circle of acquaintances, among them the famous Medicis bankers from Florence, who commissioned paintings by Van der Weyden. A hypothesis is that Memling carried out a commission for Angelo Tani, representing from 1450 till 1464 this Medicis bank in Bruges. Afterwards Tani returned several times to Bruges to complete business deals for the bank. He is represented with his wife Catharina Tanaglia on the famous *Last Judgment* (1467-71). The Tani Triptych is one of Memlings earliest paintings, showing strong affinities with the style of Roger.[17] Memling was also in touch with Tani's successors. He painted several panels for members of the Portinari family. Tommaso Portinari replaced Angelo Tani probably in 1465, the same year Memling arrived in Bruges.[18] An important Italian colony of merchants had settled in Bruges, like the 'nations' from Genoa, but also from Florence and Venice. Bernardo Bembo, a famous Venetian humanist and art lover, stayed in the Netherlands as Venetian ambassador to the Burgundian court between 1471 and 1474. He might have acquired Memling's Bembo diptych at that time.[19] Probably in 1519 the painting is mentioned in his collection in Venice or Verona. Memling's *Portrait of a man with a Roman coin* is supposed to represent Bernardo Bembo.[20]

[15] D. DE VOS, 1994, p. 90-93. N. GEIRNAERT, "Hans Memlings Kruisigingstriptiek voor Johannes Crabbe. Nieuwe gegevens over bestemming en datering", in *Brugge Stedelijke Musea Jaarboek 1987-1988*, Bruges, 1989, p. 174-183. N. GEIRNAERT, "Le Triptyque de la Crucifixion de Hans Memling pour Jean Crabbe, abbé de l'abbaye des Dunes (1457-1488). Témoignage des documents contemporains", in *Memling Studies, proceedings of the International Colloquium*, Leuven, 1997, p. 25-30.

[16] Van der Weyden visited Italy in 1450 and returned in 1451. Considering Memling's stylistic affinity with the Master it is more than probable he was working in his Brussels workshop during several years.

[17] H. LOBELLE-CALUWÉ, *Hans Memling, the success of an artist, Openbaar Kunstbezit in Vlaanderen*, Wetteren, 1994.

[18] Paintings commanded by the Portinari family: D. DE VOS, 1994, 4, p. 82-89; 9, p. 100-103, 11, p. 105-109, 79, p. 284-286

[19] J. HAND, C. METZGER and R. SPRONK, 2006, 25, p. 170-177, 292, 328-319.

[20] H. LOBELLE-CALUWÉ, "Portrait d'un homme avec une monnaie", in *Catalogue d'exposition, Bruges et la Renaissance, de Memling à Pourbus*, vol. II, *Notices*, Bruges, 1998, 5, p. 17.

Conclusion

Inscriptions on 15th century Netherlandish panel painting, especially in gothic writing, are not unusual. Several of them, executed in a typical painting style, can be attributed to the painters. This is the case for many inscriptions in Hans Memling's œuvre. But the first texts in purely humanistic letterings appear on some of his paintings. This calligraphy requires much more time and a specific technique. We suggest that Memling collaborated with professional scribes. He certainly had the opportunity to discover facets of the Italian humanistic culture during his frequent contacts with Italian residents and visitors in Bruges especially from Florence and Venice and through his connections with abbot Crabbe. As interrelationships between painters and illuminators have been illustrated, why not suggest similar relationships with professional scribes?

Hilde LOBELLE-CALUWÉ
hilde.lobellecaluwe@skynet.be

Studied Art History and Archaeology at the RUGent, Art History and Archaeology. 1975-1990: (first) head of the department: museums and archives, OCMW, Brugge. 1990: curator Stedelijke Musea Brugge, Memlingmuseum and Saint-John's Hospital, Museum of our Lady of the Poterie. Organization of different exhibitions on medieval art. Research and articles on Memling. Preparing, in collaboration with colleagues of the Scientific and Conservation Departments, the Corpus Volume on 15th c. panel painting OCMW collection Brugge.

THE ÉVORA ALTARPIECE, A PRELIMINARY REPORT

Maryan Ainsworth and Catherine Metzger

Abstract: The majestic Évora Altarpiece, one of the largest extant early Netherlandish retables, is currently being studied and restored under the auspices of the Instituto Português de Museus (IPM) with Joachim Caetano as the project director, and Maryan Ainsworth and Catherine Metzger as consultants. This discussion of the work in progress will focus on questions of the technique and execution of the thirteen main panels through preliminary results of the wood analysis, underdrawings, and paint layer structure and pigments. In particular these results will be compared with those of other major Netherlandish altarpieces made for export, such as the Cervara Altarpiece and the St. Anne Altarpiece by Gerard David and his workshop.

Commissioned for the choir of the Évora Cathedral in Portugal, the *Altarpiece of the Life of the Virgin* has long been recognized as one of the most important extant Netherlandish retables not only for its fine quality but also for its considerable size. Consisting of twelve panels measuring approximately 2 × 1.10 meters with a central panel of 2.70 × 1.6 meters, this altarpiece altogether comprises around thirty square meters. Initially assigned by Max J. Friedländer to the Master of Évora Altarpiece,[1] the work is actually a collaboration of several artists working together in the style of Hugo van der Goes and Gerard David at the beginning of the sixteenth century.

Currently a major historical and technical study of the altarpiece is underway, including the cleaning and restoration of all of the panels at the Instituto Português de Museus (IPM) in Lisbon. The research program is headed by Joachim Caetano, Director of the Évora Museums, and supported by the conservation and scientific staff of the IPM, along with the consultation of Maryan Ainsworth and George Bisacca of the Metropolitan Museum in New York and Catherine Metzger at the National Gallery of Art in Washington. The intention of this brief discussion is to provide a preliminary account of the initial findings of the research, and the varied directions this research is now taking.

It was likely the most prominent and powerful individual on the scene at the time, that is, Bishop Dom Afonso, who ordered this massive altarpiece for the main altar of Évora's cathedral.[2] There is no known record of the original configuration of the thirteen panels, nor of the six predella panels that were probably added later. At the end

[1] On the question of attribution, whether one or more artists are involved, and a review of the most pertinent literature on this altarpiece, see Didier MARTENS and Hélène MUND, "Autour de la *Pentecôte Rapaert* du Maître des Portraits Baroncelli: modèles et traditions dans la peinture brugeoise à l'aube des temps modernes", in *Jahrbuch der Berliner Museen*, 450 (2003), especially p. 30-31, notes 76-79. See also the master's thesis of Elisabeth AGIUS, *Le Retable de la vie de la Vierge de la Cathédrale d'Évora*, University of Paris-IV, Art and Archeology, June 1987.

[2] Joachim Caetano is presently carrying out in-depth research in the archives on Dom Afonso and will report on his findings at a later point.

of the seventeenth century the altarpiece was taken apart and moved to a new chapel where it was installed by the beginning of the eighteenth century.[3] Recently, Joachim Caetano has suggested a plausible layout for the altarpiece, including the added predella panels (fig. 1), based on the configuration of the paintings in the impressive retable of the *Life of the Virgin* at the church of Santa María La Major de Trujillo that was produced around 1490 by Fernando Gallego and his workshop.[4] According to this hypothesis, the series of scenes would be read in snake-like fashion from the lower left to the right on the first level, right to left on the middle level, and again left to right on the uppermost level, ending with the *Death of the Virgin*. The *Virgin in Glory* would be centrally placed, perhaps above a compartment housing the Eucharist, and beneath a sculpture of the Assumption of the Virgin.[5]

It is generally acknowledged that a number of the thirteen separate compositions of the main altarpiece are related to works by Hugo van der Goes. Some are general adaptations of known works, such as the *Death of the Virgin* related to Hugo's masterpiece in the Groeningemuseum, Bruges, or the *Adoration of the Magi* inspired by Hugo's Monforte Altarpiece in Berlin, Gemäldegalerie. Other of the Évora panels, such as the *Virgin in Glory* or the *Nativity*, are perhaps copied from lost works by Hugo.[6] Additional ones assimilate specific figures and motifs from drawings by or after Hugo, as in the *Annunciation* and the *Birth of the Virgin*.[7] Additionally, five or so are based in part on patterns found in Ghent-Bruges manuscript illumination.[8]

The art of Hugo van der Goes remained popular long after his death in 1482, and was circulated especially widely through the patterns of Ghent-Bruges manuscript illuminators. As it has been discussed recently in the 2003 exhibition catalogue *Illuminating*

[3] The original frames have all been removed, with alterations to the dimensions of each individual panel. Cleaning and restoration of the altarpiece is documented already in 1527 (report forthcoming from Joachim Caetano).

[4] Suggested by Joachim Caetano on 26 September 2006, at the Bruges conference. For a discussion and illustration of this altarpiece, see Pilar SILVA MAROTO, *Fernando Gallego*, Salamanca, 2004, p. 290-335, overall view illustrated on p. 291.

[5] This hypothetical configuration is strengthened on the basis of some technical information about the physical details of the individual panels; for example, the similar alterations in the height and width of pairs of panels as configured by Caetano (fig. 2) that could have been caused as the altarpiece was dismantled, possibly one pair at a time.

[6] The echoes of Hugo's lost work are thought to be found in works such as the *Virgin and Child with Female Saints* by the Master of 1499 in Richmond, in the Virginia Museum of Fine Arts, and the organ-playing angels in Bonkil Altarpiece shutters in the National Museum of Scotland, Edinburgh. On the *Nativity* and the sources of this work, see Lorne CAMPBELL, *National Gallery Catalogues. The Fifteenth Century Netherlandish Schools*, London, 1998, p. 254-259.

[7] For the *Annunciation*, see Stephanie BUCK, *Die Niederländischen Zeichnungen des 15. Jahrhunderts im Berliner Kupferstichkabinett, Kritischer Katalog*, SMPK, Berlin, 2001, p. 137-140. The head of Anna in her bed is taken from the head pose of the shepherdess at the far right in Hugo van Goes's drawing *The Meeting of Jacob and Rachel* in Christ Church, Oxford.

[8] For example, especially close to the scene of the Visitation in the background of the *Annunciation* is the *Visitation* by the Master of the Older Prayerbook of Maximilian I in the *Hours of Queen Isabel la Católica* (folio 115v, Cleveland Museum of Art, 63.256). For other illuminations closely related to this Visitation model, see Patrick DE WINTER, "A Book of Hours of Queen Isabel la Católica", in *The Bulletin of the Cleveland Museum of Art*, 67, December 1981, p. 380-381. Close comparisons for the background scene of Joachim greeted by the angel in *Joachim and Anna at the Golden Gate* are again by the Master of the Older Prayerbook of Maximilian I in the Book of Hours of Queen Isabel (folio 131v) and in a *Book of Hours* in Vienna (Oesterreichische Nationalbibliothek, cod. 1887, fol. 53v), both illustrated in P. DE WINTER, 1981, p. 386.

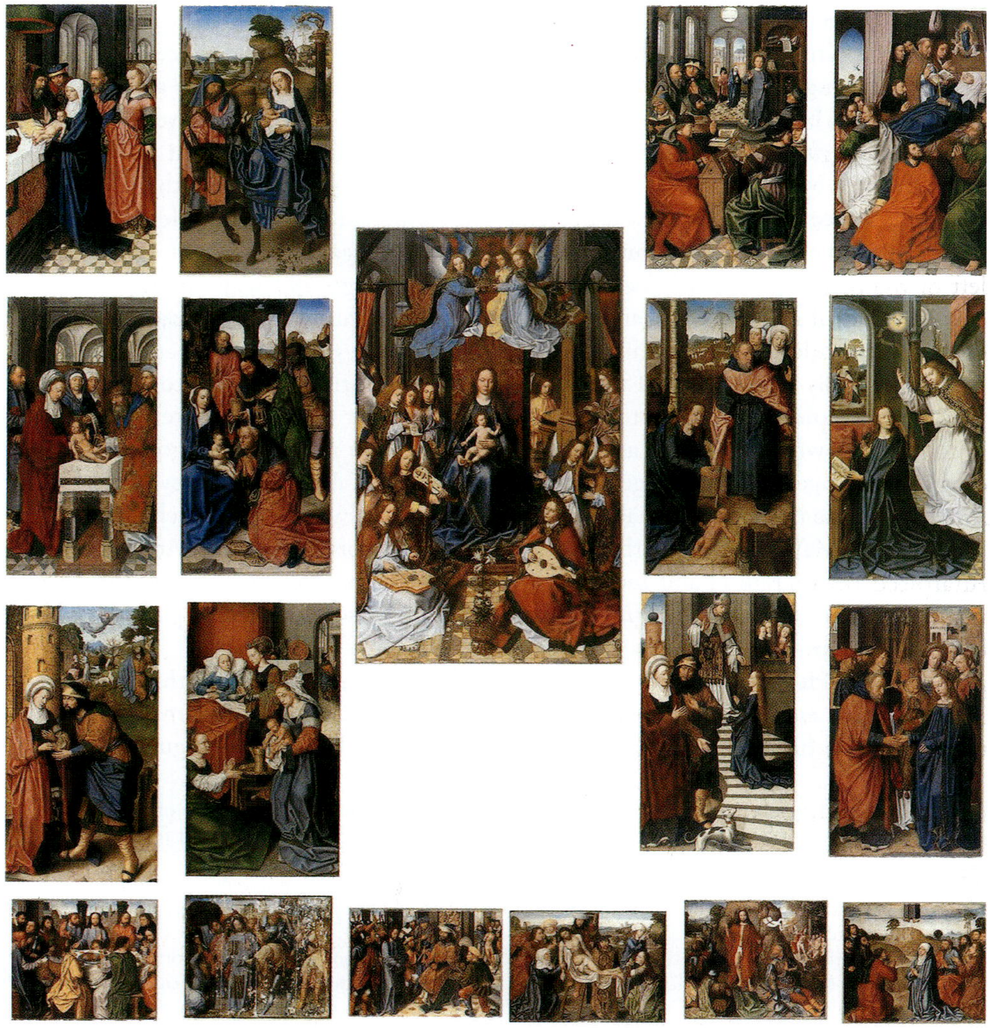

Fig. 1. Workshop or followers of Gerard David, *Évora Altarpiece*, Museum of Évora (© Instituto Português de Museus, José Pessoa).

the Renaissance. The Triumph of Flemish Manuscript Painting in Europe, these patterns most likely reached Bruges and were introduced into the workshop of Gerard David through David's friendship and collaboration with Simon Bening (son of the Ghent illuminator Alexander Bening [the Master of the Older Prayerbook of Maximilian I] and Catherine de Goes, likely the sister or niece of the painter Hugo van der Goes).[9]

[9] *Illuminating the Renaissance. The Triumph of Flemish Manuscript Painting in Europe*, exh. cat., Thomas Kren and Scot McKendrick eds., Los Angeles, 2003, p. 38-39, 44. See also Maryan W. AINSWORTH, "Diverse patterns pertaining to the crafts of painters or illuminators: Gerard David and the Bening Workshop", in *Master Drawings*, 41 (n°3, 2003), p. 240-265.

However, these borrowings (both motifs and entire compositions) from Hugo van der Goes and Ghent-Bruges manuscript illumination for the *Évora Altarpiece* are expressed in terms of the stylistic features of Gerard David's work, in particular the artist's physical types, their attitudes, and characteristic faces. This suggests that the project was carried out either in David's workshop under his aegis, or carefully monitored by one of his closest followers thoroughly practiced in his master's style. Just as in David's *Justice of Cambyses* panels of 1498, the Évora Master inserted portrait heads along with varied types, even in similar combinations of poses. Compare the grouping of heads to the left of the deathbed in the Évora *Death of the Virgin* with the arrangement and poses of portrait heads in David's *Arrest of the Unjust Judge Sisamnes.*[10] Or, consider the heads of the three male bystanders to the left in the *Marriage of the Virgin* with the three at the far left edge of the *Flaying of Sisamnes.*[11] Many of the figures, their facial types, and the emphasis on hand gestures to convey the emotional tenor of the subject have the imprint of Gerard David's art. The poses of the heads of the ladies in the background of the *Education of the Virgin* as well as the assembly of hand poses in the *Joachim and Anna at the Golden Gate* are directly related to Gerard David's drawing of *Heads and Hands* in the Louvre, Cabinet des Dessins.[12] The painters involved with the Évora Altarpiece must have had at their disposal compositional drawings, patterns of individual figural and drapery motifs and sketchbooks of studies of hands and attitudes of heads, such as those that are so well known by Gerard David. Patterns for both panel painting and illumination were readily available in David's studio; they are mentioned in the lawsuit brought against David by his assistant Ambrosius Benson in which a trunk described to have such patterns was the focus of the dispute.[13]

Beyond these stylistic features, there is also another connection to David, who around the same time (that is, at the turn of the century) was overseeing and personally involved with the execution of other large altarpieces commissioned for sites in southern Europe. These are, namely, the *Cervara Altarpiece* commissioned by Vincenzo Sauli in 1506 for the abbey church of San Gerolamo della Cervara,[14] and the *Saint Anne Altarpiece* of ca. 1510, made for export to an unknown site, possibly Mallorca in Spain.[15]

Like the *Évora Altarpiece*, the *Saint Anne Altarpiece* shows compositions in its predella paintings that were shared with miniaturists who included them in important illuminated books of the time.[16] Both altarpieces also reveal more than one hand involved in the

[10] M.W. AINSWORTH, *Gerard David. Purity of Vision in an Age of Transition*, New York, 1998, p. 58, fig. 65.

[11] M.W. AINSWORTH, 1998, p. 59, fig. 66.

[12] M.W. AINSWORTH, 1998, p. 16, fig. 14.

[13] On this dispute see R.A. PARMENTIER, "Bescheiden omtrent Brugsche schilders van de 16e eeuw. I. Ambrosius Benson", in *Annales de la Société d'Emulation pour l'étude de l'histoire et des antiquités de la Flandre*, 80, 1937, p. 92-94.

[14] See M.W. AINSWORTH, 1998, p. 179-201, and Clario DE FABIO et al., *Il Polittico della Cervara di Gerard David*, Genoa, 2005.

[15] M.W. AINSWORTH, 1998, p. 167-177. For the most recent discussion of the *Saint-Anne Altarpiece*, see Martha WOLFF, General Editor, *Northern European and Spanish Paintings before 1600 in the Art Institute of Chicago, French, German, Netherlandish, Spanish, and English. A Catalogue of the Collection,* New Haven and London, 2007, p. 73-80.

[16] P. DE WINTER, 1981, p. 401-402, 418, and John Oliver HAND and Martha WOLFF, *The Collection of the National Gallery of Art Systematic Catalogue, Early Netherlandish Painting*, Washington, 1986, p. 73-74, note 20.

production of the paintings, one of these hands suggesting a master's corrective intervention.[17] Moreover, the two altarpieces share certain stylistic features. A few examples here are typical of the common models from which the two altarpieces draw. Take, for instance, the facial type of the priest from the Évora *Marriage of the Virgin* compared with the head of *Saint Nicholas* from the *Saint Anne Altarpiece,* or the various female figures of the Évora panels that relate directly to the style and rather wooden quality of the Virgin and Saint Anne in the *Saint Anne Altarpiece.* Quite possibly some of the same painters of the predella panels of the *Saint Anne Altarpiece* may also have worked on the Évora panels.[18]

Because of the stylistic connections between the *Évora Altarpiece* and the *Saint Anne Altarpiece,* we are currently focusing on specific aspects of their manufacture in order to try to discover closer links between the two. We hope that this will ultimately help to determine where and by whom the *Évora Altarpiece* was made.

Some technical information from the *Évora Altarpiece* is already available. As with David's other panels made for export, the Évora panels were painted on oak from the Baltic region. This fact, however, does not necessarily determine the place of manufacture of the panels, for already in about 1469 late-gothic Portuguese paintings such as the *Panels of São Vicente* by Nuno Gonçalves were executed on imported northern oak.[19] So far, the tabulated dendrochronology results of Lília Esteves indicate that *Christ Among the Doctors* provides the most recent felling date of 1467.[20] Although some of the panels were later cut in length or width due to fire damage and reframing, the original width of the panels can be deduced by reference to the *Christ Among the Doctors* and the *Death of the Virgin,* which retain their original widths.[21] Although the panel supports are rectangular, the upper corners are left unpainted, and there is a partial arc of paint remaining in one or both corners of each panel. These arcs conform to a pair of circles of equal dimensions that intersect at their midpoints (fig. 2).[22] Of special interest to us concerning where the paintings were made is the unusual thickness of the planks (the thickest is 3.9 cm for the *Flight into Egypt*). Unlike Northern carpenters, Portuguese carpenters often fashioned radial cut oak boards in substantial thicknesses, as much as 4.1 cm.[23] The thick planks are joined with butt joins with rectangular wooden

[17] O. HAND and M. WOLFF, 1986, p. 69, 70, fig. 2, and M.W. AINSWORTH, 1998, p. 169-171.

[18] The connection of the Évora Altarpiece panels with the painter of the centerpiece of the National Gallery of Art *Saint Anne Altarpiece* was first suggested by Matthias WENIGER, "Provost and Portugal", in *Le Dessin sous-jacent et la technologie dans la peinture, Colloque XII (11-13 September 1997). La Peinture dans les Pays-Bas au 16ᵉ siècle. Pratiques d'atelier. Infrarouges et autres méthodes d'investigation,* Hélène Verougstraete and Roger Van Schoute eds., Leuven, 1999, p. 49, note 2. The facial types of the females in several of the Évora panels is very similar to the female faces in the St. Nicholas predella panels, for example the girls in the *Dowry* and the face of the Virgin in the *Birth of Saint Nicholas.*

[19] M. WENIGER, 1999, p. 49, note 1. In fact, 82% of Portuguese panels are painted on oak. Jacqueline MARETTE, *Connaissance des Primitifs par l'étude du bois du XIIᵉ au XVIᵉ siècle*, Paris, 1961.

[20] Verbal communication from Lilia Estevez at our joint meeting on 3/2006.

[21] Determined by George Bisacca and Miguel García at our Lisbon meeting in 3/2006.

[22] None of the panels retain their original upper edge. The diagram assumes it would have been a flat section connecting the tops of the circles (fig. 2), indicated with a dotted line. It is of course possible that the uppermost contour included an ogee arc or some other element that would have added height to each panel. A dotted line at the bottom of the diagram corresponds to the greatest current panel height, which would be the minimum original height.

[23] José MENDES, "Intervençao de Conservaçao e Restauro", in *Retábulo de Ferreira do Alentejo*, IPC, Lisbon 2004, p. 46. Jorgen Wadum gives the average thickness of northern panels as between 0.8 and

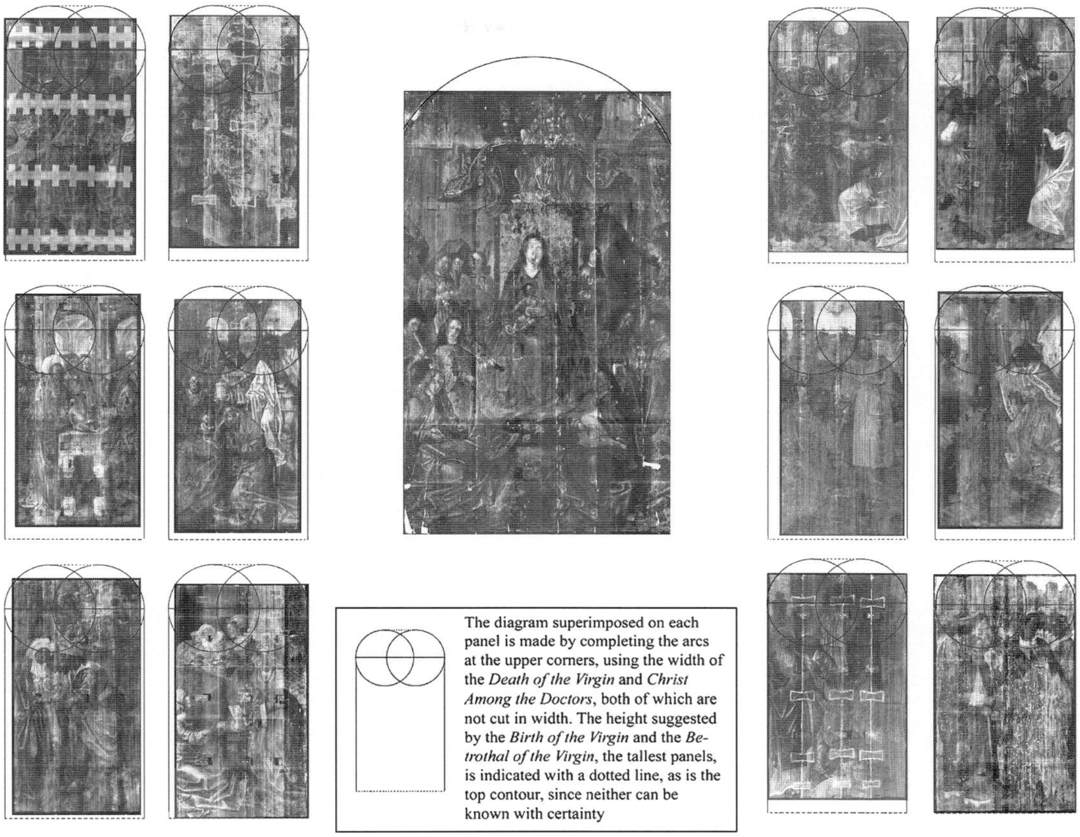

Fig. 2. Photoshop reconstruction of the Évora Altarpiece to show hypothetical original dimensions, as based on analysis of the X-radiographs. The original width of the panels is based on the widths of *Christ among the Doctors* and the *Death of the Virgin*, both of which have their original width. Continuing the radius of the unpainted arcs in the upper corners with two circles that intersect at their midpoints gives an idea of the upper profile. The minimum height is taken from the greatest panel height, assuming the top center of each panel was a flat line connecting the two circles (© X-radiographs, IPM, Divisão de Documentação Fotográfica).

tenons distributed along the length of the join. Each wood tenon is pinned with a pair of round wood pins, except in the large central panel, in which the tenons are pinned with four pins (fig. 3). The use of tenons with pins is seen only exceptionally

3 centimeters in "Historical Overview of Panel-Making Techniques in the Northern Countries", in *The Structural Conservation of Panel Paintings*, K. Dardes and A. Rothe eds., 1995, p. 154. Roger van Schoute and Hélène Verougstraete say a thickness of 0.7 and 1.5 is general, with exceptions such as Rubens' *Descent from the Cross*, twice the dimensions of the largest Évora panel, being 3.5 cm. J.R.J. van Asperen de Boer, Molly Faries and Jan-Piet Filedt-Kok give 0.5 to 1.5 as the normal thickness of Northern planks, with larger paintings sometimes on planks as thick as 2 to 3 cm, in *Kunst voor de Beeldenstorm*, Amsterdam 1986, p. 107. Peter Klein, in an email dated 4/19/07, reports that smaller Portuguese panels have about the same thickness as Netherlandish panels (around 1 cm). Bigger ones often have a thickness up to 2 cm and the large ones — as for example the Tomar altarpiece — a thickness up to 4 cm.

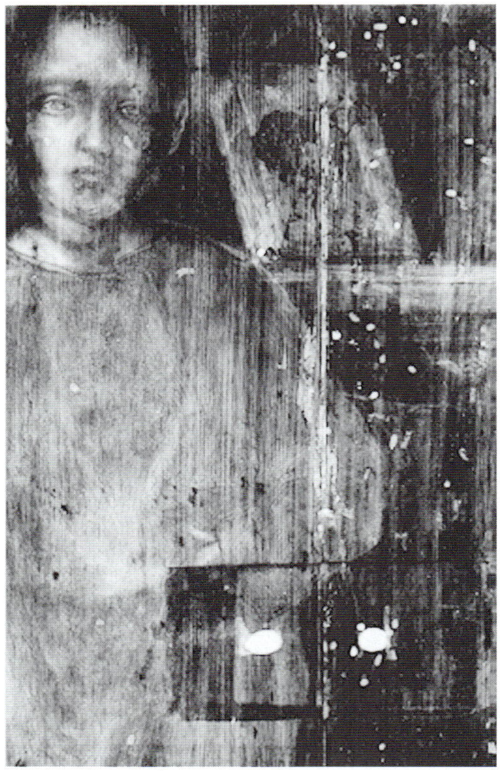

Fig. 3. Detail X-radiograph showing the pinned tenon in a join of *Christ Among the Doctors* (© IPM, Divisão de Documentação Fotográfica).

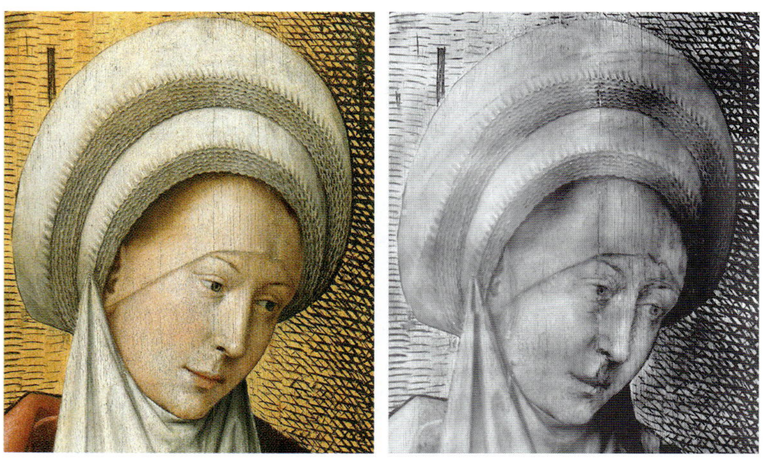

Fig. 4. Detail and underdrawing showing pouncing in the head of Anna in *The Meeting of Joachim and Anna* (© photo and IRR, IPM, José Pessoa).

in Northern carpentry, but is documented in other Portuguese-manufactured panels.[24] In most of the panels, with the exception of the *Adoration of the Magi*, the tenons are arranged in horizontal rows across the panel. Measurement of the spacing between the rows of tenons in combination with inspection of the mortise carpentry suggests that there were several carpenters involved in producing the altar panels. It is not known if these several carpenters worked in a common shop, or if the commission was distributed among a few shops working in concert. There is a partial burin mark on the reverse of *The Death of the Virgin*, but as it is cut off in the middle we cannot identify whether it is a wood-quality mark, a shipper's mark, or a forest mark.[25]

Some of the panels show unpainted edges, but none have a barbe, indicating that they were painted outside of their frames. Although the carpentry conventions intimate Portuguese manufacture, there is the question of the ground preparation. Preliminary results from samples of the white ground indicate the presence of both calcium carbonate and calcium sulfate.[26] The occurrence of both calcium carbonate and calcium sulfate in the ground has been reported in the Cervara Altarpiece[27] and is noted as well in the three panels of the Christ as *Salvator Mundi amongst Musical Angels* (the Nájera panels) by Memling in Antwerp.[28] We need to discover how much of each substance is present, and whether these are mixed or layered applications.

Among the interesting features of the Évora panels is the variety of underdrawing types, offering the possibility of various interpretations. To date we have results on only five out of the thirteen panels. Whether or not these preliminary results are representative of the entire group, we will not know until all of the data is available.[29] Two of the paintings, namely the *Meeting of Joachim and Anna* and the *Death of the Virgin*, show pouncing as well as letters for color indications (figs. 4, 5). This is notable since these two paintings show interventions by another, more accomplished hand (fig. 6). Was the pouncing carried out in order to be followed in the preliminary work up of the layers by workshop assistants? In *Joachim Meeting Anna at the Golden Gate,* the pounced pattern is not followed exactly, but changed slightly in the paint layers, as we can see with the headdress of Anna (fig. 4). Here also, the artist worked in the indications of the

[24] The panelmakers for Hugo van der Goes and Barent van Orley sometimes used square join reinforcements, but pinned them at the four corners with four pins, in Hélène VEROUGSTRAETE and Roger VAN SCHOUTE, *Scientific Examination of Easel Paintings*, 1986, p. 21. The panels made for the *Saint Anne Altarpiece*, which are also rectangular but painted with arches in the upper corners, are constructed with the more traditional dowelled butt join. The literature on Portuguese panels is scarce, but the panels made for the Charola of Tomar are joined with tenons pinned with four pins in Pedro REDOL, *Pintura da Charola de Tomar*, IPC, Lisbon, 2004, p. 33-343, ill. p. 67. The pins found in the Portuguese panels are larger in circumference than those used by Northern carpenters (author's observation).

[25] These linear scratches, seen on the far right edge of the panel as viewed verso, were put on the cut timber before it was shipped from the Baltic using '*merkynger yrons*'. Michael REIF, "Engraved Marks on Baltic Wainscot Boards", in *Constructing Wooden Images*, Brussels, 2005, p. 127-146; Tomasz WAZNY, "The origins, assortments, and transport of Baltic timber", in *Constructing Wooden Images*, Brussels, 2005, p. 115-126; Jorgen WADUM, "Historical Overview of Panel-Making Techniques in the Northern Countries", in *The Structural Conservation of Panel Paintings*, Los Angeles 1995, p. 162.

[26] Determination of the ground samples was made by the scientists at IPC.

[27] Giorgio LUCIANO et al., "Analisis chimiche su campioni prelavati dal polittico della Cervara, Palazzo Bianco, Genova", in Clario De Fabio, *Il Polittico della Cervara di Gerard David*, Genoa, 2005, p. 88.

[28] Verbal communication from Lizet Klaasen 11/2003.

[29] José Pessoa of IPC is carrying out this work.

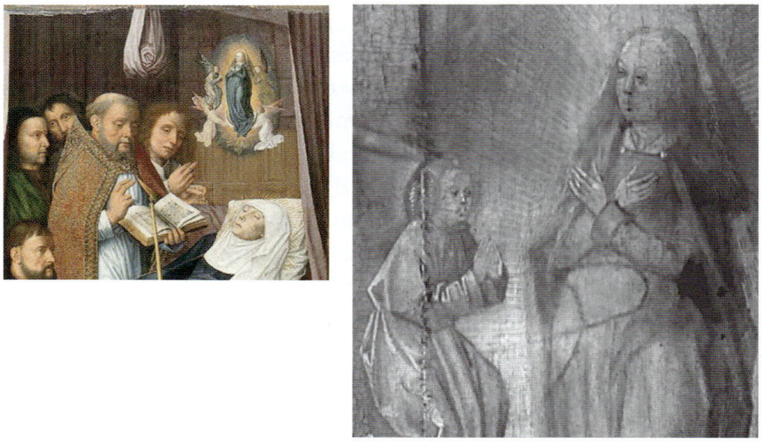

Fig. 5. Detail and underdrawing showing the letter 'P' for purple in background of *The Death of the Virgin* (© photo and IRR, IPM, José Pessoa).

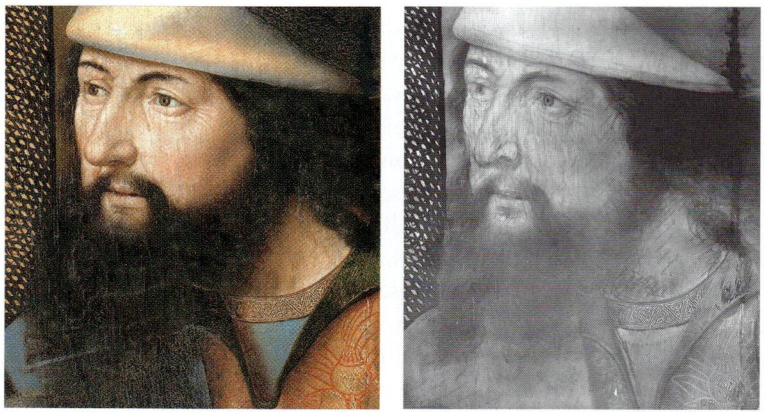

Fig. 6. Detail and underdrawing of head of Joachim in *The Meeting of Joachim and Anna* (© photo and IRR, IPM, José Pessoa).

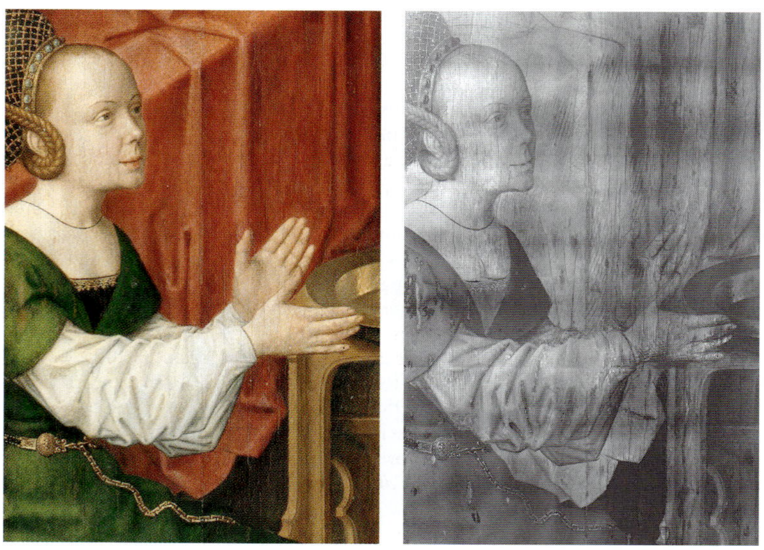

Fig. 7. Detail and underdrawing of the red drapery in *The Birth of the Virgin* (© photo and IRR, IPM, José Pessoa).

features of the face more freely. The head of Joachim is differentiated by the loose but entirely assured and confident handling worked up in black chalk, not in brush, as is the tool for most of the rest of the underdrawing in the panels (fig. 6). The underdrawing under the head of Joachim is very similar to the underdrawing under the heads in Gerard David's *Justice of Cambyses* panels, and possibly indicates an intervention here by the master of the workshop in the attention to this especially sensitively rendered head.[30]

One might also raise the question of the pouncing in these panels and their relationship to compositions copied from those of Hugo van der Goes. Does the pouncing of some of the underdrawings indicate a direct relationship to lost works by Hugo? An interesting observation here is that the one extant example where Gerard David employed pouncing on such a large scale was in the Munich, Alte Pinakothek, *Adoration of the Magi*, which is also considered a copy after a work by Hugo.[31] Did some of Hugo's patterns ultimately land in David's workshop for continued use?

The *Annunciation*, *Birth of the Virgin*, and *Virgin in Glory* do not show any signs of pouncing and also, except for one example in the *Birth of the Virgin* (fig. 7), are free of color indications. The brush underdrawing in these cases appears to be freehand and well assured. The parallel hatching in the draperies in the *Virgin in Glory* and the *Birth of the Virgin* is multi-directional, rather planar in effect, and may have been accomplished by the same hand. Here loops denote the eyes of the female figures — that is, another characteristic these few paintings have in common with each other as well as with the Washington *Saint Anne Altarpiece* panels.

Cross-sections have been taken in two campaigns; the first, to inform on the restorations, and later to categorize the painting method. A white *imprimatura* layer is found in every panel, although it is not present in all cross-sections. The distribution of the *imprimatura* may be either the result of rapid, random brushwork, or, conversely, the frequent lack of *imprimatura* under the Virgin's flesh and its ubiquitous presence under blues may be deliberate. The underdrawing is found both over and beneath the *imprimatura.* No cross-sections have been taken to locate the pouncing with respect to the *imprimatura*, or to the underdrawing where both exist on one panel. Careful mapping of the preparatory designs with respect to the *imprimatura* might elucidate the division of labor.[32] The majority of cross-sections have two or more original layers, which is like the panels of the *Saint Anne Altarpiece* completed by the workshop. Blues consist of two layers of mixed pigments over a layer of white *imprimatura*, always incorporating azurite and red lake. The particles of azurite are exceptionally large, and many of the purple-blues consist of red lake and azurite without lead white incorporated in the paint mixture, a feature noted

[30] This intervention for the head of Joachim is similar to that in the central panel of the *Saint Anne Altarpiece* where corrections are made with a dark irregular line using charcoal over the already underpainted sleeve of the Virgin, which had initially been underdrawn with a different material. Barbara BERRIE and Catherine METZGER, "Gerard David's *Saint Anne Altarpiece*: Evidence for Workshop Participation", in *Historical Painting Techniques, Materials, and Studio Practice*, Arie Wallert, Erma Hermens and Marjorie Peek eds., Leiden, 1995, p. 127-134, fig. 5.

[31] M.W. AINSWORTH, 1998, p. 47-51.

[32] It is interesting to note that when the panels are arranged as suggested by Caetano, the leftmost column and the central panel all have the *imprimatura* under the drawing, while in the remainder of panels in which the underdrawing/*imprimatura* relationship is known the *imprimatura* lies over the drawing. Three panels have no cross-sections with both *imprimatura* and underdrawing represented.

in the *Saint Anne Altarpiece* as well.[33] The expense of the ensemble is confirmed by the admixture of ultramarine blue in the upper layer of paint for the Virgin's garments. Another connection with Gerard David's shop is the use of malachite green.[34]

Based on the technical evidence that has thus far become available, it now appears most likely that the Évora Altarpiece was produced in situ. A small group of Netherlandish and locally based artists must have worked together under the supervision of a master who was trained in the style and technique of Gerard David in Bruges. All that we can say definitively is that there are several hands involved with this production. The carpentry bears a close resemblance to carpentry in altarpieces constructed in Portugal at the time. The underdrawing and the painted layers point to collaborative work, both within single compositions and among the several panels. The pigments used and the multilayered paint application point to a well funded and somewhat laborious production. Occasionally, there appears to be the intervention of the master as in the head of Joachim, the head of the Virgin in the *Virgin in Glory*, and changes in the distribution of the heads and figures in the *Death of the Virgin.* Other conclusions will be possible as more of the technical material is made available.

Maryan AINSWORTH
Metropolitan Museum of Art
European Paintings Department
1000 Fifth Avenue
USA – New York 10028
Tel: 212.396.5172
Fax: 212. 396.5052
maryan.ainsworth@metmuseum.org

Received her B.A. and M.A. degrees in the History of Art from Oberlin College and her Ph.D. from Yale University, specializing in northern Renaissance art. Since 1977 she has been employed at the Metropolitan Museum of Art in New York where she is curator, Early Netherlandish Painting. She is also Adjunct professor of Art History at Barnard College and Columbia University.

Catherine METZGER
National Gallery of Art
6th and Constitution
USA – Washington DC 20565

Senior painting conservator on the staff of the National Gallery of Art, Washington. During her sixteen years at the Gallery she has completed research and restorations on numerous Northern Renaissance paintings, including panels by Petrus Christus, Hans Holbein, and Gerard David. Together with Elizabeth Walmsley, John Delaney and Colin Fletcher, she completed initial studies into the use of solid state technology for infrared studies of painting.

[33] Barbara Berrie, Senior Scientist at the National Gallery of Art, confirms this observation. Very large particles of azurite are noted in the paintings in the Charola de Tomar by an unknown fifteenth-century Portuguese artist. P. REDOL, 2004, p. 45.

[34] L. CAMPBELL, 1998, p. 30. Only Bouts and David group paintings contain malachite.

THE PASSION OF CHRIST (1470-1490) IN THE MUNICIPAL MUSEUM OF LEUVEN. PRELIMINARY RESEARCH INTO A PAINTING BY A BRABANT MASTER

Veronique Vandekerchove

Abstract: The panel painting with scenes of the *Passion of Christ* is a recent acquisition (1998) of the Leuven municipal museum. It is an interesting example of the Passion cycle as a continuous narrative, from the entry into Jerusalem to Christ's apparitions to Peter and Mary Magdalene. The technical study provides the starting point for investigating the function and the context of the work of art, the latter focusing on the connection with late medieval Passion devotion in liturgy, literature and art. More specifically, the research deals with the spiritual pilgrimage, the Way of the Cross and the accompanying image meditation. Examination of the painting with infrared reflectography reveals that the whole composition has been underdrawn.[1] The larger scenes especially show a fine and meticulous underdrawing. Figures and faces are outlined. Diagonal hatching is used to indicate shadows and draperies. This is more refined in human figures and horses. Some details have been altered, but the general composition and the relation between the scenes seem to have been carefully studied and established before starting the drawing. The next step will be to compare the underdrawing with these of contemporary Brussels masters from the second half of the fifteenth century. This, together with the search for models, might help establish the origins of this work of art.

The *Passion of Christ* (fig. 1) belonged to a private collection and had only been shown to the public on one occasion, at the exhibition held in Liège in 1905 to mark the 75th anniversary of Belgium.[2] Presented to the Leuven City in 1998, it was acquired and joined the permanent display of late Gothic art in the Vander Kelen-Mertens Museum.

The painting is in good condition. The oak panel (101.7 × 148.6 cm) is made up of three boards. The boards are trapezoid in shape. As a result, the joins are not parallel. The reverse of the panel is smooth with slight variations in level between the boards and traces of earlier cross-members, now removed. Unpainted edges and 'barbes' appear all around. The original frame is missing.

The image is a splendid example of a *Gesammtbild* or continuous narrative: a number of actions occurring at different moments but involving much the same characters, presented together in a unified space to be observed from a single viewpoint. This creates a spatial representation of a sequence of events. Together with the time factor, this device allows the viewer to envision the Passion by moving through imaginary, depicted rooms and spaces that remind him of his own familiar surroundings. Active observation results in personally experiencing that sacred space and time.[3]

1 The Royal Institute for Cultural Heritage (IRPA-KIK) registered the full composition on site in the Leuven museum with the InfraCAM SWIR (short-wave infrared reflectography video camera) from Inframetrics (now FLIR Systems).

2 *Exposition de l'art ancien du Pays de Liège*, Liège, 1905, cat. 1002 (exhibition catalogue).

3 L. Andrews, *Story and Space in Renaissance Art. The Rebirth of Continuous Narrative*, Cambridge, 1998, p. 3-33; J. Lubbock, *Storytelling in Christian Art from Giotto to Donatello*, London, 2006, p. 282-284.

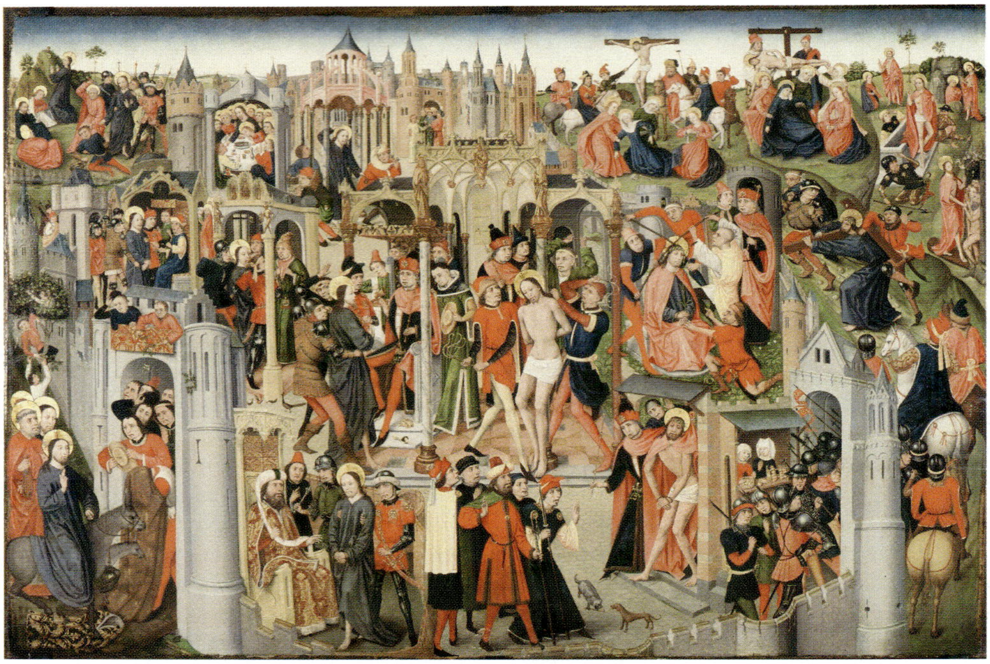

Fig. 1. Anonymous, *Passion of Christ*, oil on panel, 1470-1490, M Leuven (Municipal Museum of Leuven), inv. S/384/O (© M Leuven, photo: Eric Dewaersegger).

This Passion cycle is mainly based on a combination of the accounts given in the four gospels. It is a fine case of a late fifteenth-century narrative representation of the Passion amplified with vivid details of everyday life. The architecture is an element both of separation and union. Each event is isolated within a confined interior or exterior space. The structure of the city itself and its surroundings unifies spaces and events.[4] Jerusalem is represented as a typical western European city in the Middle Ages. This makes it easier to convey the episodes of the Passion.[5] The story unfolds across the panel. Rather than following a linear path, the action shifts back and forth. The figures move from one position to the next and from place to place as if on stage with the architecture serving as the set.[6] They move through space as well as time. The Holy City itself is a central constituent. To visit the place where Christ's suffering and sacrifice took place was the supreme goal of every Christian pilgrim. Paintings such as this helped the viewer to make an imaginary journey; entering the pictured space as in a real world, participating in the sequence of events, walking in Christ's footsteps, experiencing the Passion; each event takes on its full meaning in the context of the whole.

[4] See also: D.J. VAN DEN BERG, "Spectators in Jerusalem: urban narrative in the scenic tradition", in *Image & Narrative. Online Magazine of the Visual and Narrative*, Aug. 2001.

[5] M. BOTVINICK, "The painting as Pilgrimage: Traces of a Subtext in the Work of Campin and his Contemporaries", in *Art History*, 15, 1 (1992), p. 1-18.

[6] L. ANDREWS, 1998, p. 79.

The painting can be related to the Passion plays, to the Way of the Holy Cross and to prayer books, all of them instruments for experiencing Christ's Passion. A particular feature of the Passion painting is the gold Gothic lettering labelling each episode. It must refer to a manuscript or another medium relating to the Passion or to a guide for the completion of a spiritual pilgrimage.[7]

The entire composition is underdrawn with a brush and a liquid medium. The lines vary in width and density, with a drop at the end of a stroke. The style of the underdrawing suggests that the same hand has worked throughout the whole panel. The drawing of folds in draperies is executed in long, confident strokes, with indications of shade. In the entry into Jerusalem (fig. 2) for example, brush strokes are visible in Christ's garment and in the donkeys. The bearing of the cross also shows brushwork in the draperies. Quite thick lines can be seen in the canopy in the centre left. The artist modelled the bodies with very fine and carefully drawn lines. The most expressive example is found in the central scene with the flagellation, where Christ's body and that of one of the executioners show both thick and fine lines (fig. 4).

In the *Ecce Homo*, both Jesus and Pilate are elaborately underdrawn. The contours are either formed by a thick line or by several sketchy lines, and shadows by coarse diagonal hatching. The painter struggled with the figure of Christ. To indicate Christ's twisted torso and arms, he used abundant lines (fig. 3). He had difficulties in rendering the right arm. The result is far from perfect both in the underdrawing and the painted layer.

A series of underdrawn, not overlapping, horizontal semi-circles is visible all along the sky area. Their lower extremities are indicated with incised points, probably due to the use of proportional compasses. Two of these semi-circles show more elaborate interior detail and an extra outside semi-circle. They are probably indications of architectural details in the frame of the painting, designed to set off and complement the interior architectural canopy in the centre left of the composition. The painter did not execute his project in the paint layer.

Shadows in faces and bodies are marked with parallel hatchings, ranging from diagonal and vertical to horizontal. The underdrawing of the flagellation shows fine vertical and diagonal lines in the legs of Jesus; likewise, in the coronation, the right leg of the mocking man displays almost vertical hatching. The underdrawn modelling of Christ's body in the flagellation (fig. 4) forms the basis for the delicately painted flesh. The thick, diagonal hatchings in the architecture are less refined.

Changes were made during underdrawing but rather exceptionally during painting. The head of the donkey in the entry into Jerusalem has been modified during drawing. On the throne of Pilate, a form suggesting a coat of arms is completely underdrawn but only partially visible in the paint layer. The step, the foot of Christ and of Pilate

[7] M. SMEYERS, "Taferelen uit de Passie van Christus", in M. SMEYERS et alii eds., *Dirk Bouts (ca 1410-1475), een Vlaams primitief te Leuven*, Leuven, 1998, p. 485-488 (exhibition catalogue); V. VANDEKERCHOVE, "A Journey to Jerusalem. An unknown Passion of Christ (Leuven, Vander Kelen-Mertens Museum)", in B. CARDON, J. VAN DER STOCK en D. VANWIJNSBERGHE eds., *Als ic can. Liber Amicorum in Memory of Professor Dr. Maurits Smeyers* (B. CARDON, ed., *Corpus van verluchte handschriften, Vol. 12, Low Countries Series 9*), Leuven, 2002, p. 1429-1441.

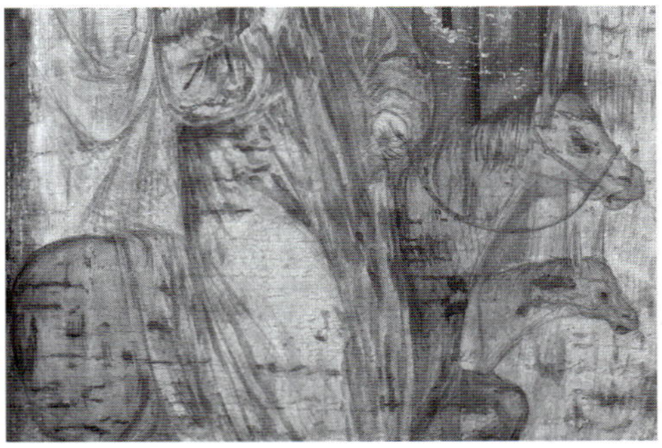

Fig. 2. Anonymous, *Passion of Christ*, IRR detail of the entry into Jerusalem with brush strokes in Christ's garment and in the underdrawing of the donkeys. The head of the adult donkey has been modified (© KIKIRPA).

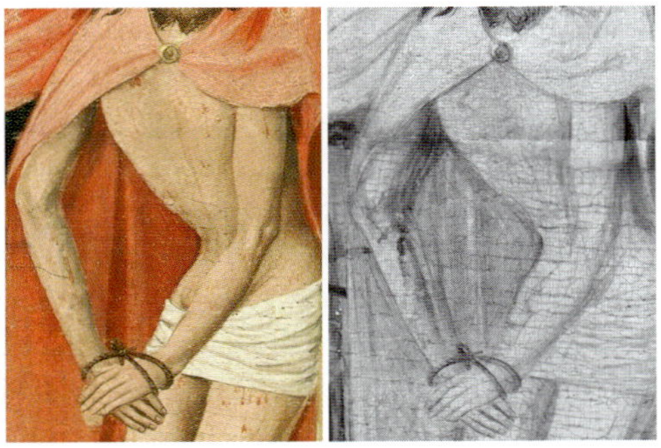

Fig. 3. Anonymous, *Passion of Christ*, details of Christ. a. Normal light. b. IRR shows that the contours of Jesus' body in *Ecce Homo* are drawn with a thick line and several sketchy lines. Shadows are indicated by coarse diagonal hatching. To indicate Christ's twisted torso and arms, an abundant amount of lines was used (© KIKIRPA).

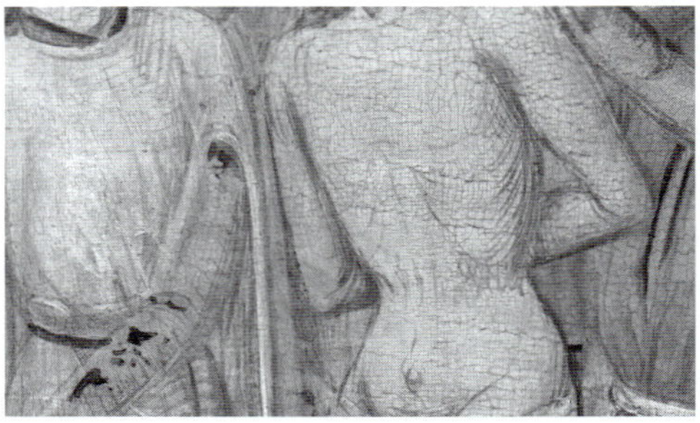

Fig. 4. Anonymous, *Passion of Christ*, detail, IRR. The modelling of Christ's body in the flagellation underlies delicately worked out painted areas. To model human bodies the artist uses both thick and fine and carefully drawn lines (© KIKIRPA).

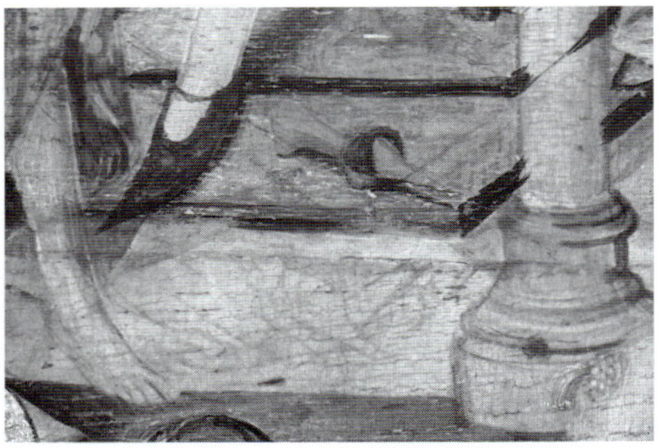

Fig. 5. Anonymous, *Passion of Christ*, detail with the throne of Pilate, IRR. Alterations can be noted at the step, the feet of Christ and of Pilate and the shoe. The document indicates the sequence of painting: first the step, then the pillar (© KIKIRPA).

and the shoe have also been altered during drawing; here, the infrared image indicates the sequence of painting: first the step, then the pillar (fig. 5). Another example of alteration during drawing is the changed position of the cane of one witness of the *Ecce Homo*.

Elements that were not underdrawn but added during the painting stage are few. The canopy over the heads of Pilate and Christ was inserted (fig. 6), clearly inorder to emphasize the separation between the coronation with thorns and the *Ecce Homo*. A painted crenelated wall is visible under the painted canopy. The dog at Pilate's feet

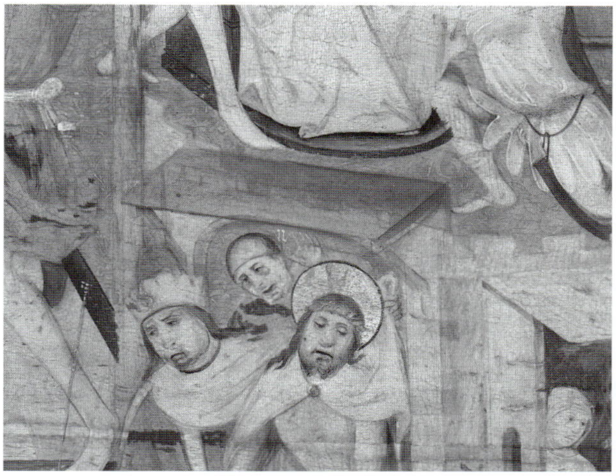

Fig. 6. Anonymous, *Passion of Christ*, IRR. The canopy over the heads of Pilate and Christ is a rare example of an element that was not underdrawn but added during the painting stage inorder to emphasize the separation between two scenes (the coronation with thorns and the *Ecce Homo*); the painted crenulated wall is plainly visible under the canopy. Thick diagonal hatchings indicate the areas of shade in the architecture (© KIKIRPA).

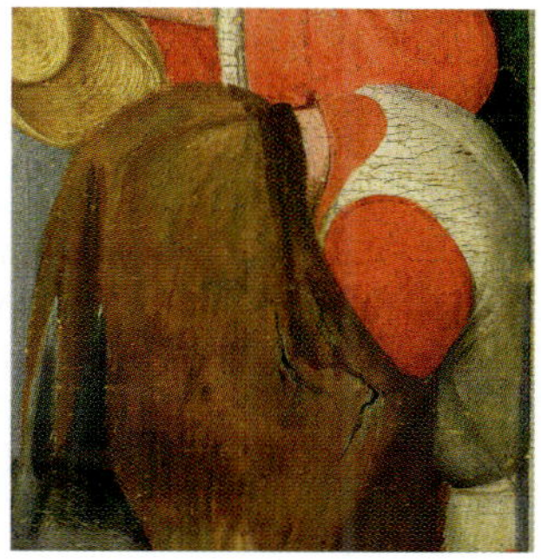

Fig. 7. The worn coat of the man animates the religious theme and brings the Passion closer to everyday life (© KIKIRPA).

is not underdrawn; it is only painted. The tree at the city gate was roughly sketched; leaves and branches were added during painting. None of the animals or vegetal elements have been underdrawn, and the same can be said for many details of clothing and headwear.

Changes both in underdrawing and painting are minor and their impact is limited. The compositional details are fully worked up in the drawing. The artist had obviously planned the composition carefully with preliminary drawings on paper, since the underdrawing is direct and confident. As was common in artists' workshops during the period, the underdrawing was probably the final stage of an elaborate preparatory process.[8] There is no evidence of mechanical copying, so it is unlikely that the panel was copied after an existing painting.

The painting was clearly commissioned as a devotional object with strong emphasis on Christ's sacrifice and with a central position for the flagellation. Creative touches are found in the very individual faces of the many figures that populate the panel and in the details of daily life such as the coat of the man who is welcoming Christ; this coat evidently needs mending. These features animate the religious theme and bring it closer to everyday life.

[8] D. BOMFORD ed., *Art in the Making. Underdrawing in Renaissance Paintings*, London, 2002, p. 37 (exhibition catalogue).

Veronique VANDEKERCHOVE
M Leuven
Willemsstraat, 7
B – 3000 Leuven
Tel: +32.16.22.69.06
Fax: +32.16.23.89.30
veronique.vandekerchove@leuven.be

The author graduated in Archaeology and Art History in 1987 at the Catholic University of Leuven. In 1989 and 1990 she carried out historical research on farm buildings at the University's Faculty of Agriculture. Since 1992 she has worked in the municipal museums of Leuven, first as a scientific collaborator and since 1998 as director. She recently started a doctoral thesis on the *Passion of Christ* (Brabant, 1470-1490) in the municipal collection. She has mainly published on archaeological research in the city of Leuven, preventive conservation of museum collections and on the collections of the municipal museums of Leuven (history, context and future).

PEETER SCHEYFVE ET AGNÈS DE GRAMME, DONATEURS DU TRIPTYQUE DE L'ADORATION DES MAGES DE JÉRÔME BOSCH (VERS 1494) CONSERVÉ AU PRADO[1]

Xavier Duquenne

Résumé: Les donateurs figurant sur l'*Adoration des Mages* de Jérôme Bosch conservée au Musée du Prado, ne sont pas, comme on l'a souvent écrit, Peeter van Bronchorst et son épouse Agnès van Boschuysen, mais bien Peeter Scheyfve, drapier et échevin d'Anvers, et sa seconde épouse Agnès de Gramme. Les personnages qui figurent sur la *Messe de saint Grégoire*, au revers, seraient Jan, le seul fils de Peeter Scheyfve, issu de son premier mariage, et Peeter de Gramme, père d'Agnès et protecteur de Jan. Le tableau date d'environ 1494, mais les personnages du revers pourraient avoir été ajoutés en 1507, après la mort de Peeter Scheyfve. La famille patricienne Scheyfve (Scheyff, Schijff, Schyf...) brilla à Anvers aux XVe et XVIe siècles et donna un chancelier au Brabant.

L'*Adoration des Mages* du Prado confondue avec deux *Adorations* perdues

En 1858, Alexandre Pinchart mit en évidence qu'au XVIe siècle, Jean de Casembroot, seigneur de Bekkerzeel, avait chez lui un triptyque des *Trois rois* de Jérôme Bosch qui portait au revers des volets les armoiries Bronchorst et Boschuysen. L'œuvre avait été confisquée lors de la répression des protestants au profit du roi d'Espagne Philippe II, d'après un inventaire officiel de 1567 (fonds *Chambre des comptes, volumes*, Archives générales du Royaume à Bruxelles)[2]. Au folio 163, le titre indique qu'on parle du mobilier trouvé le 14 avril 1567 dans une chambre à coucher de la maison de Jean de Casembroot à Bruxelles. Au folio 167 on cite «Ung tableau des trois roix faict par jeronimus bosche sevrant [ouvrant] à deux huys [ou portes, volets] ayans par dehors les armes de bronchorst et boschuyse». C'est donc au revers des volets que figuraient les armoiries.

Jean de Casembroot est né en 1525. Il était le fils d'un bourgmestre de Bruges et devint un poète apprécié. Il était surtout secrétaire et conseiller influant du comte d'Egmont, gouverneur des provinces de Flandre et d'Artois. En 1561, ce dernier démissionna avec le prince d'Orange du Conseil d'Etat en protestation contre la politique du gouvernement. Casembroot prit une part importante à la confédération des nobles, formée contre le régime espagnol en faveur d'une liberté politique et religieuse, ligue à laquelle il passait pour avoir conduit le comte d'Egmont. Les biens de Casembroot furent confisqués au profit de la couronne, mais sa seigneurie fut restituée à sa fille en 1580. Les Casembroot portaient d'*azur au chevron d'or chargé de trois quintefeuilles de gueules et accompagné de trois épis d'or, deux en chef et un en pointe.*

[1] Cet article est un condensé de l'article par le même auteur: «La famille Scheyfve et Jérôme Bosch», dans *L'intermédiaire des généalogistes*, janvier-février 2004, p. 1-19.

[2] A. PINCHART, «Archives des arts [...]», dans *Messager des sciences historiques*, 1855, p. 420, et 1858, p. 166, repris dans *Archives des arts [...]*, t. 1, Gand, 1860, p. 186 et 276.

En 1889, Carl Justi mit la découverte de Pinchart en relation avec l'*Adoration des Rois Mages* de Bosch au Prado, propriété de Philippe II au moins depuis 1574. Les armoiries du Prado furent alors considérées comme étant celles des Bronchorst et Boschuysen. Les saints patrons Pierre et Agnès prêtèrent leur prénom aux donateurs qui devinrent Pierre van Bronchorst et Agnès van Boschuysen, couple qui n'a jamais existé[3]. Justi prétendit même que Philippe II avait acquis le triptyque de Jean Casembroot en 1549, lors de son voyage aux Pays-Bas. On savait pourtant que le triptyque avait été confisqué en 1567.

Notons que les dimensions du triptyque du Prado et celles du triptyque Casembroot semblent avoir été similaires. Selon Justi, la description du triptyque faite lors de son transfert à l'Escurial en 1574 mentionne le sujet (la naissance de Notre Seigneur) et les mesures sans les volets (en hauteur 5 pieds et en largeur 3 pieds, soit respectivement 139,17 × 83,50 cm)[4]. Les dimensions de la partie centrale au Prado sont, d'après le musée, respectivement de 138 et 72 cm. La différence n'est pas grande.

Jean Casembroot avait épousé Wilhemine van Bronchorst, fille d'Andries van Bronchorst, chevalier, membre du conseil de la province de Hollande, et de Wendelmoet van Boschuysen, veuve d'Alexis de Nassau puis de Jean de Lannoy. Au revers des volets du triptyque de Jean de Casembroot, les écus étaient ceux du père de Wilhelmine (*de gueules au lion d'argent, armé, lampassé et couronné d'or*) et de sa mère (*de vair*)[5] et n'ont rien à voir avec ceux du triptyque du Prado[6].

On a prétendu également que le tableau du Prado provenait de la cathédrale de Bois-le-Duc. En 1858, Pinchart avait déjà indiqué qu'une *Adoration des Rois Mages* de Jérôme Bosch se trouvait autrefois en cette église, d'où elle fut transférée à l'hôtel de ville en 1566 à l'abri des iconoclastes. Un rapport, établi en 1608 par trois témoins pour l'historien Gramaye, mentionne qu'il y avait dans ladite église quatre tableaux de Jérôme Bosch, dont une *Adoration des Rois Mages* sur l'un des deux autels de sainte Marie[7]. Cette *Adoration* ne peut être le tableau de Philippe II signalé à l'Escurial dès 1574.

[3] C. JUSTI, «Die Werke des Hieronymus Bosch in Spanien», dans *Jahrbuch der Königlich Preussischen Kunstsammlungen,* t. 10, 1889, p. 128-129, 141-142; M. GOSSART, *Jérôme Bosch [...]*, Lille, 1907, p. 37, 47, 310; C. JUSTI, *Miscellaneen aus drei Jahrhunderten Spanischen Kunstleben*, t. 2, Berlin, 1908, p. 72-74; C. JUSTI, «Jeronimo Bosch», dans *La España moderna,* juin 1914, ne donne pas de complément notable sur le sujet; F. LAFOND, *Hieronymus Bosch,* Bruxelles, 1914, p. 37, 111; CH. DE TOLNAY, *Hieronymus Bosch*, Bâle, 1937, p. 31, 43-44, 70.

[4] C. JUSTI, 1889, p. 141-142; selon H. DOURSTHER, *Dictionnaire des poids et mesures [...]*, Bruxelles, 1840, le pied de Madrid et de toute la Castille était de 27,83 cm.

[5] J. GAILLIARD, *Bruges et le Franc*, t. 4, Bruges, 1860, p. 378-388; Casembroot, *Biographie nationale,* t. 3, 1872, *Jean de Casembroot*, par Th. Juste; A. DE BEHAULT DE DORNON, *Le château de Vilvorde [...]*, Anvers, 1922, p. 101-103; J. DE HERCKENRODE et DE VEGIANO, *Nobiliaire des Pays-Bas*, t. 1, Gand, 1865, p. 268, 320, 322; *Nieuw Nederlands biografisch woordenboek*, t. 7, 1927, *Jehan de Casembroot,* par F. BLOK; C. Douxchamps-Lefèvre (*Inventaire du fonds de Corroy-le-Château,* Bruxelles, 1962) recense des archives des couples Bronchorst-Boschuysen et Casembroot-Bronchorst, ainsi que de la fille de ce dernier, Anne de Casembroot.

[6] Les archives des époux Casembroot-Bronchorst et de leur parenté, conservées aux Archives de l'Etat, pourraient apporter d'autres renseignements. Fonds du château de Corroy-le-Château, voir n°13; Archives générales du Royaume, *Grand Conseil de Malines*: voir *Chronologische lijsten van de geëxtendeerde sententiën [...]*, t. 5, Bruxelles, 1987, n°2244, n°2607, n°2777, procès sur la succession des parents Bronchorst-Boschuysen, 1555-1560.

[7] A. PINCHART, «Archives des arts [...]», dans *Messager des sciences historiques,* 1858, f. 167, repris dans *Archives des arts [...]*, t. 1, Gand, 1860, f. 277; F. GERLACH, *op.cit.*, p. 3-4.

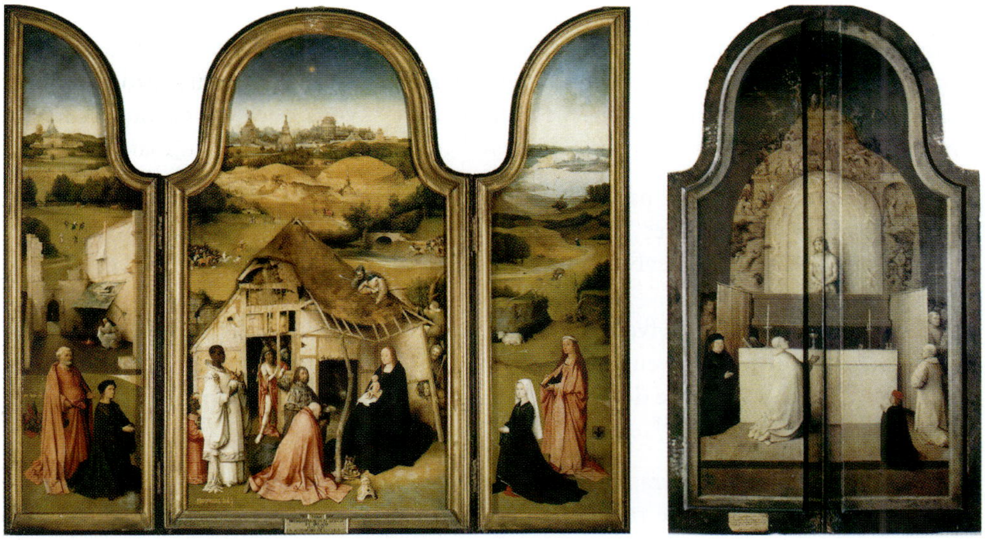

Fig. 1. a. Jérôme Bosch, *Triptyque de l'Adoration des Mages*, vers 1494, Madrid, Prado. b. Triptyque fermé avec la Messe de saint Grégoire.

Les armoiries du triptyque du Prado (fig. 1a) sont celles des familles Scheyfve et de Gramme

La tradition Bronchorst-Boschuysen eut la vie longue[8]. Un article de 1965 sur l'héraldique au Musée du Prado n'apporte guère de précisions[9]. Le catalogue de 1996[10] retient l'inexacte provenance de Casembroot. Pourtant, dès 2001, G. van Dijk affirme que le triptyque Bronchorst-Boschuysen n'est pas celui du Prado, les armoiries n'étant pas celles des deux familles[11]. L'auteur du présent article a averti dès 1964 le Musée du Prado que, d'après ses recherches, les figurants sont Peeter Scheyfve (l'oncle du chancelier) et Agnès de Gramme. L'identification erronée des donateurs a donc été contredite il y a longtemps. Dès 1893, les armoiries avaient été attribuées à la famille

[8] Notice sur Jérôme Bosch, 1983, *Biographie nationale* (de Belgique); F. Gerlach, «Wapens op schilderijen van Jeroen Bosch», dans *Brabants heem*, 1978 (reproduit dans F. Gerlach, *Jheronimus Bosch, opstellen over leven en werk*, Bois-le-Duc, 1988). *Biographie nationale,* t. 43, 1983, *Jérôme Bosch*, par R. Van Schoute et M. Van Schoute-Verboomen ; R. Van Schoute, M. Garrido et J.M. Cabrera, «Le dessin sous-jacent chez Jérôme Bosch. L'Adoration des Mages du musée du Prado à Madrid», dans *Le dessin sous-jacent dans la peinture,* Louvain-la-Neuve, 1985, p. 211-214, pl. 92-95; M. Garrido et R. Van Schoute, «El triptico de la Adoración [...]: estudio técnico», dans *Boletin del Museo del Prado,* t. 6, 1985; R. Marijnissen, *De Vlaamse schilderkunst in het Prado,* Anvers, 1989, p. 55; R. Marijnissen avec la collaboration de F. Ruyffelaere, *Jérôme Bosch, tout l'œuvre peint et dessiné*, Anvers, 2000, p. 34-35, 234-259. R. Van Schoute et M. Verboomen, *Jérôme Bosch*, Tournai, 2001, p. 191-192, signalent que l'identification Bronchorst a été mise en doute.

[9] De Montesa, «Heraldica en el Museo del Prado», dans *Arte Español*, t. 25, 1963-1967, 2, p. 84.

[10] *Museo del Prado. Catalogo de las pinturas,* Madrid, 1963 et 1996, n 2048.

[11] G. van Dijk, p. 62-63, 100.

Scheyfve[12]. Des recherches entamées vers 1900 aboutirent à deux travaux manuscrits sur la famille Scheyfve, l'un par Charles Diercxsens, l'autre, une véritable généalogie, par le baron (Constantin) de Borrekens[13]. Une généalogie semblable, mais sans la première génération, a été publiée en 1950 par Joseph De Beer[14]. Voici quelques éléments de cette généalogie pour la période qui nous intéresse[15].

I. Gillis Scheyfve, cité par son petit-fils Peeter (sous III), dans un acte scabinal d'Anvers du 24 juillet 1503[16].

II. Claus (Nicolas) Scheyfve, mort avant 1495, fils de Gillis I, père de Peeter, (qui suit en III) et de Jan.

III. Peeter (Pierre) Scheyfve, qualifié d'apprêteur de drap en 1494, doyen des drapiers d'Anvers, receveur en 1495-1504 et échevin d'Anvers en 1505[17], acheta en 1505 la seigneurie de Niel et mourut en 1506; sa succession fut partagée par ses enfants en 1509[18].

C'est ce dernier, on le verra, qui est représenté (avec sa deuxième épouse Agnès, ou Agneese, de Gramme) sur le triptyque de l'*Adoration des Rois Mages* au Prado. Peeter Scheyfve avait épousé en premières noces Barbe (Barbele) van Woelputte — morte au plus tard en 1491 — qui lui avait donné trois filles et un fils prénommé Jan. Agnès (Agneese) de Gramme, sa seconde épouse, est celle qui figure sur le triptyque du Prado; elle décéda vers 1497 sans postérité. Elle était fille de Peeter de Gramme, receveur de la ville d'Anvers en 1493-1496 et 1502-1505, et de dame (*jonckvrouwe*) Adriane J. Lenaerts. Ce Peeter de Gramme laissa toute sa fortune au seul descendant masculin de Peeter Scheyfve, Jan, issu du premier mariage[19]. La troisième épouse, Maria van der Merwede, de Dordrecht, fille de Klaas, seigneur de Baardwijk, et de Margrieta Boogaard[20], lui donna trois filles.

[12] H. HYMANS, «Le Musée du Prado», dans la *Gazette des beaux-arts*, 1893, II, p. 231-234. Vers 1934, J.G. van Gelder, directeur du musée Boymans-van Beuningen à Rotterdam adhère à cette hypothèse: lettres des 5 et 17 février 1965 de H.L. Kruimel, conservateur du *Centraal Bureau voor Genealogie*, attestant que c'est à cet organisme que van Gelder confia sa thèse.

[13] Dossier de recherche de 1901-1904 (copie au Service de Centralisation des Etudes généalogiques et démographiques de Belgique), constitué par Camille Scheyven, comprenant une généalogie Scheyfve par le baron de Borrekens en 1901, remontant à Gillis Scheyfve, une descendance du chancelier Scheyfve par Ch. Diercxens, une suite d'extraits des archives de la Cour féodale de Brabant par J.-Th. De Raadt. Sur les Scheyven: *Annuaire de la noblesse belge*, 1931-1932, t. 1, p. 205-215. *Etat présent de la noblesse belge*, 1998: *l'écu, porté déjà sous l'Ancien régime, est de gueules à six besants d'or posés 3, 2 et 1.*

[14] J. DE BEER, «Jean Scheyfve, bourgmestre d'Anvers, chancelier de Brabant, et ses deux médailles, 1575», dans *De Schakel*, 1950, p. 56-78. Une généalogie Scheyven, originaire de Tilburg au XVe siècle, a été publiée par E. VAN DEN BOSSCHE, *Scheyven-Van Thillo*, 1963.

[15] Cette généalogie est publiée *in extenso* avec des compléments substantiels dans X. DUQUENNE, 2004.

[16] Archives de la ville d'Anvers (AVA), *Schepenregisters (Sch.)*, 1503, L.G., vol. 1, f. 151; Claus est cité aussi en 1495, B.S., vol. 1, f. 113.

[17] F. PRIMS, «Jan Scheyfve (1515-1581)», dans *Antwerpiensia*, t. 6, 1932, p. 247.

[18] Archives générales du Royaume (AGR), *Cour féodale de Brabant (C.F.B.)*, 348, f. 289 (Niel); A.V.A., *Sch.*, 1490, vol. 1, f. 111; 1491, vol. 2, f. 209; 1496, B.S.A.I., f. 62; 1498, B.A., vol. 1, f. 242; 1499, B.G., f. 128; 1503, G.L., f. 151; 1509, G.L., vol. 1, f. 284, qui cite son testament du 20 janvier 1505-1506, acté par le notaire Th. Nicolas. J. DE RAADT, *[...] Niel en zijne heren*, Gand, 1889, f. 17-18.

[19] A.V.A., *Sch.*,1496, B.S., vol. 1, f. 62; 1498, B.A.,vol. 1, f. 242; *Bulletin de la propriété*, 1887, p. 67, qui cite le testament de Peeter de Gramme (notaire Th. Nicolas, 20 janvier 1505-1506) et le partage de la succession de Peeter Scheyfve en 1509.

[20] M. BALEN, *Beschrijving der stad Dordrecht*, t. 2, Dordrecht, 1677, p. 1109.

L'identification de l'écu masculin du Prado comme étant celui de Scheyfve était loin d'être certaine. La représentation de cet écu par Jérôme Bosch est imparfaite. En outre, les armoiries Scheyfve n'ont été reprises dans l'*Armorial général* de Rietstap qu'à partir de la seconde édition (1883), mais en déformant le patronyme en «Soyef», famille localisée en Belgique. Et c'est sous ce même nom déformé que les armoiries Scheyfve figurent dans le répertoire des figures héraldiques de Renesse[21].

La représentation la plus précise des armoiries Scheyfve se rapporte à Maximilien Scheyfve et fut peinte par son ami Robert van den Berghe en 1576[22]: *de gueules au chef d'or chargé de trois têtes (et col) de loup de sable,* avec comme cimier non pas la tête de loup du chef, mais une tête (et col) de lévrier (de même tonalité fer que le heaume), cimier dans un vol à l'antique, c'est-à-dire entre deux ailes, respectivement d'or et de gueules comme les lambrequins. Le chef de l'écu est parfois d'argent. Les têtes, souvent peu précises, sont parfois dites ou représentées comme têtes d'âne, de cheval ou de chien. L'écu présente parfois un signe personnel, tel qu'au champ de gueules, les initiales de Jan Scheyfve en 1536[23].

Les volets du triptyque du Prado présentent les donateurs accompagnés de leur saint patron et de leurs armoiries: ces dernières comportent, à gauche, un chef à trois têtes d'animal (fig. 3), et à droite, une fleur de lys. À gauche, l'écu est de gueules au chef noirâtre (peut-être un vert obscurci), chargé de trois têtes (et cols), peut-être de chien ou de loup, d'argent; l'écu est surmonté d'un heaume, dont le cimier est une tête comme au chef, dans un vol à l'antique de gueules et de tonalité sombre, ces deux tons se partageant les lambrequins. L'identité avec les armoiries Scheyfve n'est, à ce stade, qu'une hypothèse, car la nature des têtes est imprécise et la tonalité — peu franche — du chef et des têtes n'est pas celle des Scheyfve. Saint Pierre, identifiable par ses clés, pourrait être le patron de l'échevin Peeter Scheyfve (III dans la généalogie). La matrice de son sceau d'échevin (fig. 2) correspond bien à celui des Scheyfve[24]. L'identification se renforce par la présence, sur le tableau, au-dessus des armoiries, de la devise «*een voer al*», semblable au «ung sur tout» figurant sur la petite médaille du chancelier Scheyfve, neveu de Peeter susdit.

Grâce à l'écu et à la patronne figurant sur le volet droit, l'hypothèse Scheyfve se meut en certitude. La patronne est sainte Agnès, avec une couronne de fleurs, un livre fermé et un agneau. L'écu est *d'argent à la fleur de lys de sable.* Prénom et blason sont ceux de la deuxième épouse de Peeter Scheyfve, Agnès de Gramme, dont le père, Peeter de Gramme, receveur d'Anvers, scellait d'une fleur de lys[25].

Au revers des volets figure une *Messe de saint Grégoire* en grisaille (fig. 1b). Deux figurants — un homme d'âge mur agenouillé discrètement au fond à gauche et un très jeune homme agenouillé devant l'autel — sont peints en tonalité naturelle et semblent des surpeints[26]. Il est probable qu'il s'agisse du fils de Peeter Scheyfve, Jan, mineur à la

[21] J. RIETSTAP, *Armorial général,* 2e éd., t. 2, Gouda [1883]; TH. DE RENESSE, *Dictionnaire des figures héraldiques,* t. 1, Bruxelles, 1892, p. 47.

[22] Bibliothèque royale, *Manuscrits,* manuscrit II 6573, f. 77.

[23] A.G.R., C.F.B., 5344, déclaration du 1er décembre 1536 pour des fiefs sous Niel.

[24] A.V.A., *Matrijzen van schepenzegels,* 99.

[25] J. DE RAADT, *Sceaux armoriés [...],* t. 1, Bruxelles, 1898, p. 510.

[26] D'après une communication de M.C. Garrido rapportée par R. Marijnissen et F. Ruyffelaere, 2000, p. 240.

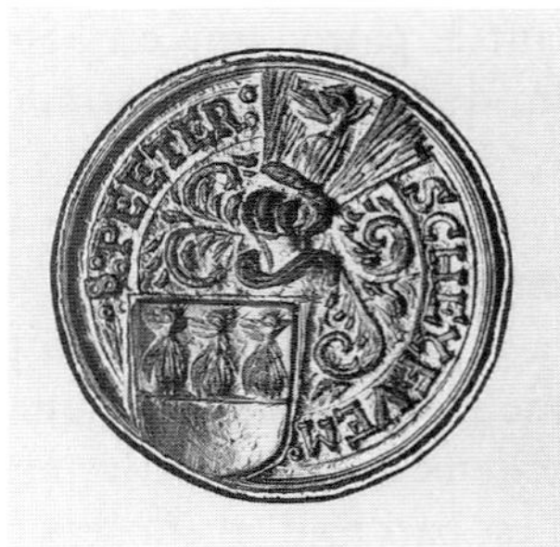

Fig. 2. Matrice du sceau de Peeter Scheyfve (© Archives de la Ville d'Anvers)

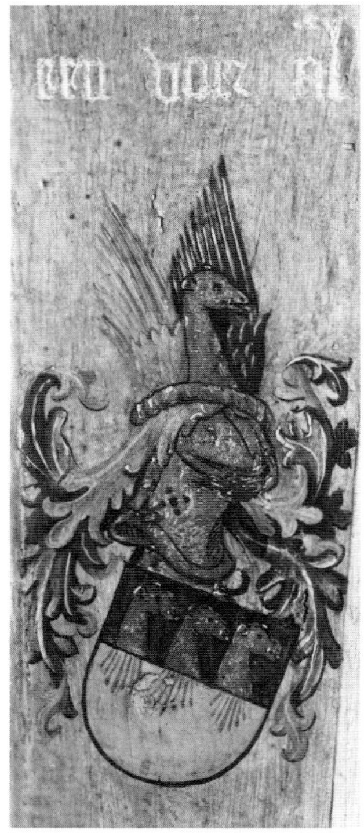

Fig. 3. Jérôme Bosch, *Triptyque de l'Adoration des Mages*, Madrid, Prado. Détail avec les armoiries et la devise de Peeter Scheyfve.

mort de son père. L'autre homme, en retrait, pourrait être un protecteur ou un tuteur: soit l'oncle Jan Scheyfve ou l'un des deux tuteurs de Jan. Tous trois consentirent à la vente de la seigneurie de Niel par le fils mineur en 1506. Il est plus probable encore qu'il s'agisse de Peeter de Gramme, beau-père de Peeter Scheyfve, dont on sait qu'il laissa en 1505 sa fortune au jeune homme. L'absence d'armoiries auprès de ces deux figurants du revers s'expliquerait par la présence de leurs armoiries à l'avers. Le tableau étant peint vers 1494, comme on l'a vu, l'ajout aurait été fait vers 1507, alors que Jan Scheyfve était encore mineur[27].

Différentes datations entre 1490 et 1510 ont été proposées pour le triptyque[28]. La présente découverte permet une plus grande précision. La première épouse de Peeter Scheyfve mourut au plus tard en 1491 et la seconde épouse en 1496, 1497 ou 1498: le tableau date donc des années 1490-1498, soit vers 1494. Les actes des échevins et autres documents des archives de la ville d'Anvers, qui sont à présent largement répertoriés, pourraient peut-être un jour préciser davantage l'époque.

27 On peut donc rejeter les essais d'identification de Fraenger et Tolnay ainsi que ceux de van Dijk. W. FRAENGER, «Der vierte Koenig des Madrider Epiphanias-Altar von Hieronymus Bosch», dans *Deutsches Jahrbuch für Volkskunde,* t. 3, 1957, p. 170-172; G. VAN DIJK, p. 53 et 63; CH. DE TOLNAY, *Hieronymus Bosch*, Baden-Baden, 1965, p. 372.

28 L. VON BALDASS, «Die Chronologie der Gemälde der Hieronymus Bosch», dans *Jahrbuch der Koenigliche Preussischen Kunstsammlungen,* 1917, p. 184; L. DEMONTS, «Deux Primitifs néerlandais au musée du Louvre», dans *Gazette des beaux-arts,* 1919, p. 12; M.J. FRIEDLÄNDER, *Die Altniederländische Malerei,* t. 5, 1927, p. 89; CH. DE TOLNAY, 1965, p. 371).

Conclusion

La destination primitive du tableau reste inconnue. Il fut peut-être peint pour l'autel de la corporation des drapiers d'Anvers, dont Peeter Scheyfve fut doyen; un autel à Niel serait une destination moins probable, car cette seigneurie ne fut acquise par Scheyfve qu'en 1505 (et vendue l'année suivante, après sa mort). Le triptyque aurait pu être déplacé lors des troubles iconoclastes qui sévirent en 1566, en particulier à Anvers. On ne peut exclure que le tableau fût placé dès l'origine dans la maison de Scheyfve, et aurait ensuite fait partie de la succession. L'hypothèse la plus probable serait que le tableau fût échu au fils. Comme celui-ci mourut en 1548, il n'est pas exclu que le tableau fût cédé peu après par sa veuve ou par sa fille unique au futur Philippe II d'Espagne. On sait que ce dernier était friand de Jérôme Bosch, dont il possédait au moins neuf œuvres, et qu'il séjourna à Anvers en 1549.

L'identification des donateurs du triptyque du Prado éclaire les questions de la clientèle de Jérôme Bosch et des relations de celui-ci avec Anvers. Le tableau n'était pas destiné, comme on l'a longtemps cru, à une famille d'ancienne noblesse, mais, signe des temps, à une famille d'hommes nouveaux, patriciens d'Anvers, et plus précisément à un marchand drapier prospère, Peeter Scheyfve.

Il reste à rechercher le sort des deux triptyques, cités dans les textes, qui ont été confondus avec celui du Prado.

Xavier DUQUENNE
Avenue Louise 214
B – 1050 Bruxelles

LE DESSIN SOUS-JACENT DU TRIPTYQUE DE LA RÉDEMPTION DE VRANCKE VAN DER STOCKT (MADRID, MUSÉE DU PRADO)[1]

Carmen Garrido, Duilio Bertani et Roger Van Schoute
avec la collaboration de Jaime García-Máiquez et Ana González Mozo

Résumé: À la suite de Georges Hulin de Loo, le *Retable de la Rédemption,* dit aussi de Cambrai, est généralement attribué à Vrancke Van der Stockt, successeur de Rogier Van der Weyden comme peintre de la ville de Bruxelles. La présente étude décrit le dessin sous-jacent du retable en vue d'établir des comparaisons avec l'œuvre de Rogier. Elle utilise le système de réflectographie à l'infrarouge à haute résolution présenté dans D. Bertani (dir.), *Oltre il visible. Indagini rifflettografiche*, Milan, 2001 et dans D. Bertani et L. Consolandi, «High Resolution. Imaging in the Near-Infrared», dans *Digital Heritage*, 2006, p. 207-234.

Le *Triptyque de la Rédemption* (Madrid, Prado, fig. 1) est attribué à Vrancke van der Stockt (1420-1495), disciple de Van der Weyden, appelé à succéder à ce dernier comme peintre de la ville de Bruxelles. Le panneau central[2] montre le Christ crucifié entre la Vierge et saint Jean-Baptiste dans le cadre architectural d'une église gothique. Les sept sacrements et des épisodes de la Passion sont représentés sous forme de petits groupes sculptés polychromes situés dans les voussures. Le volet gauche[3] représente l'expulsion d'Adam et Ève du Paradis terrestre. Dans le volet droit[4], le Jugement dernier est accompagné de petites scènes représentant les œuvres de miséricorde. L'extérieur des volets illustre le tribut de César. Les scènes s'inspirent, pour la plupart, d'œuvres de Van der Weyden. Le fait de représenter l'Ancien et le Nouveau Testament, le Jugement dernier et une grisaille d'un grand symbolisme sur le pouvoir terrestre et divin, a conduit les auteurs à mettre l'accent sur l'importance iconographique de l'œuvre plutôt que sur son incontestable qualité technique et artistique.

Conformément aux usages des Primitifs flamands, le support est en bois de chêne originaire de la Baltique. Le panneau central est constitué de 7 planches disposées à la verticale. Les volets comportent chacun quatre planches, également disposées à la verticale. Les joints sont renforcés par des tourillons. L'amincissement au revers du support central fait apparaître certaines encoches avec ou sans leur tourillon. Selon l'analyse dendrochronologique réalisée par Peter Klein, le panneau central était prêt à l'usage à partir de 1439 et les volets à partir de 1431.

[1] Concernant Vrancke Van der Stockt voir J. Dijkstra, dans *Les Primitifs flamands et leur temps*, B. de Patoul et R. Van Schoute (dir.), Louvain-la-Neuve, 1994, p. 524-531 et 643. Voir également J.R.J. van Asperen de Boer, J. Dijkstra et R. Van Schoute avec la collaboration de C.M.A. Dalderup et J.P. Filedt Kok, *Underdrawing in Paintings of the Rogier Van der Weyden and Master of Flemalle Groups*, Zwolle, 1992, p. 38-40, ill. 22-27. F. Koreny (dir.), *Dessins de van Eyck à Hieronymus Bosch*, Anvers, 2002, p. 101-122.

[2] Panneau central: 198 × 171,8 (± 0,3) cm.

[3] Volet gauche: 196,1 × 75, 5 (± 0,3) cm.

[4] Volet droit: 199,2 × 75,5 (± 0,3) cm.

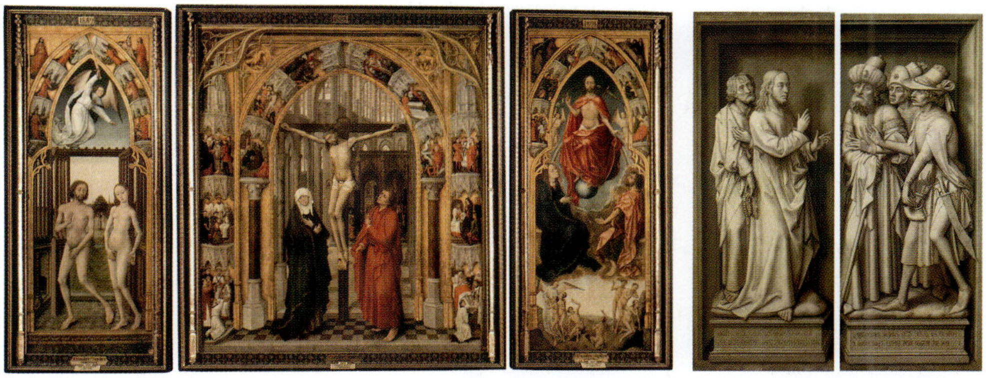

Fig. 1. Vrancke Van der Stockt, *Le Triptyque de la Rédemption* (ouvert et fermé), Madrid, Prado (© Musée du Prado).

Le dessin sous-jacent a été réalisé au pinceau avec des traits de grosseur et d'intensité variées, aussi bien pour les personnages et les architectures que pour les nuances d'ombre des différents espaces intérieurs. L'importance et l'intensité de ces traits varient selon la dimension des détails. Par endroits, de grosses lignes sont superposées à d'autres plus fines comme si le peintre avait repassé sur un projet initial. Les lignes du dessin sont longues et larges pour les figures principales; elles se brisent de manière très particulière pour signaler les draperies. Dans quelques cas, les traits se terminent en forme de crochets semblables à ceux qui apparaissent dans des dessins indépendants attribués au même artiste. Les traits marquant les ombres varient selon les zones: obliques parallèles ou croisés, verticaux ou horizontaux, etc.

Dans le panneau central, les modifications intervenant entre le stade du dessin sous-jacent et celui de la couleur sont abondantes. Le dessin des vêtements de la Vierge et de saint Jean est différent de ce qui est visible au stade pictural. Très souvent une simplification est intervenue au stade de la couleur. La position des têtes des personnages a changé. Dans un premier stade, la Vierge inclinait davantage la tête vers le centre de la scène. Le profil du visage de saint Jean a été déplacé vers la droite (fig. 2). La position de ses bras et de ses mains a changé: ses mains ont été peintes l'une plus haut, l'autre plus bas que prévu initialement. Ses pieds, séparés à l'origine, ont été rapprochés. Dans certains cas on peut observer jusqu'à cinq stades de dessin sous-jacent. Le manteau et la position du corps de saint Jean étaient différents dans le projet initial. Le dessin révèle l'existence d'une ceinture sous le ventre du saint. Le corps et la tunique sont déplacés vers l'endroit occupé par une colonne. Il est possible que tous ces changements répondent au désir de redresser le personnage pour lui donner une position plus en accord avec celle de Marie. La position initiale du personnage ressemble fort à celle de la figure en grisaille de l'un des personnages réalisés sur le volet droit de la représentation du tribut de César. Dans le personnage du Christ, signalons aussi la modification de la position des bras en relation avec le changement intervenu dans la perspective de la croix (fig. 3).

Le cadre dans lequel se déroule la Crucifixion est assez élaboré, en particulier le fond de l'église dans lequel on observe des ombres réalisées au moyen de longues lignes

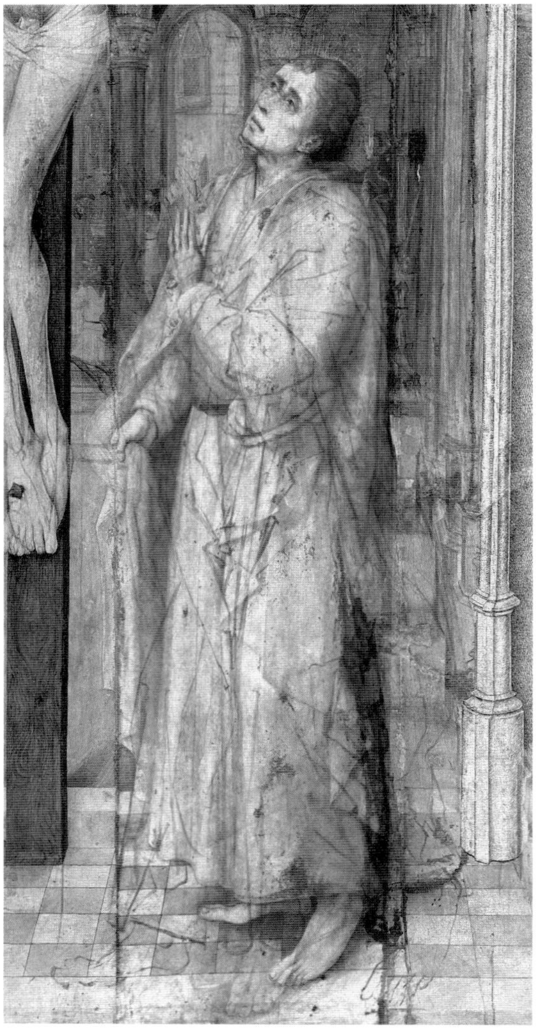

Fig. 2. Vrancke Van der Stockt, *Le Triptyque de la Rédemption*. Détail du panneau central, saint Jean, IRR (© Musée du Prado, prototype Duilio Bertani).

parallèles. De nombreux changements sont intervenus dans le dessin de l'architecture. Les colonnes sur lesquelles sont appuyées les archivoltes de l'arc central avaient des décorations végétales à la base et des décorations en zigzags dans les fûts mais ces décors n'ont pas été réalisés au stade de la peinture. Les chapiteaux qui présentaient une décoration complexe à base de cercles ont été réduits aux acanthes que l'on observe actuellement. L'horizon qui se situait plus bas que l'œil a été remonté. Les petites scènes qui se déroulent autour de la Passion et de la mort du Christ avaient été prévues dans le schéma général du dessin. Elles ont subi des changements en particulier dans la disposition des personnages comme on peut l'observer dans la Descente de croix et dans la Mise au tombeau. Dans cette dernière scène, une Marie-Madeleine a été dessinée mais elle n'a pas été retenue au stade de la couleur (fig. 4a, 4b).

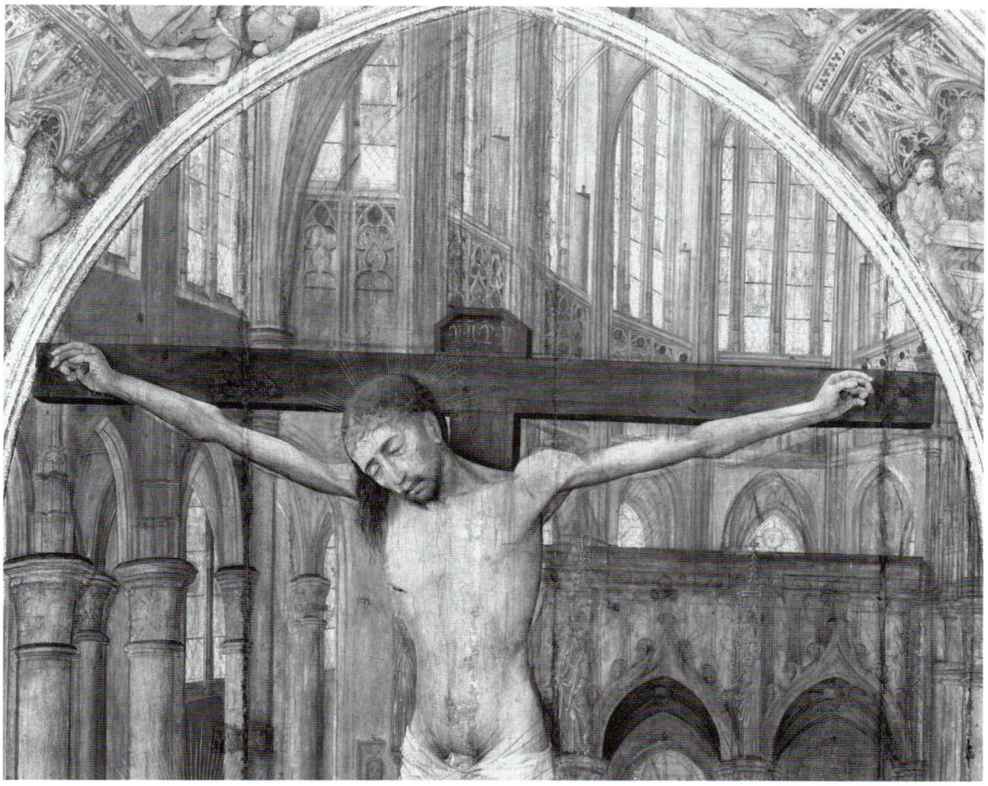

Fig. 3. Vrancke Van der Stockt, *Le Triptyque de la Rédemption*. Détail du panneau central, le Christ et architecture de l'église, IRR (© Musée du Prado, prototype Duilio Bertani).

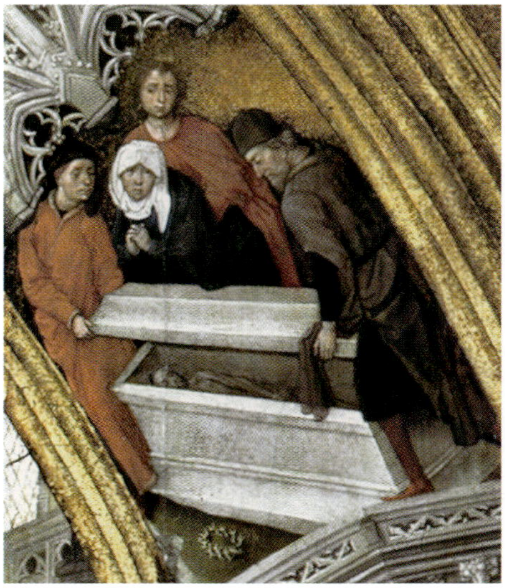

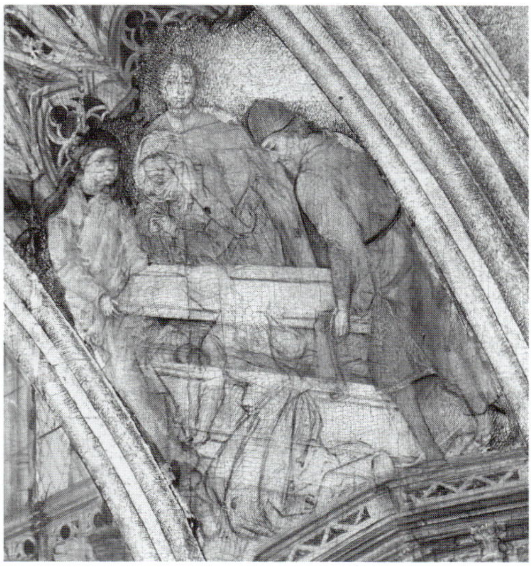

Fig. 4. a. Vrancke Van der Stockt, *Le Triptyque de la Rédemption*. Détail du panneau central, voussure avec la scène de la mise au tombeau. b. IRR (© Musée du Prado, prototype Duilio Bertani).

Il existe un dessin indépendant de la Flagellation attribué à Van der Stockt (Rotterdam, Musée Boijmans Van Beuningen, inv. A87). Dans le dessin sous-jacent du retable, le Christ et le soldat romain placé dans le fond occupent une position similaire; les deux soldats du premier plan ont été déplacés. Dans le dessin, on observe également que le soldat qui tend les cordes porte dans la bouche un objet en bois — objet visible également au stade du dessin sous-jacent du retable mais disparu dans l'exécution finale. Au stade du dessin, la dénivellation du pavement dans la partie de l'église où est représenté le sacrement de l'Eucharistie est proche de celle de la scène principale tandis que dans la peinture elle a été reculée. Il n'y a pas de dessin sous-jacent visible dans l'arc extérieur où six des sept sacrements sont représentés dans des niches. Dans l'espace où se déroulent ces épisodes — plus spécialement dans les scènes figurant l'Ordre et l'Extrême-Onction — on aperçoit une ligne de perspective en arc de cercle en face du tracé polygonal visible actuellement.

Le dessin des volets suit le même schéma d'exécution que celui du panneau central. Les corps sont dessinés au moyen de traits longs, larges et souples; certains détails sont d'exécution plus rapide comme la main gauche d'Adam (fig. 5). Dans le volet gauche, des changements significatifs peuvent être observés dans la tête d'Ève et dans les jambes d'Adam, dans la tête et les mains de l'ange ainsi que dans le tracé du sol en relation avec l'emplacement du premier dessin. La perspective du fond a été modifiée et le point de convergence des lignes de fuite abaissé lors de l'exécution finale. Dans le volet droit, le personnage de la Vierge a subi un grand déplacement vers l'intérieur de la scène par

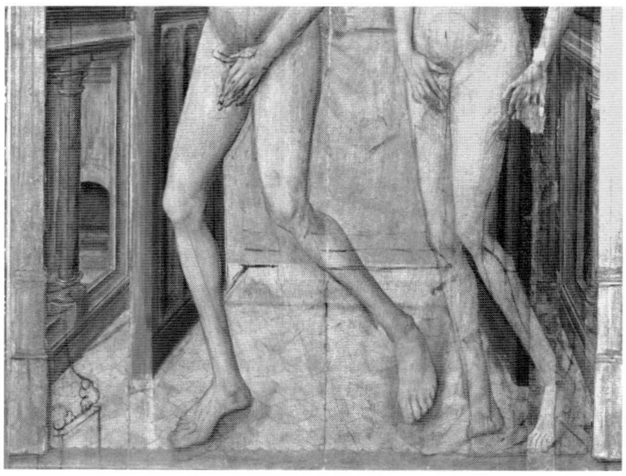

Fig. 5. Vrancke Van der Stockt, *Le Triptyque de la Rédemption*. Détail du volet gauche, jambes d'Adam et d'Eve, IRR (© Musée du Prado, prototype Duilio Bertani).

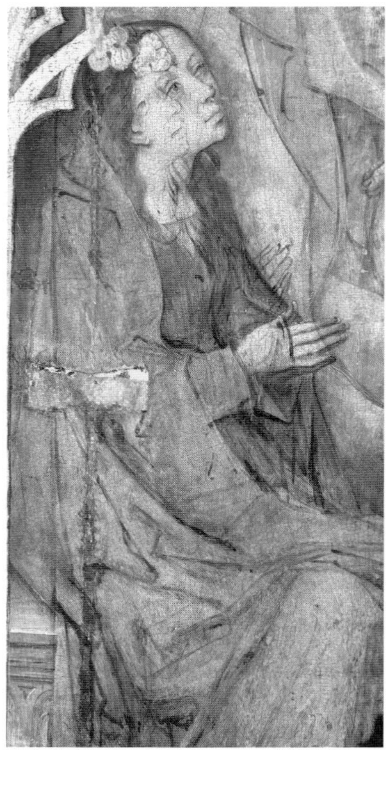

Fig. 6. Vrancke Van der Stockt, *Le Triptyque de la Rédemption*. Détail du volet droit, la Vierge, IRR (© Musée du Prado, prototype Duilio Bertani).

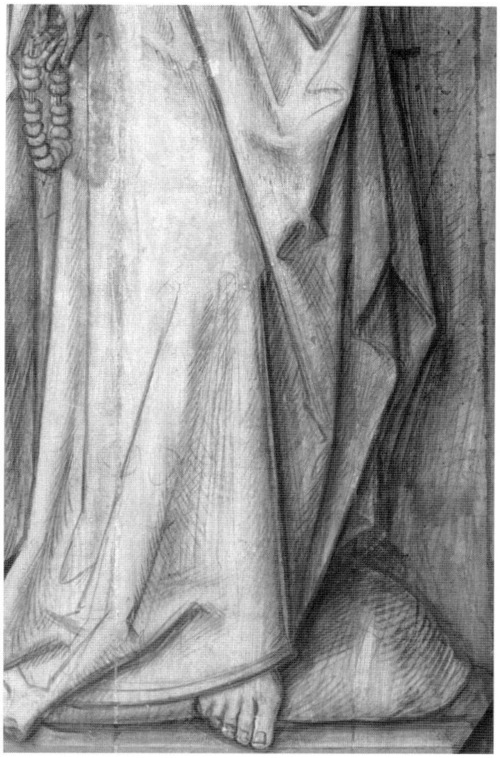

Fig. 7. Vrancke Van der Stockt, *Le Triptyque de la Rédemption*. Détail du volet gauche (revers), grisaille, le Christ, IRR (© Musée du Prado, prototype Duilio Bertani).

Fig. 8. Vrancke Van der Stockt, *Le Triptyque de la Rédemption*. Détail du volet droit (revers), grisaille, les Juifs, IRR (© Musée du Prado, prototype Duilio Bertani).

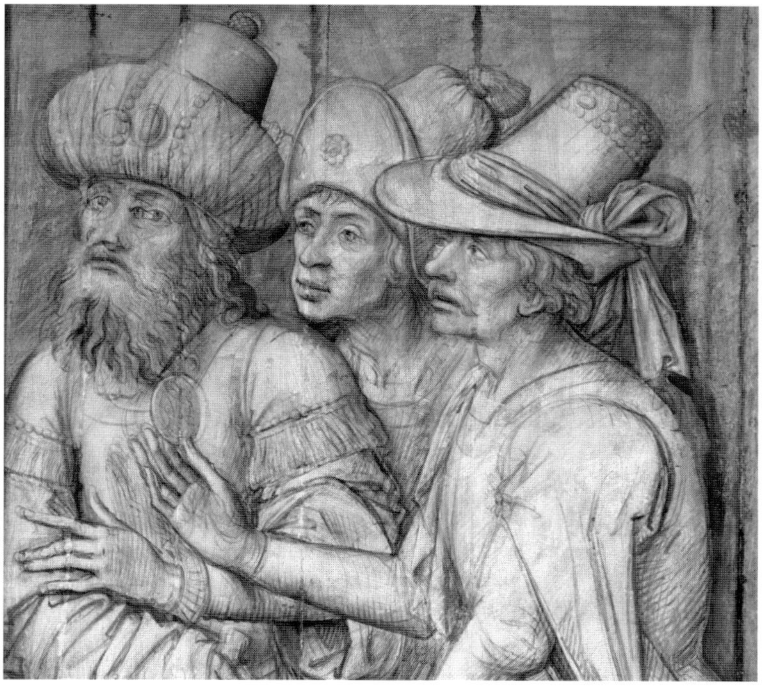

rapport au dessin; en même temps, ses mains ont été placées plus bas (fig. 6). La main gauche du Christ est rapprochée du corps. Les motifs architecturaux des deux scènes présentent des décorations et des éléments différents et on constate au niveau de la peinture un déplacement des colonnes latérales vers l'extérieur, élargissant de ce fait les deux scènes. Ces modifications de l'espace sont à mettre en relation avec celles du panneau central, qu'on observe dans le tracé des arcs, des écoinçons, des consoles, dans la mise en place de la perspective de la composition et dans le déroulement des petites scènes.

Le dessin sous-jacent des grisailles des volets est plus élaboré que celui du panneau central avec un traitement poussé des zones d'ombre (fig. 7). Les modifications et les changements y sont également importants. On signalera par exemple la rectification de la longueur des doigts du Christ et ceux des autres personnages et la modification de l'orientation de la tête et du chapeau du premier juif à droite. Ce dernier apparaissait d'abord de profil; actuellement il est vu de face. La main et la pièce de monnaie du deuxième juif ont été également déplacées pour les éloigner quelque peu du premier personnage (fig. 8).

Les différences entre le stade du dessin et celui de la peinture sont nombreuses. Nous en avons évoquées quelques-unes. Nous espérons que les observations qui précèdent constituent un premier pas dans la direction suggérée par van Asperen de Boer et *alii* en 1992 lorsqu'ils écrivaient: «*It would be interesting to pursue the study of underdrawing in the Vranke Van der Stockt Group further to see whether more of such similarities could be revealed*».

Carmen Garrido
carmen.garrido@museodelprado.es

Docteur en histoire de l'art. Spécialisée dans l'examen des œuvres d'art par les méthodes de laboratoire. Conservateur au Musée du Prado. Chef du Cabinet de documentation technique qu'elle a activement contribué à développer. Direction de projets d'investigation et de cours de spécialisation au sujet de ces matières en collaboration avec des universités et d'autres musées. Participation à un grand nombre de groupes de travail nationaux et internationaux. Nombreuses publications concernant les examens techniques des peintures et sculptures, parmi lesquelles: *Velázquez. Técnica y Evolución* (1992). *Velázquez. La técnica del genio* (1998) en collaboration avec Jonathan Brown, *El Bosco en el Museo del Prado. Estudio Técnico* en collaboration avec Roger Van Schoute et *al.* (2001) et, en collaboration, avec Peter Cherry, *Luis Meléndez: La serie de los bodegones del Príncipe de Asturias. Estudio Técnico* (2004). Commissaire de l'exposition *El trazo oculto* (Madrid, 2006).

Duilio Bertani
duilio.bertani@infn.it

1973: diplôme en Physique de l'*Universitá degli Studi* de Milan. 1975: attaché au *Centro di Studio per le cause di diparimento di metodi di conservazione delle opere d'arte*. 2000-2005: responsable de la recherche au *Laboratorio di Metrologia Ottica per la Diagnostica dei Beni Culturali* de Milan. 2006: responsable scientifique du *Centro di Rifflettografia infrarossa e Diagnostica dei Beni Culturali de l'Universitá degli Studi de Milano*. Responsable dell'*Unita Operativa* du projet pour les biens culturels du *Consiglio Nazionale della Ricerche* à l'*Istituto Nazionale di Ottica Applicata*. A enseigné les *Techniche di analisi di ottica avanzata applicata a dipinti ed elaborazione di immagini* à *l'Universitá degli Studi* de Milan. Auteur de diverses études sur la réflectographie à l'infrarouge à haute résolution.

Roger VAN SCHOUTE

Né en 1930. Licencié en philosophie et lettres (1956). Docteur en archéologie et histoire de l'art (1961). Professeur ordinaire à l'Université Catholique de Louvain (1967), Histoire de l'Art du Moyen Âge, Histoire de la peinture, Histoire de la gravure, Muséologie, Méthodes de laboratoire appliquées à l'Histoire de l'Art. Professeur à l'Institut Royal d'Histoire de l'Art et d'Archéologie de Bruxelles et Secrétaire-général de cet Institut (1968). Professeur ordinaire émérite à l'Université catholique de Louvain (1995). A fondé le Laboratoire d'étude des œuvres d'art par les méthodes scientifiques (UCL). Création du Colloque bisannuel pour l'étude du dessin sous-jacent dans la peinture avec publication d'actes (1975), ancien président du Conseil scientifique des Musées Royaux d'art et d'histoire de Belgique. Publications: les Primitifs flamands (principalement Rogier Van der Weyden, Jerôme Bosch), Pierre Bruegel et les méthodes de laboratoire appliquées à l'histoire de la peinture.

Jaime GARCÍA-MÁIQUEZ
jaime.garcia-maiquez@museodelprado.es

Licencié à l'Université de Grenade en 1997. Licencié de l'Université de Cádix en 1999. Thèse doctorale à l'Université de Cádix. Depuis 1997, il travaille dans le Cabinet de Documentation Technique du Museo del Prado pour l'interprétation des réflectographies à l'infrarouge. Il est spécialiste dans la prise, le traitement et l'interprétation des réflectographies à l'infrarouge. Il a publié des articles et donné des conférences relatives à la réflectographie à l'infrarouge, les rayons X et les analyses d'échantillons de coupes de peinture. Il a été un collaborateur direct pour la réalisation des expositions consacrées au dessin sous-jacent: *El Trazo Oculto* (Madrid, 2006) et *Dibuixos al Descobert* (Barcelona, 2008).

Ana GONZÁLEZ MOZO
ana.gonzalez@museodelprado.es

Docteur ès Beaux Arts à l'Université Complutense de Madrid en 1999. Spécialisé dans la conservation et la restauration des œuvres d'art. Entre 1990 et 1996, travaille dans le Département de Peinture et Restauration de la Faculté des beaux-arts de Madrid, avec une bourse de formation de personnel chercheur du Ministère d'Education et Science; projet: traitement digital des images appliqué à la documentation et restauration des œuvres d'art. Depuis 1996, travaille au service des Archives photographiques et documentaires du Prado, s'occupant de digitalisation dans l'optique de la conservation et de la connaissance des processus créatifs de la peinture. En 1998, elle rejoint le Cabinet de documentation technique du Prado pour digitaliser la documentation du Département et réaliser des documents techniques destinés à la recherche.

TWO DONORS WITH SAINT JOHN THE BAPTIST AND SAINT BARBARA (SAINT-MARTIN CATHEDRAL, BRATISLAVA)

Ingrid Ciulisová*

Abstract: The painting of *Two donors with Saint John the Baptist and Saint Barbara* from the Cathedral of Saint-Martin in Bratislava was recently studied and restored. The oak support results from joining two arched altar wings with a central plank of different quality. The underdrawing has given new insights both into the history of the panel and the methods involved in its execution. The coexistence of two different techniques suggests the participation of more than one hand. While shadows in the head of Saint John the Baptist are indicated by wash-like brushstrokes, most figures have only a sketchy underdrawing, typical of Colijn de Coter and his workshop. The panel in its present form seems to be the product of the Reformation. Two issues still remain unsolved: the original form of the altar and the identity of its painters.

Colijn de Coter (c. 1450/1455 Brussels? – c. 1539/1540 Brussels?) was one of the most successful masters working in Brussels at the end of the fifteenth and the beginning of the sixteenth century.[1] Three signed paintings can be reliably connected with him: *The Trinity altarpiece*, painted for the church of Saint-Bertin in Saint-Omer (now in Paris, Louvre), the *Virgin Crowned by Angels* (Düsseldorf, Herta Pape Collection) and *Saint Luke Painting the Virgin* now in the parish church of Vieure near Cosne d'Alliers. Colijn de Coter copied compositions by older masters. The aforementioned signed works are closely influenced by Robert Campin and Rogier Van der Weyden. According to Catheline Périer-d'Ieteren, the painter had a well-organized workshop.[2]

* I would like to thank Edwin Buijsen, Lorne Campbell, Maria Clelia Galassi, J.P. Filedt Kok, Micha Leeflang, Hélène Mund, Jochen Sander, Cécile Scaillierez, Sabine van Sprang, Ron Spronk and Zsuzsa Urbach for their kind assistance. Sincere thanks go to Maryan W. Ainsworth who commented on the first draft and to Catheline Périer-d'Ieteren who read the first version of the text.

[1] On Colijn de Coter see: M.J. FRIEDLÄNDER, "Bernaert van Orley", in *Jahrbuch der Preußischen Kunstsammlungen*, 29, 1908, p. 225-246; J. MAQUET-TOMBU, *Colyn de Coter, peintre bruxellois*, Bruxelles, 1937; C. PÉRIER-D'IETEREN, *Colyn de Coter et les ateliers brabançons de peinture de la fin du XVe et du début du XVIe siècle* (master thesis), Université libre de Bruxelles, Bruxelles, 1981; C. PÉRIER-D'IETEREN, *Les volets peints des retables bruxellois conservés en Suède et le rayonnement de Colyn de Coter*, Stockholm, 1984; C. PÉRIER-D'IETEREN, *Colyn de Coter et la technique picturale des peintres flamands du XVe siècle*, Bruxelles, 1985; C. STROO, P. SYFER-D'OLNE et al., *The Flemish Primitives. III. The Hieronymus Bosch, Albrecht Bouts, Gerard David, Colijn de Coter and Goosen Van der Weyden Groups*, Catalogue of Early Netherlandish Painting in the Royal Museum of Fine Arts of Belgium, Bruxelles, 2001.

[2] C. PÉRIER-D'IETEREN, "Contribution à l'étude des volets peints du retable sculpté de Pruszcz", in *Bulletin du Musée national de Varsovie*, 18, 1977, p. 55-67; C. PÉRIER-D'IETEREN, 1981; C. PÉRIER-D'IETEREN, "Aspects récents de l'étude des volets de Colijn de Coter: Marie-Madeleine affligée et Saint Jean pleurant, du Musée de Budapest", in *Bulletin du Musée Hongrois des Beaux-Arts*, 58/59, 1982, p. 107-122; C. PÉRIER-D'IETEREN, 1984; C. PÉRIER-D'IETEREN, 1985; C. PÉRIER-D'IETEREN, "Le dessin sous-jacent dans l'œuvre de Colyn de Coter", dans *Le problème de l'auteur de l'œuvre de peinture*, Colloque IV, 29, 30, 31 octobre 1981, R. Van Schoute et D. Hollanders-Favart eds, Louvain-la-Neuve, 1982, p. 131-136. See also F. GOMBERT, D. MARTENS, *Primitifs flamands. Le Maître au Feuillage brodé. Secrets d'atelier*, exh. cat., Palais des Beaux-Arts de Lille, 13 mai-24 juillet 2005, Paris, 2005; L. CAMPBELL, "Primitifs flamands. Le Maître au Feuillage brodé. Secrets d'atelier" (Paris, Réunion des musées nationaux, 2005), in *The Burlington Magazine*, 148, 2006, n°1235, p. 127-128.

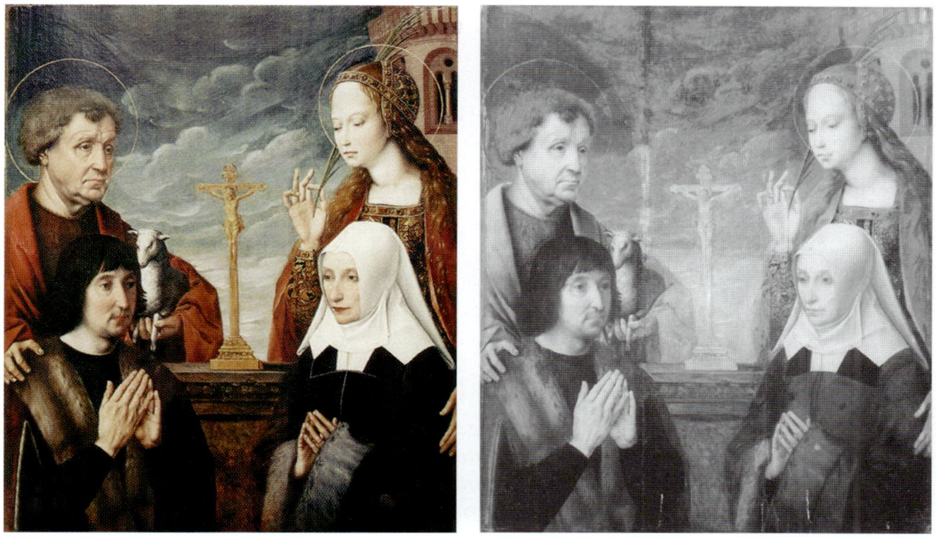

Fig. 1. a. Colijn de Coter and workshop, *Donor with Saint John the Baptist and donatrix with Saint Barbara*, Bratislava, Saint-Martin Cathedral (© Anna Mičúchová, courtesy of the archbishopric). b. IRR (© Bedrich Hoffstädter).

In 1927 Gizela Weyde, curator of the Bratislava Municipal Museum, published her article on the recently discovered *Two donors with Saint John the Baptist and Saint Barbara* from Saint-Mary chapel (*Tiefenwegkapelle*) in Bratislava (Pressburg, Pozsony) (fig. 1a). She examined its iconography and attributed the work to an anonymous Netherlandish painter of the fifteenth century. She observed that the panel resulted of joining two altar wings with a new central plank. Otto Benesch was the first to associate the wings with Colijn de Coter. He believed that the donatrix with Saint Barbara was the work of Colijn de Coter and that the donor with Saint John the Baptist might have been started by Hugo van der Goes and completed by Colijn de Coter. However, his hypothesis was not accepted by later scholars. Jeanne Maquet-Tombu, Charles de Tolnay, Catheline Périer-d'Ieteren, Micheline Comblen-Sonkes and Jarmila Vacková attributed the wings to Colijn de Coter.[3]

[3] G. WEYDE, "Die Tiefenwegkapelle", in *Grenzbote*, 23 octobre 1927, p. 2; G. WEYDE and O. BENESCH, "Zwei niederländische Altarflügel in der Pressburger *Tiefenwegkapelle*", in *Pantheon* I, 1928, p. 68-70; J. MAQUET-TOMBU, 1937, p. 73-75, fig. 31; Ch. DE TOLNAY, "Hugo van der Goes as Portrait Painter", in *The Art Quarterly*, VII, 1944, p. 181-190, p. 189, note 2; C. PÉRIER-D'IETEREN, 1981; C. PÉRIER-D'IETEREN, "Aspects récents de l'étude des volets de Colijn de Coter: Marie-Madeleine affligée et Saint Jean pleurant, du Musée de Budapest", in *Bulletin du Musée Hongrois des Beaux-Arts*, 58/59, 1982, p. 113-114, pl. 93; C. PÉRIER-D'IETEREN, 1985, p. 90, 147, 174-175, 182, note 95, pl. 195, 196 a,c; J. VACKOVÁ and M. COMBLEN-SONKES, *Collections de Tchécoslovaquie. Les Primitifs flamands. II. Répertoire des peintures flamandes du quinzième siècle*, 4, Bruxelles, Centre National de Recherches des Primitifs flamands, 1985, cat. 38, p. 97-99, pl. XLVI; J. VACKOVÁ, "Flámští 'primitivové' v československých sbírkách (Flemish Primitives in the collections of Czechoslovakia)", in *Umění* XXXIII, 1985, n°3, p. 225, note 110, plate 18; J. VACKOVÁ, "Flámští 'primitivové' v československých sbírkách (Flemish Primitives in the collections of Czechoslovakia). II", in *Umění* XXXVII, 1989, n°4, p. 323; J. VACKOVÁ, *Nizozemské malířství 15.a 16. století. [Netherlandish Paintings of the 15th and 16th Centuries] Československé sbírky*, Praha, 1989, p. 44, note 13, p. 90, pl. 9.

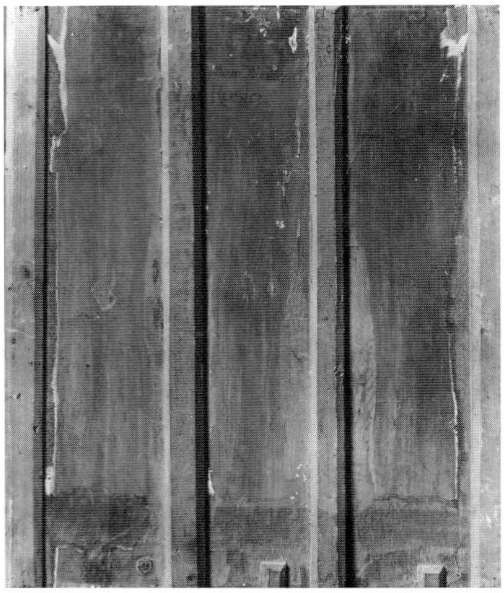

Fig. 2. Colijn de Coter and workshop, *Donor with Saint John the Baptist and donatrix with Saint Barbara*. Reverse of the panel (1954) (© Archive of the Monument Office, Bratislava).

The two kneeling donors are represented in three quarter profile in the foreground, turned in prayer towards the metal crucifix on a marble plinth of an unidentified architecture (an altar?) behind them. There are no indications of the identity of the donors. Their names were probably John and Barbara like their patron saints, Saint John the Baptist, holding the Lamb of God and Saint Barbara with the palm of martyrdom and the tower. The background of the scene is a dramatic cloudy sky.

Catheline Périer-d'Ieteren connected the Bratislava painting with two panels attributed to Colijn de Coter *Christ and the Virgin Interceding with Philip the Handsome and Joan the Mad* (Paris, Louvre).[4] There are also close relations with Colijn de Coter's paintings that were once part of the *Last Judgment* altarpiece in Cologne. The *Saint Barbara* bears a close resemblance in facial type and pose to the woman in *Saint Peter and the Blessed* (Munich, Bayerische Staatsgemäldesammlungen).[5] Details like the long thin fingers of Saint Barbara or the hands of the praying donatrix recall Rogier Van der Weyden and his *Polyptych of the Last Judgment* in Beaune.

The painting has been recently restored and extensively studied.[6] Infrared examination gave new insights both into the history of the panel and the techniques used by the painters (fig. 1b). The support (90 × 75 cm) consists of three vertical oak planks. During former restorations four boards, presumably applied in the nineteenth century, were removed, the panel was thinned and reinforced with wooden battens. The reverse

[4] C. Périer-d'Ieteren, 1985, p. 90, see esp. H. Adhémar, *Le Musée National du Louvre, Paris, I. Corpus de la peinture des anciens Pays-Bas méridionaux et de la principauté de Liège au quinzième siècle.* Centre International d'Étude de la peinture médiévale des Bassins de l'Escaut et de la Meuse, Bruxelles, 1962, cat. 88.

[5] C. Périer-d'Ieteren, 1985, fig. 223; C. Stroo, P. Syfer-d'Olne et al., 2001, fig. 174.

[6] The painting was restored in 2005 by Alena Kubová. IR photographs were taken by Bedrich Hoffstädter. All information about technique, materials and history of the previous restorations strictly refer to the restoration reports preserved at the Monuments Office archive in Bratislava.

of the added central plank originally bore the inscription *Vom Meister Rosier (or Rogier?)* and the number 123, deleted during the restoration in 1954-55[7] (fig. 2). Observation of the reverse confirmed that the wings formerly had arched tops. They were reduced into rectangles and united with the central plank bearing the crucifix. When removed from their modern frame, both wings had unpainted borders (c. 12 mm), as they were originally painted in their frames. As Alena Kubova confirmed, the unpainted edges were trimmed when the central panel was added. The edges of the painted surface bear incisions in the ground still visible at various places. All three panels are covered with a whitish chalk ground.

Infrared examination reveals that the painting is underdrawn and that there are two distinct styles. The sketchy underdrawing of Saint Barbara's head is limited to basic facial features. The eyes are indicated by simple half cercles (fig. 3a). No hatching for modelling is seen. The position of the head-dress is simply marked. By contrast, the right hand of Saint Barbara is carefully outlined, as are almost all her fingers. The underdrawing of Saint Barbara's face is comparable to the underdrawing in other works associated with Colijn de Coter. For example, the underdrawing of the face of the woman in *Saint Peter and the Blessed* (Munich, Bayerische Staatsgemäldesammlungen) shows similar contours. Also the underdrawing of the face of Saint Michael (Brussels, Royal Museum of Fine Arts) reveals similar effects, even more confident and complex.

The left side of the donor's face is heavily damaged. However, the sketchy underdrawing is still visible. No hatchings evoke a modelling. The position of the masterly painted donor's hand is indicated by swift brush strokes in form of two small crosses. Striking is that in the Bratislava painting the usual hatchings adopted for shadows are replaced by a monochrome undermodelling. IR photographs of the head of Saint John the Baptist show that the facial features are accentuated by broad brushstrokes of wash, indicating both the form and the modelling (fig. 3b). The overall effect of the undermodelling is repeated elsewhere, and fine brush strokes create a subtle transition from light to dark. The painter evidently had a special interest in modelling.

As Catheline Périer-d'Ieteren emphasized, Colijn de Coter used three-dimensional effects in his paintings. The underdrawing of the body of Christ from the Coter's *Trinity Altarpiece* for example, indicates shadows with long hatchings which vary in orientation and intensity.[8] Recent investigations by Rainald Grosshans on the Hugo van der Goes paintings in Berlin show that the underdrawing was important to indicate both the overall form and the repartition of light and shadow.[9] Shadows are indicated by

[7] The painting was restored in 1898 by István Tóth and in 1954-55 by Július Alló.

[8] C. PÉRIER-D'IETEREN, 1985, fig. 85; C. PÉRIER-D'IETEREN, *Le dessin sous-jacent dans l'œuvre de Colyn de Coter*, 1982, p. 131-136, pl. 37.

[9] R. GROSSHANS, "IRR-investigations of the panel paintings by Hugo van der Goes in the Berlin Gemäldegalerie", in *Jérôme Bosch et son entourage et autres études*, Colloque XIV, 13-15 septembre 2001, H. Verougstraete-Marcq and R. Van Schoute eds., Bruges-Rotterdam, Leuven-Paris-Dudley MA, 2003, p. 235-249; see also J. SANDER, *Hugo van der Goes: Stilentwicklung und Chronologie*, Mainz, 1992, M.F. VIANA, "Le dessin sous-jacent chez Hugo van der Goes", in *Le dessin sous-jacent et technologie de la peinture. Perspectives*, Colloque XI, 14-16 septembre 1995, R. Van Schoute and H. Verougstraete eds., Louvain-la-Neuve, 1997, p. 59; C. THOMPSON and L. CAMPBELL, *Hugo van der Goes and the Trinity Panels in Edinburgh*, Edinburgh, 1974; I. ALEXANDER, F. MAIRINGER and R. VAN SCHOUTE, "Le dessin sous-jacent chez van der Goes: le diptyque du Péché originel et la Déploration du Kunsthistorisches Museum de Vienne", in *Revue des archéologues et historiens d'art de Louvain*, novembre 1978, p. 73-83; R. VAN SCHOUTE,

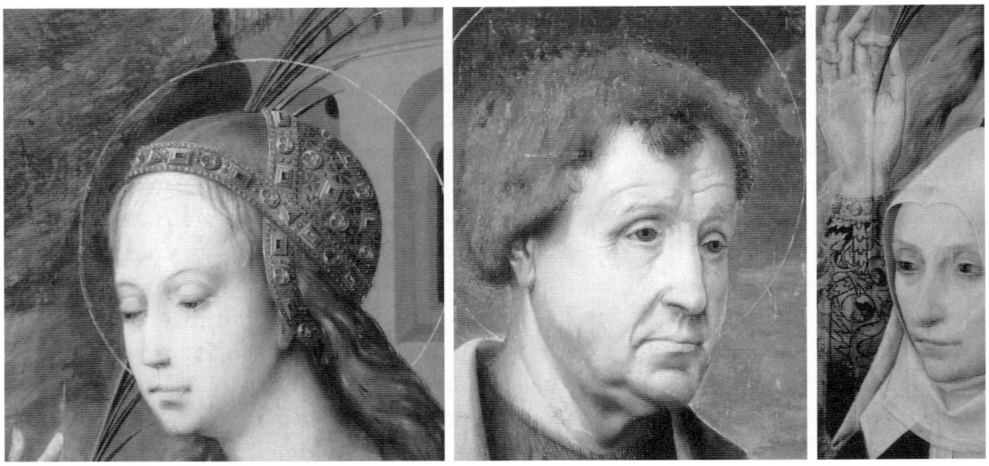

Fig. 3. Colijn de Coter and workshop, *Donor with Saint John the Baptist and donatrix with Saint Barbara.* IRR details. a. Head of Saint Barbara (© Bedrich Hoffstädter). b. Head of Saint John the Baptist (© Bedrich Hoffstädter). c. Head of the donatrix (© Bedrich Hoffstädter).

dense parallel hatching with lines of varying length. The underdrawings of Colijn de Coter and Hugo van de Goes correspond with the contemporary underdrawing practices of northern artists using parallel and cross-hatching for modelling. The underdrawing in Bratislava seems to follow the Italian practices described in Cennino Cennini's *Il libro dell'arte.* Undermodelling has also been described in sixteenth century masters like Lucas van Leyden, as documented for example in his *Last Judgement* of 1526-27.[10] Undermodelling in underdrawings is already present in Hans Memling and Gerard David.[11]

Patrons wished to determine almost every detail of their portraits. Infrared photographs revealed several practical difficulties encountered by the painter(s). For example, the nose of the donatrix was seriously altered and her head-dress was originally higher (fig. 3c). She is finally represented as long-nosed and thin-lipped. On the other hand, the head of Saint John the Baptist seems to have been painted from a carefully prepared preliminary drawing or from life. The painter contrasted the profane portrait and the sacred figure and in this case the donor is more idealized than Saint John the Baptist.

"Le dessin de peintre chez Hugo van der Goes: la *Mort de la Vierge* du Musée Groeninge de Bruges, l'*Adoration des Mages* de la Victoria Art Gallery de Bath", in *Revue des archéologues et historiens d'art de Louvain,* 1972, p. 59-66.

[10] J.P. FILEDT KOK, "Underdrawing and other technical aspects in the paintings of Lucas van Leyden", in *Nederlands kunsthistorisch jaarboek,* 29, 1978, p. 184.

[11] M.W. AINSWORTH, *Gerard David. Purity of Vision in an Age of Transition,* New York, Metropolitan Museum of Art, 1998, p. 84-85; M.W. AINSWORTH, "Minimal Means, Remarkable Results", in *Memling's Portraits,* Till Holger-Borchert ed., Ghent-Amsterdam, 2005, p. 98; concerning the role played by undermodelling and underdrawing in Italian painting technique see M. GALASSI, "Underdrawing vs. undermodelling: form construction in Tuscan painting during the first half of the 15th century", in *La peinture et le laboratoire. Procédés, méthodologie, applications,* Colloque XIII, Bruges, 15, 16, 17 septembre 1999, R. Van Schoute and H. Verougstraete eds., Leuven, 2001, p. 131-141.

This could be an example of the practice where the donor is represented disguised in apostle. On *Saint John the Baptist with Six Apostles*, which is a surviving fragment of the Colijn de Coter *Last Judgement* altarpiece, the face of one apostle has been painted on an auxiliary support glued onto the ground.[12] Also, on the reverse of the right wing of the Johanna van Maerke triptych by Colijn de Coter, the donatrix' face was originally painted on a piece of tin foil glued directly onto the ground.[13] However, the X-radiograph and ultraviolet fluorescence reveal that the face of Saint John the Baptist in Bratislava was not overpainted and that no intermediary support was used.

When Otto Benesch ascribed in 1928 the head of Saint John the Baptist and the hands of the praying donor to Hugo van der Goes, he could judge only on what he saw with the naked eye. On the basis of infrared examination we can suggest that while the right side panel with Saint Barbara seems to be a work by Colijn de Coter (1500-1510), the left side panel with Saint John the Baptist might have been completed in collaboration with another painter. The head of Saint John the Baptist has a different underdrawing style from those seen in underdrawings by both Colijn de Coter and Hugo van der Goes. The central panel is, as Gizela Weyde already pointed out, a later addition. Summing up, the Bratislava painting in its present form is a rather composite painting.

Ingrid CIULISOVÁ
Insitute of Art History
Slovak Academy of Sciences
Dúbravská cesta 9
SK – 841 04 Bratislava
dejuciul@savba.sk

The author is a senior art historian at the Slovak Academy of Sciences, Institute of Art History in Bratislava. She obtained a Ph.D. in Art History from Comenius University (1995) and was employed as a conservator at the state restoration studio. She has been awarded short-term research fellowships in the National Gallery, London, the Netherlands Institute for Advanced Study, Wassenaar, the Dutch Institute for Art History, Florence and in the American Academy in Rome. 2008: Associate Professor of Art History, Charles University in Prague.

[12] C. PÉRIER-D'IETEREN, 1985, fig. 224b; C. STROO, P. SYFER-D'OLNE et al., 2001, p. 329, note 20.

[13] C. STROO, P. SYFER-D'OLNE et al., 2001, p. 318, note 18; for other examples of portraits pasted onto the surface see R.H. MARIJNISSEN and G. VAN DE VOORDE, "Een onverklaarde werkwijze van de Vlaamse primitieven. Aantekeningen bij het werk van Joos van Wassenhove, Hugo van der Goes, Rogier Van der Weyden en Hans Memling", in *Mededelingen van de Koninklijke Academie voor Wetenschappen, Letteren en Schone Kunsten van Belgie, Klasse der Schone Kunsten*, 44, n°2, 1983, p. 41-51; R.H. MARIJNISSEN and G. VAN DE VOORDE, "Un procédé énigmatique des Primitifs flamands: annotations concernant l'œuvre de Joos van Wassenhouve, Hugo van der Goes, Roger Van der Weyden et Hans Memling", in *Le dessin sous-jacent dans la peinture. Dessin sous-jacent et autres techniques graphiques*, Colloque V, 29-30 septembre - 1er octobre 1983, R. Van Schoute and D. Hollanders-Favart eds., Louvain-la-Neuve, 1985, p. 23-34; L. CAMPBELL, "The making of early Netherlandish painted portraits", in *The image of the individual: portraits in the Renaissance*, Nicholas Mann and Luke Syson eds., London, 1998, p. 105; *Early Netherlandish Paintings. Rediscovery, Reception and Research*, Bernhard Ridderbos, A. van Buren and H. van Veen eds., Amsterdam University Press, 2005, p. 121, p. 413, note 183.

LE DESSIN SOUS-JACENT DE L'*ECCE HOMO* ET DE LA *VIEILLE FEMME S'ARRACHANT LES CHEVEUX* DE QUENTIN METSYS (MUSÉE DU PRADO, MADRID)

Pilar Silva Maroto*

Résumé: Les infrarouges aident à comprendre le processus créateur de deux œuvres de Quentin Metsys conservées au Prado. L'*Ecce Homo* présente un dessin abondant avec des rectifications, des déplacements et des changements de composition. Le dessin sous la *Vieille femme* manque de modelé, ce qui peut s'expliquer par l'influence de Leonardo, à qui l'artiste emprunte le caractéristique *sfumato*.

L'*Ecce Homo*

L'*Ecce Homo*[1] fut légué au Prado en 1940 par M. Mariano Lanuza. Au XIXe siècle, l'œuvre appartenait au marquis de Remisa. Bien qu'elle ne soit ni documentée ni signée, on considère généralement qu'elle est de la main de Quentin Metsys (Louvain 1495/1496 – Anvers 1530) et qu'elle date d'environ 1518-1520[2]. La confrontation de l'œuvre avec deux autres versions du même sujet réalisées par le même artiste à des moments différents (Musée Machado de Castro à Coimbra, documenté entre 1513 et 1517[3] et Palais ducal à Venise, ca. 1528-1529[4]) permet de constater un intérêt grandissant pour les figures de grande échelle limitées en nombre. Il s'ensuit une perte d'importance du cadre où se déroule l'action. L'artiste accentue le clair-obscur, déjà visible sur le panneau de Madrid. Finalement, il opte pour de nouvelles couleurs, comme on l'observe sur l'œuvre conservée à Venise.

Metsys conçoit la composition de l'*Ecce Homo* de Coimbra en perspective plongeante. Le Christ et Ponce Pilate se situent dans la partie supérieure du panneau, séparés par une balustrade de la violence qui éclate, plus bas, entre les juifs et les soldats. L'influence des dessins de Leonardo est perceptible dans les visages. Il n'est pas certain que Metsys voyagea en Italie et on ignore comment il eût accès, à Anvers, aux quelques dessins de Leonardo qui représentent des visages laids et déformés. Toujours est-il que, à un moment assez précoce de sa carrière, ces dessins l'incitèrent à approfondir les caractéristiques grotesques de ses propres personnages[5]. Plus tard, il parviendra à représenter

* Traduction en français, María Jesús Gozalo.

1 N° 2801, huile sur bois, 160 × 120 cm.

2 M.J. FRIEDLÄNDER, *Early Netherlandish Painting. VII. Quentin Massys*, Leyden-Bruxelles, 1971, n°10; A. DE BOSQUE, *Quentin Metsys*, Bruxelles, 1975, p. 196, fig. 63 et L. SILVER, *The Paintings of Quinten Massys with Catalogue Raisonné*, Oxford, 1984, p. 94, 95 et 220, cat. 31, fig. 86.

3 L. SILVER, 1984, p. 213, cat. 18, fig. 33. Il fait partie de l'autel de la passion donné par Manuel I de Portugal au monastère de Santa Clara de Coimbra.

4 L. SILVER, 1984, p. 228-229, cat. 46, fig. 87.

5 Metsys a connu le dessin de cinq têtes grotesques de Leonardo (Windsor Castle). Il l'utilisa pour la représentation du Martyre de saint Jean l'Evangéliste du *Triptyque de saint Jean* du Musée des Beaux-Arts d'Anvers, documenté entre 1508 et 1511 (L. SILVER, 1984, fig. 16 et 17).

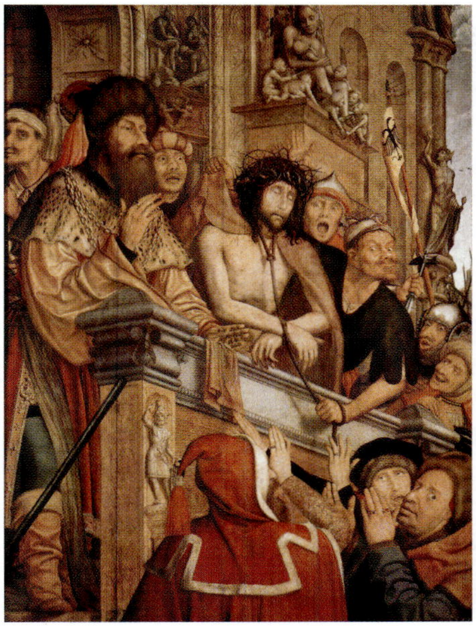

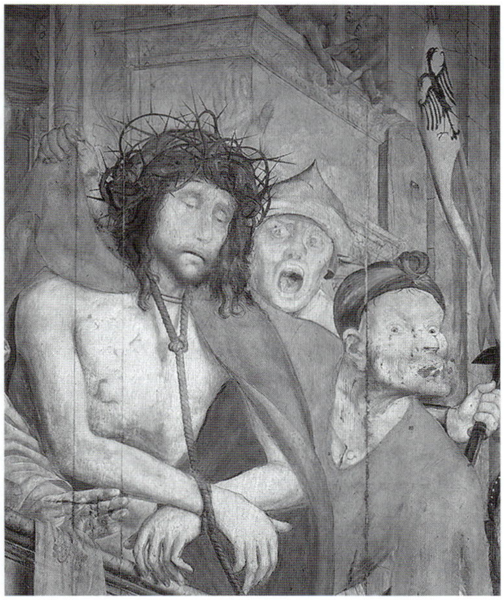

Fig. 1. a. Quentin Metsys, *Ecce Homo*, Madrid, Musée du Prado (© Prado). b. *Ecce Homo*, détail IRR (© Prado).

des personnages grotesques sans avoir encore recours aux modèles précités, tel qu'on peut le remarquer dans l'*Ecce Homo* du Prado (fig. 1a). Dans cet *Ecce Homo*, Quentin Metsys conserve la balustrade qui sépare le Christ et Pilate des juifs représentés dans la partie inférieure. Cependant, l'échelle plus monumentale des figures et le point de vue plongeant contribuent à attirer l'attention sur les fidèles qui, à hauteur des juifs, réclament la mort du Christ. Celui qui occupe le premier plan à droite dirige son regard vers le spectateur (comme à Coimbra) et l'invite à méditer sur la Passion. La disposition oblique de l'architecture produit un effet dynamique auquel contribue la violence exercée sur le Christ.

Dans l'*Ecce Homo* du Prado, de même que dans d'autres tableaux religieux, l'artiste accorde au grotesque une signification morale. Dans une conception presque manichéenne, il associe la laideur des bourreaux et des soldats au mal. La beauté quant à elle est associée à la divinité, au Christ. Bien que le peintre s'intéresse davantage aux figures qu'au cadre architectural, les personnages du Prado partagent le premier rôle avec l'architecture qui évoque Rome (la Judée était alors une province romaine) autant au travers du Prétoire de Jérusalem qu'à travers sa riche décoration sculptée[6]. Cependant,

[6] En haut, au centre, la femme nue qui allaite un nourrisson, entourée de six autres enfants, est associée à la Charité. Le vieillard barbu derrière elle pourrait peut-être inciter à l'identifier avec la Charité romaine ou l'histoire de Cimon et Pero racontée par Valère Maxime. La statue sur la colonne du côté droit du Prétoire est associée à Mercure à cause de son casque à ailes, mais l'étoile qu'il a dans la main n'est pas un de ses attributs. Il est difficile d'interpréter la scène à gauche de la Charité: un roi assis, un homme à ses pieds et un soldat debout.

à la différence d'Andrea Mantegna au Quattrocento et de ses suiveurs, la reconstruction historique ne s'applique pas aux costumes des figures bibliques.

L'*Ecce Homo* de Venise ne comporte plus de balustrade, au contraire des *Ecce Homo* de Coimbra et de Madrid. L'espace accordé à l'architecture, dans le haut, derrière les figures, est réduit pour que le fidèle puisse participer plus directement au drame religieux. La confrontation entre le bien et le mal est toujours présente, mais le nombre de figures diminue tandis que leurs dimensions augmentent.

On sait que Quentin Metsys réalisa des cartons pour tapisseries. Cependant la tapisserie de l'*Ecce Homo* du Metropolitan Museum of Art à New York[7], qui inverse la composition du Prado et introduit des variantes, n'a pas été faite au départ d'un carton de Metsys, mais plutôt à partir d'une copie qui incluait des changements dans l'architecture et dans les vêtements des bourreaux. Les coiffures, tout particulièrement, reflètent la mode de 1550-1560. Le bourreau, entre le Christ et Pilate, porte un chapeau similaire à celui de Philippe II dans ses portraits de jeunesse. L'existence même d'une copie atteste du succès de la composition de Metsys.

Les infrarouges et les rayons X de l'*Ecce Homo* du Prado[8] permettent de suivre le processus créateur. Le dessin sous-jacent est linéaire, réalisé à sec. Il se limite aux contours de l'architecture et des figures, à la définition de la plupart des traits des visages et, dans quelques cas, à des détails internes comme les plis de la coiffure de Pilate. De même que dans le reste des œuvres de Quentin Metsys influencées par le *sfumato* leonardesque, le modelé intervient au stade de la couleur. Ici aussi on observe des changements, des déplacements et des rectifications de contours[9] qui, dans quelques cas, ont eu lieu avant l'application des couleurs, mais plus souvent au stade de la couleur, et en plus grand nombre que ce qui est habituel chez l'artiste.

Étant donné la difficulté de représenter une architecture, il n'est pas étonnant qu'on trouve là quelques modifications significatives. Avant d'entamer le dessin sous-jacent, Metsys incisa des lignes obliques en haut à droite, dans l'espace couvert en partie par l'étendard, comme on peut le constater dans les rayons X (fig. 4b). D'après les infrarouges, il traça d'abord l'architecture, tout en laissant des espaces pour les figures. Au début, la disposition du socle de la Charité était plus bas et occupait un espace plus large à droite. On peut vérifier les déplacements et les rectifications de contours, aussi bien dans les parties supérieures et inférieures que dans les côtés (fig. 1b, 2a). Dans la partie supérieure, on n'avait pas réservé d'espace pour les jambes du garçon à droite. Une autre modification à mentionner dans l'architecture est la forme adoptée par le chapiteau, à gauche de la Charité, partiellement couvert par le turban du bourreau entre le Christ et Pilate, qui au stade du dessin sous-jacent présentait une forme convexe avant de devenir concave (fig. 1b). Egalement intéressant, le dessin de quelques éléments

[7] A. de Bosque, 1975, p. 257, fig. 317.

[8] Examens réalisés au Cabinet de Documentation Technique du Musée du Prado. Pour les documents à l'infrarouge le prototype de l'équipement développé par Duilio Bertani de l'Université de Milan, qui mène une investigation en collaboration avec le Musée du Prado.

[9] Dans les *Tentations de Saint Antoine* de Joachim Patinir et de Quentin Massys du Musée du Prado on peut voir des caractéristiques similaires dans les quatre femmes et dans le saint Antoine au premier plan réalisés par Metsys. Voir P. Silva Maroto, 'Paisaje con la *Tentación de san Antonio Abad.* Joaquín Patinir y Quintín Massys' dans *Patinir estudios y catálogo crítico*, éd. A. Vergara, Madrid, 2007, n°14, p. 242-255, fig. 9, p. 251.

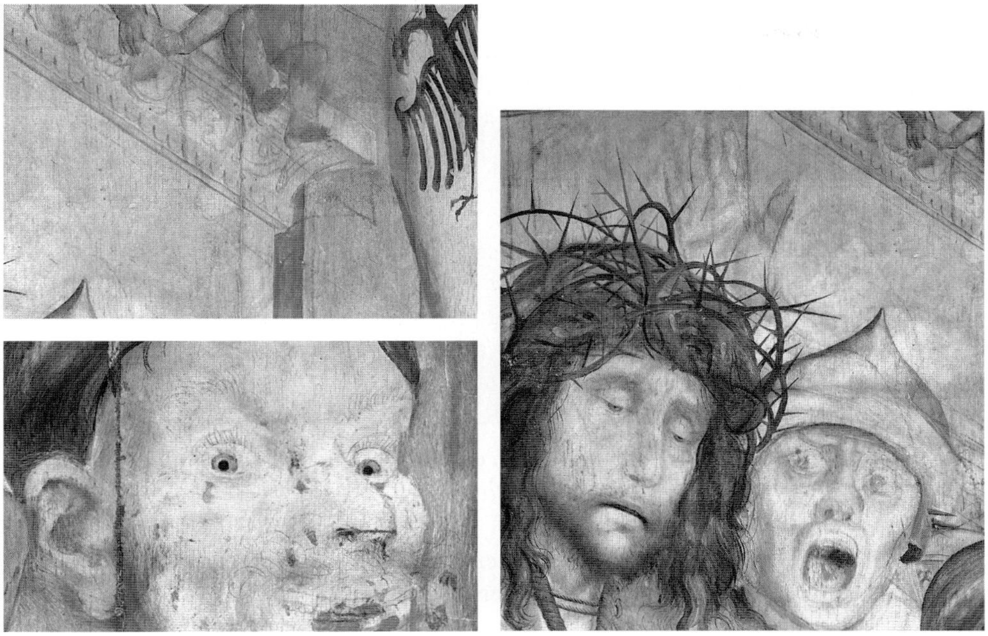

Fig. 2. a. *Ecce Homo*, IRR, détail avec le socle de la Charité, en haut à droite (© Prado). b. *Ecce Homo*, IRR, détail du bourreau qui tient la corde (© Prado). c. *Ecce Homo*, IRR, détail avec le Christ et un bourreau (© Prado).

décoratifs composant des frises et des moulures comme dans la partie supérieure du piédestal de la Charité ou les changements que Metsys fit dans l'étendard dessiné plus haut et plus à droite. Il y a aussi des rectifications dans la balustrade et dans l'extrémité de la corde tenue par le bourreau à droite du Christ.

Les changements dans les figures sont encore plus importants. Ils interviennent pour la plupart au stade de la couleur, mais parfois aussi au stade du dessin sous-jacent. Parmi les personnages qui présentent le plus grand nombre de changements, il y a le soldat en haut à droite. Il a été déplacé vers la droite, mais aussi vers le haut; les différentes positions et formes du nez, de la bouche, des yeux, de la main et du contour du casque sont mis en évidence par des traits exécutés à sec (fig. 3a). Le bourreau à turban à côté de lui, à droite[10], comporte des rectifications de contour dans le menton, la joue, le nez, les sourcils et les mains. À mentionner aussi le bourreau qui tient la corde à laquelle le Christ est attaché (fig. 1b, 2b). Quentin Metsys a dessiné tous les traits du visage, y compris les cils, les sourcils, la moustache et même les dents. À côté du bourreau se trouve un soldat qui, en un premier temps, était plus petit et se situait plus près du Christ. Son visage, son casque, ses yeux et sa bouche, de même que sa main, derrière la tête de Jésus, ont subi des rectifications de contours et des déplacements (fig. 1b, 2c). En ce qui concerne le Christ, il faut signaler les rectifications du contour des bras, des

[10] Des dégâts dans les yeux et dans la bouche, visibles aux rayons X, résultent d'actes de vandalisme qui touchent aussi le coin inférieur droit de la peinture.

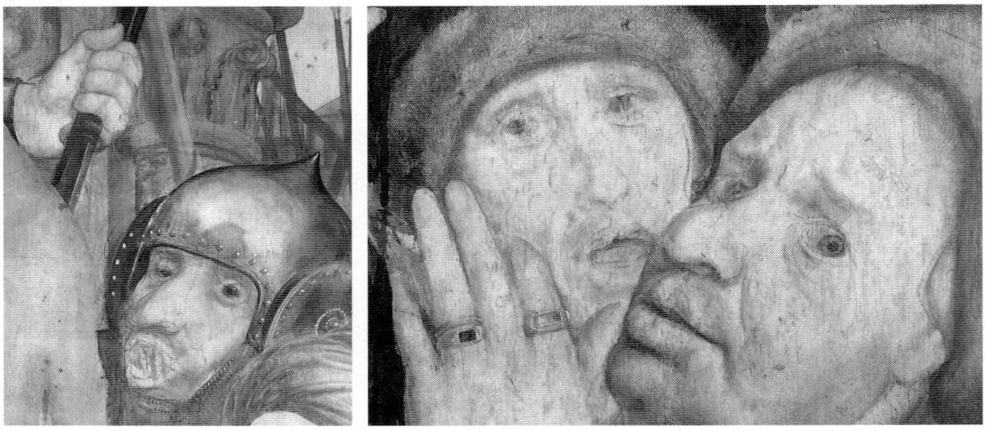

Fig. 3. a. *Ecce Homo*, IRR, détail avec un soldat, en haut à droite (© Prado). b. *Ecce Homo*, IRR, détail avec deux juifs, en bas à droite (© Prado).

mains et du manteau, le déplacement du nez vers la gauche, et la manière de tracer les cheveux (fig. 1b, 2c). Le visage du soldat entre Jésus et Pilate présente également des rectifications et des déplacements. Les rayons X révèlent la forme adoptée par ses vêtements, matérialisée en une couleur claire dans un premier stade de la couleur. Sous Pilate il y a des rectifications similaires pour le vêtement, les mains et les traits du visage. Les plis droits de la coiffe ne correspondent pas à ceux qui apparaissent en surface.

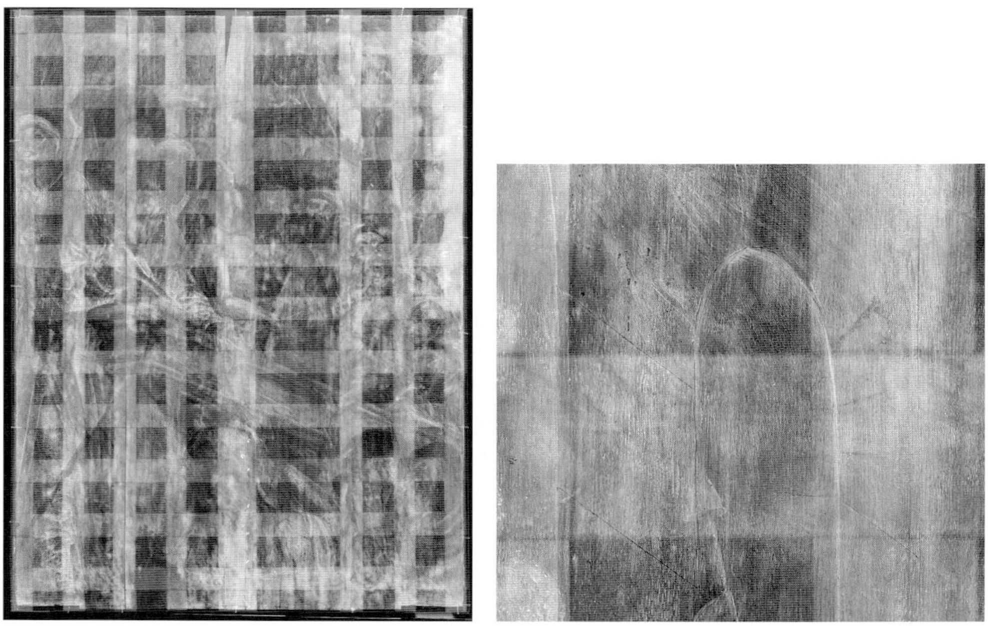

Fig. 4. a. *Ecce Homo*, RX (© Prado). b. *Ecce Homo*, RX, détail avec architecture, en haut à droite (© Prado).

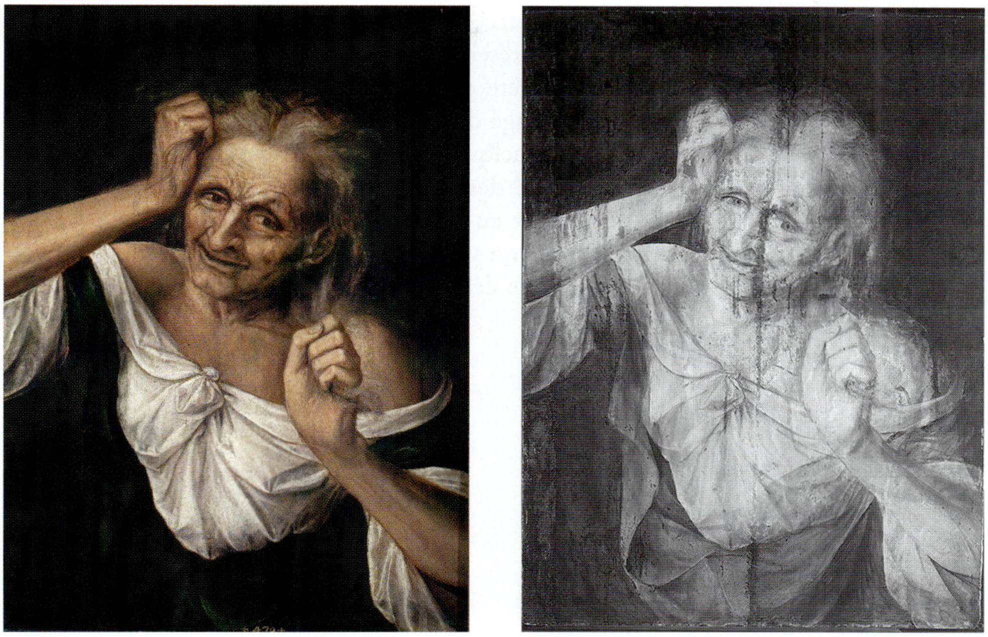

Fig. 5. a. Quentin Metsys, *Vieille femme s'arrachant les cheveux*, Madrid, Musée du Prado (© Prado).
b. *Vieille femme s'arrachant les cheveux*, IRR (© Prado).

Trois juifs sont représentés dans le bas. Celui de droite présente également des rectifications dans le contour du vêtement, de la coiffure et des mains. Plus intéressants sont les changements dans le visage, comme les yeux et surtout la bouche dont les lèvres rappellent celles d'une poupée. Celui qui vocifère à côté de lui présente un déplacement vertical particulièrement des dents et des yeux (fig. 3b), tandis que celui à gauche laisse voir des rectifications de contours similaires dans le vêtement et le profil, surtout dans le nez.

Le parquetage du support rend difficile l'interprétation de l'image radiographique. Celle-ci informe néanmoins sur l'état de conservation de l'*Ecce Homo* du Prado et sur la manière de peindre de Metsys (fig. 4a). Outre les lignes obliques incisées, déjà mentionnées, en haut à droite (fig. 4b), signalons les changements dans quelques visages qui interviennent au stade de la couleur, comme dans la zone de la bouche du bourreau qui soutient la corde attachée au Christ. Soulignons la façon de traduire les carnations ou les tissus. Avec un pinceau chargé de blanc, l'artiste définit les contours exposés à la lumière, les riches étoffes ou l'or, ou bien il détaille les traits des figures.

La *Vieille femme s'arrachant les cheveux*[11]

Achetée en 1964 par le Musée du Prado, cette peinture, qui porte au revers l'inscription MAESTRO QUINTIN, se situe dans la dernière phase de l'activité de l'artiste. Influencé

[11] N°3074, huile sur bois, 55 × 40 cm. Au XVIIe siècle il appartenait au marquis de Leganés (n°33). P. SILVA MAROTO, *Pintura flamenca de los siglos XV y XVI. Museo del Prado*, Madrid, 2001, p. 196-197.

par Leonardo et par son intérêt pour l'accentuation de traits caractéristiques — de même que pour le *sfumato* — Metsys représenta au premier plan, en raccourci, un personnage qui pourrait personnifier la colère ou la jalousie, généralement représentées sous les traits d'une vieille femme au visage déformé (fig. 5a).

Les infrarouges mettent en évidence quelques grandes lacunes surtout dans le visage, même si on lit encore presque la totalité des traits. Le dessin sous-jacent est linéaire, limité aux contours, respectés fidèlement au stade de la peinture, comme dans l'*Ecce Homo*, quoique sans les changements qu'on y observait (fig. 5b). Malheureusement, le manque de matériel comparatif pour cette dernière époque ne permet pas de connaître la technique de Metsys à ce moment[12].

Pilar SILVA MAROTO
Museo del Prado
Calle Ruiz de Alarcón 23
E – Madrid 28014
Tel: +34.91.33.02.809
Fax: +34.91.33.02.851
pilar.silva@prado.mcu.es

Docteur en Histoire de l'Art de l'Université Complutense de Madrid, où elle est professeur depuis 1972. Actuellement, professeur titulaire au Département d'histoire de l'art des temps modernes de la même université. Chef du Département de peinture flamande, des écoles du nord (1400-1600) et de peinture espagnole (1100-1500) au Musée du Prado. Se consacre à l'étude de la peinture hispano-flamande et de la première Renaissance en Castille. A réalisé sur ce sujet plusieurs publications à caractère documentaire, stylistique et iconographique. Depuis 1997, se consacre aussi à l'étude de la peinture flamande des XVe et XVIe siècles.

[12] On connait le dessin sous-jacent du *Sauveur* et de la *Vierge Marie* du Musée du Prado attribués à Quentin Metsys (n°1561 et n°1562), datés de 1529 au revers du *Sauveur*. Dans J.O. HAND, C.A. METZGER et R. SPRONK, *Prayers and Portraits. Unfolding the Netherlandish Diptych*, Washington-Anvers, 2006-2007, n°14, p. 282-283, les auteurs analysent ses caractéristiques et constatent que le dessin peu abondant est suivi de près au stade de la peinture. La ligne la plus visible apparaît entre les lèvres du Sauveur.

DENDROCHRONOLOGY AND INFRARED EXAMINATION OF JAN MASSYS' ŒUVRE

Maria Clelia Galassi*

Abstract: The œuvre of Jan Massys (ca. 1509-1573), Quinten's son, has been discussed by Leontine Buijnsters-Smets in her 1995 monograph which lists 56 paintings.[1] While Jan Massys' activity during the 1560's abounds in signed and dated paintings, his earlier production still presents problems in chronology and attribution, discussed by Buijnsters-Smets on the basis of stylistic analysis. Questions still remain, principally about the early artist's activity before his exile from Antwerp (1544-1555) and the chronology of his undated paintings. Examination of his œuvre has been carried out with dendrochronology, macrophotography, infrared photography and reflectography.[2] This paper presents some results of on-going research[3] and focuses on three topics. First, the attribution of *Saint Jerome in his Studio* (Vienna, Kunsthistorisches Museum, inv. 965) to the young Jan Massys shortly after Quinten's death. Second, the examination of some dated paintings, which provided a secure chronology to compare the dendrochronological estimations, and to delineate the evolution of Jan's underdrawing technique. Third, a new chronology for two undated paintings, *Caritas* (Genoa, Galleria di Palazzo Bianco) and *Judith* (Antwerp, Koninklijk Museum voor Schone Kunsten).

The beginnings of Jan Massys in Quinten's workshop: the *Saint Jerome* of Vienna (inv. 965)

The role played by the young Jan Massys in the workshop of his father Quinten, first as assistant, and afterwards as successor, is still not completely understood. Only a few extant paintings date before 1544, the year Jan was banished from Antwerp for religious reasons. Since Jan was born around 1509, his training must have taken place during the 1520's, and it is likely that his role as Quinten's first assistant became more important during the last years of the decade before Quinten's death in 1530. Some paintings could have been executed in collaboration with or finished by Jan; copies of Quinten's prototypes could have been produced by Jan also after 1530 (Jan enrolled in the Saint

* For allowing us to study the paintings and for their assistance, we would like to thank: Karl Schütz and Elke Oberthaler, Kunsthistorisches Museum, Vienna; Lizet Klaassen, Koninklijk Museum voor Schone Kunsten, Antwerp; Piero Boccardo, Galleria di Palazzo Bianco, Genoa; Angelo Ravasi, Pinacoteca Ambrosiana, Milan; John Hand and Sarah Fisher, National Gallery of Art, Washington; Carina Fryklund, National Museum, Stockholm; Martin Schawe, Alte Pinakothek, Munich. A special thank to Elizabeth Walmsley, National Gallery of Washington, for reading this paper and giving us helpful suggestions.

[1] L. BUIJNSTERS-SMETS, *Jan Massys een Antwerps schilder uit de zestiende eeuw*, Waanders Uitgevers-Zwolle, 1995.

[2] The University of Genoa's equipment was used by M.C. Galassi: Nikon D50 camera, objective AF Micro Nikkor 60 mm (macrophotographs); Sony Cybershot F717 camera, Hoya R72 IR filter (IR digital photographs); Hamamatsu C 2400-07 camera equipped with a N2606-26 infrared vidicon (IRR).

[3] To date, Peter Klein has analysed 39 paintings attributed to Quinten (or to his workshop) and 14 paintings attributed to Jan, a total of 119 boards.

Luke Guild in 1532), in order to perpetuate his father's legacy.[4] The recent debate about the authorship of the diptych with the *Virgin at Prayer* and the *Christ as Savior* (Madrid, Prado), most likely by Quinten with the presumed collaboration of his son, is one of the unresolved questions in this respect.[5]

Two paintings with *Saint Jerome in his studio*, associated to the Quinten/Jan Massys workshop, are housed at the Kunsthistorisches Museum in Vienna. The first (inv. 966), dated 1537, is the earliest painting that can be definitely given to Jan.[6] The second one (inv. 965) is undated and will be discussed in this paragraph. It has been described by Max Friedländer as "close to the master, possible a work by Jan", judged by Larry Silver as a copy from a lost Quinten, and given to Quinten himself by Leontine Buijnsters-Smets.[7] The latter scholar lists two nearly identical versions (Musée des Beaux-Arts of Rouen and a Dutch private collection) that she believes to be copies executed by Jan from his father's prototype.[8] The dendrochronological analysis showed that the painting could be dated during Quinten's lifetime, but that a later execution is more likely. Although 1527 is the earliest possible date for the painting's creation (that is, adding the minimum of nine sapwood rings plus two years for seasoning), a later date post-1533 (that is, adding the median of fifteen sapwood rings and two years for seasoning) is more plausible, since the tree was more than two hundred years old and the number of sapwood rings is related to the age of a tree (for an older tree, more than the minimum of sapwood rings must be added). A close inspection of the painting showed a painting technique that is similar but not completely identical to what may be observed in Quinten's last works, such as the *Ill-Matched Pair* (Washington, National Gallery, inv. 1971.55.1),[9] which exhibits very thin and transparent passages of white to model the volumes in the lights, and lacks evident brush strokes (fig. 1a). Although Saint Jerome's flesh tones have also very thin passages of white that model the forms in the lights, the texture seems quite different, with a new feeling for modeling on the surface, and it is possible to distinguish individual fine strokes of white and brown hatching on top of the paint (fig. 1b).

Infrared examination failed to detect any underdrawing in the saint's face. In contrast, the objects of the saint's cell revealed a linear and very subtle underdrawing, perhaps done in a dry medium, particularly evident where the outlines were shifted, such as in the easel on the desk. This painting and underdrawing technique is consistent with the *Saint Jerome* dated 1537 (inv. 966), which shows the same minute network of white and brown strokes to create the volumes on the paint surface and the same linear contours underdrawn in a thin dry medium, with several small shifts, especially in the saint's fingers (fig. 2).

[4] On this issue, see L. REIS-SANTOS, "Jan Quinten Massys discípulo e colaborador de seu pai Mestre Quinten Metsys", in *Belasartes*, 20, 1964, p. 3-12; V. SCARFF, "Joannes Quintini Massiis Pingebat", in *Mélanges d'Archéologie et d'Histoire de l'Art offerts au Professeur Jacques Lavalleye*, Louvain, 1970, p. 257-280.

[5] See J.O. HAND and C.A. METZGER, R. SPRONK eds., *Prayers and portraits. Unfolding the Netherlandish Diptych*, exh. cat., New Haven and London, 2006, entry 15, p. 110-115.

[6] L. BUIJNSTERS-SMETS, 1995, entry 7, p. 159-160.

[7] L. SILVER, *The Paintings of Quinten Massys with Catalogue Raisonné*, Montclair, 1984, cat. 28, p. 218-219, with previous bibliography; L. BUIJNSTERS-SMETS, 1995, p. 34-35.

[8] L. BUIJNSTERS-SMETS, 1995, entries 14 and 15, p. 168-170.

[9] L. SILVER, "The *Ill-Matched Pair* by Quinten Massys", in *Studies in the History of Art*, 1974, p. 105-123.

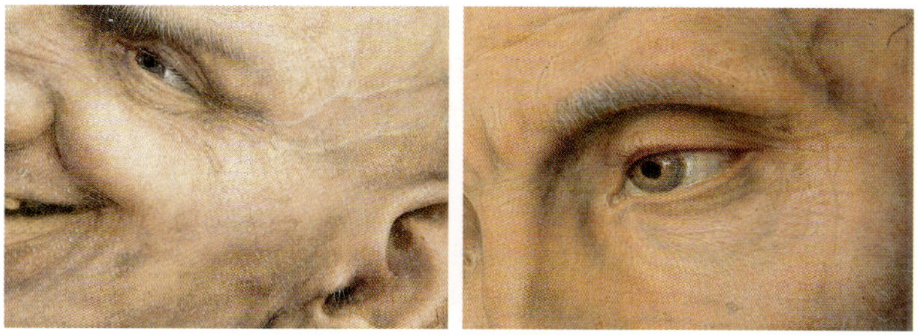

Fig. 1. a. Quinten Massys, *The Ill-matched Pair*, Washington, The National Gallery (© M.C. Galassi). b. Jan Massys (attr.), *St. Jerome in His Studio*, Vienna, Kunsthistorisches Museum, inv. 965 (© M.C. Galassi).

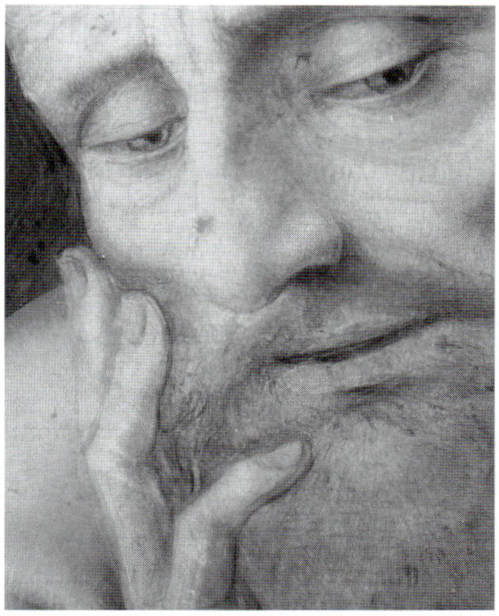

Fig. 2. Jan Massys, *St. Jerome in His Studio*, Vienna, Kunsthistorisches Museum, inv. 966. Digital IR photography (© M.C. Galassi).

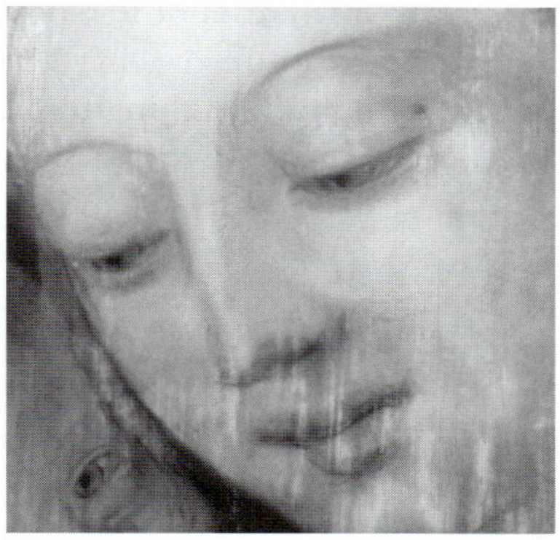

Fig. 3. Jan Massys, *The Virgin and Child*, Genoa, Galleria di Palazzo Bianco. Detail IRR composite (© M.C. Galassi).

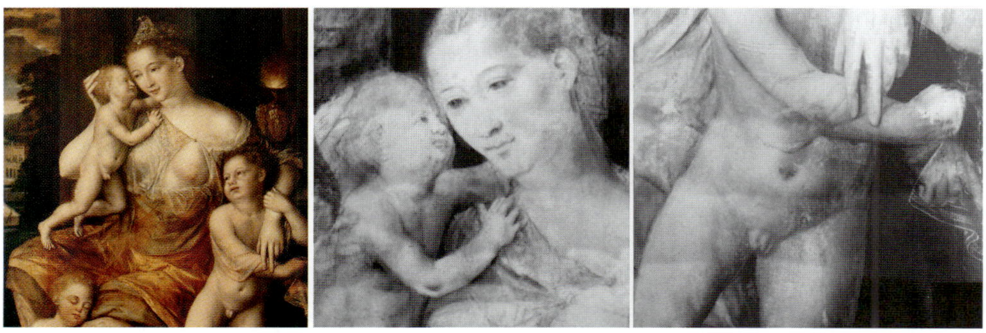

Fig. 4. Jan Massys, *Caritas*, Milan, Pinacoteca Ambrosiana. a. Detail. b and c. IRR composites (© M.C. Galassi).

Therefore, we suggest the inclusion of the undated *Saint Jerome* (inv. 965) among Jan's early works. As a young painter, Jan applied the techniques learned from his father, but at the same time he could not avoid making some interesting personal changes.

Examination of dated paintings

The œuvre of Jan Massys includes several dated paintings. A few belong to his early activity as an enrolled master in Antwerp (1532-1544) and to the exile period (1544-1555), but the majority were executed after his return to Antwerp. In 1559, the artist began to systematically sign and date his paintings. This provided a good opportunity to compare dendrochronological estimations with firm dates.

In Table 1, the difference between the youngest ring and the date of the paintings covers a range of 15-21 years, including the lost sapwood rings and the seasoning time. This means, taking into account only the minimum factors for sapwood and seasoning, that there is a range of 4 to 10 years difference between the dendrochronological dates and the dates inscribed on the paintings. However, all trees used in the analysed panels originated in the Baltic region and were more than two hundred years old. Statistically speaking, we know that 90% of Baltic oaks have a range of 11 to 21 sapwood rings and, in fact, assuming for our panels a median of 15 sapwood rings (with a range between +2 and –4 years) and 2 years for seasoning, the dendrochronological dates correspond well to the actual date of each painting.

The examination of some of these dated paintings by IRR and IR digital photography offered the opportunity to describe the achievements of Jan Massys' underdrawing technique over his entire career.

The features of Jan's early underdrawing technique, as shown by the infrared examination of his first dated painting, the 1537 *Saint Jerome* in Vienna, have been discussed above. The 1552 *Madonna and Child* in Genoa (Galleria di Palazzo Bianco), painted on walnut, belongs to the last years of his exile and for that reason exemplifies the accomplishment in style and technique that Jan reached during this undocumented period.[10] After being expelled from Antwerp in 1544, Jan probably travelled in Europe. Scholars have speculated about a stay in Italy (possibly in Genoa), Germany, or above all in France, since his style after the banishment shows an unambiguous Fontainebleau influence.[11] The use of walnut, instead of Baltic oak, could confirm an Italian or French sojourn, since this wood was often used in these countries. IR examination revealed the presence of a rough *imprimatura* done with large and irregularly oriented strokes (fig. 3), and a linear underdrawing, probably executed in a dry medium. Rather sketchy lines delineate the contours of the Virgin's face, the mouth, nose, eyes and eyelids, with some occasional hatching in the proper right cheek and eyelid. In addition, there is a *pentimento* in the Child's face, which was shifted to a higher position during the paint stage.

[10] L. BUIJNSTERS-SMETS, 1995, entry 24, p. 178-179.

[11] On the period in exile, and the relationship with the city of Genoa, see M.C. GALASSI, "Jan Massys and Artistic Relationships between Antwerp and Genoa During the Sixteenth Century", in M. FARIES ed., *Making and Marketing. Studies of the Painting Process in Fifteenth- and Sixteenth-Century Netherlandish Workshops*, Turnhout, 2006, p. 179-200, with a discussion about the underdrawing of the 1552 *Madonna*.

Table 1. Comparison between dendrochronological dates and painted dates (by P. Klein)

Title/ Museum/Inv. n°	Firm Date	Number of Boards	Last Year Ring	Earliest Possible Felling Date	Estimated Felling Date with 15 Sapwood Rings	Presumed Date for Use with 15 Sapwood Rings and 2 Years Seasoning	Difference between Firm Date and Dendrochronological Date
Saint Jerome/Vienna/966	1537	3	1517	1526	1532	1534	+ 3
Judith/Boston/121.048	1543	2	1527	1536	1542	1544	–1
Venus/Stockholm/507	1561	5	1540	1549	1555	1557	+ 4
Lot and his Daughters/Vienna/1015	1563	5	1543	1552	1558	1560	+ 3
Merry Company/Vienna/963	1564	3	1548	1557	1563	1565	–1
Saint Paul/Munich/1420	1565	3	1550	1559	1565	1567	–2
Ill Matched Pair/Stockholm/508	1566	3	1545	1554	1560	1562	+ 4

The use of dry media and fine tools for underdrawing seems to have been discarded during Jan's late career. One of the most distinctive technical features, detected in several paintings dated after 1559, is the presence of a soft and pictorial underdrawing, not always clearly distinguishable from the paint stage, created with large and liquid brush strokes, rich in shifting contours and *pentimenti*. Examples are found in the 1559 *Caritas* in Milan (Pinacoteca Ambrosiana),[12] the 1561 *Venus* in Stockholm (National Museum), the 1563 *Holy Family* in Antwerp (Koninklijk Museum voor Schone Kunsten), the 1563 *Lot and his Daughters* in Vienna (Kunsthistorisches Museum) and the 1563 *Merry Company* in Vienna (Kunsthistorisches Museum).[13] *Pentimenti* and noteworthy shifts in the placement of heads, profiles, hands, legs and feet are so frequent that it is possible to imagine the artist dealing with a repertory of single figures, the position of which were rearranged according to the general composition. This was sometimes done during the underdrawing stage, as in the feet of the *Venus*,[14] but more often during the paint stage, as in the *Healing of Tobias* (Douai, Musée de la Chartreuse), where Tobias' proper right hand was depicted at first holding a small dish, and then repainted holding the unguentarium.[15] The Milan *Caritas* (fig. 4a) shows interesting *pentimenti* both during the underdrawing and the paint stage. In particular, Caritas' proper right hand was at first drawn on the younger child's shoulder, and then redrawn on the child's head (fig. 4b); the right hand of the other child was completely painted in a lower position and then hidden by a drapery (fig. 4c).

In comparison with the earlier 1552 *Madonna*, these paintings show Jan's increased skill in using a sophisticated palette, rich in pearly and luminescent effects, and his concentrated attention to the depiction of fluent silken drapery, obtained by softly blending wet-into-wet touches of different colours. In addition, infrared examination did not reveal any evidence of the rough *imprimatura* detected in the 1552 *Madonna*.

New proposals for the chronology of two undated paintings by Jan Massys

Among the undated paintings analysed with dendrochronology (table 2), the *Caritas* in Genoa (Galleria di Palazzo Bianco) is currently placed during Jan's exile (1544 to 1555), and the *Judith* in Antwerp (Koninklijk Museum voor Schone Kunsten) is dated about 1560-65.[16] However, technical results suggest a different chronology.

According to dendrochronology, the earliest possible date for the *Judith* is 1548 (9 sapwood rings ans 2 years seasoning), although, taking into considerations the 15 sapwood rings and the seasoning time, a later date, around 1554 is more plausible. This date, which is closer to the 1552 *Madonna*, is supported by the IR examination, since both

[12] B. MEIJER ed., *Repertory of Dutch and Flemish Paintings in Italian Public Collections. Lombardy*, vol. II, entry 480, p. 61, as "signed and dated at centre right 1558 (or 1559)". With IRR, the date 1559 became clearly legible. The painting is not listed by L. Buijnsters-Smets.

[13] The paintings entries in L. BUIJNSTERS-SMETS, 1995, 35, p. 193-194; 39, p. 199-201; 43, p. 105-106; 44, p. 207-208; 45, p. 208-209; 46, p. 209-210.

[14] See M.C. GALASSI, 2006, p. 179-200.

[15] IR photography by courtesy of the Centre de Recherche et de Restauration des Musées de France, 85915. We are grateful to Françoise Baligand, curator of the Musée de la Chartreuse de Douai, for providing us with the photograph.

[16] L. BUIJNSTERS-SMETS, 1995, entry 27, p. 181-183; entry 38, p. 198-199.

Table 2. Dendrochronological dates of some undated paintings (by P. Klein)

Title/ Museum/Inv. n°	Number of Boards	Last Year Ring	Earliest Possible Felling Date	Estimated Felling Date with 15 Sapwood Rings	Presumed Date of Use with 15 Sapwood Rings and 2 Years Seasoning
Ill Matched Pair/Antwerp/566	2	1496	1505	1511	1513
Ill Matched Pair/Washington/2561	2	1509	1518	1524	1526
St. Jerome/Vienna/965	2	1516	1525	1531	1533
Merry Company/Stockholm/2661	4	1531	1540	1546	1548
Judith/Antwerp/5076*	3	1543	1546	1552	1554
Caritas/Genoa/PB285	3	1540	1549	1555	1557

* The board III of the *Judith* includes 6 sapwood rings. That means the last heartwood ring is from the year 1537.

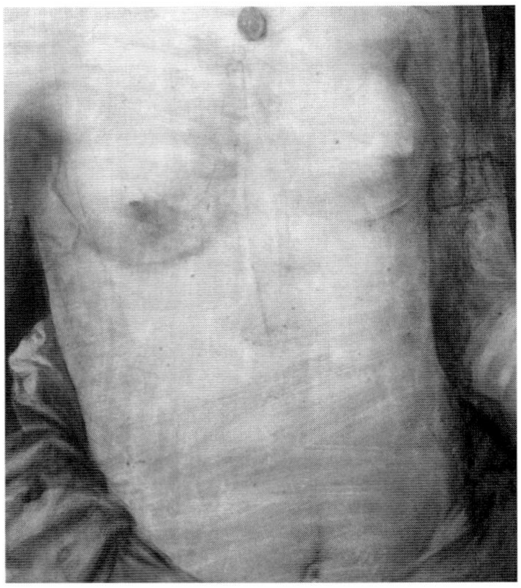

Fig. 5. Jan Massys, *Judith*, Antwerp, Royal Museum of Fine Arts. Digital IR photography (© Adri Verburg).

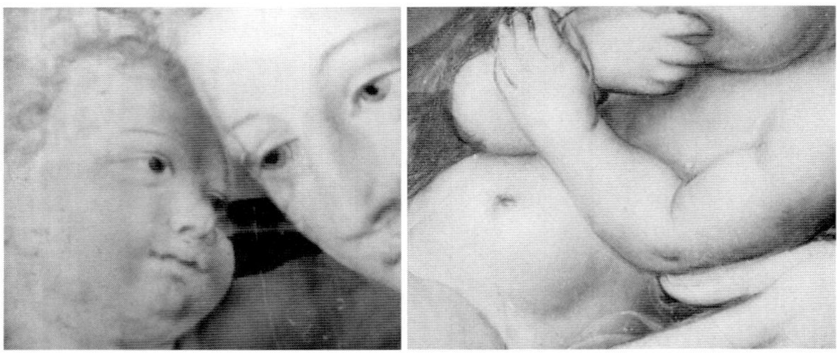

Fig. 6. Jan Massys, *Caritas*, Genoa, Galleria di Palazzo Bianco, IRR composites (© M.C. Galassi).

paintings share technical characteristics such as a prominent and rough *imprimatura* layer and a very fine underdrawing (fig. 5). Painted on Baltic oak, the Judith could be considered as one of the first works executed by Jan after his return to Antwerp, around 1555.

The earliest possible date for the *Caritas* is 1551 (9 sapwood rings and 2 years seasoning), while a later date, after 1557, is more plausible. IR examination revealed frequent shifts in the contours of the figures and a *pentimento* in the standing Child's face which was at first designed in profile (fig. 6a), and later turned to a three-quarter position. Under the luminescent colours, the underdrawing was done with liquid and large brush strokes, and with a very fluent and rounded pictorial delineation of the main contours (fig. 6b).[17]

[17] On the Genoese *Caritas*, see also M.C. GALASSI, 2006, p. 179-200.

This underdrawing is completely consistent with the underdrawings in the paintings executed around 1560 and unlike the fine, sketchy, dry underdrawing of the 1552 *Madonna*. Based on the underdrawing and dendrochronological results, we propose that the painting should not be dated to the period when Jan was in exile, but should more properly be dated around 1560.

Maria Clelia GALASSI
University of Genoa
Dipartimento di Italianistica, Romanistica, Arti e Spettacolo
Via Balbi 4
I – 16126 Genova
mcgalassi@lettere.unige.it

1977: Degree in History of Art, University of Genoa. 1982: Master in Medieval and Modern History of Art, University of Florence. 1987: Ph.D. in History of Art, University of Milan. 1988-1993: Curator, Museum G. Luxoro, Genoa. November 1990-October 1991: Temporary Professor, University of Udine. November 1998-2000: Researcher, University of Genoa. November 2000-present: Associate Professor of Methodologies in the History of Art and History of Art Techniques, University of Genoa. 2003: Paired Fellowship for Research in Conservation and the History of Art at the Center for Advanced Study in the Visual Arts (CASVA), National Gallery of Art, Washington. Research in the fields of History of Restoration, Flemish and Italian Art Techniques, and Italian and Flemish underdrawing.

LES COULEURS SIGNIFIANTES CHEZ BRUEGEL L'ANCIEN ET CHEZ SES SUCCESSEURS

Hélène Verougstraete*

Résumé: Les copies exécutées par Pierre Breughel le Jeune, son frère Jean et leurs ateliers d'après les compositions de Bruegel l'Ancien sont assez fidèles, sauf en ce qui concerne les couleurs. Les historiens de l'art ont expliqué cela par le fait que les copistes auraient disposé de dessins en noir et blanc et n'auraient donc pas vu les couleurs originales. Toutefois, certains détails apparaissent toujours dans une même couleur. Il semble bien que ces couleurs s'imposaient, soit par l'usage vestimentaire (un cardinal est vêtu de pourpre...), soit par le caractère signifiant d'une couleur. Selon les historiens spécialisés dans le symbolisme des couleurs à l'époque médiévale, le rouge est souvent associé à la richesse, le bleu délavé à la pauvreté et la juxtaposition du bleu et du rouge au contraste entre le pauvre et le riche. Le jaune habillait des personnages dont la conduite était désordonnée... Les couleurs étaient ambivalentes: elles avaient un sens positif et un sens négatif. Leur interprétation est donc compliquée mais incontestablement Bruegel et ses successeurs ont fait usage du caractère signifiant des couleurs.

La valeur signifiante des couleurs a été adoptée par Bruegel l'Ancien et ses successeurs, principalement dans les compositions allégoriques et moralisantes. Cet aspect des choses n'est pas retenu dans la plupart des monographies consacrées à ces artistes. Le présent article souhaite démontrer qu'il est indispensable de le faire.

Bruegel l'Ancien dispose souvent au centre de ses compositions, ou le long des axes principaux, les éléments les plus significatifs. Cette organisation longuement mûrie contraste avec l'exécution picturale alerte sur un dessin sous-jacent sommaire, tracé à main levée[1], et laisse supposer l'existence de dessins préparatoires indépendants fouillés. Van Mander fait état de dessins 'nets et précis' laissés en héritage par l'artiste[2]. Ce sont ces dessins qui ont dû inspirer les copies. On le soupçonnait depuis longtemps[3]. Christina Currie assoit ce fait sur des arguments solides dans son étude du Dénombrement de Bethléem dans le catalogue de l'exposition *L'Entreprise Brueghel*[4]. Un des arguments

* Je remercie Roger Van Schoute et Jacqueline Couvert pour leur relecture attentive du texte.

1 R. Van Schoute et H. Verougstraete, «Le dessin de peintre (dessin sous-jacent) chez Pierre Bruegel», dans *Scientific Examination of Early Netherlandish Painting. Applications in Art History (Nederlands Kunsthistorisch Jaarboek,* 26, 1975), Bussum, 1976, p. 259-267. J. Dunkerton, S. Foister and N. Penny, *Dürer to Veronese. Sixteenth-Century Painting in the National Gallery*, New Haven and London, p. 230, 233.

2 H. Miedema, *Karel Van Mander. The Lives of the Illustrious Netherlandisch and German Painters, from the first edition of the Schilder-boeck (1603-1604),* fol. 233v, 45-47, fol. 234, 1-2., Doornspijk, 1994, p. 192, 195.

3 G. Marlier, *Pierre Brueghel le Jeune,* (éd. posthume par J. Folie), Bruxelles, 1969, p. 69. F. Thomaes-Van Hauwaert et J. Folie, «The Cleveland Painting: a Comparison with Bruegel the Elder's Original and the Other Versions», dans *The Triumph of Death. Pieter Brueghel the Younger,* compiled by J. Corcoran, Anvers, Musée Mayer Van den Bergh, 1993, p. 15-21.

4 C. Currie, «Démystification d'un procédé: Le Dénombrement de Bethléem de Pieter Brueghel le Jeune. Une étude technique», dans *L'Entreprise Brueghel* (dir. P. Van den Brinck), Gand-Amsterdam, 2001, p. 103.

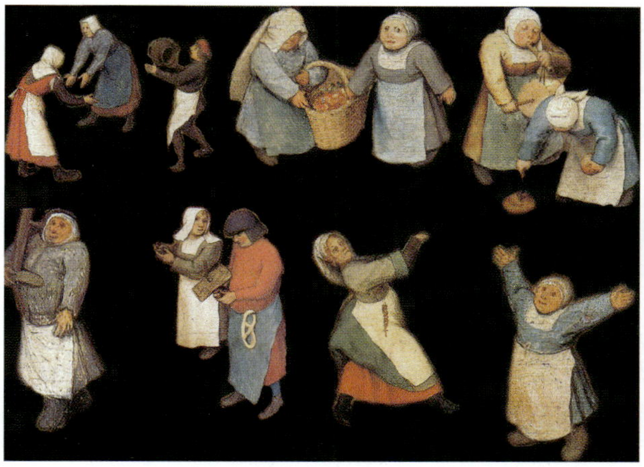

Fig. 1. Pierre Bruegel l'Ancien, détails de plusieurs tableaux, divers tabliers (© Labart, UCL).

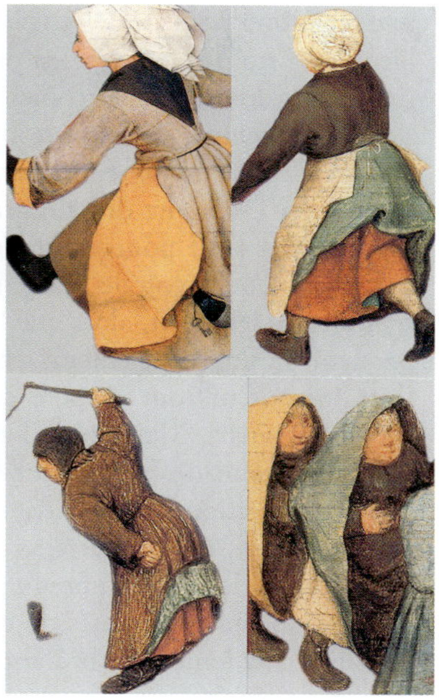

Fig. 2. Pierre Bruegel l'Ancien, détails de plusieurs tableaux, diverses jupes (© Labart, UCL).

dans ce sens: pour certains détails les copies se rapprochent plus du dessin sous-jacent que du stade pictural du tableau de Bruegel l'Ancien.

La plupart des copies sont réalisées à même échelle que la peinture de Bruegel l'Ancien. Quatre copies du *Dénombrement de Bethléem* sont, par rapport au tableau du père, à échelle de 102% à 103% et il a été suggéré que cette différence résultait d'une mise au carreau[5]. Mais cela nous semble peu probable vu le caractère laborieux du procédé qui, en outre, n'aurait laissé aucune trace, par exemple un quadrillage sous-jacent dans les copies. Il y a peut-être eu d'autres manières de report des dessins préparatoires. Dans une copie du *Triomphe de la Mort* (Bâle, Kunstmuseum cat. 199, bois, 123 × 166,5 cm), nous pensons avoir repéré un dessin au pantographe, outil peu précis en matière d'échelle (fig. 1)[6].

Les compositions de Bruegel fourmillent de détails et les copies sont nombreuses. Les comparaisons entre les diverses versions sont d'une grande complexité[7]. La couleur constitue un aspect de la question: on l'a dit, de nombreuses copies présentent des couleurs presque entièrement différentes de celles de Bruegel l'Ancien. Mais il y a des exceptions — des copies fidèles aux couleurs de l'Ancien — on les évoquera plus loin. Dans d'autres cas, par exemple dans les différentes versions des *Proverbes*, on observe

[5] C. CURRIE, 2001, p. 103

[6] H. VEROUGSTRAETE et R. VAN SCHOUTE, «Un 'dessin mécanique' au 'pantographe' pour reproduire plus vite une composition? Le dessin sous-jacent d'un Triomphe de la Mort de Pierre Breughel le Jeune», dans *Jérôme Bosch et son entourage et autres études, Actes du Colloque XIV pour l'étude du dessin sous-jacent et de la technologie dans la peinture*, Louvain-Paris-Dudley, MA, 2003, p. 298-309.

[7] C. Currie s'est attelée à cette tâche ardue dans sa thèse doctorale en voie de publication.

une répétition partielle, sélective, de certaines couleurs (fig. 2, 3). Pourtant les copies semblent bien trouver leur origine dans un dessin de Bruegel l'Ancien (et non dans la peinture) étant donné que les copies reproduisent, au moins pour un détail, un élément qui figure uniquement dans le dessin sous-jacent des *Proverbes* de Berlin considéré comme le prototype: les jambes, tendues entre les barreaux, du personnage qui joue du violon[8].

Rebecca Duckwitz dit comprendre que la huque soit bleue dans toutes les versions mais ne pas saisir pourquoi la jeune femme est toujours vêtue de rouge tout comme la personne qui ligote le diable sur un coussin. Elle souligne que souvent le rouge est reproduit fidèlement, alors que les autres couleurs varient. Elle conclut que les copistes auraient gardé «les éléments chromatiques les plus marquants de la composition»[9]. Certains auteurs ont suggéré que Bruegel a pu mettre des annotations de couleurs sur ses dessins[10]. Mais on voit mal pourquoi il aurait agi ainsi. Il a existé, il est vrai, des annotations de Bruegel sur des dessins allégoriques mais elles devaient refléter des prises de position dans les conflits socio-culturels ou religieux. Van Mander, dans le *Schilder-Boeck,* évoque certains dessins laissés en héritage à la veuve de l'artiste: «il en avait encore de très nombreux [des dessins très curieux avec des thèmes allégoriques] très nets et précis, annotés de quelques textes, dont une partie trop corrosive, qu'il a demandé à sa femme de brûler par tristesse, ou par crainte qu'elle n'ait à en souffrir, ou qu'elle n'ait à s'en justifier»[11].

Certaines couleurs s'imposaient, soit en vertu des coutumes (pourpre pour la robe du cardinal...), soit par référence à un code des couleurs. Avant d'aborder la question des couleurs signifiantes, nous évoquerons quelques aspects de la copie au sein de l''entreprise Brueghel' et quelques notions générales concernant l'usage symbolique des couleurs.

La copie au sein de l''entreprise Brueghel'

Les modèles qui ont été à la base des copies (probablement les dessins préparatoires qu'on vient d'évoquer) ont disparu. On est donc réduit à comparer ce qui est conservé. Différents auteurs ont établi un relevé des différences entre la peinture de l'Ancien et l'une ou l'autre copie d'un successeur (cela a été fait pour *Le Triomphe de la Mort, Les Proverbes, Le Dénombrement de Bethléem...*). Certaines différences semblent fortuites, sans grande signification: oublis — d'une branche d'un arbre, par exemple —, tics de métier, incompréhension d'un détail... D'autres différences modifient le sens. Parfois Pierre II ajoute un détail pour appuyer une démonstration, préférant à la manière lapidaire et condensée de son père, une façon plus explicite, voire redondante. Par

[8] R. DUCKWITZ, «C'est au fruit qu'on connaît l'arbre. Les «Proverbes flamands» de Pieter Bruegel l'Ancien et les copies issues de l'atelier de Pieter Brueghel le Jeune», dans *L'Entreprise Brueghel* (dir. P. VAN DEN BRINCK), Gand-Amsterdam, 2001, p. 76.

[9] R. DUCKWITZ, 2001, p. 71.

[10] D. ALLART, «Pieter Brueghel le Jeune a-t-il pu voir les tableaux de son père? Réflexions méthodologiques et critiques», dans *L'Entreprise Brueghel*, 2001, p. 55; R. DUCKWITZ, 2001, p. 59, 78, C. CURRIE, 2001, p. 105.

[11] *maer hadder noch seer veel (vreemde versieringhen van sinnekens) net en suyver geteykent met eenighe schriften by welcke ten dele al te seer bijtigh oft schimpich wesende hij in zijn doot sieckte door zijn huysvrouwe liet verbranden, door leetwezen, oft vreesende sy daer door in lijden quaem oft het te verantwoorden mocht hebben.* H. MIEDEMA, fol. 233v, 45-47, fol. 234, 1-2, 1994, p. 192, 195. Traduit en français par l'auteur.

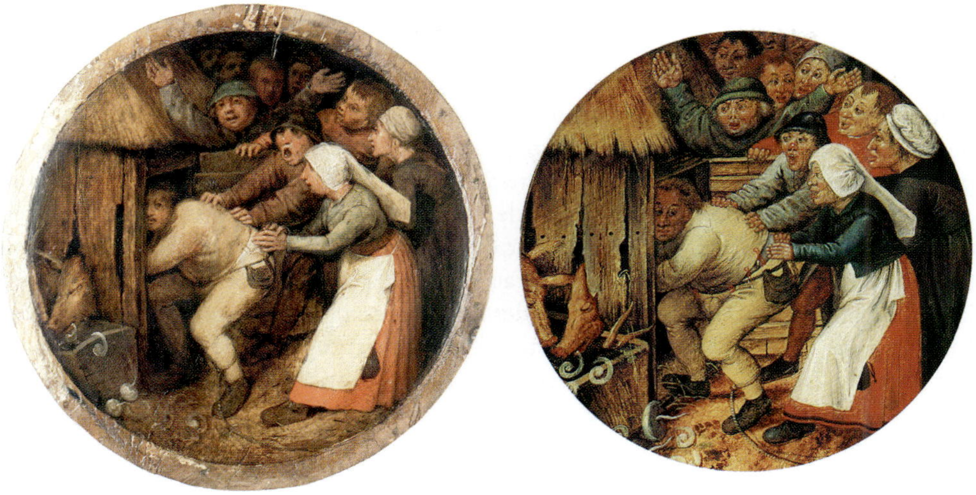

Fig. 3. a. Pierre Bruegel l'Ancien, *Cochon, à la bauge!*, signé et daté 1557, Ø 20 cm. Londres, vente Christie's, juillet 2002, lieu de conservation actuel inconnu. b. Pierre Breughel le Jeune, *Cochon, à la bauge!*, Ø 18 cm, Bamberg, Historisches Museum.

exemple, dans une copie du *Cochon, à la bauge!*[12], l'auge contient deux carottes dressées en l'air, emblèmes phalliques destinés à ceux qui n'avaient pas compris que les navets du tableau paternel (connus depuis l'Antiquité pour leurs propriétés aphrodisiaques), marquaient déjà à eux seuls que le 'cochon' flanqué à la bauge était luxurieux, et pas seulement ivrogne (deux défauts d'ailleurs souvent liés) (fig. 4a et 4b). Les copies du *Cochon, à la bauge!* diffèrent du tondo de Bruegel l'Ancien principalement par l'ajout de personnages dans le fond, ajout introduit par Jean Wierix dans une gravure réalisée en 1568, du vivant de Bruegel l'Ancien. Les copistes devaient connaître cette gravure. Dans le chef du graveur, limité par le noir et le blanc, ces personnages nombreux conféraient à la moralité une portée universelle. La notion d'universalité est exprimée dans le tondo de Bruegel l'Ancien au moyen de couleurs signifiantes: tous, riches vêtus de rouge, pauvres vêtus de bleu délavé, doivent précipiter le 'cochon' de mauvaise conduite habillé de jaune, dans la bauge aux pourceaux. Utilisées selon un code que nous évoquerons plus loin, les couleurs participent au sens de la moralité. Les copies du *Cochon, à la bauge!* cumulent deux vocabulaires: pour évoquer la portée universelle de la moralité, elles empruntent au modèle peint par Bruegel l'Ancien les couleurs symboliques et à la gravure l'entassement des personnages dans le fond.

Lorsque Pierre le Jeune était heurté par certains détails expressifs de son père, il les supprimait[13]. Dans deux copies du *Triomphe de la Mort* (anciennement Collection Peter

[12] R. VAN SCHOUTE et H. VEROUGSTRAETE, «A painted wooden roundel by Pieter Bruegel the Elder», dans *The Burlington Magazine*, CXLII, 2000, p. 140-146.

[13] Dans un mémoire de licence non publié, intitulé *Pierre Brueghel le Jeune (1564/65-1638) et la copie aux Pays-Bas,* Louvain, 1977, F. Van Hauwaert avait déjà souligné cette sensibilité particulière de l'artiste. Voir aussi F. THOMAES-VAN HAUWAERT, «La véritable signature de Pierre Breughel le Jeune: son dessin sous-jacent»,

Putnam et Galerie Corcoran, Cleveland, toile, 117 × 167 cm, signé et daté *P. BREUGHEL 1526*, lieu de conservation actuel inconnu et Bâle, Kunstmuseum cat. 199, bois, 123 × 166,5 cm), le bébé mort jeté par-dessus le cercueil tiré par les Alexiens est dessiné au stade sous-jacent — les réflectogrammes à l'infrarouge l'attestent[14] —, et omis au stade de la peinture. Ce n'est probablement pas Pierre le Jeune qui a fait le dessin (il n'aurait pas dessiné le détail cruel qu'il supprime à chaque fois au stade de la peinture), mais c'est l'atelier qui répète la composition de manière systématique à partir d'un modèle où figurait le bébé. Le dessin sous-jacent du *Triomphe de la Mort* de Bâle, œuvre que nous pensons être de la main de Pierre II au stade de la couleur, présente un dessin sous-jacent effectué d'un trait incertain, qui dérape. Il ne semble pas avoir été fait à la main (fig. 1), mais plutôt 'commandé' à distance par un pantographe (compas mathématique, 'singe', 'compas de proportion', *Storchschnabel*). En 1635, Christophe Scheiner consacre au pantographe un traité *Pantographia seu Ars delineavit quaslibet per parallelogrammum lineare*[15].

On ne dispose pas de documents aux rayons infrarouges pour un autre *Triomphe de la Mort* de Pierre Breughel II (Madrid, coll. Mallo, cat. 201, bois 50,80 × 87,60 cm), et on ne sait pas si le bébé a été dessiné, mais en tout cas, il n'a pas été peint. Au contraire de son frère, Jean de Velours n'hésite pas à peindre le bébé mort (Graz, Steiermärkischen Landesmuseum Joanneum, Alte Galerie, inv. 58, toile, 119 × 164 cm, signé). Une copie d'un successeur, débarrassée d'une signature apocryphe *Bruegel 1597* (certains y voient une production de Jean le Jeune) présente également l'enfant mort au stade de la peinture (Vaduz, collection des princes de Liechtenstein, inv. 1134, toile 119 × 162 cm)[16].

C'est dans les compositions allégoriques qu'il faut examiner les différences iconographiques entre l'œuvre du père et les copies. Ces différences, surtout celles qui semblent vouloir dévier le sens initial, aident à comprendre le sens général des œuvres de Bruegel l'Ancien. C'est ainsi que les particularités du *Triomphe de la Mort* dit de Cleveland nous ont permis de proposer une interprétation du *Triomphe de la Mort* de Bruegel l'Ancien, conservé au Musée du Prado à Madrid[17], tableau qui était jusqu'alors peu expliqué.

dans *Le dessin sous-jacent et autres techniques graphiques* (Colloque V) (R. Van Schoute et D. Hollanders-Favart éd.), Louvain-la-Neuve, 1985, p. 159-162.

[14] La réflectographie à l'infrarouge du tableau de Cleveland a été exécutée par M. Faries; celle du tableau du Kunstmuseum de Bâle par le responsable de l'atelier de conservation du musée que nous remercions ainsi que la directrice de l'institution.

[15] Ce traité dont Meder fait état, est mentionné également par C. Currie (2001, p. 112, 124, note 112) qui signale que, bien que le pantographe ne semble pas avoir été utilisé avant le 19e siècle, la possibilité de son usage est à l'étude; elle annonce la publication avec D. Allart d'expériences qu'elle mène sur ce sujet.

[16] Dans K. ERTZ, «Pieter Breughel le Jeune et Jan Brueghel l'Ancien face à l'œuvre de leur père», dans *Pieter Breughel le Jeune-Jan Brueghel l'Ancien. Une famille de peintres flamands vers 1600*, Lingen et Anvers, 1998, p. 76-79, l'auteur s'interroge également sur l'authenticité de la signature de l'exemplaire de Graz, car stylistiquement parlant, les versions de Graz et de Vaduz illustrent une manière de faire de la deuxième décennie du XVIIe siècle plutôt que la fin du XVIe.

[17] H. VEROUGSTRAETE et R. VAN SCHOUTE, «The Triumph of Death by Pieter Bruegel the Elder and Pieter Brueghel the Younger», dans *The Triumph of Death. Pieter Brueghel the Younger,* compilé par J. Corcoran, Anvers, Musée Mayer Van den Bergh, 1993, p. 35-53; ID., «The 'Triumph of Death' by Pieter Bruegel the Elder and Pieter Brueghel the Younger. *Intellegetur plus semper quam pingitur*», dans *Actes du colloque IX pour l'étude du dessin sous-jacent dans la peinture. Dessin sous-jacent et pratiques d'atelier* (R. Van Schoute et H. Verougstraete éd.), Institut supérieur d'archéologie et d'histoire de l'art, UCL, document de travail n°24, Louvain-la-Neuve, 1993, p. 213-241; ID., «Pieter Breugel d. J. Triumph des Todes, 1626»,

Les couleurs signifiantes au temps de Bruegel

Selon M. Pastoureau et R. Van Uytven[18], l'homme médiéval portait un vif intérêt aux couleurs, intérêt qui s'est perpétué jusqu'au 18e siècle. Aimant les métaphores et les allégories, il attribue à la couleur un sens lié à un usage répété dans un certain contexte. Même si la symbolique de la couleur varie selon les époques, les régions, les milieux et les techniques, quelques éléments mentionnés par les auteurs précités sont susceptibles d'éclairer la signification de certaines œuvres de Bruegel[19]. Nous les résumons ci-après. Depuis le 14e siècle, les tissus lignés ou en damier sont déconsidérés[20], surtout le jaune et le vert, mais aussi d'autres couleurs, et permettent d'identifier les vagabonds et les fous. La polychromie en jaune et vert constitue d'ailleurs une certaine redondance puisque ces couleurs ont chacune déjà une connotation négative. Toujours selon les mêmes auteurs, les couleurs du costume aident parfois à situer l'individu dans un groupe social. Des obligations ou interdits chromatiques s'appliquent aux réprouvés et exclus. Prostituées, bourreaux, lépreux, infirmes, cagots, simples d'esprit, bohémiens, hérétiques, juifs, musulmans... doivent dans certaines régions et à certaines époques porter des vêtements et des marques distinctifs et infamants. Ce phénomène, surtout urbain, est bien connu pour le 14e siècle. Le jaune et le rouge émergent au titre de couleurs obligatoires. Le rouge est normalement la couleur des chrétiens se livrant à des activités déshonorantes (prostituées, bourreaux) ou celle des infirmes (cagots et lépreux). Le jaune est la couleur de la synagogue et des non-chrétiens (juifs et musulmans) qui doivent porter un vêtement ou une partie de vêtement dans cette couleur (manteau, ceinture, bonnet, chausses, manches, gants...), mais également des laquais, des criminels, des ennemis.... Judas est vêtu de jaune.

Ambivalentes, les couleurs peuvent signifier une chose ou son contraire. Il y a un bon et un mauvais jaune, un bon et un mauvais vert. L'interprétation dépend du contexte. Entre les 12e et 15e siècles, le jaune connaît une dévaluation progressive. Il est alors moins la couleur du soleil, de la richesse et de l'amour divin (le bon jaune), que celle de la bile, du mensonge, de la trahison, de l'hérésie, de la transgression des normes, de la jouissance, de la conduite désordonnée. Au 15e siècle, dans les Pays-Bas méridionaux, la maison des parjures ou des mauvais payeurs, était peinte en jaune par décision de justice. Autant le jaune-orangé, que le jaune-citron qui tend vers le vert, sont mauvais.

dans *Jan Brueghel der Ältere. Pieter Brueghel der Jüngere. Flämische Malerei um 1600. Tradition und Fortschritt (Pieter Breugel le Jeune. Le Triomphe de la Mort, 1626, dans Breughel-Brueghel. Une famille de peintres flamands vers 1600)*, catalogue d'exposition, Kulturstiftung Ruhr Essen-Vienne-Anvers, Lingen, 1997, p. 106-111, (Anvers, 1998, p. 70-75); ID., «Bruegel et Pétrarque: une évocation de Laure dans *Le Triomphe de la Mort* de Pierre Bruegel l'Ancien», dans *Studi di Storia del'arte in onore di Maria Luisa Gatti Perer*, Milan, 1999, p. 247-251

[18] Ce chapitre s'inspire de M. PASTOUREAU, *Figures et couleurs. Etudes sur la symbolique et la sensibilité médiévales*, Paris, 1986, p. 15-57. R. VAN UYTVEN, *De zinnelijke middeleeuwen*, 3e éd., Louvain, 1999, p. 84-120. Le premier auteur s'attache principalement à l'origine et à l'aspect symbolique des couleurs, le second envisage la question des couleurs sous un angle sociologique et historique plus vaste.

[19] Soulignons encore qu'une des difficultés de l'étude des couleurs dans la peinture est d'ordre matériel: les couleurs s'altèrent et les vernis jaunis modifient la perception de la couleur originale.

[20] Cependant le vêtement peut parfois illustrer directement une partition héraldique, comme chez le donateur Philippe de Ternant, Chambellan du duc de Bourgogne Philippe le Bon, du *Retable de la Vierge* (milieu 15e s.), Ternant (Nièvre), église Saint-Roch.

[21] M. PASTOUREAU, 1986, p. 38.

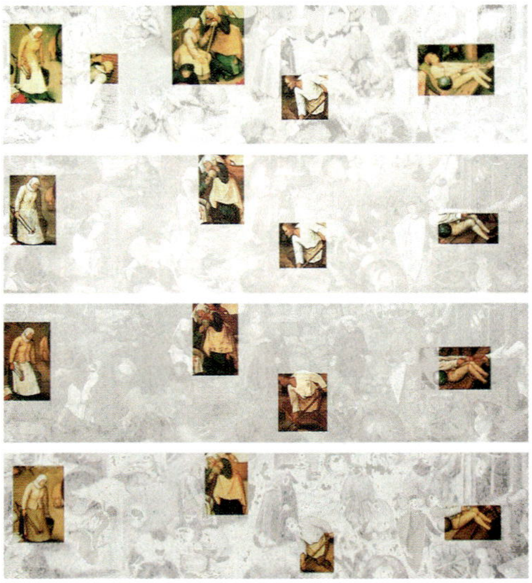

Fig. 4.

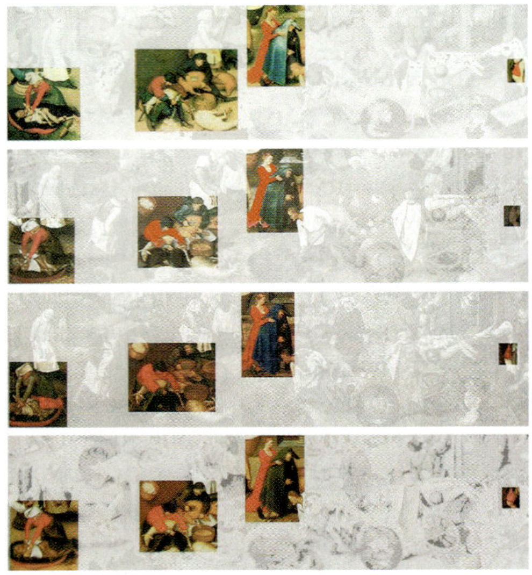

Fig. 5.

Fig. 4. Ce tableau montre des répétitions de la couleur jaune pour des détails similaires dans différentes versions des *Proverbes*. Le jaune habille les personnages dont la conduite est critiquable ou désordonnée. De haut en bas, détails repris à Pierre Bruegel l'Ancien, *Les Proverbes,* Berlin, Staatliche Museen, Preussischer Kulturbesitz. Gemäldegalerie, bois, 117,5 × 163,5 cm, signé et daté BRUEGEL, 1559. Pierre Breughel le Jeune, *Les Proverbes*, Lierre, Stedelijk Museum Wuyts-Van Campen et Baron Caroly, inv. 62, 116 × 168 cm. Pierre Breughel le Jeune, *Les Proverbes,* Anvers, Koninklijk Museum voor Schone Kunsten, inv. 5111, 120 × 170 cm. Pierre Breughel le Jeune, *Les Proverbes,* Galerie Johnny van Haeften, 1994, 119,6 × 164 cm (© Labart, UCL).

Fig. 5. Ce tableau montre des répétitions de la couleur rouge (ou l'association du rouge et du bleu) pour des détails similaires dans différentes versions des *Proverbes* (dans la même séquence que dans la fig. 4). La couleur rouge habille le riche et le bleu habille le pauvre. Dans le groupe de la huque bleue, l'inversion des couleurs est une référence au monde à l'envers ou aux abus du monde, se superposant au sens premier (normalement le rouge est la couleur du riche, le bleu celle du pauvre. A gauche, la virago 'qui lie le diable…' est en partie vêtue de rouge. A droite, la bourse du personnage 'qui tire au plus long' en s'accrochant sagement, paraît rouge dans tous les exemplaires, sauf peut-être dans celui de Lierre. La qualité des reproductions et l'état de conservation des tableaux rendent parfois difficile la lecture des couleurs (© Labart, UCL).

Le vert indique la jeunesse, l'espérance, l'amour naissant, l'insouciante liberté… mais aussi la perturbation de l'ordre établi, la perturbation sociale ou mentale, le diable, l'Islam, la jalousie, parfois la ruine. L'association du jaune et du vert indique la folie, l'impossibilité de se contrôler. Le noir, couleur du péché et des ténèbres, est moins dévalorisé qu'on ne pourrait le croire. À la fin du moyen âge, c'est le noir de la modestie et de la tempérance qui triomphe dans le costume princier et patricien. L'engouement pour le noir devient un phénomène de société qui s'étend à tout l'Occident à partir du 14ᵉ siècle. L'origine de cette mode du noir se situe dans les lois somptuaires et les décrets vestimentaires promulgués par certaines villes. Le bleu est la couleur du courage, de la loyauté, de la fidélité, mais également de la tromperie, de l'adultère.

Dans la sensibilité médiévale, une belle couleur est une couleur dense, saturée, qui résiste aux lavages[21]; elle est portée par les riches. L'artiste médiéval fera contraster les couleurs rutilantes des habits des princes et des prélats avec les couleurs ternes et délavées des 'vilains'. Le vêtement de travail porté au quotidien par la classe laborieuse est en camaïeux de gris, de bruns et de bleus délavés[22]. C'est ainsi que l'impression chromatique que laissent les Primitifs flamands est très différente de celle que laisse Bruegel. Les premiers peignent, en couleurs rutilantes, les classes privilégiées. Le second représente de nombreux personnages dans des couleurs ternes, délavées, grises et bleues, du vêtement quotidien des classes laborieuses. Il fut un temps où les tissus rouges coûtaient beaucoup plus cher que les bleus, peu saturés et éphémères. Les riches ont tendance à s'affirmer en se vêtant de rouge ou de couleurs sombres. À partir du 13e siècle, les pratiques tinctoriales permettent de teindre en bleu éclatant; le bleu est alors valorisé[23]. Les teintures tiennent mieux, les mordants sont plus efficaces. À la fin du 14e siècle, les teinturiers louvanistes produisent beaucoup plus de draps bleus que de draps rouges. Le bleu gagne en importance dans les valeurs héraldiques; mais cette couleur peut aussi être tournée en dérision, comme dans 'la nef bleue' (de *blauwe schuit*) de la gilde des plaisantins et des noceurs qui défilaient annuellement dans les villes. Selon les auteurs cités, la juxtaposition du bleu et du rouge évoque des tensions.

Les couleurs signifiantes chez Bruegel l'Ancien et chez ses successeurs

Comme on l'a dit, l'usage des couleurs signifiantes par Bruegel est une question encore peu traitée. Elle a été évoquée par W. Stechow[24]. Nous avons essayé de voir (nous basant principalement sur des écrits comme ceux de M. Pastoureau et de R. Van Uytven)[25], comment on pouvait décrypter certains messages bruegeliens par le truchement des couleurs[26]. À l'époque de Bruegel, le symbolisme des couleurs était vivant, entre autres dans le théâtre et les processions organisées par les chambres de rhétorique. Le récit de Richard Clough († 1570), témoin oculaire des cérémonies funéraires lors de la mort de Charles Quint, dans une lettre datée de 2 janvier 1559, écrite à son associé, Sir Thomas Gresham, en est un bon exemple[27]. Il décrit et explique les couleurs du cortège.

[22] *Ibid.*, p. 53.

[23] M. Pastoureau, 1986, p. 38, 53.

[24] W. Stechow voit également dans la couleur bleue dont exceptionnellement est revêtu le cardinal dans le *Triomphe de la Mort* du Prado un indice de tromperie (*Bruegel. Les grands peintres,* Paris, 1974, p. 76). Le même, *Pieter Brvegel the Elder,* New York, 1970.

[25] L'importance de la couleur pour l'homme médiéval est également soulignée dans *De Mideleeuwen in kleur, Madoc. Tijdschrift over de Middeleeuwen*, année 15, 4, 2001. Voir aussi la synthèse de J. Gage, *Colour and Meaning. Art, Science and Symbolism,* Londres, 1999.

[26] Nous avons examiné deux compositions de Bruegel l'Ancien en relation avec les copies de ses fils. La première était un *Triomphe de la Mort* signé et daté P. BREUGHEL 1626. (H. Verougstraete et R. Van Schoute, 1997, p. 106-111; Id., 1998, p. 70-75). Un *Triomphe de la Mort* est signalé pour la première fois dans la collection de Philips Van Valckenisse à Anvers en 1614, mais on n'a pas de certitude qu'il s'agisse d'un tableau du père. Pour K. Ertz, la mention *den ouden brugel* peut tout aussi bien désigner Pierre II, par rapport à son fils à lui, également peintre (K. Ertz, 1998). La seconde était le tondo *Cochon, à la bauge!* (ø 20 cm) sur lequel nous avons découvert la signature et la date *BRUEG[...] MDLVII* (R. Van Schoute et H. Verougstraete, 2000, p. 140-146).

[27] S. Schrader, «*Greater than he ever Was.* Ritual and Power in Charles V's 1558 Funeral Procession», dans *Nederlands Kunsthistorisch Jaarboek 1998, 49, Hof-, Staats- en Stadsceremonies/Court, State and City*

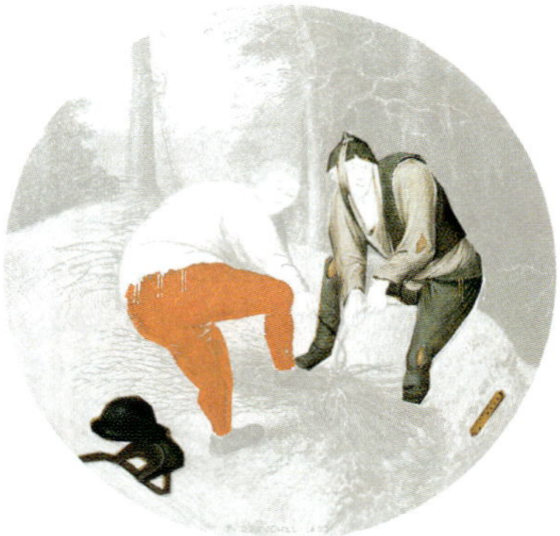

Fig. 6. *Bûcherons (?) liant un fagot,* Munich, Galerie Scheidwimmer 1984, bois, Ø 19 cm, signé et daté P.BREVGHEL, 1625. Schéma soulignant les couleurs signifiantes. La signification des couleurs est appuyée par d'autres éléments: le pauvre, en bleu, est maigre et porte des vêtements déchirés. La flûte à ses pieds, désigne sa vie dissolue. Pour avoir trop 'joué de la flûte', il est coupable de sa pauvreté. Le riche en rouge, au contraire de son misérable compagnon, présente un embonpoint signe de sa prospérité, et il a à côté de lui une besace, allusion à l'aisance dont il jouit (© Labart, UCL).

Le public, à qui les processions et le théâtre des rhétoriqueurs étaient destinés, comprenait le code des couleurs. Peut-être les écrits des rhétoriqueurs livreront-ils un jour de plus amples informations au sujet du caractère signifiant des couleurs au temps de Bruegel l'Ancien. On connaît bien les rapports de ce dernier avec les membres de la chambre de rhétorique des *Violieren*[28]. L'artiste partageait avec eux un intérêt pour des sources diverses: l'antiquité classique et l'humanisme chrétien, mais aussi le passé médiéval et la culture populaire souvent moralisatrice. C'est de ces deux derniers héritages principalement qu'est issu le langage symbolique des couleurs.

Chez Bruegel, la couleur n'est pas toujours signifiante. Certaines couleurs sont dictées par l'usage[29]. Les tabliers — portés surtout par les femmes — sont blancs ou bleus, rarement d'une autre couleur (fig. 5). Attention à ne pas confondre les tabliers et les jupes retroussées. Bruegel représente souvent les jupes du dessus retroussées, par devant ou par derrière, souvent jusqu'à la taille, quelquefois même rabattues sur la tête (fig. 6). Bruegel les représente fréquemment doublées d'un autre tissu. Les coiffes campagnardes sont blanches. Dans les tableaux descriptifs, comme dans *Le Dénombrement de Bethléem* (Bruxelles, Musées royaux des beaux-arts de Belgique, inv. 3637), beaucoup de personnages sont vêtus de gris bleu terne et délavé, couleurs portées au quotidien. Les jours de kermesse, les paysans de Bruegel revêtent leurs meilleurs atours: rouges flamboyants et bleus vifs...

Par contre, dans un contexte moralisateur ou allégorique, l'interprétation des objets et des couleurs se complique. Prenons pour exemple un objet. S'il est normal qu'on se munisse d'une bourse pour se rendre à la kermesse, dans un contexte allégorique, une

Ceremonies, Zwolle, 1999, p. 71-85. Cette procession a fait l'objet d'une gravure intitulée *La procession funéraire de Charles V en 1558* (24 cm de haut sur 1200 cm au total), gravée par Johannes et Lucas van Deutecum, publiée chez Plantin en 1559.

[28] Cette question a beaucoup été étudiée. R. VAN BASTELAER et G. HULIN DE LOO, *Pieter Bruegel l'Ancien, son œuvre et son temps,* Bruxelles, 1907, p. 88-101. W. GIBSON, «Artists and Rederijkers in the Age of Bruegel», dans *The Art Bulletin,* LXIII, 1980, 426-446. B.A.M. RAMAKERS, «Bruegel en de rederijkers. Schilderkunst en literatuur in de zestiende eeuw», dans *Pieter Bruegel. Nederlands Kunsthistorisch Jaarboek 1996*, 47, Zwolle, 1997, p. 81-105.

[29] Ce principe (le ciel et l'eau sont toujours bleus; les champs de blé toujours dorés...) est évoqué dans K. ERTZ, 1988-2000, I, p. 66.

bourse caractérise tantôt l'avare, tantôt le riche. Le sens d'un détail doit être compris dans un contexte donné. Le même principe s'applique à la couleur.

On verra d'ailleurs que certaines conventions renforcent la métaphore avancée par les couleurs. La corpulence sied au riche, vêtu de rouge saturé. Une bourse bien remplie sera parfois attachée à sa ceinture ou posée à ses pieds. Le pauvre, vêtu de bleu délavé, les vêtements en lambeaux, sera d'une maigreur qui traduit une existence précaire. Un attribut évoquant une mauvaise conduite, comme une flûte, symbole de l'adultère, est parfois posé à ses pieds (dans la sensibilité médiévale, on était enclin à le tenir responsable de son malheur). On retrouve, dans les couleurs de Bruegel l'Ancien, trace de l'origine héraldique des couleurs dont parle Pastoureau, par exemple dans les costumes des fous, souvent verts et jaunes mais l'artiste utilise la couleur signifiante dans de nombreux cas.

Un usage intensif des couleurs signifiantes peut être décrit dans *Les Proverbes* (Berlin, Staatliche Museen Preussischer Kulturbesitz, Gemäldegalerie)[30]. La tension générée par la juxtaposition du rouge et du bleu est présente dans le groupe principal: «Elle endosse la huque bleue à son mari» (*Zij hangt haar man de blauwe huik om*). L'élégante est vêtue de rouge saturé, couleur de la richesse, dans ce cas usurpée! Le bleu eut mieux convenu à cette impudente qui soutire les sous à son vieux mari (dont la richesse est attestée par la bourse qui pend à sa ceinture). Elle le recouvre de la huque bleue alors qu'il aurait dû être vêtu de rouge! C'est le 'monde à l'envers', le comble des 'abus de ce monde'. Une gravure de Frans Hogenberg (*Les Proverbes*, vers 1558, 36,5 × 56 cm, Bruxelles, Bibliothèque royale Albert Ier, cabinet des estampes) porte l'inscription *Die blau huicke is dit meest ghenaemt/maer des weerelts abuisen he[m] beter betaemt* [ceci s'appelle 'La huque bleue', mais peut de manière plus appropriée être appelé 'Les abus du monde']. Une gravure de Théodore Galle porte une inscription analogue: *Dese wtbeeldinghe wort die blavw Hvyck ghenaemt, maer des werelts abvysen haer beter betaemt.* La référence au monde renversé subsiste dans le titre qu'on donne aux *Proverbes* qui se trouvent en 1668 dans la collection de Peter Stevens: *Le Monde renversé, représenté par plusieurs Proverbes et Moralités*[31]. Dans le proverbe «Il a pissé contre la lune» (*Hij heeft tegen de maan gepist:* il s'en est mal tiré) on trouve également une inversion de couleurs. Sur l'enseigne suspendue à la fenêtre à côté de l'homme 'qui pissoyt', Bruegel l'Ancien peint une lune noire (ou bleu assombri?) sur un ciel rouge. L'inversion des couleurs représente emblématiquement un dysfonctionnement[32].

L'usage conventionnel du rouge-bleu se retrouve, cette fois sans inversion, dans «L'un tond les moutons, l'autre les cochons» (*De ene scheert de schapen, de andere de varkens*), celui vêtu de rouge tire tout le profit, l'autre, vêtu de bleu, aucun.

La couleur bleue désigne souvent la tromperie ou l'hypocrisie. J. Grauls cite l'expression «la poche bleue» (*De blauwe zak*), dans laquelle on cache de l'argent obtenu de

[30] Pour l'intitulé et le sens des proverbes, voir J. GRAULS, *Volkstaal en volksleven in het werk van Pieter Bruegel*, Anvers, 1957, p. 77-159.

[31] J. BRIELS, «Amator Pictoriae Artis. De Antwerpse Kunstverzamelaar Peter Stevens (1590-1668) en zijn Constkamer», dans *Jaarboek. Koninklijk Museum voor Schone Kunsten Antwerpen*, Anvers, 1980, p. 224 cité dans Ph. et F. ROBERTS-JONES, *Pierre Bruegel l'Ancien,* Paris, 1997, p. 211.

[32] E.M. KAVALER, «Pictorial Satire, Ironic Inversion, and Ideological Conflict: Bruegel's Battle between the Piggy Banks and Strong Boxes», dans *Pieter Bruegel. Nederlands Kunsthistorisch Jaarboek 1966*, 47, 1997, p. 159.

manière malhonnête. Ou encore: «Flanquer des fleurs bleues sur la manche de quelqu'un» (*Iemand blauw bloemekens op de mouw spelen*): être mauvais payeur[33].

La couleur rouge désigne aussi l'action violente. Des atours rouges s'imposent pour le jeune homme «qui fait tourner le monde sur son pouce» (*Hij laat de wereld op zijn duim draaien*), qui domine les autres, qui est puissant. «Elle ligoterait le diable sur un coussin» (*Zij zou de duivel op het (een) kussen binden*) se dit d'une femme difficile, une 'virago' pour laquelle le rouge s'imposait. Celui qui a les cheveux roux n'est pas comme les autres, selon une manière de voir déjà présente dans la Bible et chez plusieurs auteurs latins. Il est mauvais, cruel, mal né. Sa physionomie est souvent peu avenante et, comble de disgrâce, il est souvent gaucher. Dans le proverbe «Quand le veau est noyé, on comble le puits» (*Als t'kalf verdronken is, dempt men de put*), l'homme roux cumule tous les défauts. Son profil est disgracieux. Il bêche de la main gauche. L'accumulation de tant de défauts était synonyme de conduite désordonnée, d'incohérence[34].

Le jaune, très présent, habille ceux qui trahissent ou qui agissent sottement. «Elle porte l'eau d'une main et le feu de l'autre» (*In de ene hand water en in de andere hand vuur dragen*). Le jaune s'impose pour le mensonge, l'hypocrisie, la fausseté. «L'une charge la quenouille de ce que l'autre file» (*De ene rokkent wat de andere spint*): l'une prépare le mauvais coup, que l'autre va exécuter. Une conduite désordonnée caractérise également les personnages de «Donner de la tête au mur» (*Met het hoofd tegen de muur lopen*) et «Mal d'un pain à l'autre venir» (*Hij weet nauwelijks van 't ene brood tot 't ander te geraken*). Le premier proverbe désigne celui qui s'atèle à l'impossible, le second, celui qui ne sait pas se débrouiller, ou 'nouer les deux bouts'. Sur la table, est posé un racloir *(krabber)*. L'expression «C'est un racloir» (*Hij is een krabber*) se dit d'un homme sot et maladroit[35]. «Ils tirent au plus long» (*Si trecken omt lanxte*). Dans le tableau de Berlin, deux hommes tirent sur un 'bretzel', devenu une ficelle dans les copies. L'homme vêtu de jaune à gauche est en équilibre instable, sur le point de tomber; sa conduite est insensée. Celui qui, par contre, s'accroche à un poteau, porte une bourse rouge qui évoque la fortune meilleure qui l'attend.

La chemise blanche semble également avoir un sens allégorique. Mais dans ce cas, le sens n'est pas lié à la couleur (les chemises étaient toujours blanches). L'interprétation sera fonction du contexte: le travail se fait en chemise et c'est normal, comme dans *La Moisson* (New York, Metropolitan Museum of Art). Dans *Les Proverbes* la chemise habille de manière inachevée celui dont la conduite est désordonnée. C'est le cas de «Celui qui tire au plus long»…

Plusieurs exemples démontrent que les successeurs de Bruegel appliquaient également le code des couleurs. Certaines séries de copies dont on suspecte qu'elles sont issues d'un même atelier reproduisent toujours les mêmes couleurs. On est tenté de conclure qu'on y disposait d'une version en couleurs, restituée fidèlement. Le fait que les fils optent pour d'autres couleurs que le père ne veut pas nécessairement dire qu'ils n'ont pas vu son tableau, mais à tout le moins qu'ils ne l'avaient pas sous les yeux pendant qu'ils peignaient. Dans les copies du *Triomphe de la Mort*, la majorité des couleurs

[33] J. Grauls, 1957, p. 89-90.
[34] M. Pastoureau, 1986, p. 41.
[35] J. Grauls (1957, p. 155) ignore si l'expression existait déjà au 16e siècle.

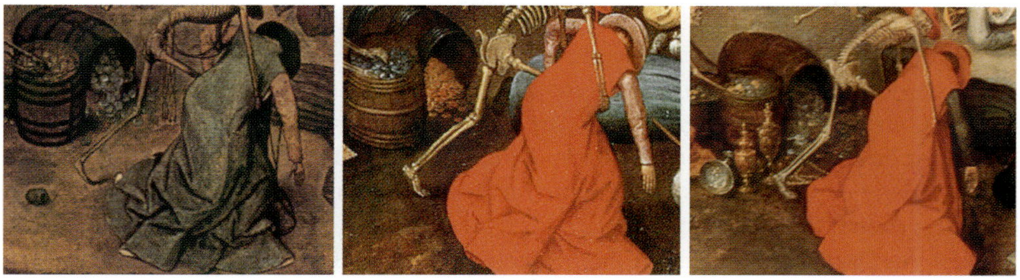

Fig. 7. Le cardinal, en bleu et en rouge pourpre dans différents *Triomphes de la mort*, détails. De gauche à droite, Pierre Bruegel l'Ancien, Madrid, Prado, inv. 1393, bois, 117 × 162 cm, détail, le cardinal vêtu de bleu. Pierre Breughel le Jeune, lieu de conservation inconnu, ancienne collection Peter Putnam (Galerie Corcoran), Cleveland, toile, 117 × 167 cm. Jan I Bruegel, Vaduz, collection du prince de Lichtenstein, signé et daté Bruegel 1597, toile, 119 × 162 cm. Notons également que les pièces d'or et de métal vil (?) sont disposées différemment dans les trois tableaux. A gauche, un squelette se sert de l'or, tandis que le métal vil est éparpillé sur le sol, dans le second cas, le squelette se sert du métal vil tandis que l'or est éparpillé. Dans le troisième cas, les pièces d'or et d'argent sont mêlées, et quelques pièces d'orfèvrerie sont représentées pour accentuer l'évocation des richesses.

est différente de celles du père. Un détail fait exception: dans toutes les versions un des fous porte un vêtement en damier rouge et blanc. Le damier, au moyen âge, reflète une idée de trouble; il revêt un personnage ambigu. Les damiers des jeux d'échecs et de dames ont longtemps été rouge et blanc[36]. Il ne faut donc pas s'étonner que ces couleurs se répètent d'un tableau à l'autre. Dans le tableau du père, le cardinal est vêtu de bleu. Les fils et les suiveurs le peignent en rouge (fig. 7). Ils ignoraient peut-être que Bruegel l'Ancien avait peint un cardinal en bleu. Pourquoi d'ailleurs peindre un cardinal en bleu? Nous avons suggéré que Bruegel fait usage du sens symbolique de la couleur bleue pour stigmatiser l'hypocrisie du cardinal Granvelle, vilipendé avec virulence par la gravure satirique de l'époque. Certains détails de la peinture paternelle, dans lesquels nous voyons une allusion au contexte politique et plus précisément à la douloureuse répression du duc d'Albe, sont absents dans les peintures des fils qui ne souhaitaient pas faire allusion à ces épisodes lointains. Les squelettes de la danse macabre dans les copies ne portent plus d'armures. Peut-être ces dernières étaient-elles absentes du dessin de Bruegel l'Ancien, qui a pu ajouter ces détails au stade de la couleur.

Dans *Le Dénombrement de Bethléem* également[37], les couleurs utilisées par les fils sont différentes de celles du père.

Pour une autre composition au moins, et peut-être pour deux, les fils reproduisent fidèlement les couleurs paternelles. Ils ont dû avoir la peinture de leur père sous les yeux en exécutant leurs copies. Ces deux cas sont: *Cochon, à la bauge!* et *Le Massacre*

[36] R. VAN UYTVEN, 1999, p. 116.
[37] C. CURRIE, 2001, p. 95.

des Innocents, (s'il est vrai que la version de Hampton Court est une œuvre du père)[38]. Dans le premier cas, il y a une raison supplémentaire de croire que les fils avaient la peinture de leur père sous les yeux en la copiant: ils ont choisi le même endroit, la bauge, pour y apposer la signature. Pourtant Wierix, dans la gravure réalisée en 1568, du vivant de Bruegel l'Ancien, avait disposé la signature ailleurs, dans le bas du tableau. Or la gravure est connue des fils, comme on l'a déjà dit, puisqu'ils y puisent les personnages regroupés dans le fond.

Pierre II, Jean de Velours et les autres copistes ont parfois mal compris un détail à défaut d'avoir vu la couleur paternelle. Ils mettent un tablier jaune à celui qui comble le trou dans *Les Proverbes*..., alors que dans le tableau de Berlin, il portait une longue chemise blanche. Les couleurs sont parfois 'normalisées'. Dans les copies du *Triomphe de la Mort*, le cardinal est à nouveau vêtu de rouge... Dans *Les Proverbes*, la lune redevient blanche dans l'enseigne *In den Maene* («À la lune»). Probablement les fils n'ont-ils pas réalisé que leur père avait peint une lune noire sur fond rouge, couleur du monde à l'envers. Parfois les copistes mettent une couleur qu'ils estiment convenir à l'allégorie. Le squelette qui tue le pèlerin dans les copies du *Triomphe de la Mort* est vêtu de jaune dans les copies, couleur qui convenait à la cruauté du bourreau. Bruegel l'Ancien avait couvert le bourreau d'une cotte de mailles. Le contraste rouge-bleu de ceux qui tondent le mouton ou le cochon, est présent dans toutes les copies des *Proverbes*. Ces couleurs s'imposaient. Dans certaines copies, Pierre II représente à côté du tondeur de moutons un panier plein de laine, pour souligner sa richesse. Il cède alors à son penchant pour la redondance. De même, il est fait usage du rouge-bleu dans toutes les copies, pour les couples à la huque bleue. D'une manière générale, le jaune caractérise les conduites désordonnées des mêmes personnages que ceux du père (fig. 2). Ceux qui 'tirent au plus long', tirent sur une ficelle dans les copies, plutôt que sur un 'bretzel' comme dans la peinture de l'Ancien. Le jaune est également présent pour les deux mégères de «l'une charge la quenouille de ce que l'autre file». Chez les fils comme chez le père, le bleu caractérise l'infidélité, l'hypocrisie. La femme infidèle, dans le coin droit du *Triomphe de la Mort*, est vêtue de bleu dans l'original du Prado. Cette couleur est bleue ou rose dans les copies, selon le maintien ou non de la notion d'infidélité pour ce groupe que semble bien à chaque fois être constitué de portraits. Dans les copies, les flûtes — allégories de l'infidélité — sont tantôt présentes tantôt omises aux pieds de l'homme, et dans une version, Jean corrige la connotation d'infidélité en ajoutant au pied du couple un chien, symbole de fidélité. Notons enfin que les jeunes femmes autour de la table dans le coin inférieur droit, où figure une scène de bordel, sont souvent habillées de jaune et de vert, couleurs du désordre et de la folie.

Un autre exemple montre que les fils appliquaient le code des couleurs. On conserve plusieurs copies d'après la grisaille de Pierre l'Ancien, représentant *Le Christ et la femme adultère*, 1565 (huile sur bois 24,1 × 34,4 cm Londres, Courtauld Institute of Art, Count Antoine Seilern Collection). On sait que la grisaille a été conservée par Bruegel

[38] Voir au sujet du *Massacre des Innocents* C. CURRIE, «Découverte de la réutilisation d'un panneau chez Pierre Brueghel le Jeune: premier témoignage de recyclage économique chez cet artiste», dans *Revue belge d'archéologie et d'histoire de l'art*, LXXI, 2002, p. 47-57. Pour l'auteur «une version du *Massacre des Innocents*...Bruxelles, inv. 8726 a été vraisemblablement copiée d'après un modèle disparu de Martin Van Cleve».

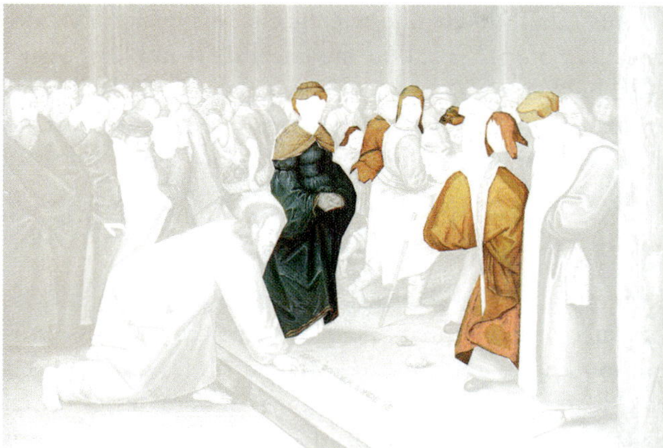

Fig. 8.

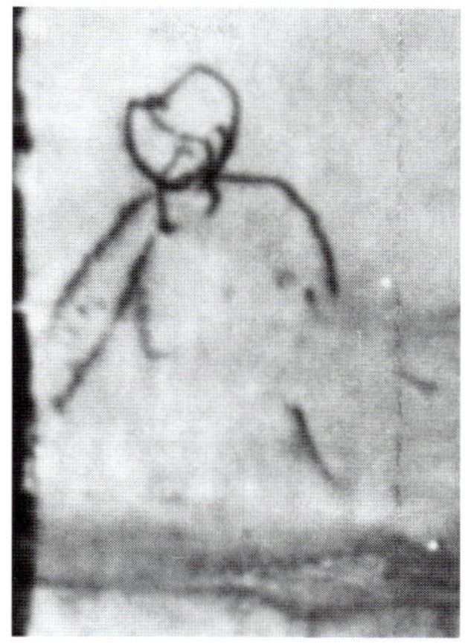

Fig. 9.

Fig. 8. *Le Christ et la femme adultère* de Pierre Breughel le Jeune, Philadelphia Museum of Art, John G. Johnson Collection, bois, 26,7 × 39,7 cm, signé P. Breughel. Schéma des couleurs signifiantes, ici le jaune et le bleu.

Fig. 9. Pierre Breughel le Jeune, *Le Triomphe de la mort,* Bâle, Kunstmuseum, dessin sous-jacent d'un détail dans le fond du tableau, réflectogramme à l'infrarouge. Le caractère très maladroit du dessin de ce tableau évoque l'usage d'un pantographe.

jusqu'à sa mort et que son fils Jean en a hérité. Les fils en font des copies, en grisaille, mais également en couleurs. Ils semblent n'avoir eu aucun problème de traduire dans les couleurs appropriées la composition de leur père. Dans une des copies en couleurs (signée P. BREUGHEL, Philadelphie, Museum of Art, John G. Johnson Coll., Acc. Nr J. 423, bois 26,7 × 39,7 cm) (fig. 8)[39], les juifs de la synagogue sont vêtus de jaune (ou, à tout le moins, portent un bonnet jaune), couleur parfois associée à du vert. Un juif est roux, ce qui était péjoratif également. La femme inconstante porte le bleu de l'adultère, avec un petit foulard jaune rayé qui pourrait également marquer sa condition. Ces couleurs correspondent au code cité précédemment.

On retrouve l'opposition entre le rouge du riche et le bleu délavé du pauvre dans une composition moins connue, représentant deux bûcherons (?) qui lient un fagot (fig. 9). Seules des versions par les fils sont conservées, mais les auteurs pensent que Pierre l'Ancien inventa le thème[40]. K. Ertz en mentionne douze versions, mais n'explique pas le sujet[41]. Deux versions en couleurs sont reproduites dans son ouvrage[42]. La référence au code des couleurs peut mettre sur la voie de l'interprétation de la scène. Un des bûcherons est corpulent (riche) et vêtu de culottes rouges. Une bourse et un chapeau sont posés à ses pieds. Le pauvre porte des vêtements bleu délavé en lambeaux. Il est maigre.

[39] K. ERTZ, 1998, I, p. 382, 385, 445, ill. 274.
[40] K. ERTZ, 1998, I, p. 151-154, 159, 212-213.
[41] K. ERTZ, 1998, I, p. 151-156.
[42] (München, Galerie Scheidwimmer, signé et daté P. BRVEGHEL 1625 et Enschede, Rijksmuseum Twenthe, signé P. BRVEGH...). K. ERTZ, 1998, I, p. 163-164.

Une flûte est posée à son côté. Ici aussi, la flûte est une référence à l'expression «vendre des flûtes» (vendre des récits trompeurs, tromper… la flûte faisant office d'emblème visuel pour l'acte sexuel). La pauvreté du bûcheron est liée à son usage abusif de la flûte… On a déjà signalé que les flûtes aux pieds du couple dans le coin inférieur droit du *Triomphe de la Mort* du Prado devaient être interprétées dans le même sens.

Conclusion

Dans ses tableaux allégoriques ou moralisateurs, Bruegel l'Ancien a adopté le code des couleurs signifiantes. Ses fils et ses successeurs connaissaient eux aussi ce code des couleurs. Dans de rares cas les copistes ont vu les couleurs de l'original; ils les ont répétées. Dans d'autres cas, plus nombreux, leurs couleurs sont totalement différentes et on en conclut qu'ils ignoraient les couleurs du modèle. Mais dans d'autres cas encore, les couleurs choisies par eux se répètent pour certains personnages ou certains détails. La raison de cette répétition est que ces couleurs s'imposaient pour leur caractère signifiant dans un certain contexte. Ces couleurs signifiantes n'étaient d'ailleurs pas toujours les mêmes que celles que Bruegel l'Ancien avait choisies pour les mêmes détails. Bruegel l'Ancien, dans le langage pictural extrêmement élaboré qui était le sien, avait parfois fait un choix inattendu ajoutant une signification différente de celle à laquelle on pouvait s'attendre, par exemple en peignant un cardinal vêtu de bleu plutôt que de rouge, ou encore en inversant les couleurs comme il convenait dans un 'monde à l'envers' (une lune noire plutôt que blanche…). Dès lors le sens symbolique d'une même composition peut, d'une génération à l'autre, présenter des variantes ou des nuances nombreuses.

Hélène VEROUGSTRAETE
Université catholique de Louvain
Laboratoire d'étude des œuvres d'art
par les méthodes scientifiques
Rue Cardinal Mercier 53
B – 1348 Louvain-la-Neuve
Tel: +32.10.47.48.97
Fax: +32.10.47.48.95
helene.verougstraete@uclouvain.be

Docteur en Histoire de l'Art et Archéologie (UCL). Thèse: cadres et supports dans la peinture des Pays-Bas méridionaux aux 15e et 16e siècles. 1967-1969: formation en restauration des œuvres d'art, KIKIRPA. 1987-2005: professeur à l'Institut royal d'Archéologie et d'Histoire de l'Art de Bruxelles. 1992-: professeur à l'UCL. 1997-: également professeur à la KULeuven. 1992-: direction du Laboratoire d'étude des œuvres d'art par les méthodes scientifiques. 1975-: avec R. Van Schoute, organisation du présent colloque et publication d'actes. Travaux d'expertise, notamment pour les tribunaux. Publications: Primitifs flamands, Pierre Bruegel (e.a. une interprétation du *Triomphe de la Mort).* Examen d'œuvres restaurées par Jef Van der Veken et préparation avec l'équipe du Laboratoire de l'exposition et du catalogue qui lui sont en grande partie consacrés: *Restaurateurs ou faussaires des primitifs flamands. [Fake not Fake. Het verhaal van de restauratie van de Vlaamse primitieven]*, Cat. d'exp. Bruges Groeningemuseum 26 novembre 2004 – 28 février 2005, Gand (Ludion), 2004.

THE *LAST SUPPER* COMPOSITION BY WILLEM KEY (c. 1515-1568)

Linda Jansen

Abstract: The article focuses on Willem Key's *Last Supper* in the Dordrecht Museum, a painting reflecting the famous composition by Pieter Coecke, known in many versions (e.g. in the Royal Museums of Fine Arts of Belgium, Brussels). IRR revealed an extensive underdrawing with many alterations, pointing to a creative genesis of the composition. The figure in the lower right corner is supposed to be a self-portrait. Several versions produced in Key's workshop (present locations unknown), do not represent this figure. A triptych for Gilles de Smidt (Royal Museum of Fine Arts, Antwerp) presents another version of the composition with *The Last Supper* placed on the outer wings. Often cited as one of the most important emulations of Key's composition, it is painted by Adriaen Thomasz. Key, presumably a distant relative and pupil. Adriaen made important alterations to the composition, making it an invention of his own. He signed 'ADRIANUS THOMAE. KEII FECIT 1575' and included a portrait in the lower right corner. As said before, the pictorial tradition of Willem Key's *Last Supper* can be traced back to his master Coecke. Adriaen Key's *Last Supper* can be traced back to his master Willem Key. How was this pictorial tradition passed on from master to pupil? Can we speak of an original composition that lies at the heart of this tradition, or are all these compositions, with the alterations made by the individual artists, inventions in their own right? The underdrawing and the alterations are indicative of the way the composition came into being. Since the underdrawing of the *Last Supper* by Willem Key is the first one by him ever revealed, it is the first indication of his workshop routine and therefore an important starting point for further comparison.

The *Last Supper* by Willem Key[1] at the Dordrechts Museum in Dordrecht (fig. 1)[2] is the first of Key's works to be examined with infrared reflectography.[3] The present study explores Key's underdrawing in relation to his working method and workshop organization. The infrared image will also be used to examine the hypothesis that one of the figures represents a self-portrait. The painting is neither signed nor dated. Wescher attributed it to Key,[4] grounding his opinion on stylistic similarities with Key's signed paintings. Correspondences with contemporary *Last Supper* compositions, notably by

[1] Unless otherwise stated, the name "Key" will point to Willem Key and not to Adriaen Thomasz. Key, a distant relative and pupil.

[2] Willem Adriaenz. Key, *Last Supper*, panel, 150 × 234 cm, Dordrechts Museum, Dordrecht.

[3] IRR examination at the request of the Dordrechts Museum. Further results will be published in the Museum Bulletin by this author. In the new presentation of the Museum's collection, starting from 2009, more focus will be given to Key's *Last Supper*. The IRR examination was carried out by Margreet Wolters (RKD), with the equipment of the RKD in The Hague: Hamamatsu C 2400-07 with a N2606 IR vidicon, a Nikon Micro-Nikkor 1:2.8/55 mm lens, a Heliopan RG 850 (or RG 1000) filter, with a Lucius & Baer VM 1710 monitor (625 lines). Digital documentation was done with a Meteor RCB framegrabber, 768 × 574 pixels, colorvision toolkit (Visualbasic). The reflectograms were assembled in Panavue and Photoshop by Margreet Wolters and Linda Jansen.

[4] P. WESCHER, "Zu Anthonis Blocklandt", in *Oud Holland*, 45 (1928), p. 276. The painting had been attributed to Anthonis Blocklandt from Utrecht, probably because of Van Mander's reference to another *Last Supper* in Dordrecht by the painter (fol. 254).

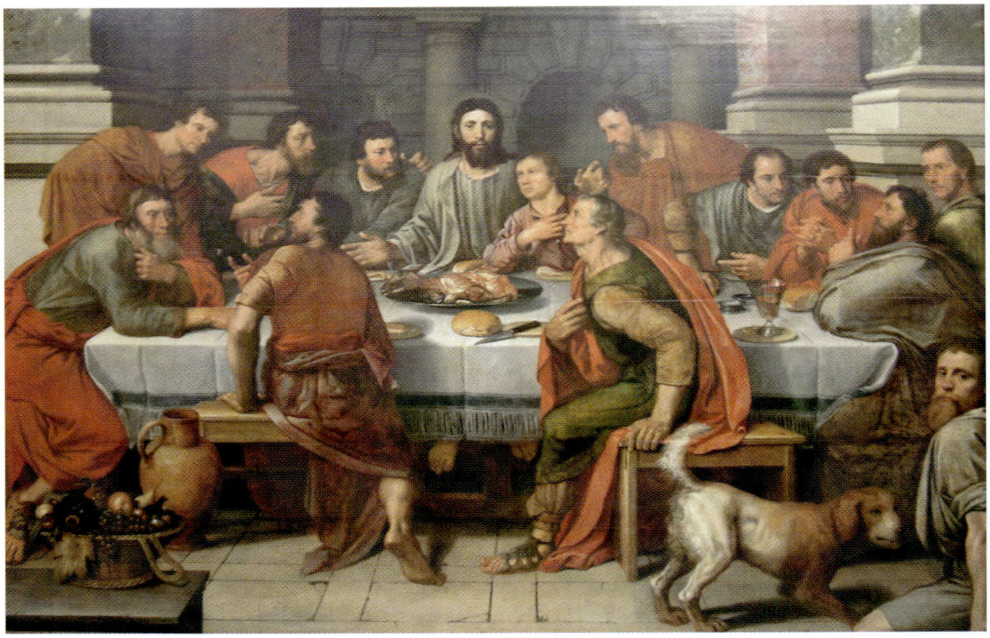

Fig. 1. Willem Key, *Last Supper*, Dordrecht, Dordrechts Museum (© Jansen, with the courtesy of the Dordrechts Museum).

Pieter Pourbus, allow to date the painting around 1560.[5] Before 1572, Key's *Last Supper* was in the *Grote Kerk* in Dordrecht in the Chapel of Saint John the Evangelist, for which it was probably commissioned.[6] The donors are unknown. After Protestant troubles the painting was transferred in 1572 to the City Hall.

Willem Adriaensz. Key: a short introduction

Key descended from an artists' family in Breda. He left for Antwerp in 1529 to receive his initial training from Pieter Coecke van Aelst.[7] After his training he may have continued to work for this master or he may have worked with his brother Wouter, master in Antwerp since 1531. Around 1538, he continued his training with the Liège painter Lambert Lombard. He returned to Antwerp in 1542 and became a master in

[5] Exh. Cat. Amsterdam, 1986, *Kunst voor de Beeldenstorm. Noordnederlandse kunst 1525-1580*, Amsterdam/The Hague, 1986, cat. 209, p. 331. Particularly relevant are two *Last Supper* compositions from Pieter Pourbus, dating from 1559 and 1562.

[6] Exh. Cat. Amsterdam, 1986, p. 331.

[7] About the Key family in Breda, see D. TILLEMANS, "Willem Key – Biografische gegevens", in *Gentse Bijdragen tot de Kunstgeschiedenis*, vol. XXV, (1979-80), p. 63-90; on Key's training with Coecke, see L. JANSEN, "Considerations on the size of Pieter Coecke's workshop: apprentices, family and journeymen. A contribution to the study of journeymen on micro level", in N. PEETERS ed., *Invisible hands? Role and Status of journeymen in artist's and craftsman's workshops in the Low Countries ca 1450-1650*, Louvain, 2007, p. 83-103.

the Guild of Saint Luke.[8] From the enrolment lists of the Guild it is known that he took on two apprentices: Hansken Spruyt and Dirk Aertsen. However, he presumably had even more apprentices or assistants, among them Adriaen Thomasz. Key. Adriaen was never officially registered as an apprentice to Willem, but the closeness in style between the two artists undoubtedly reveals a master-pupil relationship.[9] Adriaen frequently signed as Adriaen Thomasz. Key, as can be seen in his interpretation of the *Last Supper* in Antwerp.[10] It is not certain whether he was a relative, or whether he simply took over his master's name as a brand-name when he founded his own studio in 1568, the date of Key's death.

Guicciardini, Lampsonius and Van Mander all refer to Willem Key as a portrait painter and history painter. Today he is primarily known as a portrait painter. His œuvre of history paintings remains small and still undefined. Only two signed and dated paintings have been preserved: a *Suzanna and the Elders* at Pommersfelden (1546/K/WFA)[11] and a *Lamentation* (1553) in a private collection (W. Keien).[12] Based on stylistic correspondences with these paintings, a varying number of other works have been attributed to Key. His style is often compared to the monumental Italianate style of Frans Floris, his fellow student in Lombard's workshop.[13] Key's technique is more refined and his style is more restrained in movement and colour compared to Floris' technique. Another typical aspect of Key's style are the portrait-like features of the heads.

For the *Last Supper*, Key was clearly inspired by Pieter Coecke's much copied composition of the same subject,[14] for example by Pieter Pourbus. However, Key, as a pupil of Coecke, must have had first hand access to his master's invention. Key took over the strongly horizontal composition with the table parallel to the picture surface. The gesticulating apostles are placed behind the table, with the exception of two of them, seated in front of the table in twisted positions. The dog and the fruit basket are also strongly inspired by the master. However, despite similarities with Coecke's *Last*

[8] P. ROMBOUTS and T. VAN LERIUS, *De Liggeren en andere historische archieven der Antwerpse Sint Lucas Gilde*, vol. 1, Antwerpen, 1864-1876, p. 143. Van Mander incorrectly reports that Key became a master in 1540.

[9] C. VAN DE VELDE, in J. TURNER ed., *Dictionary of Art*, vol. 18, New York, 1996, p. 6-7; K. Jonckheere, "Wars van decorum. Een interpretatie van het Laatste Avondmaal van Adriaen Thomasz Key", in K. VAN DER STIGHELEN ed., *Munuscula amicorum – Contributions on Rubens and his Colleagues in Honour of Hans Vlieghe*, Turnhout, 2006. K. JONCKHEERE, *Adriaen Thomasz. Key (c.1545-c.1589). Portrait of a Calvinist painter*, Turnhout, 2007.

[10] Adriaen Thomasz. KEY, pair of wings with *The Last Supper*, Koninklijk Museum voor Schone Kunsten, Antwerp, inv. 230/231; both panels 181 × 118 cm and signed 'ADRIANUS.THOMAE. KEII. FECIT. 1575'.

[11] This should be read as 'Key Willem Fecit Antwerpiae'; M.J. FRIEDLÄNDER, *Early Netherlandish Painting, Anthonis Mor and his contemporaries*, vol 13, Leiden/ Brussels, 1975, p. 52.

[12] M.J. FRIEDLÄNDER, "Über Willem Key", in *Pantheon*, vol. 3 (Jan), 1929, p. 254-256; G.J. HOOGEWERFF, "De werken van Willem Key", in *Belgisch Tijdschrift voor Oudheidkunde en Kunstgeschiedenis*, XVII (1947-48), p. 48-49; FRIEDLÄNDER, 1975, p. 52-53.

[13] M.J. FRIEDLÄNDER, 1975, p. 53; K. Jonckheere is preparing an article on the relation between Willem Key and Frans Floris; I would like to thank him for sharing his views on Key with me.

[14] Numerous copies have been preserved, such as the one in the Royal Museum of Fine Arts of Belgium in Brussels. For Coecke's series, see L. JANSEN, "The *Last Supper* as a Starting Point for the Study of the Workshop Practices in the Group Pieter Coecke van Aelst", in H. VEROUGSTRAETE and R. VAN SCHOUTE eds., *Le dessin sous-jacent et la technologie dans la peinture. Colloque XIV: Jérôme Bosch et son entourage et autres études (2001)*, Louvain, Paris, Dudley, 2003, p 165-174.

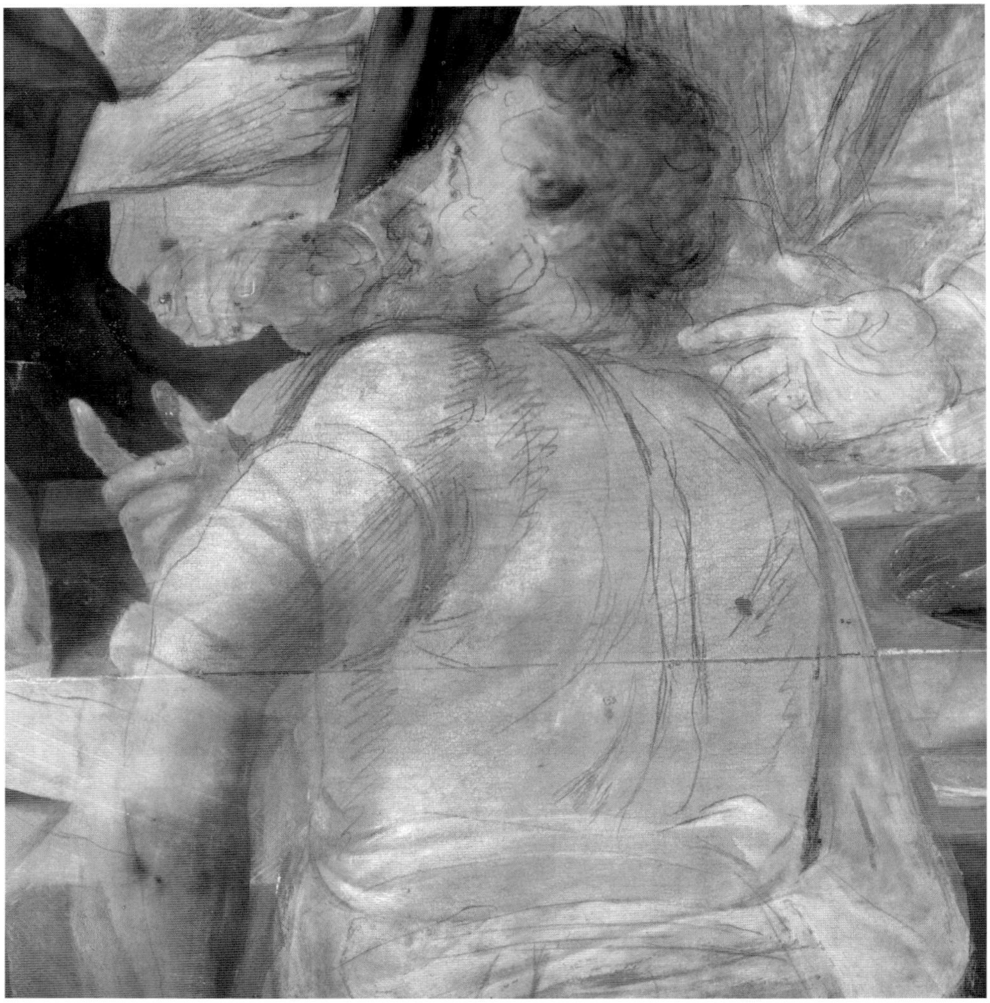

Fig. 2. Willem Key, *Last Supper*, Dordrecht, Dordrechts Museum. Detail, left apostle sitting in front of the table, digital IRR assembly (© RKD, digital assembly Wolters/Jansen).

Supper, Key's composition is an original invention. The monumental scale of the figures, their individualized characters and the severe architectural setting are all typical of Key's style. Unlike Coeke's composition, produced in series, Key's was an exclusive product made on commission, as his working procedure will show.

Key's working procedure in the *Last Supper*

Infrared reflectography reveals a complex and creative working procedure in several stages, with many changes in design. The broad brush strokes of a preparatory layer are visible throughout the painting. The underdrawing is applied on top of this layer in a

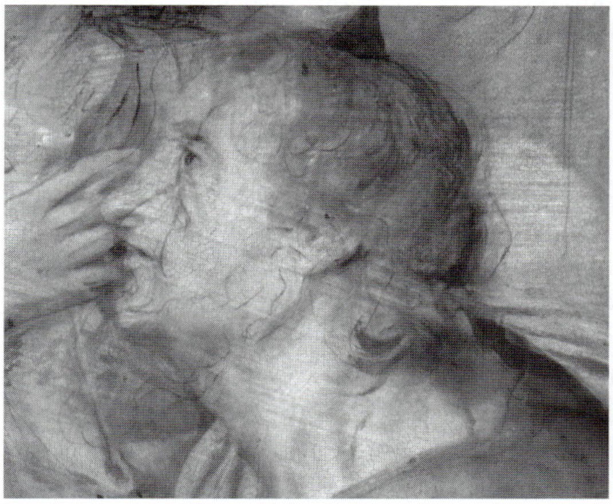

Fig. 3. Willem Key, *Last Supper*, Dordrecht, Dordrechts Museum. Detail, head of Judas, digital IRR assembly (© RKD, digital assembly Wolters/Jansen).

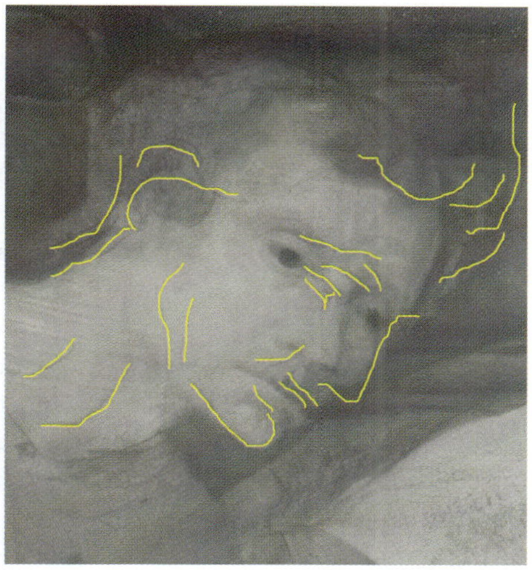

Fig. 4. Willem Key, *Last Supper*, Dordrecht, Dordrechts Museum. Detail, head of the apostle left digital IRR assembly. Reinforcement of the underdrawn lines in yellow by Jansen (© RKD, digital assembly Wolters/Jansen).

seemingly dry material. The line is uneven — sometimes thin, then wider — and it is interrupted by the preparatory layer. It is not certain whether the underdrawing was made in chalk, or with a dry-brush. The overall character of the underdrawing is sketchy, as in the apostle sitting to the left in front of the table (fig. 2). The lines are loose and sometimes repeated, as if Key were searching for the right forms. Besides contour lines, the artist used both zigzag and parallel hatchings. These roughly indicate the shadows rather than the volumes.

There are numerous changes in the paint layers. In several places the underdrawn folds have been simplified in their final painted form, and therefore have a more striking appearance. In the apostle sitting to the left in front of the table, the garments were not extensively underdrawn. Key focused on the structure of the muscles in the back of the figure. He broadened the outlines of arms and back, making the figure more monumental.

Extensive changes were made in the heads of all the apostles. The eyes, nose and cheeks of an apostle in front of the table were initially placed lower in the underdrawing and in the first stage of painting. The final face was painted over the background. The eyes and nose of Judas were also positioned lower in the underdrawing (fig. 3). Judas originally had a pointed beard. The underdrawn head looks more archaic, resembling the Judas-figures of the first half of the century, particularly the one in the *Last Supper* by Pieter Coecke. Key seemed to individualize heads in the final paint layers. The apostles on the far left and right have undergone significant changes (fig. 4). They were planned and painted in profile. In the final paint stage, Key decided on a three quarter pose.

How can we interpret these alterations with regard to Key's working procedure? The sketchiness of the underdrawing and the alterations in the paint layers are indications of continuous creative activity. It is nonetheless likely that for a complex composition such as this Key also made a compositional sketch or model drawing prior to underdrawing. This procedure is confirmed by comparing Key's *Last Supper* with the versions made after it.

Key's composition is known in four more versions (present whereabouts unknown)[15] of smaller dimensions, with different faces and clothes and considerably inferior in quality. They most likely are later copies rather than workshop replicas. Certain variations with regard to Key's composition appear in all the copies. For instance, in the copies Judas has a beard and the apostles at the far left and right are still depicted in profile. In these details, the copies seem to follow the underdrawing of Key's *Last Supper* rather than the painted surface. It is unlikely, however, that they based themselves on the underdrawing. They must have had access to the same source Key himself used for his underdrawing — his compositional sketch or model drawing.

The servant as Key's self-portrait

The servant in the lower right corner (fig. 5a), the only figure looking out of the plane of the painting, has portrait-like characteristics. Hoogewerff believed him to be a self-portrait by Key.[16] He compared the features with those of a much debated self-portrait from 1566. The features of the servant do resemble the engraved portrait of Key by Wierix in Lampsonius' *Pictorium aliquot celebrium* [...] from 1572.[17] Portraits are usually not, or little underdrawn,[18] and sometimes added in the final paint stages.[19] Indeed, the alleged self-portrait of Key differs in lay-out and execution from the apostles (fig. 5b). The figure is not underdrawn although its position was determined at an early stage. The underdrawing of the robe of the apostle sitting behind the servant does not extend under the servant. In the copies of Key's composition, the servant is absent. This most likely means that the figure was not included in Key's original design of the composition.

Another argument in favour of the hypothesis of the servant being a self-portrait is the appearance of the figure in Adriaen Thomasz. Key's *Last Supper* in Antwerp. Both servants are placed in the lower right corner, but their features are strikingly different. As an assistant of Willem, Adriaen Thomasz. Key must have been familiar with the

[15] RKD database: Follower of Key, *Last Supper*, private collection, The Netherlands, panel 114 × 177 cm (RKD art number 53717); Follower of Key, *Last Supper*, whereabouts unknown, measurements unknown, (RKD art number 53718); Follower of Key, *Last Supper*, London, Sotheby's, 28 October 1999, lot. 313, panel 124.5 × 178.5 cm (RKD art number 61711); Follower of Key, *Last Supper*, London, Sotheby's, 5 July 1989, lot. 172, panel 104 × 168,5 cm.

[16] G.J. HOOGEWERFF, 1947-48, p. 48.

[17] I would like to thank Koenraad Jonckheere for sharing his views on this topic with me.

[18] M.W. AINSWORTH and M. FARIES, "Northern Renaissance Paintings. The Discovery of Invention", in *The Saint Louis Art Museum Bulletin*, 18/1 (1986), p. 19-28.

[19] For instance, in the self-portraits by Joos van Cleve, see M. LEEFLANG, "Joos van Cleve's Self-Portraits: New Insights", in H. VEROUGSTRAETE and R. VAN SCHOUTE eds., *Le dessin sous-jacent et la technologie dans la peinture. Colloque XIV: Jérôme Bosch et son entourage et autres études*, Louvain, 2003, p. 175-182.

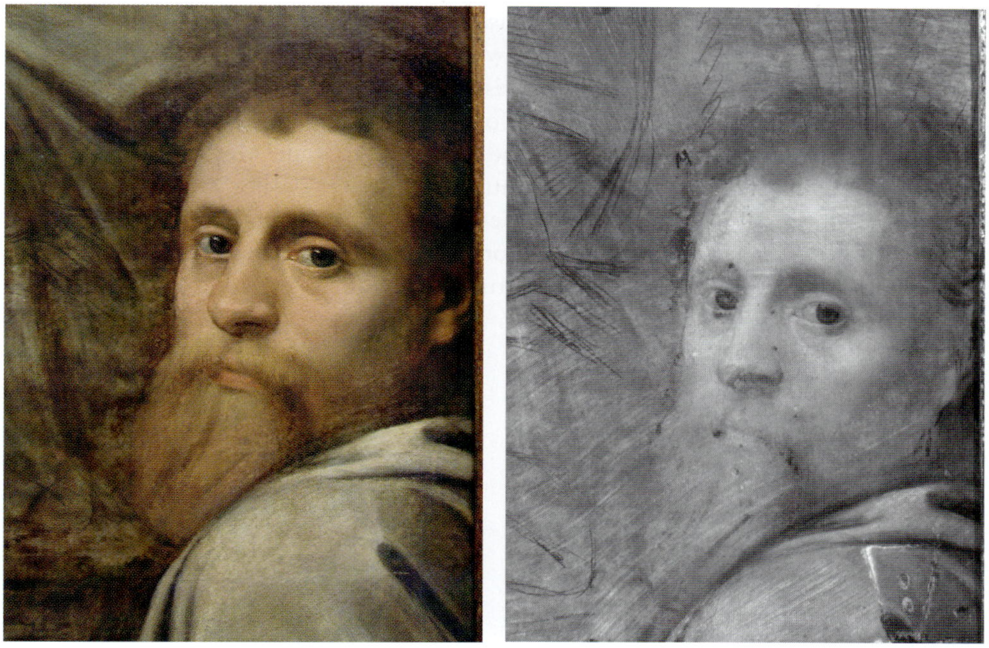

Fig. 5. a. Willem Key, *Last Supper*, Dordrecht, Dordrechts Museum. Detail, head of a servant (© Linda Jansen, with courtesy of the Dordrechts Museum). b. Willem Key, *Last Supper*, Dordrecht, Dordrechts Museum. Same detail, digital IRR (© RKD, digital assembly Wolters/Jansen).

stratagem of using the servant as a self-portrait. If the figure were a merely a servant, it would not have been necessary to change his features. Since Adriaen's *Last Supper* is often seen as an emulation of his master's composition, it is tempting to presume that Adriaen Key borrowed this personal 'wink-of-the-eye' from his master to honour him, and to 'sign' his own invention.

Conclusion: quality over quantity?

The analysis of the *Last Supper* gives a valuable first insight into the working procedures of Key. The *Last Supper* presents three or four different phases:

1) the composition was prepared in a compositional sketch or model drawing
2) based on the preparatory drawing, the artist loosely worked out an underdrawing on the panel
3, 4) he worked out the composition in paint in at least two stages

Key made alterations at every stage. In addition to the addition of his self-portrait, these alterations testify to a highly personal working method. To verify this conclusion, of course, the underdrawing and painting technique must be compared with the two signed paintings by Key and with other paintings that are reliably attributed to him. But for now the *Last Supper* shows a certain exclusiveness in working procedure which can be representative for the way he organised his workshop. No standardized production methods can be found, unlike what we see in large contemporary workshops at that time. Although two pupils and one assistant were more then average at that time,[20] Key probably had a small-scale workshop, focusing mainly on prestigious commissions. The number of his assistants will never be known since some were unregistered like Adriaen Thomasz. Key. However, the paintings themselves do not present evidence of collaboration of many assistants, nor do the archival or contemporary sources mention a large workshop such as that of Frans Floris. His presumed small-scale management of the workshop proved to be a success formula providing Key wealth and esteem.

Linda JANSEN
University of Groningen
Postgraduate Research Centre
Grote Rozenstraat, 38
NL – 9712 TJ Groningen
Tel: 00.31. (0)50.363.62.36
l.jansen@let.vu.nl

Studied History of Art and Architecture at the University of Groningen (1993-1998); thesis about the Bruges painter Lanceloot Blondeel. In 1999 she worked as an IRR assistant for M. Faries at the University of Groningen. Since July 2000, she has been working as *Assistent in Opleiding* in the NWO sponsored project at the University of Groningen (2000-2005): *Painting in Antwerp before Iconoclasm (c. 1480-1566): a socio-economic approach,* (dir. M. Faries and M.P.J. Martens). The title of her Ph.D. thesis is *Pieter Coecke van Aelst (1502-1550) and his Workshop: the Production, Distribution and Consumption of his œuvre.* She is currently affiliated to the Postgraduate Research Centre in Groningen.

[20] M.P.J. MARTENS and N. PEETERS, "Artists by Numbers: Quantifying Artists'Trades in Sixteenth-Century Antwerp", in M. FARIES ed., *Making and Marketing. Studies of the Painting Process in Fifteenth- and Sixteenth-Century Netherlandish Workshops*, Turnhout, 2007, p. 216-218. In the period 1500-1579, 67-68% of the masters never took on apprentices and only 7,48% took on two.

RESILIENCE AND ENTERPRISE. THE FRANCKEN-BROTHERS IN THE YEARS BETWEEN THE ICONOCLASM AND THE SPANISH FURY (1566-1576)

Natasja Peeters

Abstract: This paper concentrates on a lesser-known period of Antwerp painting: the turbulent years between the first wave of Iconoclasm and the Spanish Fury (1566-1576). After the boom of the 1560s, both the market for luxury goods and private commissions suffered from the economical, political and religious instability in the Southern Netherlands. Artists keenly felt this. Based on the case study of three members of the Francken family, this paper explores the opportunities and limitations of the painter's craft in a time of crisis.

Hieronymus (1540-1610), Frans (1542-1616) and Ambrosius Francken (1544/45-1618) left their small native town of Herentals between 1555 and c. 1560 after their initial training with their father Nicolas (c. 1520-1596) and went to Antwerp.[1] They completed their training in the studio of the renowned Frans Floris (1519/1520-1570), Antwerp's most famous history painter. Carel van Mander mentions them as disciples or journeymen with an intellectual stature. The dates of their stay are known only approximately. They had come to Antwerp to prove their luck at a time the metropolis was booming.[2] They were not the only ones to do so: before 1560, seventy per cent of registered and non-registered master painters were immigrants. The greatest influx took place between c. 1500 and 1560.[3]

Carel van Mander estimated the number of journeymen in Floris's workshop at 120, although this figure is difficult to corroborate.[4] Possibly, it enhanced a painter's reputation to have been in Floris's orbit, even if it were only for a day or two. The precise activities of the brothers in the workshop remain a matter of conjecture. No literal Francken copies of Floris's work are known to exist, and none are attributed to them, but an often published document of 1576 refers to a copy which Frans Francken painted around 1565 with *Cain and Abel* after an original by Floris.[5] The copy of study heads (*tronies*) was also something Floris possibly delegated to his journeymen.

[1] For specific references, see N. PEETERS, *Tussen continuiteit en vernieuwing. De bijdrage van Frans en Ambrosius Francken I, en de jonge generatie Francken, tot de historieschilderkunst te Antwerpen ca 1570-1620*, unpublished PhD-thesis, Vrije Universiteit Brussels, 2000 (publication forthcoming).

[2] M.P.J. MARTENS and N. PEETERS, "Artists by numbers. Quantifying Artist's Trades in 16th century Antwerp", in M. Faries ed., *Making and Marketing: Studies of the Painting Process in Fifteenth- and Sixteenth-Century Netherlandish Workshops*, Turnhout, 2006, p. 211-222.

[3] N. PEETERS, "Zugewanderte Künstler in Antwerpen in der Frühen Neuzeit", in K.J. Bade, P.C. Emmer, L. Lucassen and J. Oltmer eds., *Enzyklopädie Migration in Europa vom 17. Jahrhundert bis zur Gegenwart*, Paderborn, 2007, p. 1108-1110.

[4] P.F. ROMBOUTS and T. VAN LERIUS, *Liggeren en andere historische archieven der Antwerpsche Sint-Lucasgilde,* The Hague, 1864-1876, 2 vols., Amsterdam, 1961, vol. 1, p. 216; C. VAN DE VELDE, *Frans Floris (1519/20-1570), Leven en werken*, Brussels, 1975, p. 441.

[5] C. VAN DE VELDE, 1975, S 89 and p. 472-473, referring to City Archives, Antwerp (hereafter quoted as: S.A.A.), Notary D. Van den Bossche, n°3638, 1576-1577 fol. 2v° (7/1/1576).

While the brothers were still working in Floris's workshop, religious troubles and a deep economic crisis darkened Antwerp's horizon. This context was determining in the early careers of the Francken brothers. Their endurance was to be severely tested. In August 1566, the first wave of iconoclasm swept over the Netherlands; many altarpieces and other works of art were irreparably damaged or destroyed,[6] and many people fled Antwerp. During the years thereafter, altarpieces and works of art in public and religious buildings were repaired or replaced, which especially benefited painters like Frans Floris, Willem Key, Adriaan Thomasz. Key, Michiel Coxcie and Maarten de Vos. But what about the new generation of history painters?

On average, after an approximately four-year training period, an apprentice spent between one and eleven years as a journeyman before acquiring a master's title. An apprentice had around a one in five chance of becoming a master. How did the Franckens fare?

Hieronymus probably left Floris's workshop for Paris before or in 1566, possibly as an art-dealer.[7] Frans moved on before 1567-1568 and Ambrosius most likely left a year later. They did not leave the workshop without having sharpened their skills in history painting and certainly not without a secure network of contacts spanning various countries. After Floris's death in 1570, many of his former journeymen kept in contact, even after migrating to neighbouring countries.

Frans got his master's title in 1567-1568,[8] became an Antwerp citizen (*poorter*) at the end of March 1568,[9] married and started his career as an independent master painter. Different signs point towards Frans having a workshop at that time, where he employed journeymen. Clearly Frans had work to offer. Interestingly, he took in an assistant of Floris: Herman van der Mast (c. 1550-1610), who came to live with Frans shortly after Floris's death, before travelling on to Paris and Bourges.[10] He also took in Gortzius Geldorp (1553-1616), who had been apprentice to Frans Francken's former co-assistant Frans Pourbus around 1570-1571.[11]

Van Mander mentions that Frans Francken was indeed active in the 1570s, and that, shortly after Floris's death, his workshop was involved in producing copies after him — Frans Francken had already done so during the 1565s — thus cashing in on the demand for copies after his famous masters' paintings.[12]

Frans relied not only on the network of Floris connections (who had the right skills to copy), but also on his brothers, and his brothers on him. The altarpiece of the *Adoration of the Magi*, dated 1571, formerly attributed to Floris, of which the provenance is unkown, bears a double monogram '*FF HF*' (fig. 1).[13] It reminds of Floris in

[6] C. Van de Velde, *Het kunstpatrimonium*, in W. Aerts ed., *De Onze Lieve-Vrouwekathedraal van Antwerpen*, Antwerp, 1993, p. 187; D. Freedberg, *Iconoclasm and painting in the revolt of the Netherlands 1566-1609*, New York, (1972) 1988.

[7] P.-M. Auzas, *Hiérosme Francken dit Franco peintre du roi Henri III et du roi Henri IV*, Brussels, 1968.

[8] P.F. Rombouts and T. Van Lerius, 1961, vol. 1, p. 233.

[9] S.A.A., Vierschaar, 149, Poorterboeken (1566-71), f° 200 r° (unpublished).

[10] C. Van de Velde, "Le portrait de la famille Francken, une œuvre de Herman van der Mast", in *Ars auro prior. Studia Ioanni Bialostocki sexagenario dicata*, Warsaw, 1981, p. 341-348.

[11] H. Miedema ed., *Karel van Mander, Lives of the Illustrious Netherlandish and German Painters*, Doornspijk, 1994, vol. 1, p. 378-79, f° 282 r°-v°.

[12] H. Miedema ed.,1994, vol. 1, p. 230, f° 243r°.

[13] Royal Museums of Fine Arts of Belgium, Brussels (central panel and backs of wings) and London, Brompton Oratory (front sides of wings).

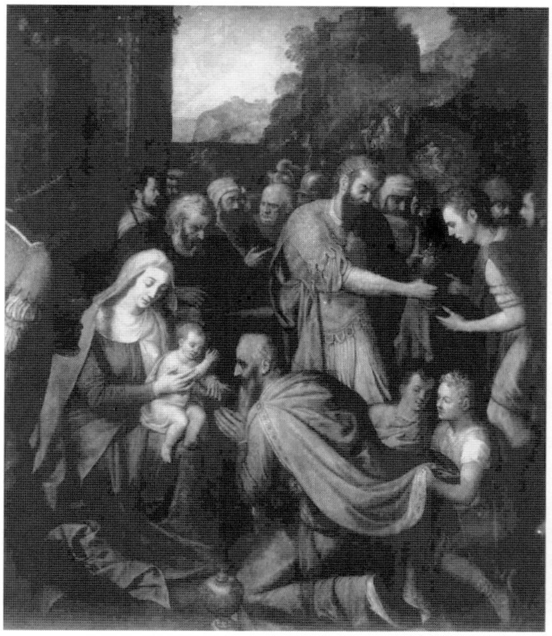

Fig. 1. Hieronymus and Frans Francken, *The Adoration of the Magi*, Brussels, Royal Museums of Fine Arts of Belgium. Central panel (© KIKIRPA).

Fig. 2. Anonymous print, *Pieter-Pots-abbey and surroundings*, late 17th century, Antwerp (© City Archives).

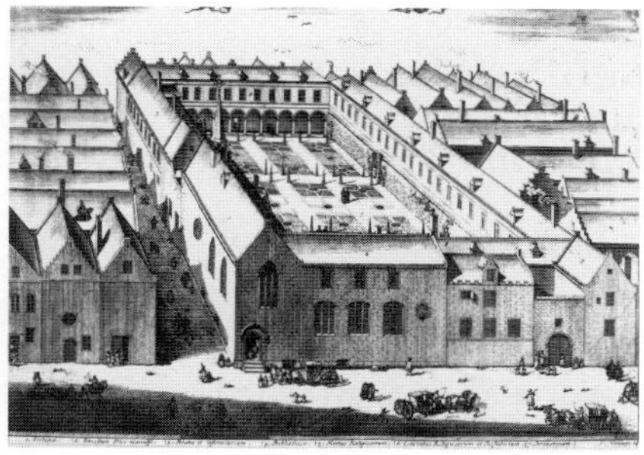

composition and style. Hieronymus inserted a self-portrait in profile on the left and Frans did the same on the right.[14]

Hieronymus, recorded as a master painter in Paris in 1570,[15] was clearly not thinking of coming back to Antwerp permanently: he received his Paris' citizenship in September 1572 and married a French woman.[16] Still, the contacts between the Franckens remained close: Hieronymus's administrative affairs were dealt with by his brothers, and he returned from time to time to Antwerp for short periods.

Around 1572, Frans came to live in Pieter-Potstraat (fig. 2)[17] near Hoogstraat, in a middle class house he rented for 60 guilders from the religious order of the Pieter-Pots-brothers.

In 1569, Ambrosius, the youngest brother, went to Tournai for a short time, as did other Floris apprentices and relatives such as Cornelis Floris II and the latter's brother, the glazier Jacob Floris († 1581).[18] Van Mander (who also worked in Tournai, as did his master Pieter Vlerick) mentions him in the service of bishop Gilbert d'Ognies, but

[14] N. PEETERS, "Brothers in Art: Hiëronymus and Frans Francken I and their family: portraits and crypto-portraits in the second half of the sixteenth century", in E. Griffey ed., *Envisioning Self and Status. Self-representation in the Low Countries 1400-1700*, Hull, 1999, p. 80-102.

[15] J.J. GUIFFREY, "La communauté des maîtres peintres et sculpteurs dite Académie de Saint-Luc depuis son origine en 1391", in *Archives de l'art français*, 1915, 3, p. 6.

[16] J.J. GUIFFREY, *Histoire générale de Paris. Artistes parisiens des XVIe et XVIIe siècles*, Paris, 1915, p. 33-44.

[17] S.A.A., Privilegiekamer 665, Rekwestboeken (1586), f° 88v°.

[18] D. ROGGEN and J. WITHOF, "Cornelis Floris", in *Gentsche Bijdragen* (1942) 8, p. 167; P. ROLLAND, "La Famille Floris à Tournai", in *Jaarboek van de Koninklijke Oudheidkundige Kring van Antwerpen*, 1950, p. 294-95 and p. 311-312; J. VAN DAMME, A. HUYSMANS and C. VAN MULDERS, *Cornelis Floris: 1514-1575: beeldhouwer, architect, ontwerper*, Brussels, 1996, p. 111-113.

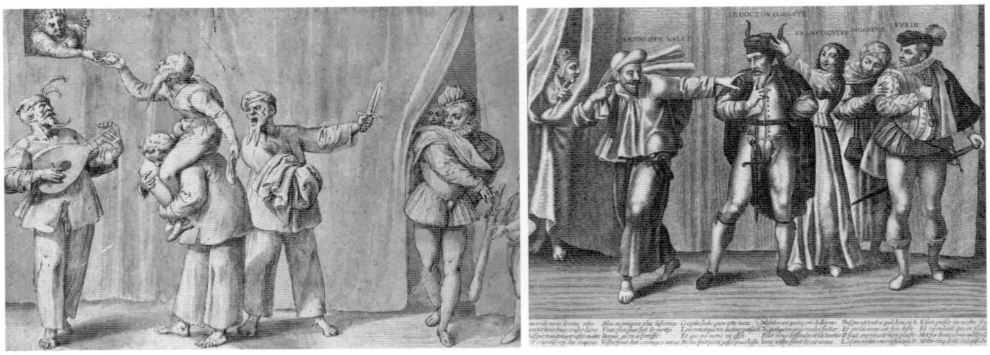

Fig. 3. a. Ambrosius Francken, *Scene from the commedia dell'arte*, drawing, Amsterdam, (© Amsterdam, Historisch Museum). b. After Ambrosius Francken, *Scene from the commedia dell'arte*, print, Paris, Bibliothèque nationale (© Paris, Biblothèque nationale).

does not explain anything further.[19] The town had been much vandalized by the Iconoclasts, and thus attracted a fair number of artists.[20] But Ambrosius did not stay there for long. He might have left Tournai with Cornelis Floris III, son of Cornelis II, who is recorded in Paris in 1568 at the age of 18 years, where he was an apprentice to Hieronymus Francken. Ambrosius is mentioned on May 27 in 1570 at Avon, the little parish church near the palace of Fontainebleau.[21] It is not clear why he made the trip to Paris. He may have gone there to check out the market, to secure trade-contacts and contracts or to design prints. A series with figures from the *Commedia dell arte* (figs. 3a, 3b), very popular in France at that time, seem the earliest works of art pointing to Ambrosius' stay in Paris, as its iconography is closely related to a troupe of players performing in Paris back then.[22] We should not forget that Ambrosius was not yet a free master and he could not work for his own account; maybe he worked in the studio of Hieronymus. We also have to bear in mind that France, and Paris in particular, was the stage of troubles as well, especially the Night of Saint Bartholomew (24 August 1572). The date of Ambrosius' return is not known, but he became a free master in Antwerp in the autumn of 1573.[23]

During the 1570s, the economic, political and religious crisis of the Netherlands deepened and was keenly felt in Antwerp.[24] Due to the religious troubles, altarpiece commissions were at a low point. Antwerp nevertheless remained an important centre

[19] H. MIEDEMA ed., 1994, vol. 1, p. 229, f° 242 v°; E. CORTYL, *Un évêque du 16e siècle, Pierre Pintaflour, évêque de Tournai 1575 à 1580*, Lille, 1893, p. 99-100.

[20] M. VAN VAERNEWIJCK, *Van de beroerlijke tijden in de Nederlanden en voornamelijk in Gent (1566-1568)*, Hasselt, 1966, p. 150.

[21] E.-H. DE GROUCHY, "Extraits des registres paroissiaux d'Avon", in *Annales de la Société historique et archéologique du Gâtinais*, 1890, p. 5. See also: N. PEETERS, "Den 'Quaden tyt'? The career of the young Antwerp artist Ambrosius Francken before 1585", in *Oud Holland* (forthcoming 2008).

[22] N. PEETERS, 2000, p. 92ff.

[23] P.F. ROMBOUTS and T. VAN LERIUS, 1961, vol. 1, p. 250.

[24] H. VAN DER WEE, *The growth of the Antwerp Market and the European economy (fourteenth-sixteenth centuries)*, The Hague, 1963, 3 vols., vol. 2, p. 205, 207, 209-220.

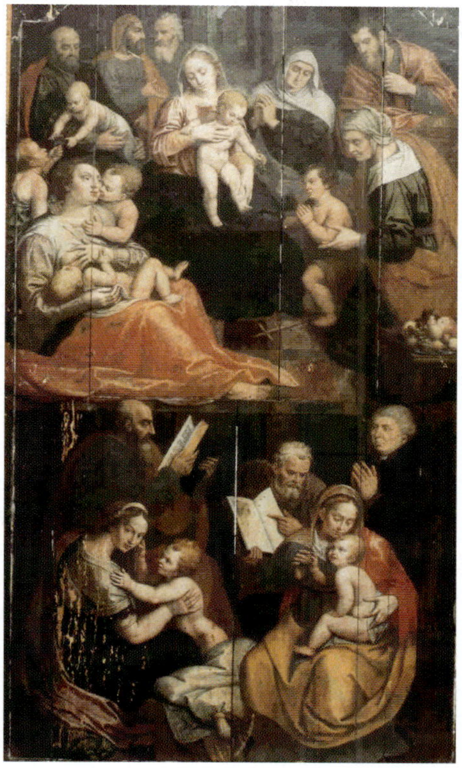

Fig. 4. Frans Francken, *Altarpiece of Sainte Anne* (partial view) during restoration, Herentals, Church of Saint Waldetrudis (© KIKIRPA).

for the production of paintings and prints.[25] While Frans Francken was active as a painter, Ambrosius, back in Antwerp, seems to have hedged his bets on the printing business as an *inventor* of images, something he was already involved in during his stay in Paris. Floris had been active in this sector as well. In this respect it is also interesting to mention that Floris made his disciples execute drawings after his works for prints, as is mentioned by Van Mander.[26] Ambrosius possibly discovered his talent as a creator of images for prints then. Printers and editors in the 1570s benefited from the high demand of religious pamphlets, moralising series of prints and liturgical books. Other painters, like Crispijn van den Broeck, Maarten de Vos and the elderly Michiel Coxcie were also active in this sector. Like them, Ambrosius worked for important printing dynasties: the Collaerts, Wierixes, Galles, De Jodes and Verhaechts. Another early series by Ambrosius is the *Fable of the father, the son and the donkey*, a moralising tale about stupidity, engraved by Carel de Mallery and published by Johannes Galle.[27]

After the *Adoration of the Magi*, another early painting of Frans Francken exists, an altarpiece of *Saint Anne* from 1575 (Herentals, Church of Saint-Waldetrudis) (fig. 4)

[25] L. VOET, *De gouden eeuw van Antwerpen. Bloei en uitstraling van de metropool in de XVIe eeuw*, Antwerp 1974, p. 236-238; C. VAN DE VELDE, 1996, p. 191.

[26] H. MIEDEMA ed., 1994, vol. 1, p. 214.

[27] N. PEETERS, 2000, p. 93-94; F.W.H. HOLLSTEIN, *Dutch and Flemish etchings, engravings and woodcuts ca. 1450-1700*, Amsterdam, 1954, vol. 11, p. 160, 125-128.

was probably made for the Brotherhood of Saint Anne from his native town Herentals, which he had left more than 15 years before.[28] It shows his strong adherence to the style and manner of Floris, both in composition and detail. The crisis in religious commissions was apparently less felt in smaller villages and towns such as Herentals. A further testimony to Frans' early activities is a reference in the alderman's registers from July 1576 mentions a contract between Frans Francken and panel-maker Peter Brauws, about whom nothing further is known.[29]

Between 1566 and 1576, the three brothers only needed to look around them to see countless painters who never quite succeeded. Knowing that an aspiring artist had little chance of becoming a master, the Franckens seem to have done rather well.[30] They worked as journeymen for ten years or a little less before acquiring their masters' titles, which seems to have been the average.[31] This phase allowed them to earn money, get experience and try out the market. They postponed their independence which was no asset in a time of crisis. The Francken brothers' years in Floris's workshop, and the next decade, were thus put to profitable use in every respect. Making the best of bad situation, they tried to keep themselves afloat in various ways between 1566 and 1576. First, they were flexible enough to accept diverse commissions, paint copies, design prints and migrate to Paris and Tournai in the hope for better positions. Second, they took advantage from the network of social and professional relations inherited from their time in Floris's studio. Third, they also exploited their family connections, something they could rely on time and again. The fact that all three were married and independent masters by 1573 indicates the had enough self-confidence in their capabilities to support their families. Already by 1575 they were slowly moving up the economic ladder. The following decades show the Francken brothers growing into a close-knit, enterprising family, a ultimately a successful dynasty.[32]

Natasja PEETERS
Curator of paintings and sculptures, Royal Military Museum, Brussels
Parc du Cinquantenaire 3
B – 1000 Brussels
Tel: +32.(0)2 737.79.18
natasja.peeters@klm-mra.be

[28] N. PEETERS, "De Maagschap van de heilige Anna", in D. Allard ed., *S.O.S. Oude Schilderijen. Redding en behoud van 20 werken op paneel, S.O.S. Peintures anciennes. Sauvegarde de 20 œuvres sur panneau*, cat. exhib., Brussels, 1996, p. 95-100; N. PEETERS, "Het paneel 'De maagschap van de H. Anna' in de Sint-Waldetrudiskerk te Herentals", in *Historisch Jaarboek van Herentals*, 2000, p. 24-28.

[29] S.A.A., Schepenregisters 345 AM II (1576), f° 301, 13 July 1576 (unpublished).

[30] N. PEETERS and J. DAMBRUYNE, "Artists of the twilight zone. Some introductory remarks on assistants in artists' workshops in the Southern Netherlands c.1450-c.1650", in N. Peeters ed., *Invisible hands? Role and Status of journeymen in artists' and craftsmen's workshops in the Low Countries c.1450-c.1650*, Louvain, 2007, p. ix-xxiv for artists' prospects.

[31] N. PEETERS and J. DAMBRUYNE, 2007, *passim*.

[32] N. PEETERS, "Frans I and Ambrosius I Francken, painters of the metropolis Antwerp, and their altarpieces in the years just after the fall of Antwerp (1585-1589)", in *Jaarboek van het Koninklijk Museum voor Schone Kunsten van Antwerpen*, 2003, p. 69-91; N. PEETERS, "Marked for the Market? Continuity, collaboration and the mechanics of artistic production of history painting in the Francken-workshops in Counter-Reformation Antwerp", in *Kunst voor de Markt, Nederlands Kunsthistorisch Jaarboek*, 1999, 50, p. 59-80.

Ph.D. in Art History (*De bijdrage van Frans en Ambrosius Francken I, en de jonge generatie Francken, tot de historieschilderkunst te Antwerpen c. 1570-1620*) and a BA in History (Vrije Universiteit Brussels), 2000 and 1994. She was recipient of a grant of the Fonds for Wetenschappelijk Onderzoek (Vlaanderen) from 1994-1998. Between 2000 and 2003 she was a Post-doc at the Rijksuniversiteit Groningen (project: 'Painting in Antwerp before iconoclasm (c. 1480-1566): A socio-Economic Approach', sub-project 'Ownership of paintings in 16th century Antwerp'). In 2004, she was curator of the exhibition *De uitvinding van het landschap. Het Vlaamse landschap van Patinir tot Rubens, 1520-1700*, Koninklijk Museum voor Schone Kunsten, Antwerp. From 2004-2006 she was involved in the project "Scientific study of the 'Rubens group' in the Royal Museums of Fine Arts of Belgium, Brussels, with catalogue raisonné". Since 2006, she is curator of paintings ans sculptures at the Royal Military Museum, Brussels.

UNE TÊTE DE VIEILLARD SUR PARCHEMIN (JORDAENS?)

Louise Decq, Jacqueline Couvert et Hélène Verougstraete,
avec la collaboration de Marie Geraerts

Résumé: Une *Tête de vieillard* (17e siècle; 38,6 × 27,8 cm, fig. 1) peinte à l'huile sur parchemin marouflé sur bois a été étudiée au Laboratoire d'étude des œuvres d'art de Louvain-la-Neuve. Le support de parchemin provient d'un manuscrit du 14e ou du 15e siècle. Il a sans doute été plus grand à l'origine et a pu inclure un deuxième portrait; l'usure de ses bords, a incité à le maroufler sur bois probablement au début du 18e siècle. Une restauration abusive incluant de nombreuses retouches intervient le siècle suivant. Une dernière intervention a encore été entreprise au 20e siècle. Cette esquisse au style assuré et spontané évoque la manière vigoureuse de Jordaens.

Les résultats présentés ici résultent d'un travail d'équipe auquel ont participé Jacqueline Couvert pour la réalisation de l'analyse des pigments par microfluorescence X, Marie Geraerts pour un état de la question au sujet des têtes esquissées par les grands maîtres du 17e siècle ainsi que Louise Decq et Hélène Verougstraete pour l'observation, la description matérielle de l'œuvre et la réalisation des documents. Les différentes méthodes d'analyse (fluorescence d'ultraviolet, réflectographie à l'infrarouge, radiographie, microfluorescence X et microscope stéréoscopique) ont permis de reconstituer l'histoire matérielle de cette œuvre qui se rattache à la meilleure production artistique du 17e siècle.

Le support en parchemin provient de deux pages d'un manuscrit, toutes deux incomplètes (fig. 2). Elles sont collées l'une à l'autre en se chevauchant légèrement. La partie principale, à gauche, comporte deux colonnes de texte. À droite, l'étroite bande de parchemin ne comprend qu'une partie d'une colonne. On peut s'étonner que la peinture ait été exécutée sur deux feuilles jointes alors qu'une feuille eut suffi à la représentation entière. Cette particularité sera un argument, parmi d'autres, pour fonder l'hypothèse d'un support plus grand à l'origine sur lequel auraient figuré deux ou plusieurs têtes.

Le parchemin

Le parchemin comporte un texte manuscrit en latin visible à l'œil nu mais plus clairement lisible en réflectographie à l'infrarouge. Ce texte est inversé de haut en bas par rapport à la peinture et exécuté à la plume en gothique cursive, écriture adoptée entre le 14e et le début du 16e siècle. Les majuscules situées au début d'une nouvelle section sont indiquées au moyen d'une boucle rouge un peu plus grande.

La colonne centrale est entière, ce qui permet de déchiffrer des phrases complètes. Les professeurs Paul Deproost et Baudouin Van den Abeele de l'UCL nous ont apporté leur aide dans la transcription et dans l'interprétation du texte. Nos collègues concluent que le texte est certainement postérieur à 1153, année de la mort de saint Bernard dont un passage est cité. Il y est évoqué un «martyre non sanglant», thème présent dans la

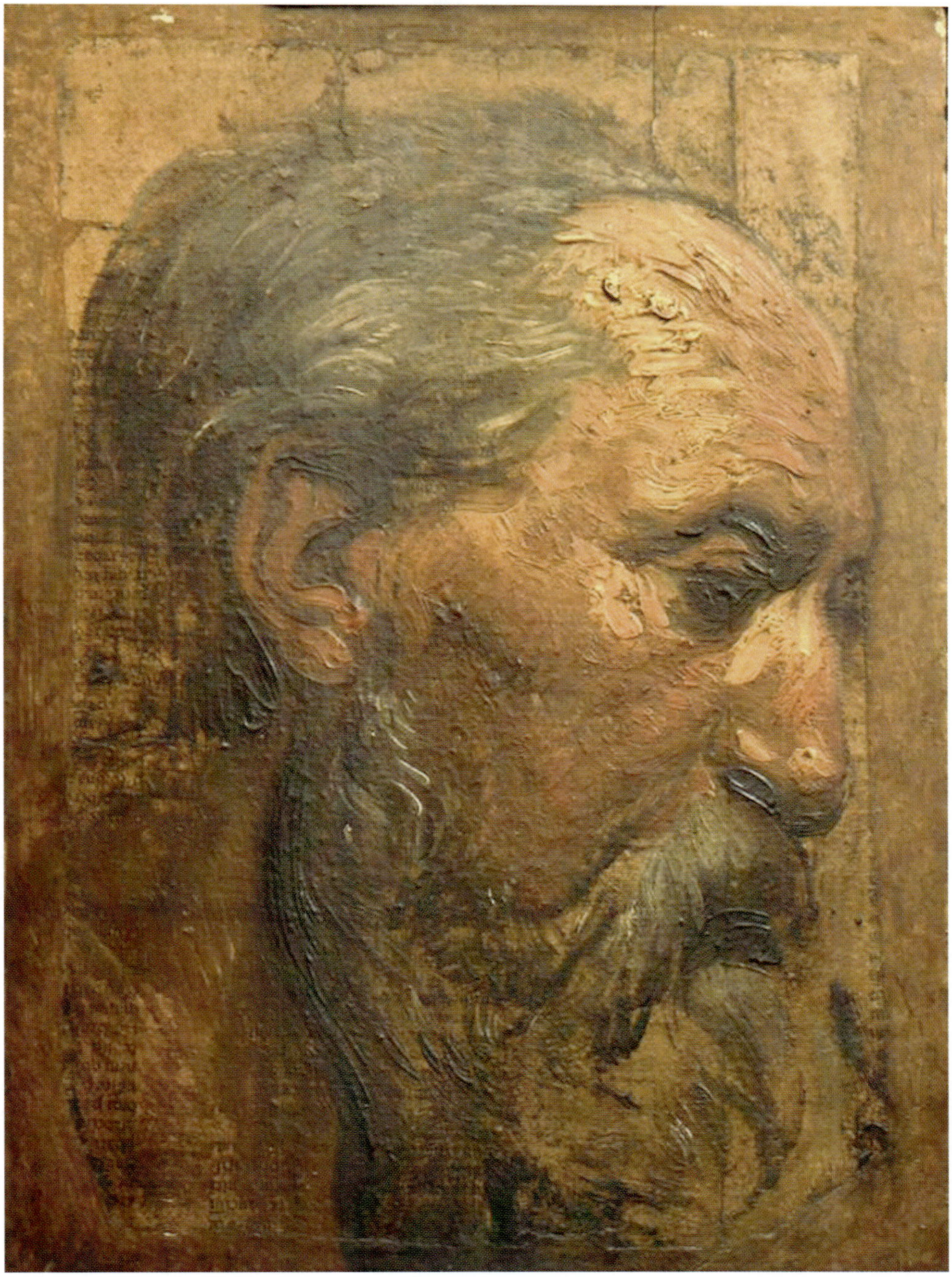

Fig. 1. *Tête de vieillard*, XVIIe siècle, huile sur parchemin marouflé sur chêne, 38,6 × 27,8 cm, collection privée.

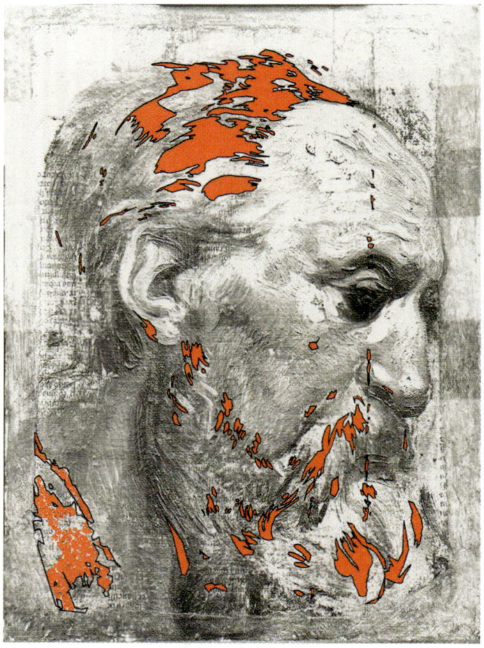

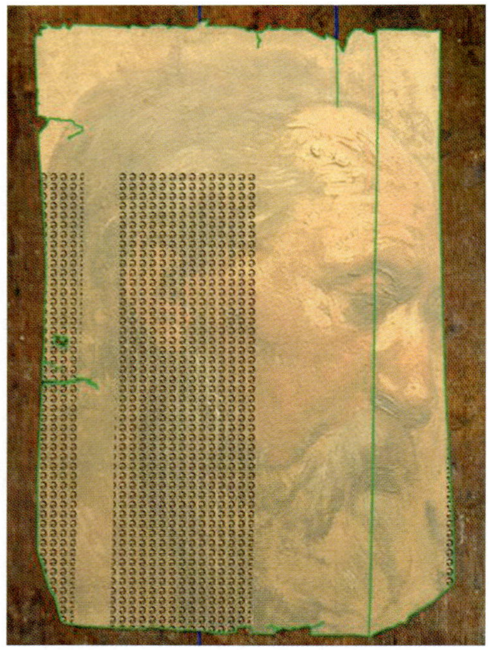

Fig. 2. *Tête de vieillard*. a. Schéma montrant, en rouge, les surpeints. b. Schéma montrant, en vert, les déchirures, les lacunes et le contour du parchemin et, en noir, la disposition du texte.

spiritualité monastique. D'autres passages émanent des *Lettres* et des *Homélies sur l'Evangile de saint Jean* de saint Augustin, des *Sentences* de Bernard de Clairvaux et des *Homélies sur les Evangiles* de Grégoire le Grand.

L'usure des bords du parchemin, visible sur la radiographie, surtout dans le haut et le bas, révèle que l'œuvre était endommagée avant marouflage. Cette usure pourrait être le résultat d'un usage intensif de l'esquisse, à titre de modèle, dans l'atelier d'un peintre. C'est notamment pour remédier à cet état que le parchemin aurait été marouflé sur bois. Les côtés latéraux du parchemin, moins endommagés, font penser qu'à l'origine l'esquisse était peut-être plus large et aurait pu comporter deux ou plusieurs têtes juxtaposées, comme on le rencontre fréquemment dans les esquisses de Van Dyck ou de Jordaens. De plus, le bord à droite est situé très près de la tête et coupe même une petite partie du sourcil, peut-être pour épargner une hypothétique seconde tête, qui aurait pu se trouver à droite. L'adoption de deux feuilles de parchemin serait alors justifiée. Elles auraient été nécessaires à l'exécution de deux ou plusieurs têtes, que l'on a dissociées ensuite.

Marouflage sur parchemin

C'est un parchemin affaibli par l'usage qui a été marouflé sur support de bois. Le panneau est constitué de deux éléments réunis par un joint vif renforcé de deux tourillons. Le bois est débité sur quartier ou sur faux-quartier, comme il convenait pour éviter le gauchissement. Les deux éléments présentent le bois de cœur vers les bords extérieurs. Le panneau est soigneusement chanfreiné sur les quatre côtés. Ces caractéristiques techniques, et le soin avec lequel le support a été construit, indiquent qu'il s'agit d'un panneau ancien, dont il conviendrait de réaliser l'analyse dendrochronologique en gardant à l'esprit que l'esquisse et le bois ne sont pas nécessairement tout à fait contemporains.

Les dimensions du parchemin sont inférieures à celles du panneau. C'est pourquoi des bandelettes de papier et de tissu imprégné ont été soigneusement coupées suivant le contour irrégulier du parchemin et collées autour de l'esquisse. Ensuite, elles ont été recouvertes d'une couche d'enduit et de peinture. L'aspect aujourd'hui quelque peu négligé des bords résulte de restaurations abusives. Il contraste avec le soin initialement accordé à la découpe et au collage des bandelettes latérales.

Couches picturales

La séquence des couches disposées sur le parchemin se présente comme suit. Une couche de couleur brune assez foncée a d'abord été apposée, probablement dans le but de recouvrir le texte de même couleur. Ensuite, une *imprimatura* grise (comme on en trouve souvent dans les tableaux du 17e siècle) a été brossée rapidement et présente un aspect strié horizontal, ce qui pourrait également plaider en faveur de l'existence de deux ou plusieurs têtes à l'origine, les stries se disposant de préférence dans le sens de la plus grande largeur du support.

Visage et fond ont réagi très différemment à un nettoyage: le second semble s'être assez facilement dissout. Le fond aurait été exécuté avec un liant aqueux, tandis que le liant du visage est huileux. Par ailleurs, des traces de couleur sombre, visibles sur les bandelettes le long des bords, font penser que les bandes de papier et de soie enduitées ont été peintes pour se rapprocher de la couleur du fond de l'œuvre qui devait être grisâtre à l'origine (peut-être déjà noirci par la saleté au moment du marouflage). La figure 2 présente une reconstitution avec les sous-couches d'origine. Les pigments détectés dans les zones non retouchées sont le blanc de plomb, le vermillon, des sels de cuivre et du jaune de plomb et d'étain. Tous ces pigments sont compatibles avec une datation ancienne.

Le portrait est réalisé au moyen de touches épaisses, assurées et spontanées, avec un pinceau rond assez souple. Certaines touches sont courtes et vives tandis que d'autres sont longues et souples. Sur la joue, le peintre a disposé des touches parallèles assez caractéristiques. Il a parfois utilisé la couleur grise du fond pour réaliser localement la tonalité, comme c'est le cas pour certaines zones de la barbe ou l'ombre de la joue.

L'état de conservation

Pour pallier un assombrissement causé par l'âge, un agent chimique aqueux a du être utilisé à des fins de nettoyage au 19e siècle, ce qui expliquerait que les couches de fond brune et grise peintes avec un liant aqueux, plus fragiles, aient été enlevées à ce moment, de même que la couleur sombre des bords. Le nettoyage a d'ailleurs dû être assez vigoureux, la preuve en est l'usure de la partie supérieure, plus claire, et les lacunes de la couche picturale. Le parchemin présente même des traces d'abrasion.

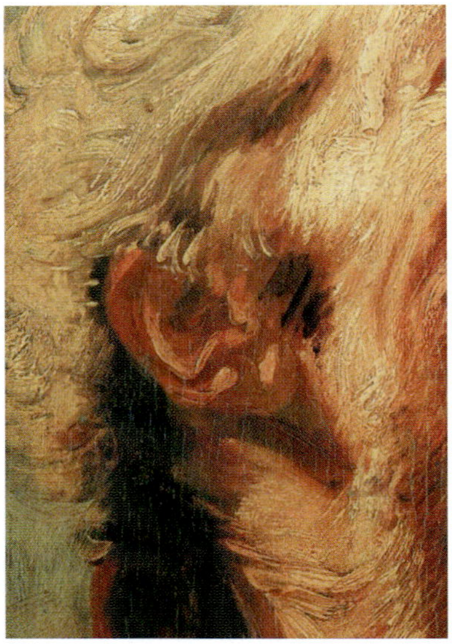

Fig. 3. J. Jordaens, *Tête de Grapheus*, ca. 1620-1621, Douai, Musée de la Chartreuse, inv. 198. Détail, l'oreille.

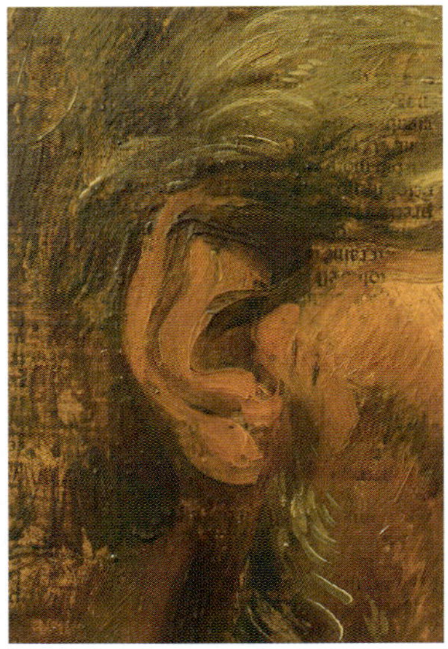

Fig. 4. *Tête de vieillard*, collection privée. Détail, l'oreille.

Les retouches qui interviennent probablement peu après le nettoyage se limitent aux aux zones les plus usées: les cheveux et la barbe. Ces retouches contiennent du blanc de zinc, on peut donc les situer après 1834, date de commercialisation de ce pigment. L'aspect blanchâtre du vernis sous la lumière ultraviolette indique qu'il s'agit probablement d'un vernis ancien, appliqué probablement lors de la même intervention.

Un autre vernis sans grande fluorescence (20e siècle?) a été disposé au dessus de retouches plus récentes, localisées notamment au niveau du joint des deux feuilles de parchemin, sur le coin inférieur gauche et sur les bandelettes latérales.

L'artiste

L'esquisse se rattache aux nombreuses études de têtes des plus grands maîtres du 17e siècle. Ces portraits peints 'sur le vif', appelés *tronies* dans les textes flamands de l'époque, servaient de modèles aux peintres qui s'en inspiraient souvent pour leurs grandes compositions. Jordaens, par exemple, fit de nombreuses esquisses notamment de la tête expressive de Grapheus, dont il se servit dans plusieurs peintures. Dans *L'Allégorie de la fécondité* (Bruxelles, Musées royaux des beaux-arts de Belgique), Grapheus apparaît sous la forme d'un satyre. Rubens, et surtout Jordaens et Van Dyck, ont exécuté des *tronies* sur papier. Notons que la distinction avec le parchemin est parfois difficile à faire. Les mentions «sur papier» que l'on rencontre dans la littérature n'excluent pas à notre avis qu'il s'agisse de parchemin.

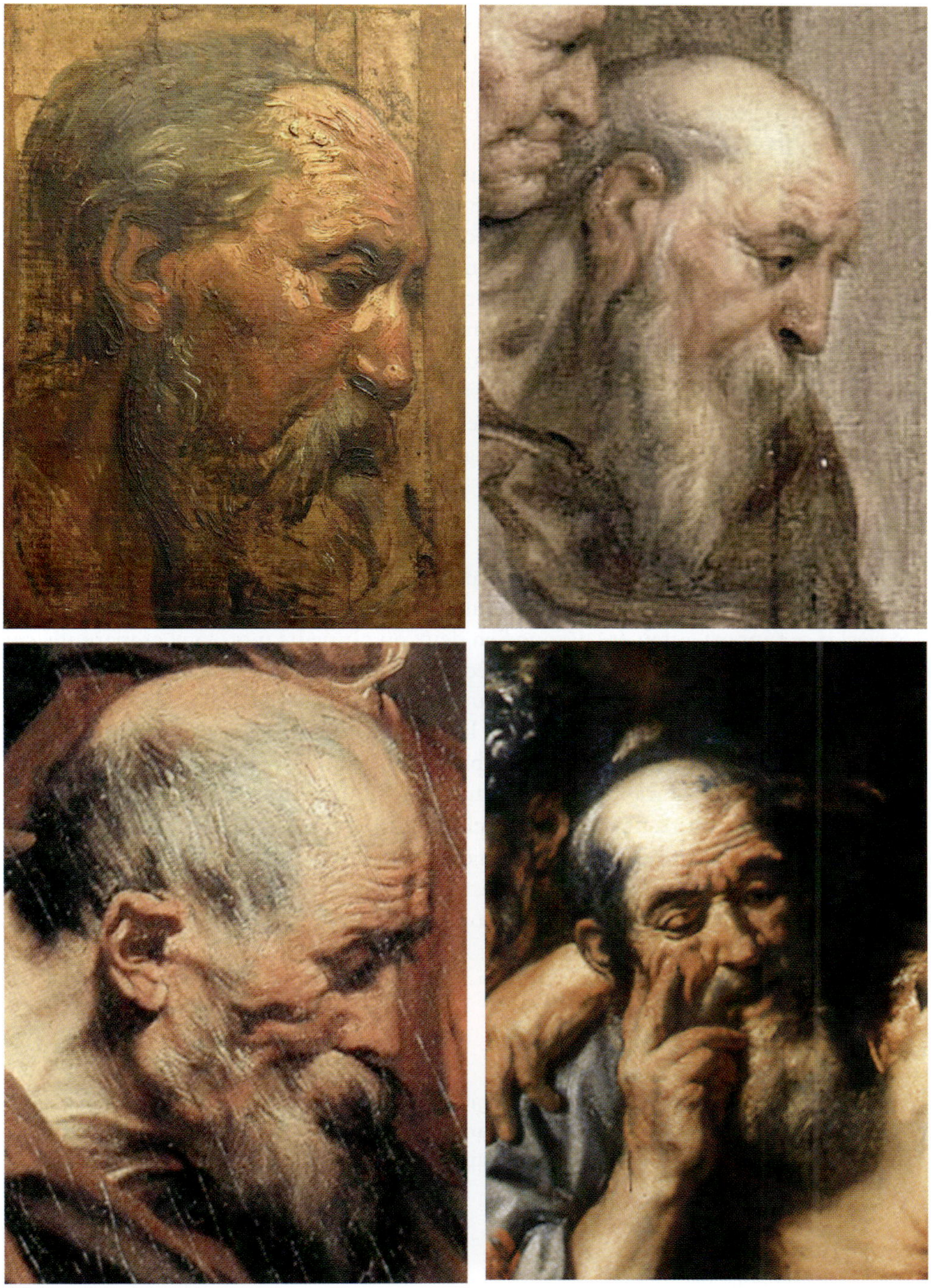

Fig. 5. a. *Tête de vieillard*, collection privée. b. *Jésus chassant les marchands du temple* (détail), Paris, Musée du Louvre. c. *Les disciples au tombeau* (détail), Dresde, Gemäldegalerie (inversé). d. *Héraclite et Démocrite* (détail), collection privée.

Si beaucoup de questions restent non résolues au sujet de l'attribution de ces portraits exécutés 'sur le vif' par ces peintres, des éléments de comparaison peuvent être relevés. On constate que la touche d'Antoine Van Dyck (surtout en début de carrière) et de Jacob Jordaens se rapproche de celle que l'on observe dans notre esquisse. On y retrouve les accents rouges et les petits traits caractéristiques.

La touche picturale vigoureuse de l'esquisse examinée ici nous fait penser davantage à Jordaens qu'aux autres grands maîtres de son époque. On peut comparer l'exécution de l'oreille de Grapheus de Jordaens avec celle de la tête de vieillard que nous présentons ici (fig. 3 et 4). On peut dès lors penser que le peintre aurait pu intégrer la tête de vieillard dans l'une ou l'autre composition. On peut suggérer, par exemple, qu'un vieillard assez semblable est représenté dans *Jésus chassant les marchands du temple*, en haut à gauche. Un personnage assez similaire se retrouve dans *Les disciples au Tombeau* ainsi que dans une composition avec *Héraclite et Démocrite* (collection privée) dont l'attribution à Jordaens ne fait pas de doute vu que les deux philosophes sont peints sur une esquisse préparatoire à la *Fécondité* (fig. 5).

Conclusion

L'étude de cette *Tête de vieillard* a donc permis d'en reconstituer la vie mouvementée. À l'origine, deux portraits ont pu être peints sur des pages d'un manuscrit du 14e ou du 15e siècle; ce support aurait été découpé en deux ou plusieurs morceaux et marouflé sur bois, à la fin du 17e ou au début du 18e siècle, peut-être pour donner à l'objet un caractère négociable. Ensuite, probablement au cours de la deuxième moitié du 19e siècle, une restauration abusive a été entreprise: les sous-couches brune et grise ont été enlevées; la couche picturale elle-même et le parchemin sont devenus lacunaires. Le texte, complètement caché à l'origine, est devenu apparent à certains endroits. Des retouches sont alors effectuées surtout dans la chevelure et la barbe. Lors d'une dernière restauration, des retouches parfois grossières sont disposées surtout sur les bords et le coin inférieur à gauche, sans enlever le vieux vernis du 19e siècle.

Cette esquisse de qualité soutient la comparaison avec les visages réalisés par Jordaens.

Louise DECQ

Née en 1982 à Louvain (Belgique). Bioingénieur à la Katholieke Universiteit Leuven (2000-2005). En 2005 et 2006, travaille au Laboratoire d'Etude des œuvres d'Art par les méthodes scientifiques (Université Catholique de Louvain).

Jacqueline COUVERT
Université catholique de Louvain
Laboratoire d'étude des œuvres d'art
par les méthodes scientifiques
Rue Cardinal Mercier 53
B – 1348 Louvain-la-Neuve

Tel: +32.10.47.48.97
Fax: +32.10.47.48.95
jacqueline.couvert@uclouvain.be

Licenciée en Histoire de l'Art et Archéologie (UCL, 1997) et graduée en chimie industrielle. Depuis 1999, assistante de recherche puis collaboratrice technique au Laboratoire d'étude des œuvres d'art de Louvain-la-Neuve (analyse des pigments par microfluorescence X, réalisation de radiographies et de réflectogrammes à l'infrarouge). Examen de tableaux et réalisation de dossiers d'expertise.

Hélène VEROUGSTRAETE

Voir Hélène VEROUGSTRAETE, *Les couleurs signifiantes chez Bruegel l'Ancien et chez ses successeurs.*

Marie GERAERTS
Tel: +32.477.48.76.11
Fax: +32.11.58.75.06

Née le 5 août 1962 à Genk. Etudes d'Architecture et de Restauration de monuments, ensuite restauration de peintures. Après avoir obtenu son diplôme de restaurateur, elle étudie l'Histoire de l'Art à l'Université libre de Bruxelles. Depuis, chercheur scientifique (Vénus dans l'œuvre de Pierre-Paul Rubens) pour le *Corpus Rubenianum Ludwich Burchard* au *Centrum voor de Vlaamse Kunst van de 16de en de 17de eeuw* (Rubenianum) à Anvers.

MORE ON JOSEPH VAN DER VEKEN (1862-1964)

Louise Decq

Abstract: The contributions of the highly skilful restorer Jef Van der Veken are often regarded as falsifications considering his radical way of restoring and the perfection of his copies and pastiches. Occasionally he endeavoured to create 'Flemish Primitives' of his own. Two works are presented here in which his hand is recognizable. In the first case, Van der Veken over-restored a late 15th century *Madonna and Child* still in its original frame by a follower of Van der Weyden. The other painting, a *Saint Anna* owned by a close descendant of the restorer, is entirely new, though painted on an ancient panel. Rather minor in quality, the painting was probably an experiment since it tries out various unusual techniques. Incompatible paint materials were used to achieve a speckled texture in the clothes and different aging techniques were experimented.

The *Madonna and Child*

In the autumn of 2005, a *Madonna and Child* (oak: 44.8 × 30.6 cm) was presented at the Laboratoire d'étude des œuvres d'art (UCL) for technical examination (fig. 1). It turned out to be a drastically restoration by Van der Veken of a central part of a late 15th century triptych by a follower of Van der Weyden. The painting was executed on an oak panel, assembled out of two quarter sawn elements. In the original oak frame, carvings intended for the placement of hinges, now lost, testify to the former existence of wings. The frame's profile, its black and gold polychromy and the insertion of the panel in a groove of the frame, indicate that the frame originated at the end of the 15th century. The tenon and mortise joint in the corners of the frame is common for this period.[1] In this case, the tenon is apparent in the lateral sides of the frame. Normally, a tenon and mortise joint is reinforced with a pin. Here the system is different. In the tenon, a rather deep diagonal groove was carved. In this groove a wooden wedge was inserted, forcing the tenon to expand after assemblage of the frame. This rare construction is part of the original tenon and mortise assemblage. No further reference was found for this assembly technique in the period.

At first view, the painting seems in good condition, but its appearance today is the result of a large scale restoration. On the diagram (fig. 2), one can see that less than one quarter of the surface dates back to the 15th century. The other areas are painted by Van der Veken. In the hatched areas, mainly in the background, the hair and the clothes, the original paint has been scratched off down to the naked wood. No lead white was used in the renewed parts. The outline of these areas can be seen as a rupture on the radiography. Pigment analysis by XRF confirmed the results of the examination

[1] H. VEROUGSTRAETE and R. VAN SCHOUTE, *Cadres et supports dans la peinture flamande aux 15e et 16e siècles*, Heure-le-Romain, 1989.

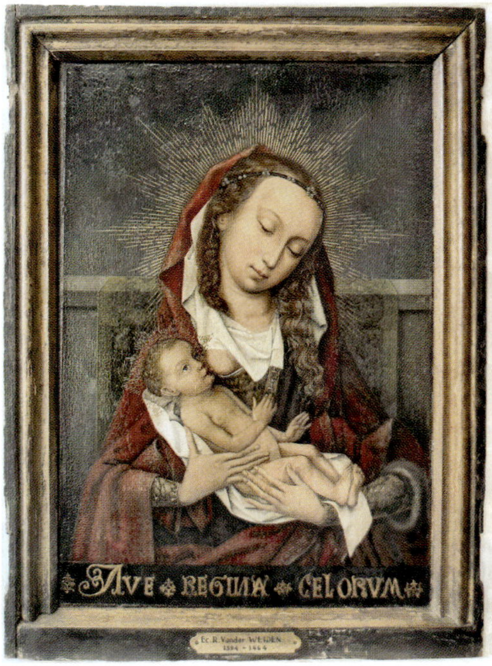

Fig. 1. Anonymous (15th c.) and Joseph Van der Veken, *Madonna and Child*, 44.8 × 30.6 cm, private collection (© Labart, Louise Decq).

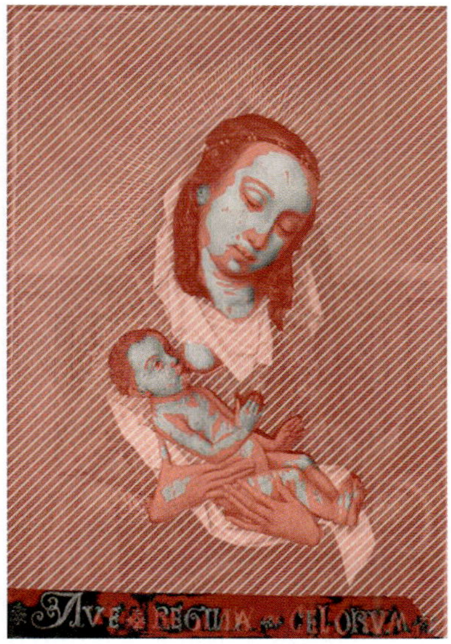

Fig. 2. *Madonna and Child.* Diagram of damaged and restored areas (in red) (© Louise Decq).

carried out with stereoscopic microscope and radiography. Infrared reflectography shows a well executed underdrawing beneath the original areas, for example in the Child's curly hair, his eyes and his navel, as well as in Mary's face and hair.

The restorer used a wide range of techniques to make his additions look old. The new areas show artificial cracks, which can be recognized by their trembling pattern. Similar cracks have been described in other restorations by Van der Veken, for example when the Laboratoire d'étude des oeuvres d'art (UCL) investigated works restored by Van der Veken for the exhibition *Fake or not Fake* in Bruges in 2004.[2]

While in the pale new areas the cracks were filled with ink, inorder to appear as dark contrasting lines, the cracks in the darker, more colourful new areas, such as in the blue background, are left empty and colourless. In those areas Van der Veken wore the edges of the cracks by rubbing them, causing the white preparation layer to show through.

[2] H. Verougstraete, R. Van Schoute and T.-H. Borchert with contributions of E. Bruyns, J. Couvert, R. Pieters and J.-L. Pypaert, *Fake or not fake. Restaurateurs ou faussaires des primitifs flamands, [Het verhaal van de restauratie van de Vlaamse primitieven]*, Ghent (Ludion), 2004. Exh. cat. *Fake or not fake* in the Groeningemuseum in Bruges, November 26, 2004 - February 28, 2005. See also H. Verougstraete and R. Van Schoute, "La Madone Renders et sa restauration par Joseph Van der Veken (1872-1964)", in *La peinture et le laboratoire. Actes du Colloque XIII pour l'étude du dessin sous-jacent et de la technologie dans la peinture*, H. Verougstraete and R. Van Schoute eds., Louvain-Paris-Sterling, 2001, p. 7-28.

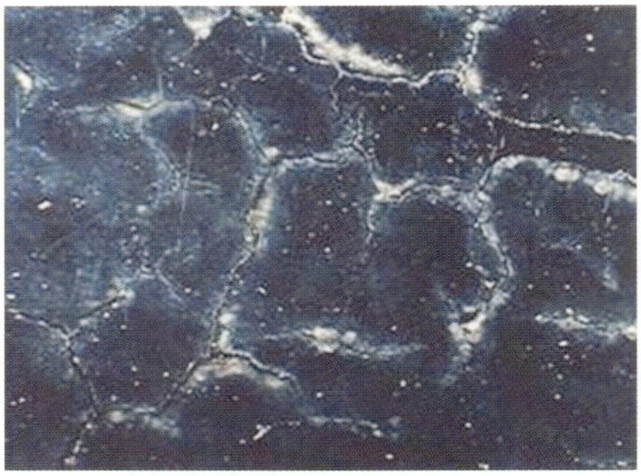

Fig. 3. *Madonna and Child.* Microphotography (×10). The cracks in the blue background are left empty, but their edges are worn artificially, exposing the preparation layer. The contrast between the worn cracks and the surrounding blue colour imitates a typical feature in ancient paintings (© Labart, Louise Decq).

In doing so, he achieved a contrast between the cracks and the surrounding colour, imitating a feature typical in ancient paintings (fig. 3). In the overpaintings over old areas, and in the transition zones between old and new, fake cracks were drawn with a pencil or scratched with a pen, often in continuity of the original ones.

Many lacunae exist in the paint layers, even in the new areas. Most of the lacunae are filled and retouched. Retouchings were also found on top of the new varnish as can be seen under UV-fluorescence. Van der Veken seems to have restored his own work. This addition of an extra stage in the pictorial execution is likely to be part of the falsification.

The invasive but skilful interventions, the use of aging techniques, the making of a convincing crack pattern and the care adopted in the depiction of details are all very similar to what was found in other restorations by Van der Veken. The modern pigments identified here, such as zinc white, chrome green, cerulean blue and barium white, have already been identified in other work by Van der Veken.[3] All this leads us to recognize his hand in this extensive restoration.

The bottom limit of the painting, the only original border that remains, presents a 'barbe' corresponding to the frame. The painting was made when the panel was inserted in the frame, as is usual for this period. Van der Veken made his own 'barbes', but they fail to comply with the ridges of the frame.

Jef Van der Veken started his career as a restorer around 1914. Although he restored paintings of various centuries, his overwhelming passion for the Flemish Primitives led him to work mainly on 15th century paintings. While he was quite reserved in restoring

[3] J. COUVERT, "La Madone Renders. Analyses par spectrométrie de microfluorescence X", in *Actes du Colloque XIII pour l'étude du dessin sous-jacent et de la technologie dans la peinture. La peinture et le laboratoire*, H. VEROUGSTRAETE and R. VAN SCHOUTE eds., Leuven, 2001. *Merciful Christ with a benedictine* (Flemish school, follower, New York, Metropolitan Museum of Art), see H. VON SONNENBURG, "A Case of Recurring Deception", in *The Changing Image. Studies in Paintings Conservation*. The Metropolitan Museum of Art Bulletin, Winter 1993-1994, 1994, p. 9-19; and a *Saint Bernard with donator* (Flemish school, follower, National Gallery of Washington, Chester Dale Collection) see J.O. HAND and M. WOLFF, *Early Netherlandish Painting. The Collections of the National Gallery of Art. Washington Systematic Catalogue*, Washingon-Cambridge, 1986, p. 88, note 6.

works in public collections, he used to have a more radical approach for paintings in private collections. The technique of scratching off the paint down to the wood seems to date from the first part of his career. After the 1920's he seems to have preferred reusing old paint layers, cracks included, after rubbing off the coloured surface. Since the restoration of the *Madonna and Child* includes a scratching off of certain areas, it can probably be dated in the early 1920's.

The *Saint Anna* (fig. 4)

The second work (oak, 30.0 × 25.1 cm), owned by a descendant of Jef Van der Veken, depicts Saint Anna with Mary and Child. It is entirely new. On the reverse, traces of a severely damaged ancient green marbling can be seen. Clearly, an old panel was reused here.

Most interesting, however, is the remarkable number of different techniques experimented by the painter. It looks like Van der Veken used this little panel to refine his techniques.

Under the surface of the painting a blue layer was applied, probably for optical reasons. In some areas, the colourful upper layer is intentionally scratched off, making the underlying blue layer visible. The resulting light blue lines help to soften the transition

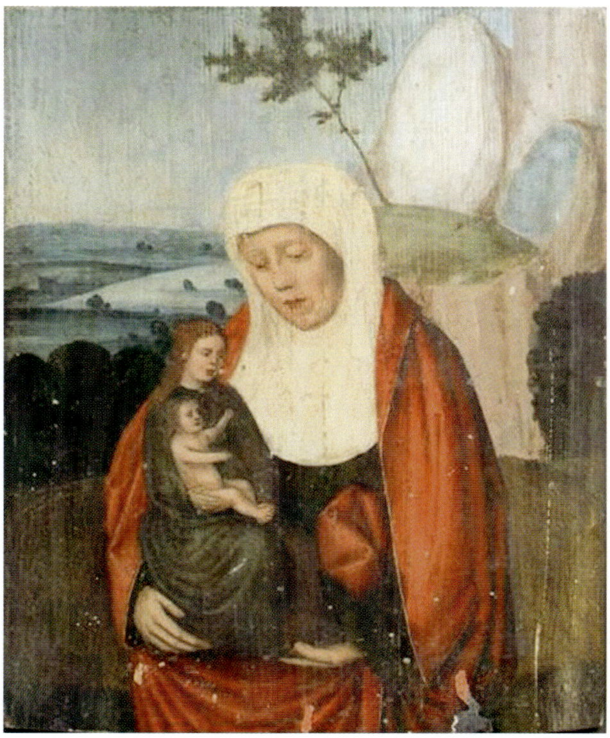

Fig. 4. Joseph Van der Veken, *Saint Anna*, 30 × 25.1 cm, private collection (© Labart, Louise Decq).

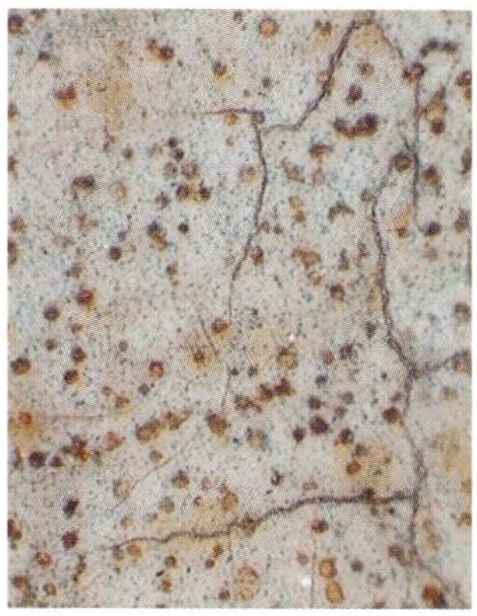

Fig. 5. Joseph Van der Veken, *Saint Anna*, private collection. Microphotography (×10). Small craters, probably made by a metal brush and filled with patina (© Labart, Louise Decq).

between two colours. Another type of scratches was used for Saint Anna's nose and mouth. Here, Van der Veken scratched the fresh paint to achieve a softer effect.

In several places stained patterns were observed. Van der Veken experimented with incompatible materials. He used a piece of fabric to apply a special, maybe greasy paint onto the surface, sometimes rather dry, sometimes diluted. Anna's and Mary's clothes are completely done like this. Once dry, these droplets were difficult to cover with subsequent layers of paint, causing small gaps in the top layer. For the grass field in the bottom left corner, a material was applied to which the patina, used for aging, did not adhere. As a result, it concentrated in the depths and formed large stains.

Compared to the relatively rough execution of the painting itself, Van der Veken paid remarkable attention to giving it an old appearance. Cracks were made in the direction of the wood grain, probably by exposing the painting to dramatic changes in relative humidity. Today, the panel is warped, but the cracks are still narrow. It is possible that the panel was already warped before the creation of the cracks. The cracks were filled with a liquid brown material; they look very natural.

After this artificial aging by climatic changes, a patina was applied. We found patterns of a brush, fabric and even fingerprints. Elsewhere, small craters, probably made by a metal brush, were filled with the patina (fig. 5). This method in particular is a convincing aging technique. Excess material was removed by intensive rubbing, causing a scratched and worn look. A wax finish was then applied on the painting, which makes the surface look old and satin-like. The work has not been varnished. Finally, some speckles were added.

The *Saint Anna* panel must have been a test panel for Jef Van der Veken. It illustrates how the restorer experimented techniques and materials to perfect his craft. It is possible that the observed techniques will be found in other works of this restorer, and they could help in attributing other paintings to him.

Louise DECQ

See Louise DECQ, Jacqueline COUVERT et Hélène VEROUGSTRAETE, *Une tête de vieillard sur parchemin (Jordaens?).*

UN FAUX 'PRIMITIF FLAMAND' AU MUSÉE DES BEAUX-ARTS DE BILBAO: ANALYSE TECHNOLOGIQUE, CRITIQUE DE SOURCES ET ESSAI D'ATTRIBUTION

Didier Martens et Ana Sánchez-Lassa de los Santos*

Résumé: Le triptyque du *Mariage mystique de sainte Catherine* conservé au Musée des beaux-arts de Bilbao a été longtemps considéré comme une œuvre flamande de la fin du XVe siècle, proche de Hans Memling. Les examens de laboratoire permettent toutefois de conclure à un faux moderne. Ils ont révélé de nombreuses anomalies, tels l'usage d'un support en noyer et la présence, dans la couche picturale, de plusieurs pigments modernes, dont le blanc de baryum et le bleu de Prusse. Quant à la composition peinte sur les trois panneaux, elle constitue un véritable collage d'emprunts, principalement à Hans Memling, mais aussi à Jean van Eyck. Plusieurs de ces emprunts peuvent également être reconnus dans un tableau qui, vers les années 1950, se trouvait sur le marché d'art américain. Ce tableau a été attribué à l'un des plus redoutables faussaires des 'Primitifs flamands': le Maître du Triptyque Palmer. Il semble évident que ce dernier a également réalisé le triptyque de Bilbao.

Depuis 1952, le Musée des beaux-arts de Bilbao possède un triptyque dont la forme générale et le style des figures évoquent, de prime abord, l'art des 'Primitifs flamands'[1] (fig. 1). Il représente, sur le panneau central, le *Mariage mystique de sainte Catherine*, sur le volet gauche, un *Donateur avec saint Jean-Baptiste*, sur le volet droit, une *Donatrice avec saint Georges*. L'œuvre, léguée par une collectionneuse basque, fut cataloguée en 1969 par Crisanto de Lasterra comme «école flamande, XVe siècle», une classification reprise en 1978 par Javier de Bengoechea[2]. En 1995, Xesqui Castañer proposa d'attribuer l'ensemble à un artiste «proche de l'atelier de Hans Memling», en raison des nombreux emprunts au grand maître brugeois reconnaissables dans les trois panneaux[3]. Elle signala en particulier l'influence conjointe exercée par le triptyque Donne de la National Gallery de Londres et par celui du *Mariage mystique de sainte Catherine*, conservé à l'Hôpital Saint-Jean à Bruges. Le lien avec cette dernière œuvre avait déjà été relevé par Crisanto de Lasterra. Aucun des auteurs cités ne semble avoir nourri le moindre doute quant à l'ancienneté du triptyque de Bilbao. Pourtant, en 1984, lors

* Les auteurs tiennent à remercier tout particulièrement Tomás Antelo et Araceli Gabaldón (Madrid, Instituto del Patrimonio Histórico Español), pour l'aide apportée dans l'examen du dessin sous-jacent, ainsi que le père Robert Godding (Bruxelles, Société des Bollandistes), pour l'identification du texte apposé sur le cadre du triptyque.

[1] Bilboko Arte Eder Museoa / Museo de Bellas Artes de Bilbao, inv. 69/321; legs Doña Sofía Gil Iturriaga, Bilbao. Mesures avec cadre: 76,5 × 48,7 cm (fermé) et 76,5 × 98,6 cm (ouvert); mesures sans cadre: 60 × 37 cm (panneau central) et 66 × 24,5 cm (chaque volet). La moulure à la face des volets a été sculptée dans la masse et forme donc corps avec les deux panneaux. Il n'y a pas de figures sur les revers.

[2] C. DE LASTERRA, *Catálogo descriptivo. Sección de Arte antiguo. Museo de Bellas Artes de Bilbao*, Bilbao, 1969, p. 139-140; J. DE BENGOECHEA, *Museo de Bellas Artes. Bilbao*, Bilbao, 1978, p. 97.

[3] X. CASTAÑER, *Pinturas y pintores flamencos, holandeses y alemanes en el Museo de Bellas Artes de Bilbao*, Bilbao, 1995, n°11. Du même auteur, «La pintura del siglo XV en el Museo de Bellas Artes de Bilbao: formas, temas, modelos», dans *Kobie*, 8, 1991, p. 13.

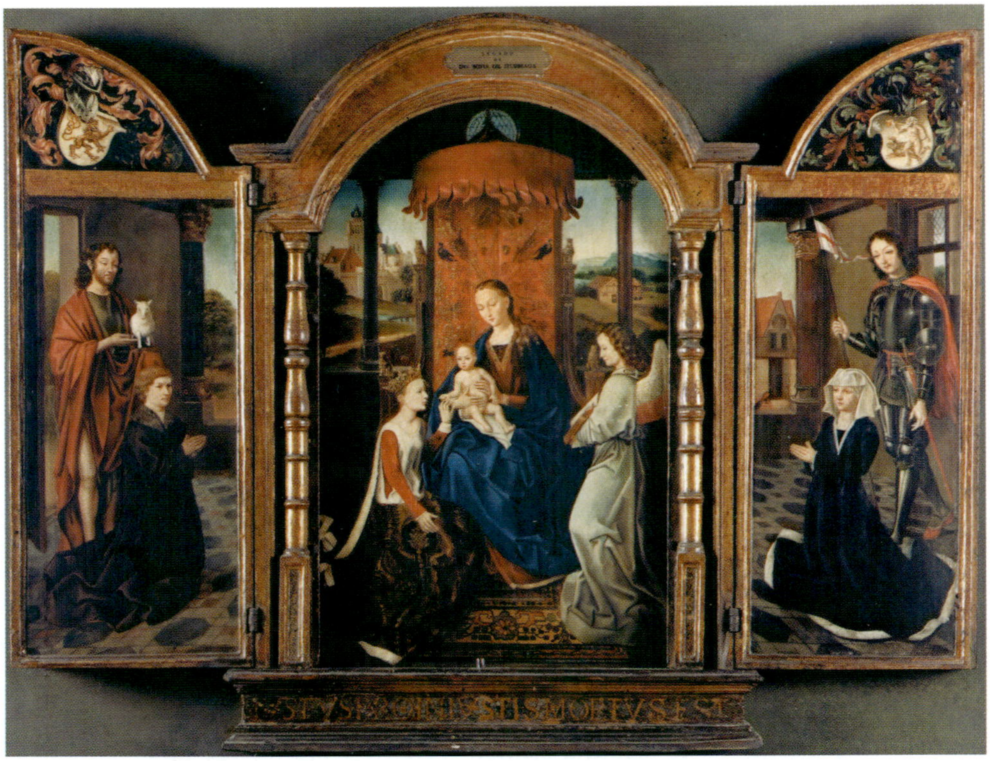

Fig. 1. Faussaire moderne, *Triptyque du Mariage mystique de sainte Catherine*, Bilbao, Museo de Bellas Artes (© Museo de Bellas Artes, Bilbao).

d'une visite de prospection dans le cadre de la préparation de l'exposition Europalia *Splendeur d'Espagne et les villes belges*, Matías Díaz Padrón et le regretté Ignace Vandevivere ont contesté son authenticité. L'œuvre fut retirée peu après des cimaises et mise en réserve. En 2005-2006, le triptyque a été soumis à une série d'examens de laboratoire. En voici les résultats.

- Le bois utilisé, tant pour les volets que pour le panneau central, est du noyer. Ce point mérite d'être relevé dans la mesure où, comme on le sait, les 'Primitifs flamands' utilisaient normalement le chêne.
- Le support semble ancien. Le revers du panneau central a été aminci sur deux bandes horizontales, de manière à permettre la fixation d'un double renfort. Le bois mis à nu à cette occasion présente de nombreuses galeries creusées par des insectes xylophages. Les volets paraissent également d'un âge respectable. Sur les revers, on remarque une couche de peinture usée et craquelée, visiblement plus ancienne que celle recouvrant les faces.
- L'ancienneté des panneaux ne saurait pour autant être invoquée comme argument en faveur de l'authenticité des peintures. Un triptyque ancien a fort bien pu être réutilisé. C'est ce que suggère notamment l'inscription en capitales romaines *JUSTUS PRO*

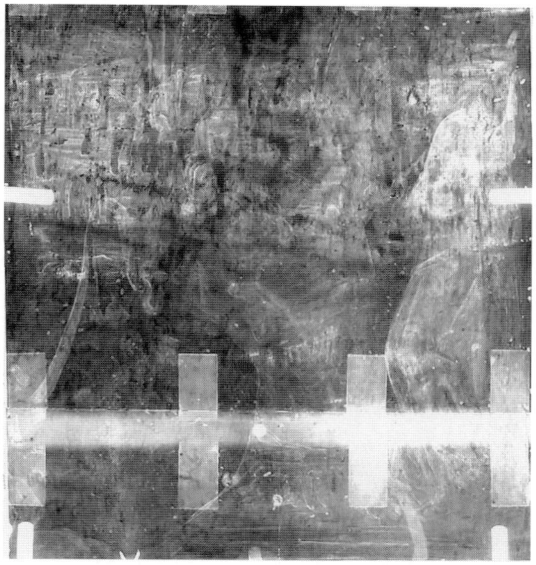

Fig. 2. Faussaire moderne, *Triptyque du Mariage mystique de sainte Catherine.* Détail du panneau central en radiographie (© Museo de Bellas Artes, Bilbao).

INJUSTIS MORTU(U)S EST. L'expression, qui provient de la première *Épître de saint Pierre*, fait référence au sacrifice sanglant du Christ, au 'Juste mort pour des injustes'[4]. L'encadrement aurait été conçu à l'origine pour une scène de la Passion ou pour un *Christ en croix*, pas pour un *Mariage mystique de sainte Catherine.*

- La radiographie n'a pas permis de relever des changements de composition significatifs dans les volets. Par contre, elle accrédite l'hypothèse d'un réemploi pour le panneau central. Sous les figures du *Mariage mystique*, on observe une composition tout à fait différente (fig. 2). Il s'agit apparemment d'un paysage avec des maisons, peint dans une technique peu soignée.
- La réflectographie à l'infrarouge a révélé un dessin sous-jacent extrêmement limité (fig. 3). De simples tracés de mise en place ont pu être détectés dans les festons du baldaquin, dans le visage de la Vierge et du donateur (fig. 4), ainsi que dans la colonne de droite du panneau central. Aucune préparation du modelé à l'aide de hachures n'a été observée. Ce dessin sous-jacent sommaire n'évoque en rien celui, bien plus complexe, de Memling et de ses disciples[5]. On notera que les festons du baldaquin présentaient à l'origine l'aspect de pointes et étaient situés plus haut.
- La technique d'exécution diffère aussi de celle des 'Primitifs flamands'. À gauche du manteau de sainte Catherine, le travail du pinceau est bien visible et se signale par son caractère étonnamment expressif. La couleur a été posée de manière rapide, contrairement à l'usage en vigueur au XVe siècle dans les anciens Pays-Bas.

[4] «*Quia et Christus semel pro peccatis nostris mortuus est, justus pro injustis, ut nos offerret Deo* (...)» (I Pt, 3, 18).

[5] Voir, à ce sujet, M.W. AINSWORTH, «Hans Memling as a Draughtsman», dans *Hans Memling. Essays*, Bruges, 1994, p. 78-87.

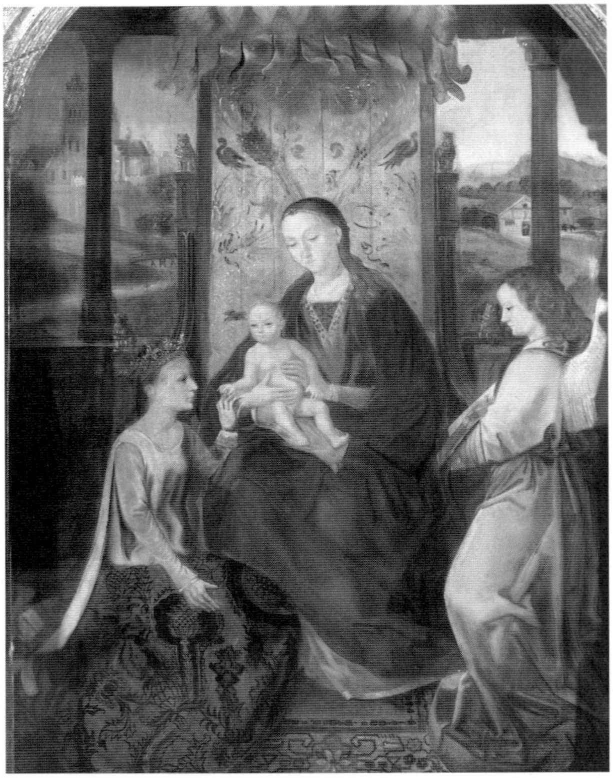

Fig. 3. Faussaire moderne, *Triptyque du Mariage mystique de sainte Catherine*. Panneau central, IRR (système utilisé: VARIM) (© Instituto del Patrimonio Histórico Español, Laboratorio de Estudios Físicos, Madrid).

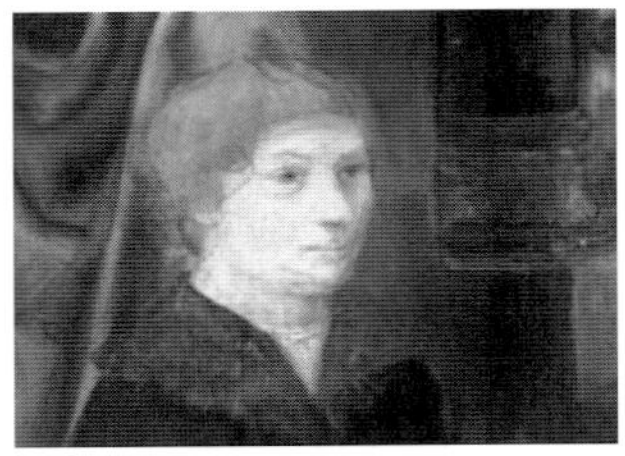

Fig. 4. Faussaire moderne, *Triptyque du Mariage mystique de sainte Catherine*. Détail de la tête du donateur; IRR (système utilisé: VARIM) (© Instituto del Patrimonio Histórico Español, Laboratorio de Estudios Físicos, Madrid).

- Plusieurs zones du triptyque sont dépourvues de craquelures. Celles-ci se concentrent dans certaines parties de l'image, notamment dans l'aube de l'ange, dans la coiffe de la donatrice et dans sa robe noire. Pour le reste, l'état de conservation de la couche picturale est anormalement bon pour une œuvre du XVe siècle, comme le confirme la photographie dans l'ultraviolet.
- Des échantillons de peinture, prélevés principalement sur les bords du panneau central, dans des zones non retouchées, ont été analysés[6]. La stratigraphie observée n'est pas toujours conforme à celle des œuvres flamandes du XVe siècle. Ainsi, dans le blanc provenant de la bordure du manteau de sainte Catherine, on a identifié une préparation non pas à base de carbonate de calcium, mais à base de plâtre et de terres. En outre, il apparaît que les pigments utilisés sont de fabrication industrielle et que certains ne sont pas attestés avant le XIXe siècle. Dans la couche blanche proprement dite du manteau de sainte Catherine, par exemple, on a retrouvé du blanc de baryum,

[6] Cette analyse a été effectuée par l'entreprise ARTE-LAB, S.L., de Madrid.

du blanc de zinc et du bleu d'outremer synthétique en basse concentration. On a détecté du bleu de Prusse dans un échantillon prélevé dans le ciel tandis que, dans le manteau de la Vierge, ont été découverts du bleu d'outremer synthétique, du blanc de plomb, du blanc de baryum et du blanc de zinc. Ce dernier échantillon a en outre révélé la présence, dans le niveau inférieur de la stratigraphie, d'une couche constituée de pigments plus anciens, tels le blanc de plomb, le vert-de-gris ou le jaune de plomb ou d'étain. Par-dessus, on observe une seconde couche, appliquée sur la peinture primitive par l'auteur des figures de style 'memlingien'. Elle comporte une double préparation, composée, d'une part, de carbonate de calcium et magnésium (dolomite), d'autre part, de sulfate de calcium, terres et noirs d'os. Sur cette double préparation ont été posées les couleurs (fig. 5).

L'analyse combinée du bois, de la technique d'exécution, de la stratigraphie de la couche picturale et des pigments utilisés permet de conclure que le triptyque de Bilbao, pris en tant qu'image, est bien une imitation moderne, même si le support est ancien.

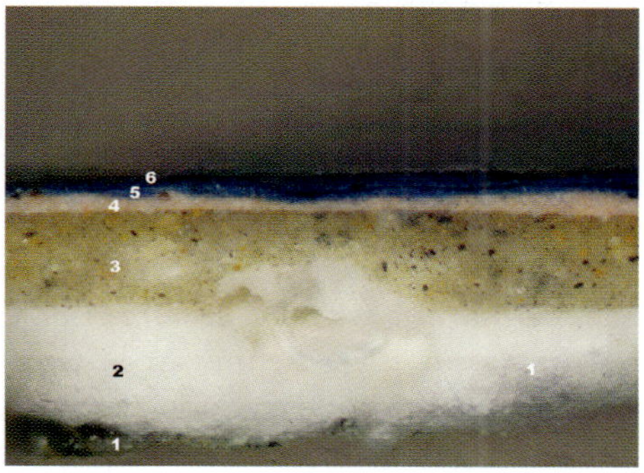

Fig. 5. Faussaire moderne, *Triptyque du Mariage mystique de sainte Catherine*, échantillon du bleu du manteau de la Vierge (© Museo de Bellas Artes, Bilbao). 1. Restes d'une couche de peinture verdâtre composée de blanc de plomb, de carbonate de calcium, de vert-de-gris, de jaune de plomb et d'étain, de terres (en très basse concentration). 2. Stuc blanc composé de carbonate de calcium et de magnésium (dolomite). 3. Préparation brune composée de gypse, de terres et de noir d'os. 4. Couche de peinture rosée composée de blanc de plomb, d'un colorant rouge, de gypse (en très basse concentration) et de blanc de zinc (idem). 5. Couche de peinture bleue composée de bleu d'outremer synthétique, de blanc de plomb, de blanc de baryum et de blanc de zinc. 6. Vernis brun composé de résine de colophane.

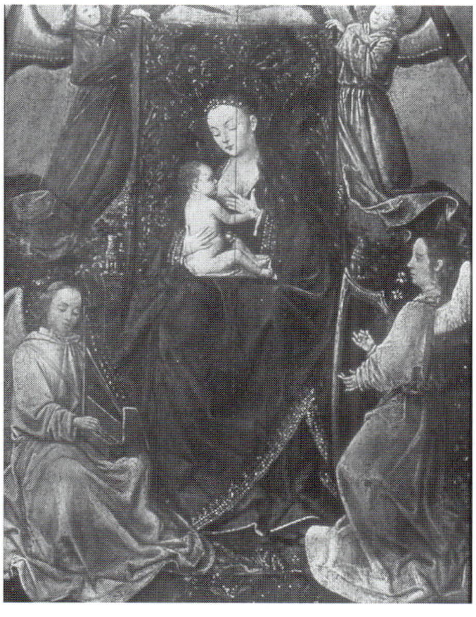

Fig. 6. Faussaire moderne, *Vierge à l'Enfant en trône*, localisation actuelle inconnue (© Centre d'étude de la Peinture du XVe siècle dans les Pays-Bas méridionaux et la Principauté de Liège, Bruxelles).

Le faussaire avait visiblement l'intention de réaliser un 'Primitif flamand' mais, plutôt que de prendre le risque d'inventer de toutes pièces une composition dans le goût du XVe siècle, il a préféré réaliser un 'collage' à partir de motifs empruntés à des œuvres authentiques de cette époque.

Ainsi, du triptyque Donne, il reprend le cadre architectural et l'arrière-plan des deux volets, ainsi que la figure de saint Jean-Baptiste. L'habit du donateur du triptyque de Bilbao est presque entièrement calqué sur celui de Heinrich Von Werl, le donateur d'un des volets de Madrid traditionnellement attribués au Maître de Flémalle. Quant à la donatrice, elle porte la robe de Caterina Tanagli, l'épouse du commanditaire du triptyque du *Jugement dernier* de Memling conservé à Dantzig. La coiffe blanche provient en revanche d'une œuvre de Rogier de la Pasture: le triptyque de la *Crucifixion* de Vienne. La donatrice représentée par le maître tournaisien sur le panneau central arbore un couvre-chef quasi identique.

Le saint protecteur figurant sur le volet droit du triptyque de Bilbao constitue, à nouveau, un emprunt à Hans Memling. Cette fois, c'est le saint Georges du petit tableau votif de la National Gallery de Londres qui a été pris pour modèle. La sainte Catherine du panneau central se révèle, quant à elle, un étonnant montage. Deux sainte Catherine similaires de Hans Memling ont été fondues en une seule figure: celle du *Mariage mystique* de New York et celle du monumental triptyque de l'Hôpital Saint-Jean. À la première, le faussaire a emprunté le détail du pouce écarté de la main droite et le motif de la main gauche redressée. À la seconde sainte Catherine, la plus connue, il a repris en grande partie les motifs végétaux du brocart de la robe.

Dans la Madone trônant, le faussaire combine des motifs tirant leur origine des deux mêmes *Mariage mystique de sainte Catherine* 'memlingiens'. L'Enfant est représenté dans l'attitude qui est la sienne sur le tableau de New York mais il tourne la tête vers la gauche, comme sur le panneau central du triptyque brugeois. Quant à Marie, si elle penche le visage vers la gauche, comme sa consœur de la version new-yorkaise, la bordure ornée de pierreries de son décolleté en pointe a probablement été reprise à son autre consœur du triptyque de l'Hôpital Saint-Jean.

L'ange présentant un livre est également le résultat d'une greffe combinant deux figures apparentées du corpus 'memlingien': la partie supérieure du corps a été empruntée au *Mariage mystique de sainte Catherine* de l'Hôpital Saint-Jean, la moitié inférieure et les ailes au panneau central du triptyque Pagagnotti. Ce dernier modèle a également inspiré au faussaire le motif du baldaquin dont les festons sont agités par le vent, ainsi que l'*oculus* inscrit dans l'axe de symétrie vertical de la composition. Quant aux deux colonnettes de la loggia et à leurs chapiteaux, ils proviennent de la *Madone assise sur un banc* de Berlin, autre création de Memling étudiée par notre faussaire.

L'arrière-plan du panneau central trouve son origine dans deux paysages peints par le maître brugeois: celui de la *Madone Pagagnotti*, dont le faussaire a notamment repris le moulin à eau et la moitié gauche de la ville fortifiée, et celui du *Martyre de saint Sébastien*, œuvre appartenant aux Musées royaux des beaux-arts de Bruxelles. À ce second tableau ont été empruntés le petit pont surmontant une rivière et la moitié droite de la ville fortifiée.

Outre Hans Memling, Rogier de la Pasture et le Maître de Flémalle, Jean van Eyck fait aussi partie des artistes dont l'auteur du triptyque de Bilbao est débiteur. Le trône aux quatre lions et la partie supérieure du manteau de la Vierge, du côté droit, ont été copiés sur la fameuse *Madone de Lucques* conservée à l'Institut Städel de Francfort, tandis que le drap d'honneur reprend, sous une forme simplifiée, celui de la *Madone à la fontaine* du Musée des beaux-arts d'Anvers.

Enfin, pour les armoiries des donateurs, ce sont les gravures héraldiques d'Albrecht Dürer qui ont servi de modèles. L'écu de l'homme, avec son haume, a pour origine le célèbre cuivre au Lion et au Coq (Meder 97). Quant à l'écu de l'épouse, qui comporte l'image du Chasseur maudit, il a été inspiré par celui de Johann Tscherte (Meder 294), tandis que le haume reproduit celui de Lorenz Staiber (Meder 293).

Il faut reconnaître que la suture de ces nombreux emprunts — nous ne prétendons d'ailleurs pas les avoir tous identifiés — a été opérée avec une habileté étonnante. Bien que constituée d'éléments de provenance aussi diverse, la composition du triptyque paraît équilibrée. À l'instar de Memling, le faussaire recherche un effet global de symétrie, tout en évitant d'enfermer ses personnages dans une structure géométrique par trop rigide. Si l'on excepte le couple de donateurs en prière, les attitudes se signalent même par une grande variété.

Le triptyque a d'ailleurs suscité des commentaires plutôt élogieux. Javier de Bengoechea observe une «prédominance de la verticalité dans la composition pour renforcer le caractère sublime de la scène» et note que «les lumières lointaines des paysages du fond enveloppent de mystère l'événement représenté»[7]. Quant à Xesqui Castañer, elle estime suffisante la qualité de la peinture que pour envisager, au moins à titre d'hypothèse, sa réalisation dans l'atelier de Hans Memling[8].

La composition, incontestablement, 'tient'. On ne relève dans le triptyque que peu de ces erreurs typiques des 'faussaires-collagistes', erreurs imputables à la fragmentation des figures prises pour modèles ou au fait qu'elles ont été arrachées à leur contexte original. Seules trois incohérences, peu visibles, sont à signaler. Le faussaire a pris le parti de ne pas représenter les pieds du donateur. Il a toutefois conservé, dans le drapé de sa robe, un 'arc' correspondant au pied droit de Heinrich Von Werl. De même, dans l'aube de l'ange au livre du panneau central, on discerne un pli motivé par l'existence d'une marche sur le modèle 'memlingien', la *Madone Pagagnotti*. Or, dans le triptyque de Bilbao, une telle marche est absente. Le pli ne se justifie donc plus. Enfin, sur le volet droit, la main libre de saint Georges semble tenir un objet invisible. Dans le modèle utilisé, le panneau votif de Hans Memling conservé à Londres, le pieux chevalier porte effectivement un chapeau dans la main gauche. Le faussaire a jugé bon d'éliminer ce détail, sans pour autant prendre la peine de redessiner la main.

On sait que la méthode de travail des faussaires présente parfois des analogies frappantes avec celle des historiens d'art. Ce principe s'applique parfaitement au triptyque de Bilbao. Comme un historien d'art, son auteur a étudié avec une attention toute particulière les répétitions partielles ou complètes de figures dans l'œuvre de Hans Memling. Dans les *Mariage mystique de sainte Catherine* conservés à Bruges et à New York, il a notamment relevé les relations étroites unissant les deux pieuses vierges d'Alexandrie et les deux Enfant Jésus. De même, il a remarqué le lien entre l'ange au livre du grand triptyque de l'Hôpital Saint-Jean et le harpiste ailé de la *Madone Pagagnotti*.

[7] J. DE BENGOECHEA, 1978, p. 97.
[8] X. CASTAÑER, 1995, p. 48.

En général, les historiens d'art fondent sur de tels rapprochements soit l'attribution d'une œuvre, soit sa datation. Ce sont là des questions qui, normalement, ne préoccupent guère les faussaires. L'intérêt de l'auteur du triptyque de Bilbao pour la variante chez Hans Memling répond sans doute à un tout autre souci: celui de s'écarter du modèle, sans pour autant courir le risque de commettre une erreur historique ou esthétique. La manière la moins périlleuse de réaliser une nouvelle figure dans le style de Memling ne consiste-t-elle pas à combiner deux variantes autographes? Le faussaire demeure ainsi dans les limites mêmes de la production conservée du peintre qu'il cherche à imiter, tout en se permettant d'innover...

Peut-on déterminer l'auteur du triptyque de Bilbao? Dans un premier mouvement, l'historien d'art hésitera à répondre par l'affirmative. C'est que, par définition, les œuvres d'un faussaire-collagiste ne comportent aucune marque stylistique personnelle. Pour peu qu'il soit abouti, le faux constitué de fragments copiés présente exclusivement les caractéristiques stylistiques des auteurs plagiés. Si l'approche formaliste apparaît donc peu prometteuse pour l'étude de ce type de falsifications, il existe néanmoins une méthode rendant possible, dans certains cas, l'attribution de faux réalisés sur le mode du collage: l'étude systématique des sources figuratives utilisées. On imagine difficilement, en effet, que deux contrefacteurs différents aient eu à peu près simultanément l'idée d'emprunter le même détail au même modèle, ou de combiner dans une seule œuvre les mêmes emprunts. Aussi, quand on rencontre des reprises de motifs identiques dans deux peintures d'ancienneté douteuse, la probabilité est très grande qu'elles soient dues à un seul et unique 'fabricant'.

Semblable raisonnement a permis récemment de rassembler une série de falsifications unies par un réseau étroit de références communes. En 2001, Maryan Ainsworth a constitué un premier groupe de huit faux 'Primitifs flamands' autour d'un triptyque moderne de style 'memlingien', conservé au Palmer Museum of Art de l'Université de Pennsylvanie à University Park[9]. Le Maître du Triptyque Palmer était né. Deux ans plus tard, Didier Martens en a étoffé le catalogue, portant à vingt le nombre de ses productions présumées[10]. Parmi les modèles favoris du faussaire, on citera notamment: la Vierge de l'*Annonciation* du polyptyque de l'*Agneau mystique* (deux occurrences), l'Elisabeth Borluut du même ensemble (quatre occurrences), la Vierge à l'Enfant du triptyque de Vienne de Hans Memling (cinq occurrences) et le Guillaume Moreel du triptyque de Bruges, par le même artiste (deux occurrences).

Au nombre des œuvres qui peuvent être attribuées au Maître du Triptyque Palmer figure une *Vierge à l'Enfant en trône* qui, dans les années 1950, serait passée entre les mains de l'antiquaire new-yorkais Wildenstein[11] (fig. 6). Il s'agit, une fois encore, d'un collage d'emprunts. L'attribution repose sur l'indice suivant: la partie inférieure du drapé de la robe de la Vierge a été reprise au triptyque de Vienne de Hans Memling.

[9] M.W. AINSWORTH, «*Caveat emptor.* An Early Twentieth-Century Workshop for Flemish Primitives», dans *Apollo*, 153, juin 2001, n°472, p. 20-29.

[10] D. MARTENS, «Les frères Van Eyck, Memling, Metsys *et alii* ou le répertoire d'un faussaire éclectique», dans *Wallraf-Richartz-Jahrbuch*, 64, 2003, p. 253-284.

[11] D. MARTENS, 2003, p. 268-269. Dans un courriel adressé à Didier Martens le 19 septembre 2006, Jennifer Van Etten (New York) signale qu'elle n'a pu retrouver la moindre trace de l'œuvre dans les archives de la Galerie Wildenstein.

Si le panneau Wildenstein est bien l'œuvre du faussaire redécouvert par Maryan Ainsworth, la même attribution peut être avancée pour le triptyque de Bilbao. En effet, dans les deux peintures, le trône aux lions de la *Madone de Lucques* et une partie de son manteau se retrouvent associés au drap d'honneur de la *Madone à la fontaine* et à l'ange de profil du triptyque Pagagnotti. De plus, dans le panneau Wildenstein, son aube présente également le pli motivé par la présence d'une marche 'invisible'. Ces coïncidences dans le choix des modèles sont trop étroites pour être fortuites. Des sources identiques ont été combinées de manière similaire. C'est donc, à l'évidence, le même faussaire qui, dans le triptyque de Bilbao comme dans le panneau Wildenstein, a mis à profit les mêmes études réalisées d'après Jean van Eyck et Hans Memling.

Didier MARTENS
Université libre de Bruxelles
Faculté de Philosophie et Lettres
Avenue Franklin Roosevelt 50, C.P. 175
B – 1050 Bruxelles
Tel: +32.2.650.24.19
didier.martens@ulb.ac.be

Licencié en Philologie germanique (1982) et en Histoire de l'Art de l'Université Libre de Bruxelles (1984), Didier Martens est titulaire de la Chaire d'Histoire de l'Art de la fin du Moyen Âge et des Temps modernes dans cette même université et travaille principalement, à l'heure actuelle, sur les relations artistiques entre Flandres et Péninsule ibérique aux XVe et XVIe siècles.

Ana SÁNCHEZ-LASSA DE LOS SANTOS
Bilboko Arte Eder Museoa / Museo de Bellas Artes de Bilbao
Museoko Plaza, 2
E – 48011 Bilbao
asanchezlassa@museobilbao.com

Diplômée en Conservation/Restauration de l'Escuela central de Artes aplicadas a la Restauración de Madrid (1973) et licenciée en Histoire de l'Art de l'Universidad Complutense de Madrid (1977), Ana Sánchez-Lassa a eu en charge le Département de Conservation/Restauration du Musée des beaux-arts de Bilbao (1975-2005). Elle est devenue depuis lors conservateur de ce musée et dirige le Département des Collections. Elle travaille principalement sur la peinture flamande des XVe et XVIe siècle dans les collections publiques basques.

PAINTING TECHNIQUE IN THE LATE WORKS OF GIOTTO: INFRARED EXAMINATION OF SEVEN PANELS FROM ALTARPIECES PAINTED FOR SANTA CROCE

Maria Clelia Galassi and Elizabeth Walmsley*

Abstract: The long career of the Florentine painter Giotto di Bondone (c. 1266-1337) spans a period of approximately forty years that was crucial for the stylistic and technical development of Italian painting in fresco and panel painting. Giotto's own development of a more naturalistic depiction of figures, imbued with a more convincing three-dimensionality, marks the introduction of the Renaissance style. Although several important studies of Giotto's early panel painting technique have been published, questions about the development of Giotto's style during the last decades of his career and the role of his many workshop assistants remain open. According to scholars, the Peruzzi altarpiece (NCMA, Raleigh) was executed by Giotto and assistants during the 1330's for the church of Santa Croce. This paper presents results of infrared examination. The underdrawing and undermodeling technique are discussed and compared to the early works and to other late works, specifically two panels thought to have come from a dismantled polyptych also painted for Santa Croce: the Goldman Madonna and Child (NGA, Washington) and the Saint Stephen (Horne Museum, Florence).

As Hayden Magginis has written, "Perhaps no painter has been, and remains, so controversial as Giotto. Perhaps no art-historical debate has been more divisive than of attempting to define Giotto's artistic personality".[1] About a dozen surviving panel paintings can be attributed to Giotto's workshop, but only one — the great *Ognissanti Maestà* of about 1310 — has been universally accepted as a work by the master himself.[2] Among the frescoes attributed to Giotto, only the cycle in the Arena Chapel, in Padua, dating to about 1305, is firmly recorded as Giotto's work. The famous essay by Richard Offner, "Giotto, Non-Giotto," published in 1939, represented an important contribution to this question, by using strict, stylistic criteria to differentiate works that might be 'autograph Giotto'. But Offner's methodology was soon found to be inadequate for understanding the complexity of Giotto's production. In 1967, Giovanni Previtali offered a new approach in his book, "Giotto and his workshop". He focused on the reality of a large workshop.[3] The role that artists associated with Giotto could have played

* We are very grateful to the Center for Advanced Studies in the Visual Arts, Samuel H. Kress Foundation, and Heinemann Fund for funding this research. For allowing us to examine their paintings, we would like to thank Elisabetta Nardinocchi, Museum Horne; Antonio Paolucci, Soprintendenza Speciale per il Polo Museale Fiorentino, Florence; Antonio Natali and Alessandro Cecchi, Galleria degli Uffizi, Florence; William Brown, Noelle Ocon, and David Steel, North Carolina Museum of Art.

[1] H.B.J. Magginis, "In Search of an Artist", in *The Cambridge Companion to Giotto*, A. Derbes and M. Sandona eds., Cambridge, 2004, p. 10.

[2] J.I. MILLER and L. TAYLOR-MITCHELL, "The Ognissanti Madonna and the Umiliati Order in Florence", in A. DERBES and M. SANDONA, 2004, p. 157.

[3] R. OFFNER, "Giotto, non-Giotto", in *The Burlington Magazine*, LXXIV (1939), p. 259-268; LXXV (1939), p. 96-113; G. PREVITALI, *Giotto e la sua bottega*, Milan, 1967. For a discussion on the critical

in the workshop is still not completely understood. As Magginis noted, it seems more likely that "Giotto headed a workshop that was more a joint-venture company than a studio entirely determined by the relationships of master to apprentices and followers".[4] In addition to Bernardo Daddi, Maso di Banco, and Taddeo Gaddi, there were undocumented students, collaborators, and assistants, including the so-called *Parente di Giotto* (Giotto's relative), the painter that scholars recognize as his most important collaborator.

Since the 1990's, technical examination, particularly infrared reflectography, has been used to study paintings attributed to Giotto in order to determine the working method and to obtain new information about Giotto's workshop. Fundamental studies of Giotto's painting technique have focused only on his very early career, that is, the end of 13th century through the first decade of the 14th century.[5] This paper presents the results of our investigation into Giotto's late panel paintings to show the technical achievements of the Giottesque workshop at the pinnacle of its success in Florence and to determine if there were different hands at work.

We carrried out infrared studies of seven panels thought to belong to a group of four altarpieces which, according to the sculptor Lorenzo Ghiberti, were executed by Giotto for the church of Santa Croce in Florence between 1320-1335. Only the Baroncelli altarpiece is still in place, as well as Giotto's fresco cycles in the Peruzzi and Bardi Chapels. We studied the five panels known as the 'Peruzzi altarpiece' now at the North Carolina Museum of Art in Raleigh: the Blessing Christ, with the Virgin, Saint John the Evangelist, Saint John the Baptist, and Saint Francis.[6] We also examined the so-called Goldman Madonna and Child at the National Gallery of Art in Washington,[7] and the Saint Stephen at the Horne Museum in Florence.[8] Infrared reflectography of an eighth painting, the early *Ognissanti Maestà*[9] at the Uffizi, circa 1310, was carried out to determine the hallmarks of Giotto's technique, since it is the only panel painting that has been unanimously attributed exclusively to Giotto.

debate on Giotto see G. BONSANTI, "La bottega di Giotto", in *Giotto. Bilancio critico di sessant'anni di studi critici e ricerche*, exh. cat., A. TARTUFERI ed., Florence, 2000, p. 55-73.

[4] H.B.J. MAGINNIS, 2004, p. 25.

[5] See M. SCUDIERI ed., *La Croce giottesca di San Felice in Piazza: storia e restauro*, Venice, 1992; D. BENZATO ed., *La Croce di Giotto. Il restauro*, Milan, 1995; M. CIATTI and C. FROSININI eds., *La Madonna di San Giorgio alla Costa di Giotto. Studi e restauro*, Florence, 1995; F.R. PESENTI, "Del modo di colorire le incarnazioni e i vestiri in Giotto, dopo alcuni restauri", in *Scritti e immagini in onore di Corrado Maltese*, S. MARCONI ed., Rome, 1997, p. 149-157; M. CIATTI and M. SEIDEL eds., *Giotto. La Croce di Santa Maria Novella*, Florence, 2001 (Italian), 2002 (English); E. MOGNETTI, "*De visu...* Observations sur la peinture italienne des XIVe et XVe siècles", in *Techné*, 15, 2002, p. 11-19.

[6] Accession number GL.60.17.7, 105.7 × 250.2 cm. See also W. BROWN with contributions by D. FINLEY, B. WOJCIK and M. MARCH, *Peruzzi Altarpiece: Italian Catalogue Examination Form*, June, 2001, NCMA Conservation Department files; F.R. SHAPLEY, *Catalogue of the Italian Paintings of the National Gallery of Art*, Washington, 1979, p. 219-221; F.R. SHAPLEY, *Italian Paintings XIII-XV Century from the Samuel H. Kress Collection*, London, 1966, p. 21-22.

[7] Accession number 1939.1.256. 85.4 × 61.6 cm. F.R. SHAPLEY, 1966, p. 20-21.

[8] Inv. 52. 84.3 × 54 cm. See also M. BOSKOVITS, "Santo Stefano", in A. TARTUFERI, 2000, p. 182-186.

[9] Inv. 1890, 8344. 325 × 204 cm. For additional technical studies, see *La Madonna d'Ognissanti di Giotto restaurata, Gli Uffizi. Studi e ricerche*, 8, Florence, 1992, in particular p. 87-123.

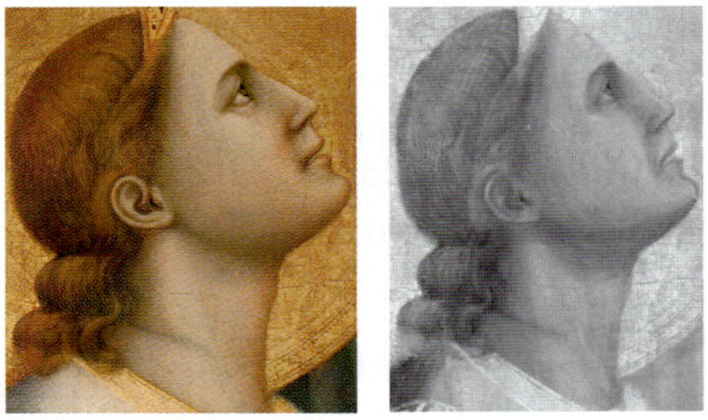

Fig. 1. *Ognissanti Maestà*, Florence, Uffizi. a. Detail, angel. b. IRR composite (© the authors).

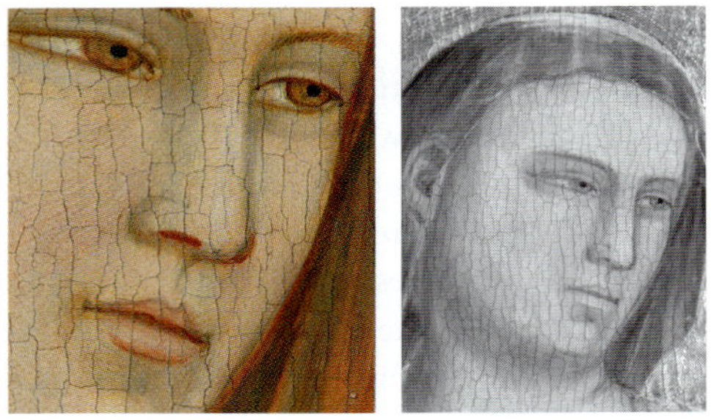

Fig. 2. *Peruzzi Altarpiece*, Raleigh, North Carolina Museum of Art (NCMA). a. Detail, the Virgin (© the authors). b. IRR composite (© Noelle Ocon).

One of the most distinctive features in the infrared reflectograms of the *Ognissanti Maestà*[10] is the presence of undermodeling in the fleshtones.[11] For example, infrared reflectography of the angels' faces (figs. 1a, 1b) revealed large areas of shadow, made with multiple brushstrokes using a liquid medium, much larger than in the finished painting, as seen in the shadows of the neck. This feature is described by Cennino Cennini in his *Libro dell'arte*, our main technical source for the Giottesque age.[12] Before beginning to paint, the artist outlined the compositional forms in charcoal to allow for mistakes and revisions. Next, the contour lines were fixed in black ink using a pointed brush. Volumes were given bold shadows with more of the black ink using a soft, cropped brush, the *pennello mozzetto*, with the bristles held by the fingers into a fan-shape. Once this was completed, the first paint layers were applied: the ink washes in the shadows were reinforced with *verdaccio*, a brownish or greenish mixture of pigments that also can register in infrared reflectograms. The strongly contrasting *chiaroscuro* created

[10] IRR details of the painting *in situ* by the authors using the University of Genoa's Hamamatsu C2400-03 camera with N2606-26 vidicon tube, Nikon 50 mm 1:2.8 lens, RG1000 filter, Toshiba Satellite 2540 CDS laptop, USB video capture device mosaics with Adobe Photoshop on a Macintosh computer. For previously published IRR details, see M. SERACINI, "Diagnostica per immagini dell'opera", in *Gli Uffizi. Studi e ricerche*, 1992, p. 99-104; M.C. GALASSI, *Il disegno svelato: progetto e immagine nella pittura italiana del primo Rinascimento*, Nuoro, 1998, especially p. 99-100.

[11] For discussion of undermodeling in other works, see P. BRACCO and O. CIAPPI, "Technique of execution, state of conservation and restoration of the Crucifix in relation to other works by Giotto. The painted surface", in M. CIATTI and M. SEIDEL eds., 2002, p. 343.

[12] C. CENNINI, *The Craftsman's Handbook. Il Libro dell' Arte,* translated by D.V. THOMPSON Jr., New York, 1933. See also M.C. GALASSI, "Underdrawing vs. Undermodelling. Form Construction in Tuscan Painting during the first half of the 15th Century", in *La Peinture et le Laboratoire. Procédés. Méthodologie. Applications. Le dessin sous-jacent et la technologie dans la peinture. Colloque XIII, septembre 1999*, R. VAN SCHOUTE and H. VEROUGSTRAETE eds., Leuven, 2001, p. 131-141; R. BELLUCCI and C. FROSININI, "Painting Flesh Tones", in C.B. STREHLKE and C. FROSININI eds., *The panel paintings of Masolino and Masaccio: the role of technique*, Milan, 2002, p. 32-33.

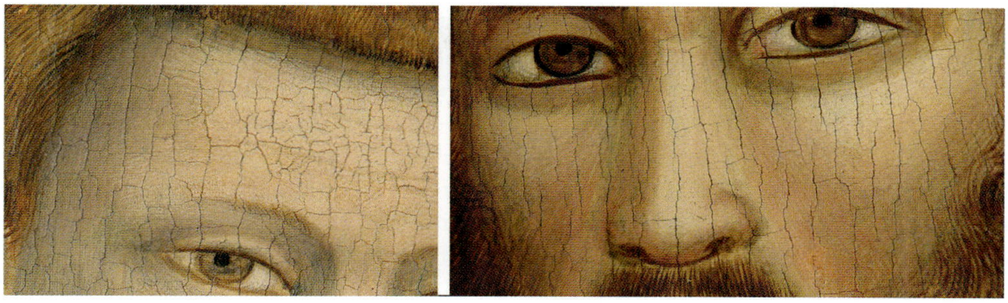

Fig. 3. *Peruzzi Altarpiece*, NCMA. a. Detail, Saint John the Evangelist. b. Detail, Saint John the Baptist (© the authors).

during the undermodeling stage appears to be typical of Giotto's working method for both panel and mural paintings.[13] Only in the paint stage were the volumes modulated in softer transitions between light and shadow.

The Peruzzi altarpiece may be dated in the 1330's. In the 20th century, the Christ Blessing came onto the market separately from the other panels. William Suida was the first to recognize it originally belonged to the same altarpiece, formerly in the Peruzzi chapel. The attribution to Giotto is still under debate and many scholars have speculated about a possible attribution to Giotto's assistants. According to Ferdinando Bologna, the Christ Blessing, Saint John the Baptist, and Saint Francis were painted by Giotto himself, while the Virgin and Saint John the Evangelist were by the workshop.[14] Our examination suggests a slightly different grouping. Although the five panels have a similar underdrawing technique (linear underdrawing and washed undermodeling) (figs. 2a, 2b),[15] we found two different approaches to the creation of fleshtones during the paint stage, thus suggesting the possibility of two different hands at this stage. The faces of the Virgin, the Evangelist (fig. 3a), and Christ were built up with pink and white hatchmarks of opaque paint applied with a pointed brush, following the contours, and creating a compacted surface. Translucent gray shadows were created by scumbles of light paint over the washed undermodeling.[16] Other shadows were made by leaving small areas of the transparent gray-green underpaint visible. The faces of Saint Francis and the Baptist (fig. 3b) were modeled by exploiting the color relationships of green and pink paint, and by layering transparent and opaque colors. More of the transparent, greenish underpaint was left visible than in the other three panels. Opaque white paint applied in areas of highlight was followed by translucent pink brushstrokes, creating a transition between the white highlights and the darker shadows.

[13] M.C. Galassi has noted a similar working method in the now-damaged Peruzzi Chapel fresco cycle, and G. Bonsanti in the Capella di San Nicola, Assisi (A. TARTUFERI, 2000, p. 62).

[14] Noted in Bonsanti, 1992, p. 70. F.R. SHAPLEY, 1966, with previous bibliography.

[15] IRR of the altarpiece by Noelle Ocon and M.C. Galassi using the NCMA's FLIR-SwirCAM PtSi camera, 1.5-1.8 micron filter, Foresight Imaging I-50 frame grabber; mosaics by N. Ocon using Adobe Photoshop on a Windows 2000 computer.

[16] This effect is recommended by Cenino Cennini for "fusing and softening like a puff of smoke": C. CENNINI, 1933, p. 94.

The Goldman Madonna has been attributed to Giotto's assistant, Bernardo Daddi, but is now generally accepted as the work of the master and is dated 1320/1330 (fig. 4a).[17] Infrared reflectograms[18] reveal the two phases of underdrawing — the outlining of the forms and the undermodeling — described by Cennino, but with a remarkable difference in the way the undermodeling was carried out. The modeling of the Madonna and Child's faces relies, in fact, more on the build-up of fine, black hatchmarks following the curves of the face than on the *pennello mozzetto* washes (fig. 4b). Using a stereomicroscope,[19] we found that the hatchmarks were applied during both the underdrawing and the paint stages.

Attributed in the past to Bernardo Daddi, to Taddeo Gaddi, and to Giotto's workshop, the Saint Stephen is now given to Giotto (fig. 5a).[20] Dated circa 1330-1335, it is thought to have formed a part of the same altarpiece as the Goldman Madonna. This last hypothesis has been rejected by some scholars because the two panels have a different bole under the gold-leaf, green in the Saint Stephen and red in the Madonna.[21] Nonetheless, the two paintings share strong similarities in the way the underdrawing and paint layers were applied. Digital infrared photography[22] of Stephen's face (fig. 5a) reveals a minor shift in the position of the eyes at the linear underdrawing stage, similar to other changes in the Goldman Madonna. There are *pennello mozzetto* washes in the shadows, and the modeling was again emphasized with hatchmarks (fig. 5b). Color details show the paint applied in overlapping hatchmarks of varying widths in translucent pinks, opaque white and black.

During the paint stage, the fleshtones in the seven late panels were modeled with the addition of black to the palette of pink and white, by incorporating the underdrawing and underpaint layers, and by reinforcing some of the contours. Green-tinged shadows in the figures' faces were made by allowing narrow glimpses of the transparent green earth layer beneath the striations of pink and white paint. As seen in the fingers of the Goldman Madonna (figs. 6a, 6b), the contours were softened by covering the black underdrawing with opaque fleshtones or were sharpened by allowing the underdrawing to remain visible and then further emphasized with a second black or red outline added as a final touch. The red outline varies among the seven panels: a deep transparent red or sometimes a more opaque dark pink was used in the Goldman Madonna, an orange-red in the Saint Stephen, and a bright red was used to reinforce the contours of the hands of the Evangelist (fig. 7), Virgin, and Christ. The Evangelist's hands are

[17] F.R. SHAPLEY, 1966, p. 20-21, with previous bibliography.

[18] IRR by E. Walmsley using the NGA's FLIR Vis/InGaAs camera, 1.5-1.8 micron filter, Nikon 1:1.2 lens, pk-11 extension ring, Dell Inspiron 600m laptop, National Instruments 1422 IMAQ PCI frame grabber, Indigo IRVista v2.5; mosaics using Adobe Photoshop on a Macintosh computer. Sarah Feinstein and Bart Devolder assisted the authors with earlier IRR examinations of this painting.

[19] Examination by E. Walmsley at the NGA, Washington.

[20] M. BOSKOVITS, 2000, p. 186, with previous bibliography.

[21] D. BOMFORD, J. DUNKERTON and D. GORDON, *Art in the Making. Italian Painting Before 1400*, London, 1990. D. GORDON, "A Dossal by Giotto and His Workshop. Some Problems of Attribution, provenance and patronage", in *The Burlington Magazine*, CXXXI, 1037, 1989, p. 524-531. It should be noted that the Peruzzi panels also have a green bole.

[22] Digital infrared photographs of the painting *in situ* by the authors using the University of Genoa's Sony Cybershot F717 ccd camera in nightshot mode and Hoya R72 IR filter.

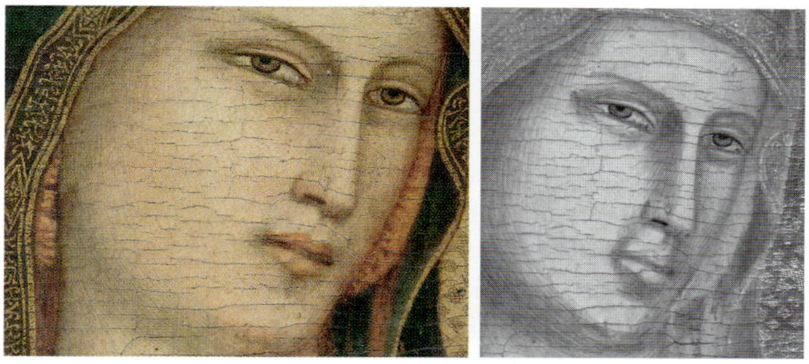

Fig. 4. *Goldman Madonna*, Washington, National Gallery of Art (NGA). a. Detail of face. b. IRR composite (© the authors).

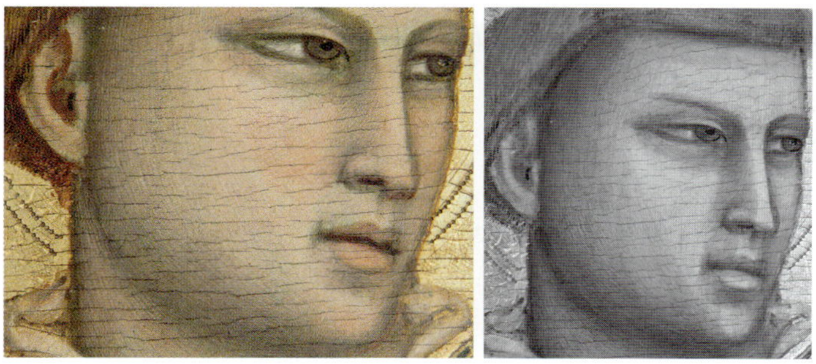

Fig. 5. *Saint Stephen*, Florence, Horne Museum. a. Detail of face. b. Digital infrared photograph (© the authors).

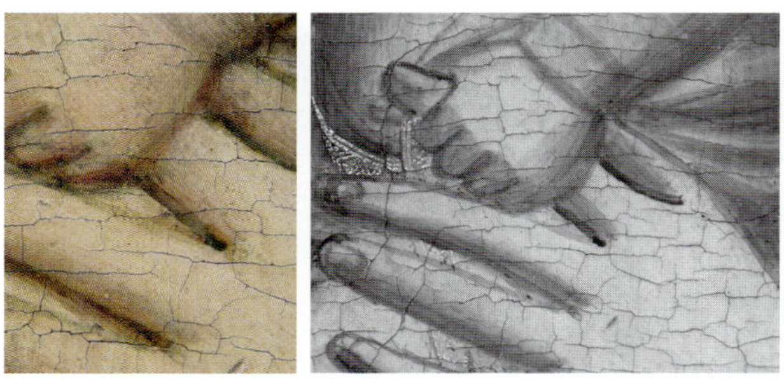

Fig. 6. *Goldman Madonna*, NGA. a. Detail of hands. b. IRR composite (© the authors).

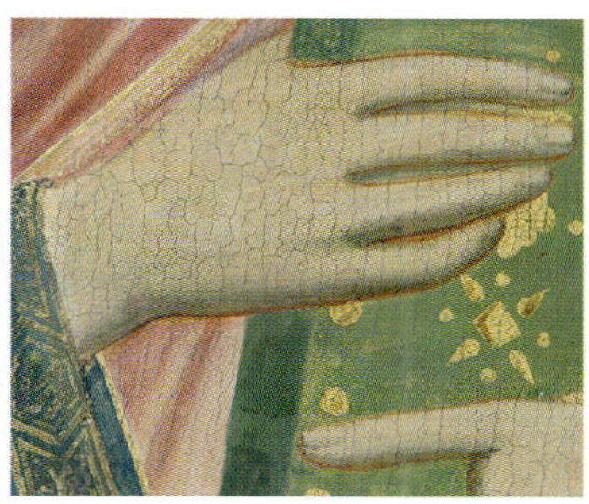

Fig. 7. *Peruzzi Altarpiece*, NCMA. Detail with Saint John the Evangelist (© the authors).

almost completely outlined, regardless of where the shadows lie. The hands of the Baptist and of Saint Francis do not rely on a narrow outline of strongly contrasting color. In keeping with the broader modeling of their faces, Saint Francis' drapery has a nearly vibrating red underlayer. The Peruzzi Virgin has a red line along the side of her face and accentuating her nostrils. Her eyes, as well as the eyes of the other figures in the Peruzzi Altarpiece, are delineated in varying shades of brown.

In conclusion, we found that the infrared reflectograms of the five Raleigh panels reveal a similar approach to the underdrawing and undermodeling, one that is consistent with the technique we observed in Giotto's *Ognissanti Maestà*. However, the two different approaches to the creation of fleshtones during the paint stage suggest that two different painters could have completed the panels: one the Evangelist, the Virgin and the Christ, and a second the Baptist and Saint Francis. The Goldman Madonna and the Horne Saint Stephen were created with similar techniques at the underdrawing stage and in the painting of the fleshtones; this suggests they could have been executed by the same hand, lending support to the theory that they are from the same altarpiece. The greater reliance on black hatchmarks, at both the underdrawing and the paint stages, to create shadows and to model the faces is substantially different from the working method noted in the *Ognissanti Maestà* and the Raleigh panels.

Maria Clelia GALASSI

See Maria Clelia GALASSI, *Dendrochronology and Infrared Examination of Jan Massys's Œuvre.*

Elizabeth WALMSLEY
National Gallery of Art
2000-B South Club Drive
Landover
MD 0785 USA
e-walmsley@nga.gov

She is a painting conservator on the staff of the NGA, Washington. She began the present research on Giotto's underdrawings with Maria Clelia Galassi during a CASVA Paired Fellowship in 2003. She studied underdrawings of Italian paintings in the NGA's collection during a Kress Fellowship at the American Academy in Rome in 2001. She received her undergraduate education at Dartmouth College and a Masters degree in art history/art conservation from the State University College at Buffalo.

THE *SAINT GILES* PANELS IN THE NATIONAL GALLERY OF LONDON

Lorne Campbell*

Abstract: The Master of Saint Giles is named after *Saint Giles and the Deer* and *The Mass of Saint Giles*, believed to have been painted in Paris in about 1500 and now in the National Gallery in London. The underdrawings, executed in more than one technique, are untidy; a great number of changes have been made at all stages of the underdrawing and painting processes. Many of the alterations are radical: in *The Mass of Saint Giles*, for example, the retable of Charles the Bald is painted over an underdrawn altarpiece representing the Crucifixion. The layer structures are often complex, partly because of the changes made during the course of painting. The alterations seem to indicate that the compositions were not fully worked out in preliminary studies but were fully resolved only at the painting stage (the grisailles on the reverses are less complex and the changes there are less radical). The Master's style, techniques and methods are very different from what appear to be standard Netherlandish practices: which raises doubts about the generally accepted theory that the master was an emigrant from the Low Countries who settled in France.

The Master of Saint Giles was named after two paintings: *Saint Giles and the Deer*, acquired for the National Gallery in 1894, and the *Mass of Saint Giles*, acquired for the National Gallery in 1933. On the reverses are grisaille images of an unidentified bishop saint and Saint Peter (fig. 1).[1] On the latter is a mark often interpreted as the Master's signature. It would be strange, however, if the Master had chosen to sign this painting, manifestly inferior to all the others and apparently the work of an assistant. With two companion paintings now in Washington, the Saint Giles panels evidently came from the wings of a large altarpiece in which episodes from the lives of several saints were illustrated. All the painted surfaces are about 61.5 cm high. Peter Klein has established that the wood of the panels is oak grown in western Europe, that the last ring of the London panels was formed in 1463 and that the last ring of the Washington panels was formed in 1462.[2]

Both faces of both the London panels were cleaned in 1974-5 and were the subject of an important article by my colleagues David Bomford and Jo Kirby, published in 1977 in the first issue of the *National Gallery Technical Bulletin*.[3] They have been re-examined by Rachel Billinge and myself during our work on the catalogue of sixteenth-century French and Netherlandish pictures in the National Gallery.

* I am extremely grateful to Rachel Billinge, who made the reflectograms and helped me to interpret the images, and to Peter Klein, who has kindly permitted me to cite his reports on the two panels.

[1] M. Davies, *National Gallery Catalogues. The Early Netherlandish Schools*, 3rd ed., London, 1968, p. 107-13.

[2] J.O. Hand and M. Wolff, *The Collections of the National Gallery of Art, Systematic Catalogue. Early Netherlandish Painting*, Washington, 1986, p. 162-176; in the National Gallery dossier are reports by Peter Klein dated 13 April 1993, on the Washington panels, and 16 December 1993, on the London panels.

[3] D. Bomford and J. Kirby, "Two Panels by the Master of Saint Giles", in *National Gallery Technical Bulletin*, 1, 1977, p. 46-56.

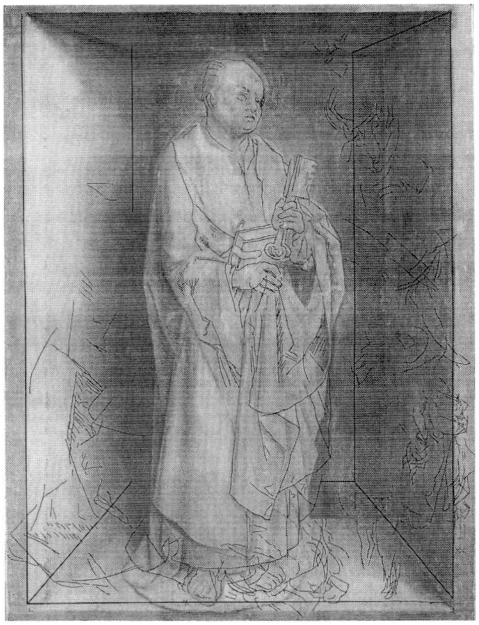

Fig. 1. Master of Saint Giles, *Saint Peter*, London, National Gallery. IRR composite, some underdrawn lines reinforced in black (© National Gallery).

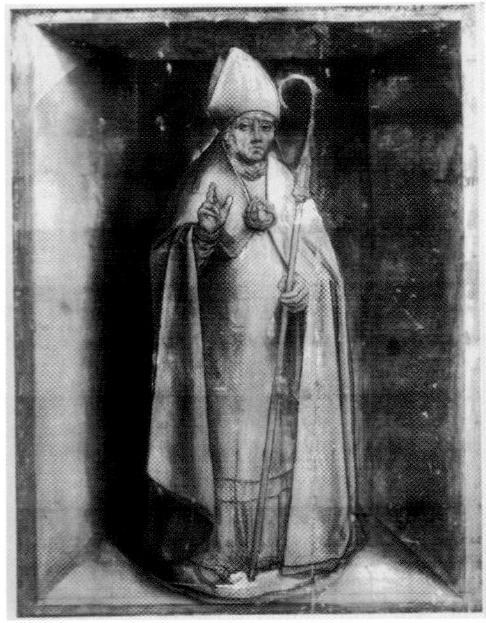

Fig. 2. Master of Saint Giles, *Bishop Saint*, London, National Gallery. IRR composite (© National Gallery).

The legend of Saint Giles seems to have very little basis in fact. The Master and his assistants have followed the much-copied *Life of Saint Giles* thought to have been written early in the tenth century by a monk of the abbey of Saint-Gilles in Provence.[4] In the first panel, Saint Giles has rescued from the royal hunt the deer which has been feeding him with her milk. The two men in the foreground are Flavius, King of the Goths and the Bishop of Nîmes. In the other panel, Saint Giles is securing for Charlemagne remission of a sin too terrible to confess. It was widely believed that this miracle had taken place in the Cathedral at Orléans; but the Master has set the incident in the basilica of Saint Denis. The town in the background of the first panel is often identified as Pontoise but this is far from proved.[5]

A great deal of underdrawing is visible to the naked eye, under the microscope and in infra-red reflectograms. Some of the visible underdrawing does not register in reflectograms: which confirms the impression that more than one technique was used in the execution of the underdrawing. There are enormous numbers of changes. The reflectograms, the X-radiographs, the paint samples taken during cleaning and examination under the microscope show that these changes were made at all stages of the underdrawing

[4] E.-C. JONES, *Saint Gilles. Essai d'histoire littéraire*, Paris, 1914.

[5] W.M. HINKLE, "The Iconography of the Four Panels by the Master of Saint Giles", in *Journal of the Warburg and Courtauld Institutes*, XXVIII, 1965, p. 110-144.

and painting processes. I can mention only the most obvious alterations and those that seem to me most significant.

In the grisailles on the reverses, various alternative shapes have been tried for the architectural elements. The bishop saint once stood under an arch; the word *cruce* is written near the top of the right edge (fig. 2). Much of the drawing beneath Saint Peter bears no relation whatsoever to the painted image (fig. 1).

In *Saint Giles and the Deer* (fig. 3), the saint was drawn lower and further to our right and the deer too was different in the underdrawing. A red circle opposite the head of the deer is now concealed under the grey of Giles's habit and may have been the first idea for the serious wound that he suffered in his efforts to protect the deer. The hunting dog that occupies the centre of the foreground in the underdrawing was never painted. The painted archer, who seems just to have shot the arrow that has wounded the saint, is not underdrawn and covers parts of a horse's head that were at least underpainted. Underdrawn lines beneath and to our right of the archer can be resolved into a man wearing a large hat and raising his right hand to the level of his face. He was on a bigger scale than the painted archer and seems to have been touching his hat rather than shooting an arrow or even bending his bow. There is no trace of an underdrawn bow or arrow.

King Flavius's head was drawn to our right of his painted head; the thumbs of his praying hands were crossed; his drapery fell towards the lower edge of the panel; and

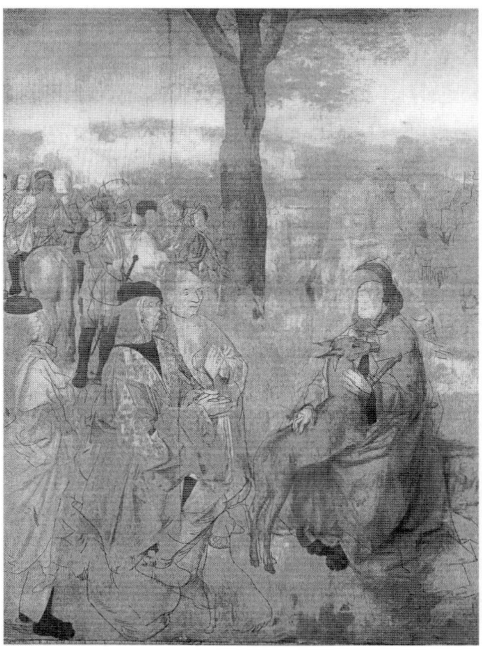

Fig. 3. Master of Saint Giles, *Saint Giles and the Deer*, London, National Gallery. IRR composite, some underdrawn lines reinforced in black (© National Gallery).

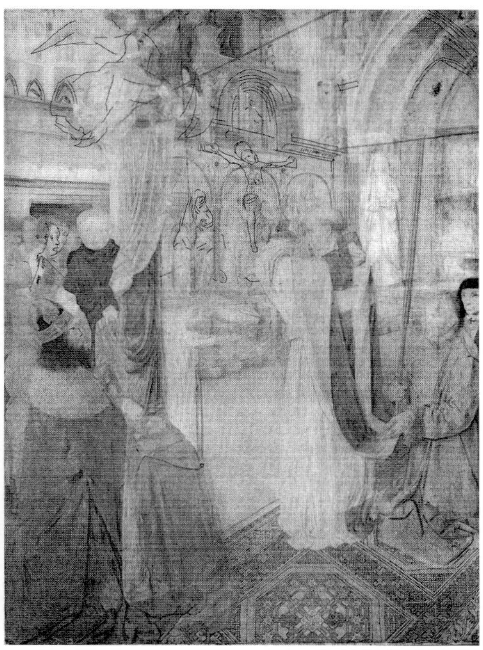

Fig. 4. Master of Saint Giles, *Mass of Saint Giles*, London, National Gallery. IRR composite, some underdrawn lines reinforced in black (© National Gallery).

his right leg was drawn on a diagonal — presumably so that it would not be concealed behind the dog. The underdrawing beneath the young man behind the bishop has yet to be interpreted. Many of the plants in the foreground, including the large mullein in the lower right corner, are neither underdrawn nor reserved. The central tree seems also to have been an afterthought, painted on top of the blue sky and apparently also on top of the landscape. It is difficult or impossible to reconstruct the exact sequence of these additions and changes. Reserves were left for three of the buildings in the town, while rocks are drawn under the strip of landscape on the extreme right.

In the *Mass of Saint Giles* (fig. 4), the alterations are perhaps still more dramatic. The cross of Saint Eligius, the platform supporting Saint Louis's shrine and the altar of Charles the Bold are not underdrawn but are painted over a drawing of an altarpiece with a raised semicircular top. It represented the crucified Christ between the Virgin and, presumably, Saint John the Evangelist. Saint Giles's head was drawn to our right of the painted contour. In the underdrawing, the flying angel in the top left corner was contained within the frame; his right wing was lower and his draperies floated further to our left. Charlemagne's head was drawn at less of an angle and his hands were joined in prayer. The subsidiary figures have been much altered. The head of the man on our left of the altar is drawn and reserved on a different scale and at a different angle. Between him and the left edge, two underdrawn heads have been suppressed while the fair-haired youth is not underdrawn or reserved. The man on the extreme right has also been added at a late stage, on top of the paint of the tomb.

The tomb of Dagobert was there in the underdrawing, though it was smaller and was seen from a different angle. Dagobert's tomb was, and remains, near the high altar of Saint Denis but the most obvious references to the basilica — the cross of Saint Eligius and the altar of Charles the Bald — were added after the underdrawing had been completed. Such changes may be interpreted in many different ways. It does not appear, however, to have been the patron who changed his mind about the ways and the settings in which the narratives were to be presented. The Master, it seems, was accustomed to working out his ideas on his panels and to making radical alterations as he changed his mind at every stage of his work.

In that respect, the two Saint Giles panels resemble the Washington *Baptism of Clovis* but differ totally from the fifteenth- and indeed the sixteenth-century Netherlandish pictures in the London collection. Jan van Eyck, working towards the compositional perfection of his *Portrait of Giovanni (?) Arnolfini and his Wife*, did not make such drastic alterations.[6] Rogier Van der Weyden, apparently persuaded by his patrons to introduce portraits into the *Exhumation of Saint Hubert*, did not greatly change the central narrative.[7] Even painters as apparently spontaneous as Bruegel worked things out in advance and rarely made major alterations.[8]

Careful and disciplined planning seems to have been characteristic of Netherlandish painting but certainly not of the Saint Giles pictures. The techniques, moreover, are

[6] L. CAMPBELL, *National Gallery Catalogues. The Fifteenth Century Netherlandish Schools*, London, 1998, p. 182-184.

[7] L. CAMPBELL, 1998, p. 411-414.

[8] D. BOMFORD, *Art in the Making, Underdrawings in Renaissance Paintings* (exhibition catalogue, National Gallery, 2002), London, 2002, p. 176-180.

demonstrably unlike those of, say, Van der Weyden or van der Goes. The idea that the Master of Saint Giles was a Netherlander who was trained in the Low Countries before emigrating to work in France should be questioned.[9] He was perhaps a foreigner, who was educated in a less rigorously disciplined tradition, went for a short period to the Low Countries and there acquired a veneer of fashionable Netherlandish ideas. He took them mainly from van der Goes and his followers. The Master of Saint Giles, however, never emulated Hugo's exacting discipline of method.

Lorne CAMPBELL
The National Gallery
Trafalgar Square
UK – London WC2N 5 DN
Tel: (0)20.7747.2481
Fax: (0)20.7747.2472
lorne.campbell@ng.london.org.uk

He is George Beaumont Senior Research Curator at the National Gallery in London. Among his publications are *Renaissance Portraits* (1990) and *National Gallery Catalogues. The Fifteenth Century Netherlandish Schools* (1998). He is currently working on the catalogue of sixteenth-century Netherlandish and French paintings and is President of the Scientific Committee organising the exhibition *Rogier Van der Weyden, Master of Passions*, which is to take place in the Museum at Leuven in 2009.

[9] See, for example, P.-G. GIRAULT, "De Bruges à Blois. Gauthier de Campes et le Maître de saint Gilles", in *Bulletin des Amis du Château et des Musées de Blois*, 33, décembre 2002, p. 15-23.

TWO PAINTINGS OR ONE?
THE BAPTISM OF CHRIST BY JAN VAN SCOREL*

Arie Wallert and Anne van Oosterwijk

Abstract: Two paintings attributed to the workshop of Jan van Scorel have been recently investigated. The largest painting depicts the *Baptism of Christ* (Rijksmuseum, Amsterdam, inv. nr. SK-A-1636). The subject matter of the other painting is not very clear. It presently is owned by Johnny van Haeften (London) under the title *People at the Bank of a River*. The two paintings look remarkably similar in several aspects: the same colours are used, the figures show great resemblance and — most of all — the landscapes continue from one painting to the other. Both paintings may therefore have been part of the same composition. Scientific examination has been carried out to gain insight into this problem (stereo-microscope, infra-red reflectography (IRR), X-radiographs). A small number of paint samples were analysed with the microscope and with microchemical, spectrometric and chromatographic methods, inorder to identify inorganic pigments, organic colorants and binding medium, as well as to understand the paint stratigraphy. Dendrochronology has provided useful evidence. The study of these paintings gives us insight into Van Scorel's workshop practices. One of the artist's most famous autographic works, the *Baptism of Christ* (Frans Hals Museum, Haarlem) is dated 1527-1529. The two workshop paintings resemble the Haarlem painting, not only in theme but also in the motives, and give us more information about the development of a popular theme in the workshop of Jan van Scorel.

In 2005, an unpublished oil painting on panel representing *People at the Bank of a River* by the workshop of Jan van Scorel showed up on the art market (fig. 1).[1] The absence of a well-defined subject matter and cutting marks on top and bottom of the panel strongly suggest that this painting is a fragment. For several reasons, *The Baptism of Christ* by the workshop of Jan van Scorel (Rijksmuseum Amsterdam, inv. nr. sk-a-1636) seems strongly related to it (fig. 2). There is evidence that both panels were once one and the same painting. What are the arguments?

Two paintings or one?

In the *People at the Bank of a River* twelve men gather near a river. Some of them direct their attention to something on the left, outside of the picture plane. Others are looking to the sky, again to something outside of the picture plane. Some kneel down with their hands folded, while others raise their hands in awe. At the upper- and lower side of the panel, cutting marks are visible, indicating that the painting originally may have been larger. The cutting marks and the focus of attention of the persons strongly suggest that a significant part of the painting is missing at the left. This missing part may well have been the Rijksmuseum *Baptism of Christ.*

[1] The painting is at the moment with Johnny van Haeften, London.

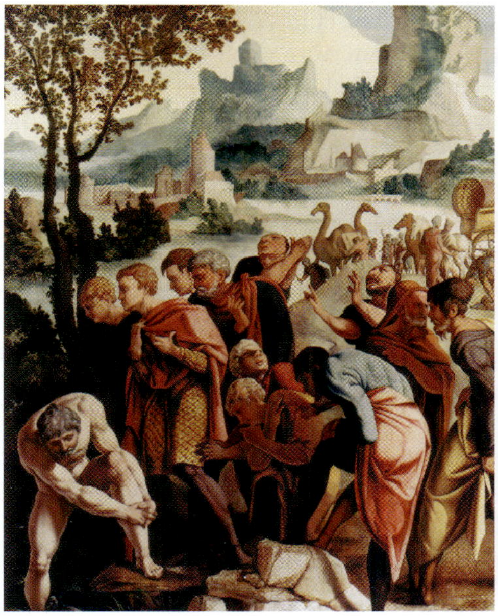

Fig. 1. Jan van Scorel, *People at the Bank of a River*, oil on panel, 84 × 66.7 cm, London, Collection Van Haeften.

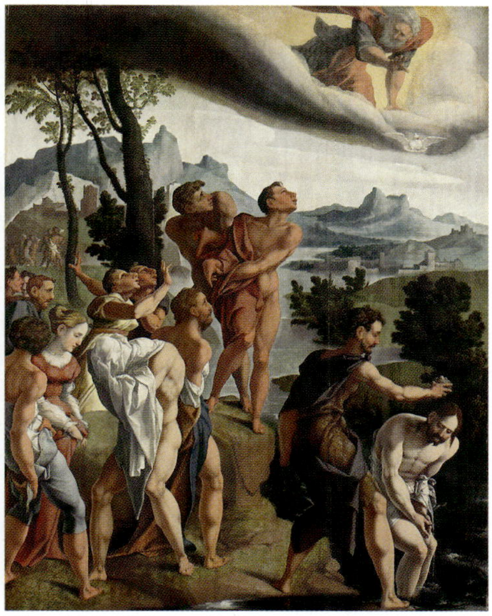

Fig. 2. Jan van Scorel, *Baptism of Christ*, oil on panel, 127.8 × 98.3 cm, Amsterdam, Rijksmuseum, inv. sk-a-1636.

In the story of the *Baptism of Christ* it is told how, near the river Jordan, Saint John the Baptist was preaching that baptism was the way to salvation. Many people, among them Jesus, came from Jerusalem, Judea and the region of the Jordan to be baptized by him. While Jesus was baptized a dove appeared in the sky and a voice was heard saying "Thou are my Son, my Beloved".[2] Many elements of this story can be found in *People at the Bank of a River*. The scene takes place near a river. In the background people are approaching to be baptized. The tilt-cart in the background suggests that some may have travelled long distances.[3]

Jan van Scorel painted a well-known *Baptism of Christ* in 1527-1530, now in the Frans Hals Museum in Haarlem, that served as model for many artists.[4] One of the

[2] Mark 1:1-11, Matthew 3:1-17, Luke 3: 1-22 and John 1:19-34. The quote is from Mark 1:11, taken from *The New English Bible, New Testament*, Oxford/Cambridge, 1961.

[3] The motive of a man stepping in the water with his hands around his knees is also used in a *Baptism of Christ* (formerly in the collection Streit), anonymous, circle of Maarten van Heemskerck. In this painting the figure is depicted mirrored and represents Jesus. The painting is only known to us by a picture in R. GROSSHANS, *Maerten van Heemskerck, die Gemälde*, Berlin, 1980, fig. 152.

[4] Most important literature about this painting with further references is: M.A. FARIES, *Jan van Scorel, his style and its historical context*, unpublished dissertation, Bryn Mawr 1972, p. 81-90; M.A. FARIES, "Jan van Scorel *Baptism of Christ*", in *Art before Iconoclasm*, J.P. Filedt Kok and W. Kloek (eds.), Amsterdam, 1986, II, p. 184-5, cat. 62; T. VAN BUEREN, *Tot Lof van Haarlem*, Hilversum, 1993, p. 339-342, cat. B4, including a photograph of the painting. The painting was often exactly and freely copied by artists working in or outside his workshop. The most important examples are preserved by the Museum of Art Philadelphia (John G. Johnson collection, nr. 414, published in *Paintings from Europe and the Americas in the Philadelphia Museum of Art. A concise catalogue*, Philadelphia, 1994, p. 96, as workshop Jan van Scorel),

works inspired by the Haarlem painting is the Rijksmuseum *Baptism of Christ*.[5] The general consensus is that the Rijksmuseum work is not autograph but made in Van Scorel's workshop. The composition is somewhat out of balance because Jesus and Saint John are placed in the front, quite close to the right. While Jesus bows his head to the right and keeps his folded hands before his legs, Saint John is standing with his back to the spectators. Spectators, in the background, are undressing themselves to be baptized. Some are looking at the appearance of God the Father on a cloud, accompanied by a dove. Others are looking at the baptism. When the two paintings are placed next to each other, it becomes clear how close they are related. Not only do they share a subject matter, they also show great resemblance in style, figures, landscape and use of colour. Most striking, is the fact that the Amsterdam painting contains exactly those elements that are missing in *People at the Bank of a River*. The people at left watching towards the sky must be looking at the baptism of Christ and the appearance of God the Father. This strongly suggests that both panels originally were part of the same picture. A tentative reconstruction was made. Such a reconstruction, of course, does not provide any evidence, but only suggests what the painting possibly may have looked like, if the two panels were really part of a whole (fig. 3). This still needs to be convincingly demonstrated.

Arguments pro

More similarities between the two panels came to light after scientific examination. Infrared reflectography revealed an extensive underdrawing in a dry black material in both paintings.[6] These underdrawings show strong similarities, although the underdrawing in the *People at the Bank of a River* may appear sketchier at places. Both underdrawings are characterized by long contour lines and extensive hatching to indicate the shadows (fig. 4). These hatchings are sometimes quickly applied with zigzags and occasionally cross the contour lines. The underdrawings on both panels are also characterized by summary lines indicating the profiles with a sharp angle for the noses, which are seen from below (fig. 5). The locks of hair of the spectators are indicated by quite idiosyncratic 'commas'.[7] The similarities in style and method of drawing indicate that the underdrawings were made by the same hand.

the Gemäldegalerie, Berlin (inv. II 153, published in *Gemäldegalerie Berlin. Gesamtverzeichnis*, Berlin, 1996, p. 112, as follower of Jan van Scorel), now in a private collection in the United States (pictured in A. DE HEVESY, "Quelques tableaux inconnus de Jean Scorel", in *Oud Holland*, 49 (1932), p. 142, fig. 4).

[5] Inv. Sk-a-1636. The painting caught little attention in the literature, see G.J. HOOGEWERFF, "Jan van Scorel", in *Peintres de la Renaissance Hollandaise*, Den Haag, 1923, p. 140, nr. 6; G.J. HOOGEWERFF, *De Noord-Nederlandse Schilderkunst*, IV, Den Haag, 1941/42, IV, p. 124, n. 1; *Jan van Scorel*, Utrecht, 1955, cat. 22, p. 41; P.J.J. VAN THIEL, *All the Paintings of the Rijksmuseum in Amsterdam*, Amsterdam, 1976, p. 512; M.A. FARIES, "Two additional panels from the Jan van Scorel's workshop", in *Le Dessin sous jacent*, Colloque IV, Louvain-la-Neuve, 1982, p. 121-130.

[6] Infrared reflectography (IRR) was done by Margreet Wolters of the RKD. The equipment used is a Hamamatsu C2400-07 with a N2606 IR vidicon, a Nikon Micro-Nikkor 1:2.8/55mm lens, a Heliopan RG850 (or RG 1000) filter, with an Lucius & Baer VM 1710 monitor (625 lines). Digital documentation with a Meteor RCB framegrabber, 768 × 574 pixels, colourvision toolkit (Visualbasic). The assemblies are made with PanaVue ImageAssembler and Adobe Photoshop 5.5 by Margreet Wolters and Anne van Oosterwijk.

[7] The term 'commas' is used for the pointed shapes that are used to indicate hair locks. The expression was introduced by Molly Faries in her article "Underdrawings in the workshop production of Jan van Scorel. A study with infrared reflectography", in *Nederlands Kunsthistorisch Jaarboek*, 1975, p. 177.

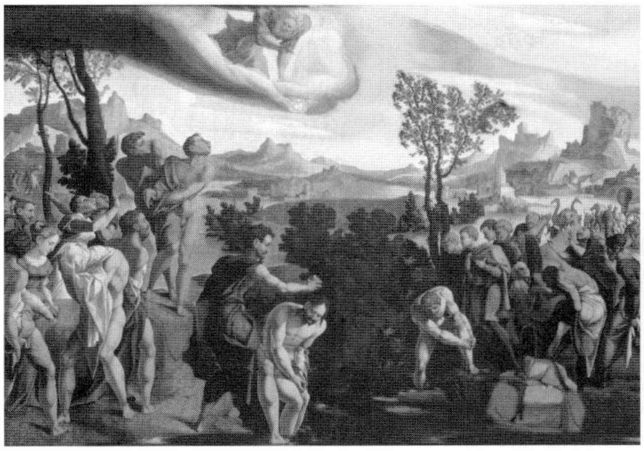

Fig. 3. Reconstruction. The *Baptism of Christ* and the *People at the Bank of a River* combined into one image.

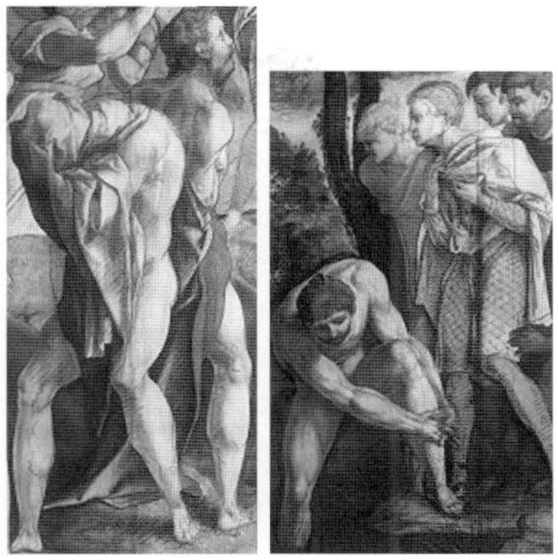

Fig. 4. IRR-details. a. Detail of The *Baptism of Christ*. b. Detail of *People at the Bank of a River*. IRR-assemblies by Margreet Wolters and Anne van Oosterwijk (© RKD, The Hague).

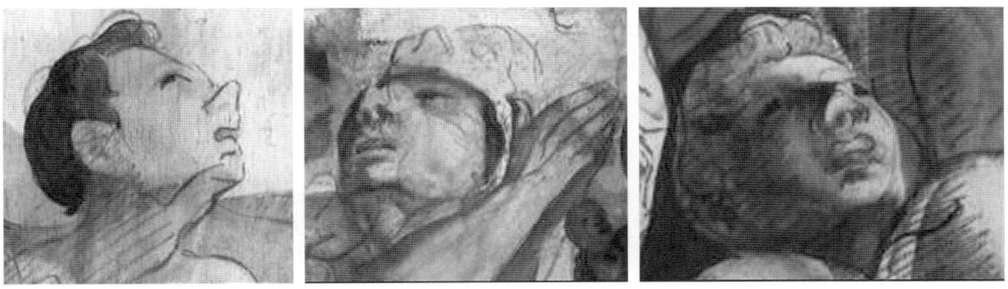

Fig. 5. IRR-details from sharp-angled noses. a. Detail of the *Baptism of Christ*. b and c. Details of *People at the Bank of a River*. IRR-assemblies by Margreet Wolters and Anne van Oosterwijk (© RKD, The Hague).

Faries discussed the *Baptism of Christ* in 1981.[8] At that time, she pointed to strong similarities between the underdrawing in the *Baptism* and the underdrawings securely attributed to Jan van Scorel. However, if the underdrawing of the *Baptism of Christ* should be attributed to Jan van Scorel, then the underdrawing for the *People at the Bank of a River*, must also be by him. In our opinion, the underdrawing on both panels can convincingly be attributed to Jan van Scorel. However, the rather poor quality of the paintings, makes it highly unlikely that Jan van Scorel painted them himself. Thus the paintings and the underdrawings would have been executed by a different hand. This seems quite exceptional in Van Scorels œuvre. According to Faries, only two times before has it been found that the underdrawing was done by Van Scorel while

[8] M. FARIES, 1982, p. 122-130.

the paint layers were applied by someone else.[9] As both panels appear to share this very same peculiar production procedure, the suggestion of them being part of one and the same painting becomes more convincing.

Not only stylistic similarity in the underdrawing ties the two paintings together. Also the manner in which the paint layers have been applied support a common origin. Examination of the red draperies on both panels show a similar build-up. The red lake pigment appears directly applied onto the calcium carbonate ground,[10] without any intermediate isolating priming layer. Such a construction seems quite uncommon in the production of Scorel's studio, where almost all paintings present a continuous intermediate layer between ground and paint. Faries et al. have convincingly argued that the three-fold function of such a layer (isolating the ground, adding luminosity, and providing 'tooth' for the underdrawing) was important to the efficacy of Scorel's painting technique.[11] The absence of this layer in two panels from Scorel's workshop can hardly be discounted as mere coincidence. Elemental analysis of the underdrawing material in both samples showed — not surprisingly — a similar composition.[12] The black chalk, a carbonaceous shale, is characterized by the presence of Si, C, Al, K, some Ti, and Fe and some Sr.[13]

The dark blue for the dress of the man on the foreground on the Amsterdam *Baptism of Christ* is applied in separate layers of mixtures of azurite and lead white, with an admixture of red lake in the lower layer.[14] This build-up is the same for the blue on the London *People at the Bank of a River*.[15] The average particle size of the azurite

[9] The two paintings are *Madonna with the Wild Roses* (Centraal Museum, Utrecht) and *Christ Blessing*. For the former see: *The Madonna's of Jan van Scorel 1495-1562: serial production of a cherished motive*, M.A. Faries and L. Helmus (eds.), Utrecht, 2001, cat. B and for the latter: M. FARIES, 1982, p. 123-5.

[10] Observation of microscopic cross sections were done with a Leitz Orthoplan polarizing microscope. Objective magnifications ranged from 100×, 200× to 500×.

[11] M. FARIES, C. STEINBÜCHEL and J.R.J. VAN ASPEREN DE BOER, "Maarten van Heemskerck and Jan van Scorel's Haarlem Workshop", in *Historical Painting Techniques, Materials, and Studio Practice*, preprints to the symposium University of Leiden 26-29 June 1995, A. Wallert, E. Hermens, and M. Peek eds., Los Angeles, 1995, p. 135-139.

[12] Scanning electron microscope (SEM) examinations with energy dispersive spectrometry of x-rays (EDX) were performed on a JOEL JXA-840A electronprobe microanalyzer, usually at 10nA., 25kV, with a 39 mm working distance. Samples were in low vacuum mode. EDX analyses were performed at various points throughout the cross section by measuring the emitted X-rays with a Noran Vantage EDS-system with Pioneer Norvar detector.

[13] A. WALLERT, J. DIK, G. DAVIES and I. JOOSTEN, "Black Chalk; Material for Painting, Drawing, and Underdrawing".

[14] Azurite was identified by polarized light microscopy (PLM) as pale blue, highly birefringent mineral grains with specific anomalous extinction, and refractive indices higher than 1.66. Polarized light microscopy (PLM) of pigment dispersions in transmitted light was done with the Leitz Orthoplan microscope. The mounting mediums for the microscope slides were Cargille melt mount (nD = 1.662), and Permount (nD = 1.539). Microchemical analyses (MCA) appeared quite helpful. The formation in a sample of the blue of a salmon red product in reaction to $K_4Fe(CN)_6$ reagent was taken as a positive identification of Cu.

[15] The admixture of red lake particles to azurite blue is not uncommon in his workshop. It was also found in the *Deposition of Christ* (Centraal Museum, Utrecht, inv. 6450), made in Scorel's studio around 1540, and in Scorel's *Madonna with the Roses*, of 1529. J.R.J. VAN ASPEREN DE BOER, "Natuurwetenschappelijk onderzoek van schilderijen uit de groep Jan van Scorel", in *Jan van Scorel in Utrecht*, cat. Centraal Museum, 1977, 51-55, p. 52.

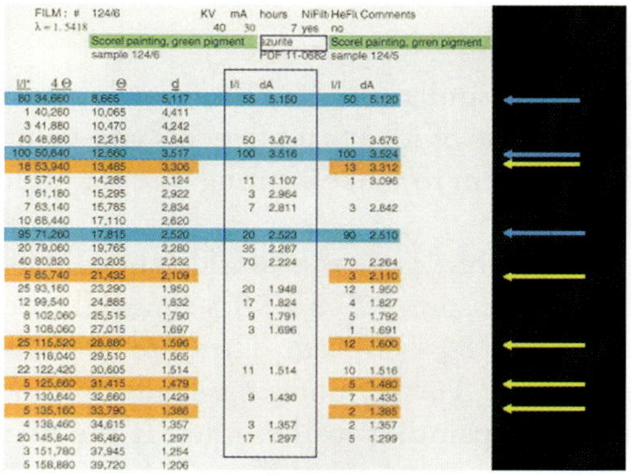

FILM: # 124/6 — λ = 1.5418 — KV 40, mA 30, hours 7, NiFilt yes, HeFl no, Comments

Scorel painting, green pigment, sample 124/6 | azurite, PDF 11-0682 | Scorel painting, grren pigment, sample 124/5

I/I*	4Θ	Θ	d	I/I	dA	I/I	dA
80	34,660	8,665	5,117	55	5.150	50	5.120
1	40,260	10,065	4,411				
3	41,880	10,470	4,242				
40	48,860	12,215	3,644	50	3.674	1	3.676
100	50,640	12,660	3,517	100	3.516	100	3.524
18	53,940	13,485	3,306			13	3.312
5	57,140	14,285	3,124	11	3.107	1	3.096
1	61,180	15,295	2,922	3	2.964		
7	63,140	15,785	2,834	7	2.811	3	2.842
10	68,440	17,110	2,620				
95	71,260	17,815	2,520	20	2.523	90	2.510
20	79,060	19,765	2,280	35	2.287		
40	80,820	20,205	2,232	70	2.224	70	2.264
5	85,740	21,435	2,109			3	2.110
25	93,160	23,290	1,950	20	1.948	12	1.950
12	99,540	24,885	1,832	17	1.824	4	1.827
8	102,060	25,515	1,790	9	1.791	5	1.792
3	108,060	27,015	1,697	3	1.696	1	1.691
25	115,520	28,880	1,596			12	1.600
7	118,040	29,510	1,565				
22	122,420	30,605	1,514	11	1.514	10	1.516
5	125,660	31,415	1,479			5	1.480
7	130,640	32,660	1,429	9	1.430	7	1.435
5	135,160	33,790	1,386			2	1.385
4	138,460	34,615	1,357	3	1.357	2	1.357
20	145,840	36,460	1,297	17	1.297	5	1.299
3	151,780	37,945	1,254				
5	158,880	39,720	1,206				

Fig. 6. X-ray diffraction (XRD) patterns of green paints, mixed of azurite and a yellow component.

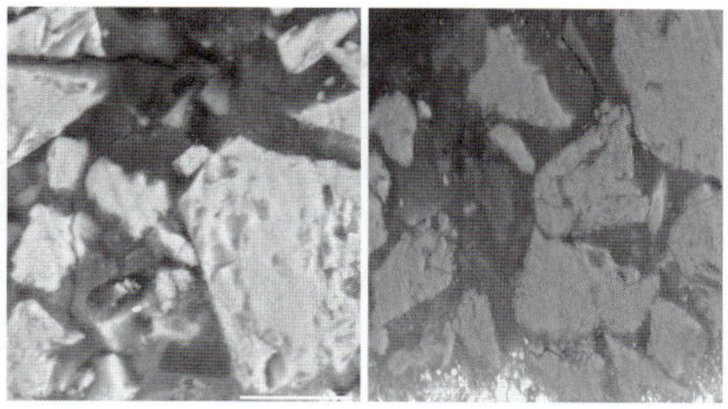

Fig. 7. Scanning electron (SEM) micrograph of azurite with pitted surface. Two different paint samples with azurite based green. Azurite has the same particle morphology in both samples. Chemistry of admixture the same in both samples. Scanning electron microscope. Back scattered electron image, aquired in low vacuum mode, scale bar = 10 μm.

seemed fairly similar on both panels. Again, no priming layer was found between ground and paint layers.

We also found striking similarities in two samples of green paint for the foliage: both were made of a mixture of azurite and some yellow component, rather than the expected verdigris.[16] The yellow components could not be identified.[17] The diffraction patterns in both samples for the yellow components and for the azurite (PDF 11-0682) appeared so remarkably similar that they must stem from the same paint mixture (fig. 6). In the scanning electron microscope analyses (SEM), the unidentified yellow particles appeared as mineral fragments with lower molecular weight than that of the associated azurite particles.[18] Elemental analyses showed the presence of Si, Al, Mg, K, and Ca. The chemistry of the admixture was the same for both samples. Examination with SEM of the azurite-based green showed that the azurite has the same peculiarly 'pitted' particle morphology in both samples (fig. 7).[19]

[16] The pigments were identified with PLM, MCA and x-ray diffraction (XRD). Debye-Scherrer powder diffraction patterns were obtained with 57.3 mm cameras with Gandolfi mounts, with the sample in cedar oil on the tip of a glass spindle. Exposures varied from 4-7 hours. CuKx radiation (x = 1.542Å) was used at 40kV, with tube current of 30 mA. All intensities were estimated visually.

[17] A yellow sample from the Amsterdam panel, subjected to XRD analyses showed a diffraction pattern that provided a perfect match with PDF file 24-589 for lead tin yellow Pb_2SNO_4. No such agreements could be found for the yellow components in the green mixtures.

[18] In the azurite-rich layer some small (ca. 10 μm) barite particles were found. The finding of barite ($BaSO_4$), being paragenetic with azurite in only a limited number of locations, may allow us to trace the geographical origin of Scorels pigments.

[19] This green mixture seems to be at variance with the pigment compositions commonly found in paintings from Scorel or his workshop. Most greens seem to be malachite-based compositions. Green verditer was found in the Lochorst triptych (Centraal Museum, Utrecht, inv. nr. 256) of 1526, while the greens on Scorel's 1540 altarpieces for the Abbey of Marchiennes (Douai, Musée de la Chartreuse) were identified as posnjakite $[Cu_4(SO_4)(OH)_6.(H_2O)]$, J.R.J. VAN ASPEREN DE BOER, *Natuurwetenschappelijk onderzoek*, 1977, 54.

Both panels under our investigation share that green anomaly.

Arguments contra

The iconographical correspondance, the stylistic similarities in underdrawing and painting, in procedures and working methods, the use of identical pigments and paint mixtures, the identical build-up of paint layers, all seem to make a strong case for both panels to belong to one and the same painting.

However, there remain unresolved problems. Notably, there is a considerable difference in scale. The Rijksmuseum's *Baptism of Christ* measures 127.8 × 98.3 cm, while the London *People at the Bank of a River* is measuring only 84 × 66.7 cm. There are no obvious signs of cutting in the Rijksmuseum painting.[20] Furthermore, the landscapes of both paintings do not cross over exactly from one painting to the other. If the two panels really once belonged together, then a small strip in the middle must be lost.

Although both panels are made up of three planks of oak, the average widths of the planks of the *Baptism of Christ* is 32.7 cm,[21] whereas the average width for the planks of the *People at the Bank of a River* is 22.2 cm.[22] While it is not uncommon for 16th century panel planks to vary in width, a panel maker would likely have alternated the narrower planks with the broader ones.

Also the age of the wood for both panels varies. Dendrochronological examination has shown that the wood for the *Baptism of Christ* was obtained from two trees with an earliest possible felling date in 1466. Counting with sapwood rings, shipping and processing time, it seems probable that this painting was created anytime from 1474 onwards. The tree that provided the wood for the *People at the Bank of a River*, was cut down in 1516 at the earliest. The creation of that painting could only have been possible from 1524 onwards.[23] Although both paintings were probably painted ca. 1540 — when Van Scorels workshop had his glory days — the differences in the dating of the wood make their use in one single product problematic at best.[24]

X-radiographs and inspection during the restoration of both paintings also confirmed that the two panels could not originally have been constructed as part of the same board.[25] The planks of the *People at the Bank of a River* are butt-joined and the seams are reinforced with linen strips. The panel for the *Baptism of Christ* in the Rijksmuseum, was assembled with dowels. It is clear that the two panels were not manufactured as a whole.

There is a marked difference in scale between the figures on the *People at the Bank of a River* and those on the *Baptism of Christ*. In the latter painting the people appear larger.[26] A possible explanation for this may be that the artist tried to suggest that the people in the other painting are further away. Their smaller size would result from

[20] There is, however, no conclusive evidence that the painting was not cut.

[21] From left to right the planks have widths of 31.8, 35.2 and 31.2 cm respectively.

[22] From left to right these planks measure 24.5, 19.1 and 23.1 cm.

[23] Dendrochronological examination by Peter Klein, Hamburg University.

[24] Molly FARIES, "Jan van Scorel as Painter/Entrepreneur", in *Le dessin sous jacent colloque XII. La Peinture dans les Pays-Bas au 16e siècle*, p. 13, n. 28. According to Faries it is possible that Van Scorel had some well-seasoned wood at his disposal.

[25] The painting of the Rijksmuseum was restored in 1977. *People at the Bank of a River* in 2005.

[26] The figures standing in front of the picture plane of the Rijksmuseum painting measure about 52-62 cm, while the figures of the art market painting measure 40-45 cm.

perspective. However, closer inspection of the two paintings shows that the two men in front of the *People at the Bank of a River* are placed at the same distance from Jesus and Saint John as the two men standing on the rock in the Rijksmuseum *Baptism of Christ*, disproportionally small to fit properly in a single 'combined' piece.[27] An accomplished artist (or artistic entrepreneur) like Jan van Scorel would not have made such a mistake.

Further arguments may be gathered from the X-radiographs. The *Baptism of Christ* was painted with great accuracy. The lines from the underdrawing are followed very neatly; different colours do not overlap and sometimes even leave some space between them. No significant *pentimenti* or notable deviations from the initial underdrawing could be found. The painter of the *People at the Bank of a River*, however, seems to have worked much less precisely, sometimes with double outlines for the figures. He quite often corrected himself. In that respect there is a difference in the execution of both paintings.

On the left edge of the *People at the Bank of a River* the lines, hatchings and shapes of the underdrawing are clearly interrupted as a result from the cutting down of the panel. On the right hand edge of its purported counterpart, the *Baptism of Christ*, no such interruptions could be observed. All forms have been entirely worked out. There is no reason to suspect that the panel was cut down.

The differences in the construction of the panels, in the date of the wood, in the scale of the figures and in the occurrence of pentimenti, make strong arguments *against* the hypothesis that the paintings were once united.

Conclusion

The similarities between the two paintings in underdrawing style, stratigraphy of the paint layers, combination of the pigments, paints and characteristic pigments are striking. It is only in the presence of the *Baptism of Christ*, that the *People at the Bank of a River* can tell its story and become meaningful. The *Baptism of Christ* provides *People at the Bank of a River* with the iconographical necessities. Yet, it can not be concluded with certainty that these two paintings once belonged together. It seems that the strongest arguments *contra* can be gathered from the panels, while the strongest arguments *pro* focus on the upper areas of both paintings: underdrawing, paints and pigments. This has led to the suggestion that a panel maker assembled some new wood with an older panel to make it into one large panel. This assembled construction would then have been delivered to the painters' studio. In that studio Jan van Scorel would then have made the preparatory underdrawing and have left the execution in paint to his assistants. Such a scenario would explain the differences between the panels and the similarities between the underdrawing and the paint layers. However, it does not explain the differences in execution and scale of the figures.

[27] An old copy of the *People at the Bank of a river* (formerly in the collection Städtischen Kunstsammlungen Haus zu Cavazzen, Lindau am Bodensee) shows a larger part at the bottom. It shows that the figures standing in the bend of the river are looking to the other side of the river. It is very likely that this must also have been the case in our *People at the Bank of a River.*

Other hypotheses have also been suggested, sometimes appealing, but none was entirely convincing. Arguments are diverse and opinions are divided.[28] It seems that, for the time being, the problem will remain unresolved.

Arie WALLERT
Rijksmuseum Amsterdam
Tel: +31.(0)20.674.7283
a.wallert@rijksmuseum.nl

MA, Ph.D., studied in Utrecht and Groningen. He worked at the Department of Mediaeval Studies of Groningen University, 1987-91, and as museum scientist at the J. Paul Getty Museum in Malibu, California, 1991-96. In 1996 he was appointed curator of paintings and advisor on the scientific examination of art works at the Rijksmuseum, Amsterdam. His interests include the identification of historic pigments and organic colorants, the study of historical painting techniques and technology.

Anne VAN OOSTERWIJK
Tel: +31.(0)20.674.7139
annevanoosterwijk@gmail.com

She studied Art history in Groningen (1999-2006) and wrote her master thesis about Jan Gossaert. She is now a student of the Museum curator master program at the Free University of Amsterdam, writing a master thesis about the present subject. As part of this master program she was in 2005/2006 an intern in the Rijksmuseum Amsterdam.

[28] There is a difference in opinion between the two authors of this paper. One author (AvO) finds the arguments *contra* the strongest, while the other (AW) favours the *pro* arguments.

UN *PORTRAIT DE FEMME* PEINT PAR THÉODORE CHASSÉRIAU (fig. 1)

Jacqueline Couvert

Résumé: Un *Portrait de femme* (coll. privée) signé et daté «Th^{re} Chassériau, 1845» a été étudié au Laboratoire des œuvres d'art de Louvain-la-Neuve. La peinture, de belle facture, est caractéristique de la sensibilité romantique de l'artiste.

Théodore Chassériau est né dans l'île de Saint-Domingue en 1819. Sa famille s'installe à Paris en 1822. À l'âge de 12 ans, Chassériau intègre l'atelier de J.D. Ingres. Ses véritables débuts datent de 1839 lors du Salon où deux de ses œuvres seront exposées. En 1840, l'artiste entreprend un voyage de sept mois en Italie. C'est à cette époque que Chassériau se détache de son maître Ingres. Un voyage en Algérie en 1846 permettra à l'artiste de s'orienter ensuite vers un romantisme teinté d'exotisme proche de la sensibilité de Delacroix. Théodore Chassériau meurt à Paris en 1856 à l'âge de 37 ans. En dehors des œuvres peintes, Chassériau a également produit de nombreux dessins. Les portraits réalisés par l'artiste — on en recense près d'une centaine — furent exécutés surtout dans les années 1840-1841 et 1846. Si l'on a souvent considéré Chassériau comme la synthèse d'Ingres et Delacroix, cet artiste a cependant développé un art particulier. Ses personnages féminins notamment, à la longue silhouette souple, au visage idéalisé et au regard mélancolique n'appartiennent qu'à lui.

Ce type de visage se retrouve notamment dans le *Portrait de femme* (coll. privée) examiné au Laboratoire. Réalisé dans des tons sobres, l'œuvre représente une jeune femme en buste sur un fond vert, vêtue d'une robe noire découvrant largement ses épaules. Le vêtement est orné de dentelles blanches et d'une broche dont les motifs (une mère et son enfant) sont réalisés dans des couleurs vives. Une longue cordelière dont le pendentif n'est pas visible pend à son cou. Un châle de couleur sombre enveloppe le corsage et cache les bras du personnage. De sensibilité romantique, la peinture est exécutée avec une touche nerveuse, précise et parfois épaisse. Le blanc presque pur modèle le front et les épaules. La dentelle du corsage est exécutée en relief indiquant un geste rapide de l'artiste à cet endroit. La radiographie rend compte du mouvement du pinceau traçant des stries dans la matière picturale encore fraîche. Le peintre n'a pas pris le temps de changer de pinceau, on remarque des traces de couleur foncée dans la couleur blanche. La même rapidité du mouvement créatif se remarque aussi pour la broche. Les 'anglaises' et le châle sont aussi traités avec un coup de pinceau assez vif. Par contre, le visage, empreint de beaucoup de douceur, est peint d'une manière plus paisible. Mais cette apparente sérénité est démentie par le jeu important des ombres et des lumières. Le *Portrait de femme* (coll. privée) s'inscrit bien dans l'œuvre de Chassériau à l'époque de son exécution (1845). La ressemblance avec le *Portrait du R.P. Dominique Lacordaire de l'ordre des prêcheurs* (Paris, Louvre) antérieur pourtant de 4 à 5 années est remarquable. Les deux visages présentent les mêmes caractéristiques (fig. 2a, 2b): grands yeux de même couleur au regard lointain animé de petites touches blanches horizontales, long nez

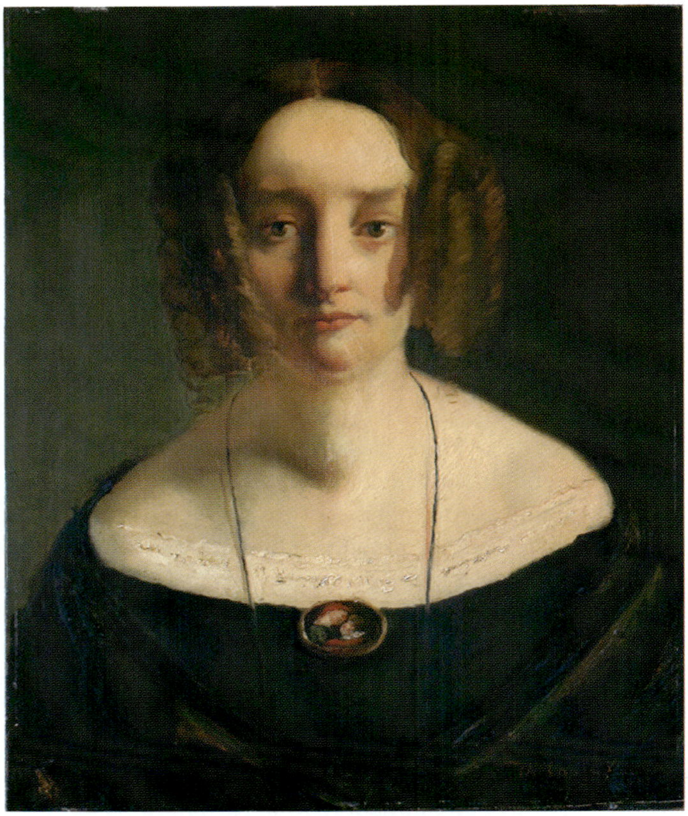

Fig. 1. Théodore Chassériau, *Portrait de femme* (coll. privée).

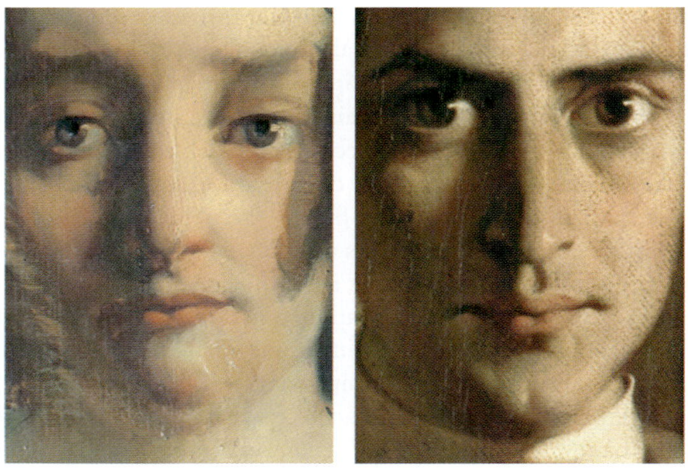

Fig. 2. a. Théodore Chassériau, *Portrait de femme* (coll. privée), détail. b. Théodore Chassériau, *Portrait du R.P. Dominique Lacordaire de l'ordre des frères prêcheurs* (Paris, Louvre), détail.

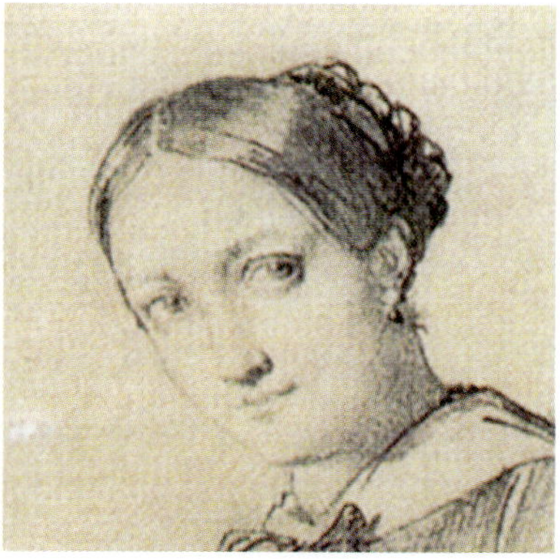

Fig. 3. Théodore Chassériau, *Portrait de Madame Jules Amiel* (coll. A. Bomberg), détail.

dont l'arête est marquée d'une ligne blanche verticale, bouche un peu pincée. L'ombre est également placée presque identiquement sur la gauche des deux visages. Parmi les nombreux dessins exécutés par Chassériau, un portrait d'une jeune femme peut se comparer avec *Le portrait de femme* (coll. privée). Malgré une coiffure différente, on peut discerner une ressemblance certaine entre les visages de Madame Amiel (fig. 3) et de la jeune femme représentée sur la peinture étudiée ici. Jules Amiel était probablement un condisciple de Chassériau dans l'atelier d'Ingres[1]. Les deux portraits sont d'ailleurs exécutés en 1845.

La peinture est exécutée sur toile marouflée sur un support de bois (contreplaqué) dont les dimensions (60,5 × 50,5 cm) sont légèrement supérieures à celles de la toile, coupée quelque peu irrégulièrement. Le marouflage a vraisemblablement été motivé par l'état de conservation du châssis original aujourd'hui disparu et/ou de la toile. On observe quelques boursouflures de la toile sans doute occasionnées par cette opération. La toile a probablement conservé ses dimensions originales malgré son découpage en vue du marouflage. L'emplacement de la signature et de la date, situées près du bord inférieur droit, confirme cette hypothèse.

La signature «Th^re^ Chassériau, 1845» (fig. 4a, 4b, 4c), prend la forme adoptée par Chassériau au moins à partir de 1937. En effet, à cette époque, le peintre traçait le premier 's' de son patronyme en majuscule[2]. On observe que les traits de la signature et la date sont parcourus de craquelures d'âge se poursuivant dans la couche de peinture sur laquelle ces inscriptions ont été tracées ce qui permet d'affirmer que la signature et la date sont contemporaines de l'exécution du tableau.

Une couche de préparation blanchâtre a été étendue sur la toile du *Portrait de femme* (coll. privée). Elle est principalement visible le long des bords. Quelques lacunes dans la couche picturale sont à mentionner. Elles apparaissent sous la forme de zones noires sur la radiographie (fig. 5). Elles sont localisées à gauche de la tête, dans la chevelure,

[1] S. GUEGAN, V. POLMAREDE et L.-A. PRAT avec une biochronologie par B. CHENIQUE et la participation de C. PELTRE et P.B. MILLER, *Chassériau. Un autre romantisme* [Cat. exp. Grand Palais, Paris 26/2-27/5/2002], Paris, 2002, p. 256.

[2] L.-A. PRAT, «Théodore Chassériau: œuvres réapparues», dans *Revue de l'art,* 125/ 3, 1999, p. 71-77.

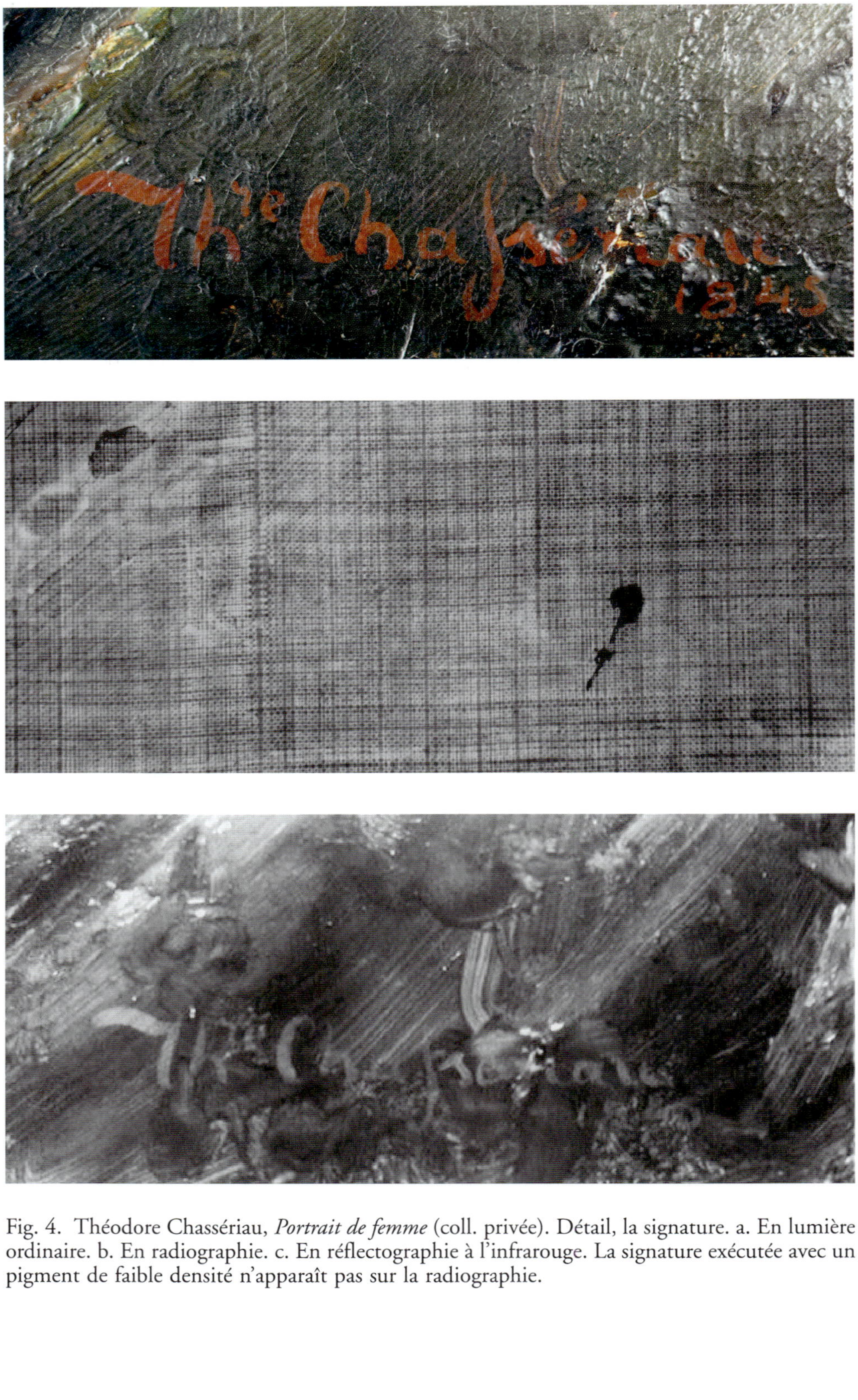

Fig. 4. Théodore Chassériau, *Portrait de femme* (coll. privée). Détail, la signature. a. En lumière ordinaire. b. En radiographie. c. En réflectographie à l'infrarouge. La signature exécutée avec un pigment de faible densité n'apparaît pas sur la radiographie.

Fig. 5. Théodore Chassériau. *Portrait de femme* (coll. privée). Radiographies.

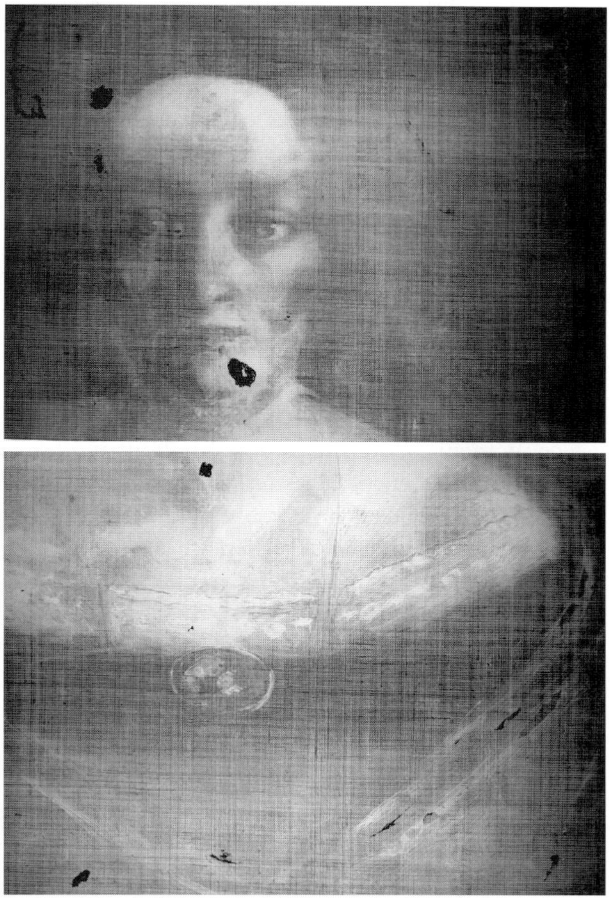

Fig. 6. Théodore Chassériau, *Portrait de femme* (coll. privée). Réflectographie à l'infrarouge, détail, le visage et les épaules (montage digitalisé de 30 réflectogrammes).

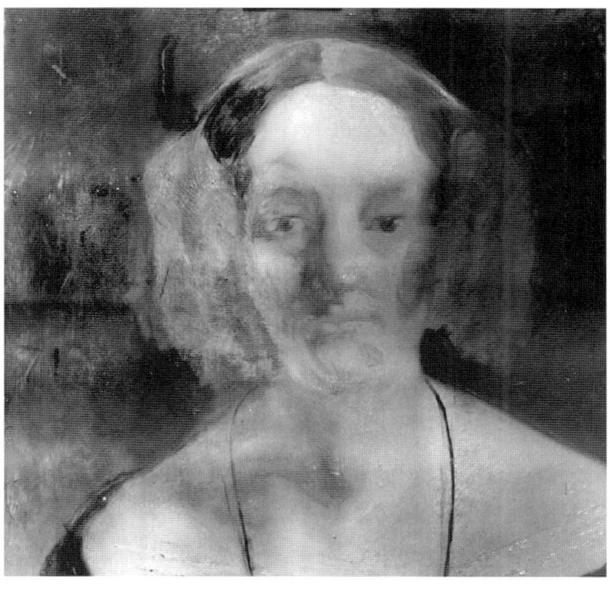

sur le menton et le cou, sur le châle et dans l'angle inférieur droit et plus précisément entre les lettres r et i de la signature disposée à cet endroit. En radiographie, les lettres et chiffres de la signature et de la date n'apparaissent pas (fig. 4b), ce qui s'explique par la faible densité des éléments constituant les pigments utilisés pour ces inscriptions. Les zones originales analysées par microfluorescence X révèlent la présence de blanc de plomb, de blanc de zinc, de terres, de vermillon, et d'azurite. Le blanc de zinc, commercialisé en 1834, permet de situer l'exécution de la peinture après cette date. Les autres pigments ne constituent pas un critère de datation puisqu'ils ont été employés de tout temps et sont encore utilisés aujourd'hui (ou l'étaient il y a peu). On remarque que le blanc de zinc ne se retrouve pas dans les zones de carnation ni dans les zones blanches du corsage. Ce pigment a été employé notamment en mélange avec la couleur brune de l'iris des yeux. Le fond original est composé d'azurite, de terres et de blanc de plomb. La radiographie visualise la présence de blanc de plomb. L'artiste utilise ce pigment de densité importante pour modeler le visage et les épaules et marquer ainsi les ombres et les lumières. La dentelle du corsage est aussi constituée de blanc de plomb. Les touches de matière picturale parfois très épaisses sont bien discernables. La fluorescence d'ultraviolet présente une surface relativement homogène. Un vernis d'apparence laiteuse masque les retouches sous-jacentes. La réflectographie à l'infrarouge ne met en évidence que peu de dessin sous-jacent (fig. 6). On observe des traits délimitant la position et la forme des narines, des yeux ainsi que l'iris de l'œil droit. La réflectographie à l'infrarouge permet de rendre compte des surpeints abusifs (recouvrant inutilement les matières picturales originales) comme par exemple la zone retouchée dans les cheveux et la déchirure la jouxtant à gauche. On remarque également un fond perturbé. Malgré les lacunes et les nombreux surpeints, l'état de conservation de l'œuvre demeure cependant bon en ce qui concerne les parties les plus importantes notamment le visage et la signature.

Le *Portrait de femme* (coll. privée), caractéristique du courant romantique, a été exécuté par Théodore Chassériau dans le style à la fois précis et nerveux. Il est représentatif de l'art de cet artiste dans les années 1840.

Jacqueline COUVERT

Voir Louise DECQ, Jacqueline COUVERT et Hélène VEROUGSTRAETE, *Une tête de vieillard sur parchemin (Jordaens?).*

THE MULTIPLE VIEWS OF ITALY BY DANISH GOLDEN AGE ARTISTS: REALITY OR MANIPULATION?

Jørgen Wadum, Kasper Monrad and Mikkel Scharff*

Abstract: During the Danish Golden Age in the early 19th century, Danish artists often travelled abroad, notably to Rome.[1] They drew sketches on site, and later used these drawings as models for their paintings. This paper discusses to what extent the paintings were faithful to the initial drawings. Infrared images of the paintings reveal that the artists made underdrawings. These differ in certain details from the initial drawing — and also from the final painted image — thus offering a unique insight into the artistic creative process.

This paper presents the initial results of an ongoing project examining how Christoffer Wilhelm Eckersberg (1783-1853) and his pupils created images of Italy. Using infrared imaging,[2] we examined several paintings from the large collection of the Danish Golden Age in the Statens Museum for Kunst in Copenhagen. Knowing that Eckersberg and his pupils made an underdrawing on the canvas before painting, we wanted to investigate whether the underdrawing was followed faithfully during the painting process or if the artist added, deleted or altered elements. We wanted to know whether the artist sought to create a realistic picture or instead, a more idealistic view influenced by aesthetic considerations. Eckersberg himself taught his students to observe nature very carefully. However, he did not want them to mimic nature in every detail. He wanted them to show motifs as they would be viewed under ideal circumstances.[3]

The existence of preparatory drawings for many of the paintings gave us an opportunity to examine how the scenery recorded 'on site' was later repeated or copied onto the canvas. In certain cases, we were able to assess the creative process in three phases: the drawing after nature, the drawing transferred on the canvas and finally the painted image.

* Acknowledgements: Henrik Bjerre, Head of the Jørn Rubow Centre and Jakob Skou-Hansen, Head Photographer, both Statens Museum for Kunst.

[1] Kasper MONRAD et al., *The Golden Age of Danish Painting*, Exhibition catalogue, Los Angeles County Museum of Art, Los Angeles and Metropolitan Museum of Art, New York (Hudson Hills Press), 1993.

[2] Infrared imaging was carried out by Jørgen Wadum and Mikkel Scharff with an Artist PRO® camera (Art Innovation, Hengelo, The Netherlands) mounted with a CCD progressive scan image sensor (1360 × 1036 pixels) and a Schneider Kreuznach Xenoplan 23 mm F/1.4 CCTV lens in near infrared 2 with a long wave pass filter 1000 nm. The images were captured with Artist software (release 1.2) and stitched with vips/nip2 (ver.7.10.20) (figs. 2, 6, 8, 12). Other images were captured by Mikkel Scharff and Henrik Bjerre with the FLIR IRR camera with a 256 × 256 pixel PtSi-array, lower threshold 1100 nm (Inframetrics InfraCam SWIR®) equipped with a bandpass filter 1500 to 1800 nm and as source of IR radiation two tungsten bulbs. The 8-bit images were transferred to a computer where the images were subsequently processed in the image processing software vips/nip2 (GNU) by means of mosaicing, further image analysis and enhancement.

[3] Kasper MONRAD in Catherine JOHNSTON et al., *Baltic Light: Early Open-Air Painting in Denmark and North Germany/Im Lichte Caspar David Friedrichs*, Exhib. cat., National Gallery of Canada, Ottawa, 1999, 4.

The drawings, most of which are kept in the Statens Museum for Kunst, have in some cases been squared before their transfer to canvas. How accurately the artist would copy the details from the drawing into the painting would depend not only on his artistic filtering of disturbing elements but also on how carefully he made the transfer.

Eckersberg, the 'Father' of Danish painting

Eckersberg was trained at the Royal Academy of Fine Arts in Copenhagen from 1803 until 1809. After studying another three years in Paris, he travelled to Rome where he continued his studies from 1813 until 1816. He introduced open-air painting into Danish art[4] and produced a large number of views after antique ruins and medieval churches in Rome. His paintings should be considered as finished paintings, not as sketches. After his return to Denmark, he became in 1818 professor at the Academy and he exerted a considerable influence on almost all Danish artists during the following 35 years. Notably, he made open-air painting a part of the tuition. Our question is to what extent can the influence of Eckersberg be identified in his pupils' paintings?

Porta Angelica and Part of the Vatican

One of the first paintings Eckersberg executed in Rome is a view of the *Porta Angelica and Part of the Vatican* (fig. 1) dated 1813.[5] In the underdrawing (fig. 2), the composition has undergone a few minor changes compared to the compositional drawing (fig. 3). Although many auxiliary perspective lines are found in the architectural elements, Eckersberg lowered the top of the Porta Angelica slightly in the underdrawing, preventing it from crossing the line of the wall in the distance. By doing this, Eckersberg strengthened the horizontal lines in the picture and thus adjusted reality in accordance with his compositional principles. Other minor details in the architecture were similarly altered.

It is notable that the vegetation on the architectural monuments appears much richer in the painted version compared to the drawing — not to mention the loose arabesques indicating vegetation in the underdrawing. By adding so much vegetation, the artist stressed the picturesque character of the view. The Porta Angelica was probably painted entirely in the studio.

View through the three Arches of the Colosseum

Eckersberg's *View through the three arches of the Colosseum* (fig. 4) was executed after 1814, when he had started painting outdoors in front of the subject. During the transfer of the drawing (fig. 5) to the ground of the painting, the composition was only

[4] Kasper MONRAD, "Eckersberg and Open-Air Painting", in Philip CONISBEE et al., *Christoffer Wilhelm Eckersberg*, Exhibition catalogue, National Gallery of Art, Washington, D.C. 2003, 13-25.

[5] C.W. ECKERSBERG, *Porta Angelica and Part of the Vatican*, (Rome, 1814), oil on canvas, 31,7 × 41,3 cm (Statens Museum for Kunst, inv. KMS7383); CONISBEE, 1993, cat. 20, drawing in private collection, 244 × 350 mm. Peter Michael HORNUNG and Kasper MONRAD, *C. W. Eckersberg – dansk malerkunsts fader*, Copenhagen, 2005, 134.

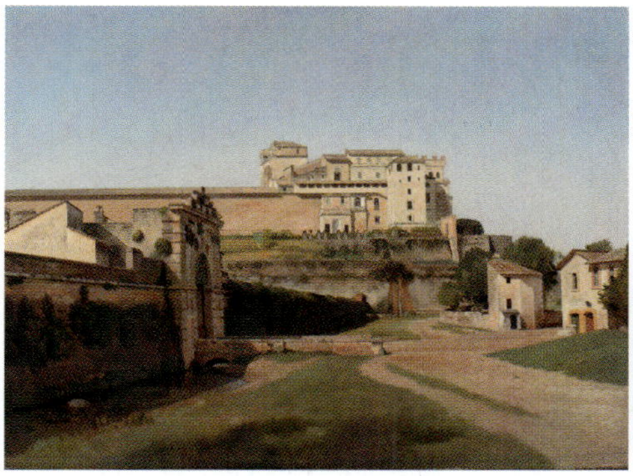

Fig. 1. C.W. Eckersberg, *Porta Angelica and Part of the Vatican*, 1813, oil on canvas, 31.7 × 41.3 cm, Statens Museum for Kunst, inv. KMS7383 (© Statens Museum for Kunst).

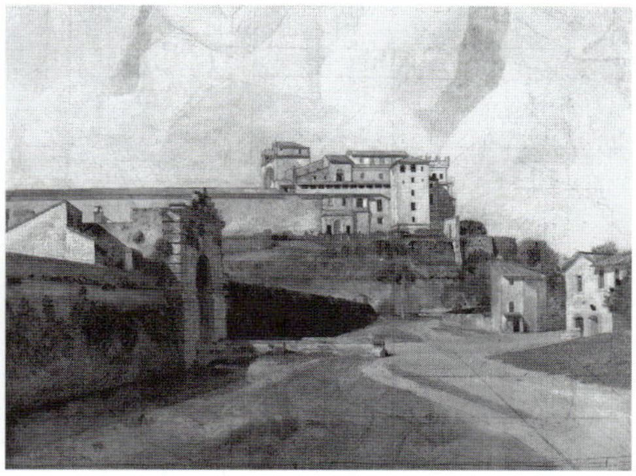

Fig. 2. C.W. Eckersberg, *Porta Angelica and Part of the Vatican*, 1813, oil on canvas, 31.7 × 41.3 cm, infrared image, underdrawing (© Statens Museum for Kunst).

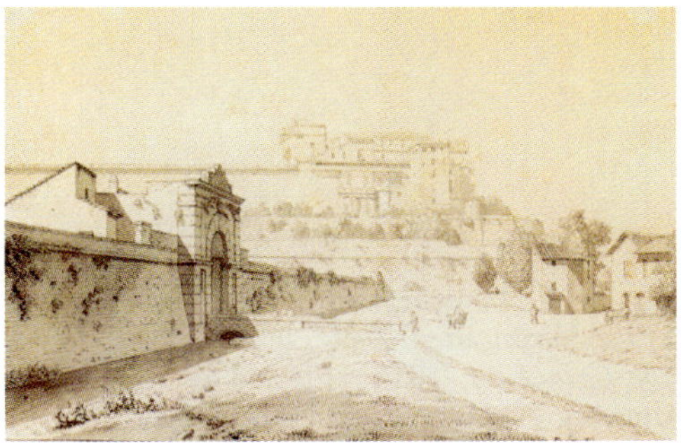

Fig. 3. C.W. Eckersberg, *Porta Angelica and Part of the Vatican*, 1813, pen and wash on paper, 24.4 × 35.0 cm, private collection.

slightly altered.[6] Superimposition of the drawing on the painting shows that the central opening is faithfully copied from the drawing, while the left and right semicircular openings facing the city of Rome are raised a little. This causes the vaults to appear thinner and lighter compared to the drawing. Furthermore, the right-hand opening has been shifted a few millimetres to the right.

In the far right column, Eckersberg painted the large square block at a slightly lower level than in the drawing and in the faithful underdrawing (fig. 6). Apart from this, there are only small architectural changes or alterations in the distant cityscape and its vegetation.

In his preparatory drawing, Eckersberg used a ruler and compass to construct the convincing spatial foreshortening of the foreground architecture. These were set down in

[6] C.W. Eckersberg, A View through Three of the North-Western Arches of the Third Storey of the Colosseum. A Thunderstorm is Brewing over the City (Rome, 1815), oil on canvas, 32,0 × 49,5 cm (Statens Museum for Kunst, inv. KMS3123); CONISBEE, 1993, cat. 25, drawing in pencil on paper, 337 × 529 mm (inv. KS2001-1) — measurement of the drawing itself: 318 × 490 mm. HORNUNG and MONRAD, 2005, 164.

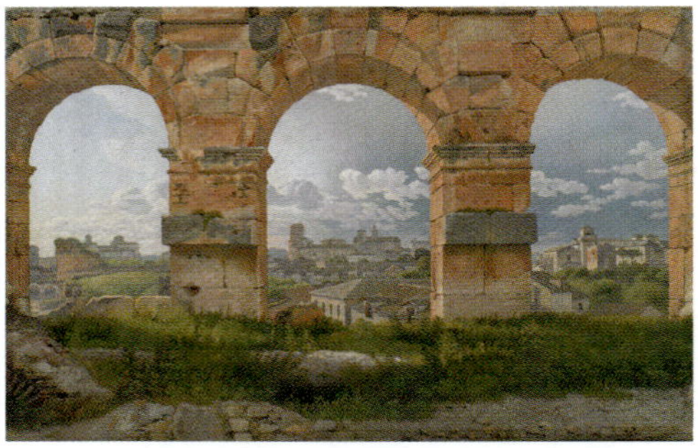

Fig. 4. C.W. Eckersberg, *View through Three of the Northwestern Arches of the Third Storey of the Colosseum. A Thunderstorm is Brewing over the City*, 1815, oil on canvas, 32.0 × 49.5 cm, Statens Museum for Kunst, inv. KMS3123 (© Statens Museum for Kunst).

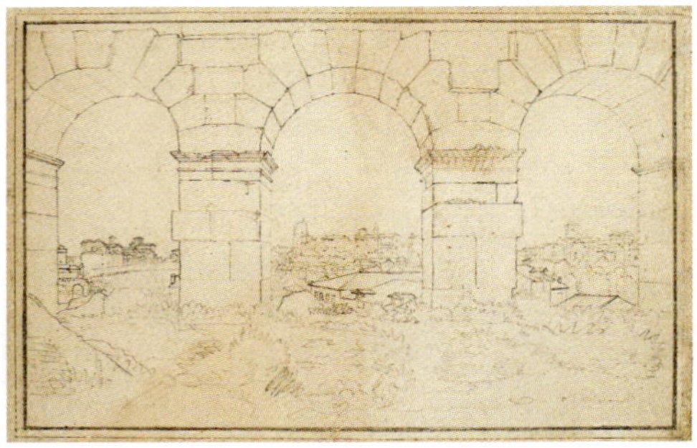

Fig. 5. C.W. Eckersberg, *View through Three of the Northwestern Arches of the Third Storey of the Colosseum*, 1815, pencil on paper, 31.08 × 49.0 cm (size of drawing), 33.7 × 52.9 cm (size of the paper), Statens Museum for Kunst, inv. KKS2001-1 (© Statens Museum for Kunst).

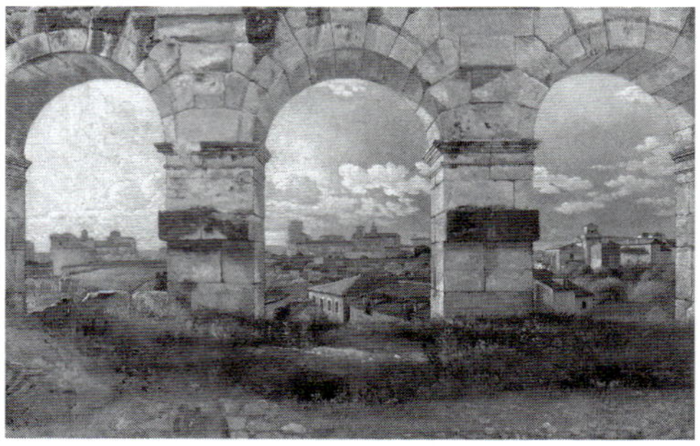

Fig. 6. C.W. Eckersberg, *View through Three of the Northwestern Arches of the Third Storey of the Colosseum*, 1815, infrared image, underdrawing (© Statens Museum for Kunst).

a thin, probably hard pencil. Over these thin lines, the drawing was firmly reinforced using a darker, softer pencil whose lines remain clearly visible. This drawing was finally transferred to the canvas for the painting of the *Colosseum*. It is, however, in the painted version of the architecture, stones and bricks, that Eckersberg deviated from the preparatory drawing: each of the vaulted openings have been given an extra architectural element, namely a carved ridge along the edge of the vaults.

Superimposition of the drawing and the painting reveals that they are of equal size, with the exception of a few millimetres in the vertical direction. However, the thin construction lines in the drawing are not precisely copied in the underdrawing. While preparing the underdrawing, Eckersberg did not merely trace or copy the preliminary drawing, but made a completely new construction on the painting's ground layer using rulers and a compass. This is revealed by images from the two infrared imaging systems used in this study.

The documented underdrawings in Eckersberg's two paintings mentioned above were superimposed in the VIPS/nip2 programme in order to understand how the scale of the images, the architectural details and the construction lines were copied. While there are few changes in the Colosseum painting, some distortion took place in the

Porta Angelica. In the latter, certain details were intentionally distorted in order to fit the image to the size of the canvas, which had another height to width ratio than the paper used for the sketch. Also, minor changes were observed in the placement of architectural elements and construction lines after superimposition of the images of the initial drawing and the painting's underdrawing revealed in the infrared image.

Successors of Eckersberg

In the work of Eckersberg's pupils we sometimes see an equally careful copying from preliminary drawing to underdrawing. However, contrary to the master's work, we also notice a certain loosening up of the strict rules of perspective and a lack of faithfulness to the recorded image. Unlike their teacher, the pupils often executed an oil sketch as an intermediate stage between the drawing and the final painting. These oil sketches were always painted on paper, most likely in front of the subject.

These characteristics are found in drawings and paintings by Christen Købke (1810-48). During his stay in Italy (1838-40), he travelled with fellow painter Constantin Hansen to Naples, Pompeii and Capri where he painted in the open air.

Castel dell'Ovo in Naples

In Købke's drawing (fig. 7) of *Castel dell'Ovo in Naples* (1839/40), the detailed execution reveals a thorough observation of the rock-like, almost circular fortification.[7] Careful vertical hatching indicates the shadows on the left parts of the scene. There are numerous rowing and sailing boats in the gulf, and in the far background the mountains show their rugged contours above the low horizon.

The underdrawing (fig. 8) in the oil sketch is of a much more sketchy nature. The outlines are steeper or located slightly more towards the left compared to the drawing. The left outline of the main tower was moved to the left in the underdrawing, but shifted back during painting in accordance with the original preparatory drawing. There is meticulous hatching in the shaded areas. Across the bulk of the fortification we observe only a few bluntly applied diagonal underdrawn lines. In the drawing, the stones in the water along the pier in the foreground give individual forms to the boulders whereas in the underdrawing they are hardly indicated at all. However, the lack of detail and the use of simplified notations in the underdrawing point to the fact that the artist only needed raw indications for his oil sketch (fig. 9) as he could get inspiration from the preparatory drawing during the painting process. There is no underdrawing for the mountains in the far background to the right of the fortress, the artist using the preliminary drawing for guidance. The most prominent change after the preliminary drawing is that the bridge in the far left corner of the drawing has been omitted in the oil sketch, both in the underdrawing and in the paint layer. The small bridge did not disappear just as a matter of simple cropping prior to transfer of the image to

[7] Christen KØBKE, *Castel dell'Ovo in Naples* (1839/40), oil on paper on canvas, 25,0 × 36,0 cm (Statens Museum for Kunst, inv. KMS3467); Hans Edvard NØRREGÅRD-NIELSEN and Kasper MONRAD eds., *Christen Købke 1810-1848*, Exhibition catalogue, Statens Museum for Kunst, Copenhagen, 1996, cat. 139. Drawing in pencil and framed in pencil, 223 × 355 mm (Statens Museum for Kunst, inv. KKSgb2908).

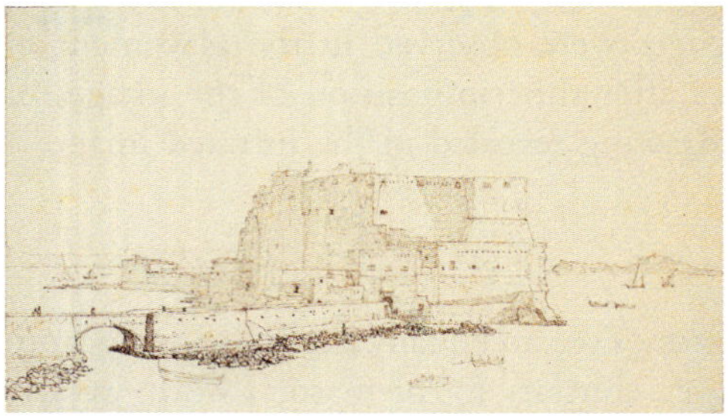

Fig. 7. Christen Købke, *Castel dell'Ovo in Naples*, 1839-40, pencil on paper, 22.3 × 35.5 cm, Statens Museum for Kunst, inv. KKSgb2908 (© Statens Museum for Kunst).

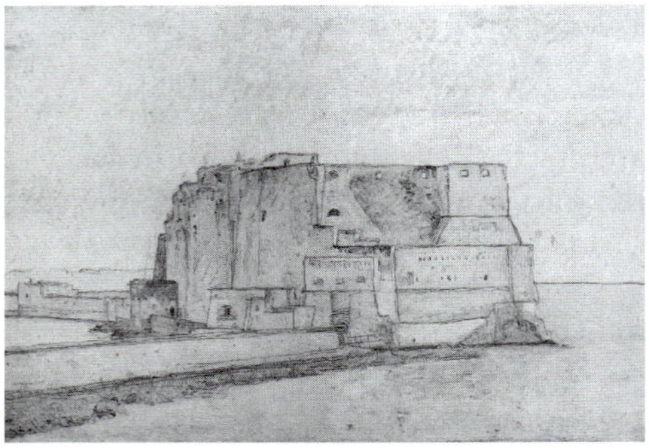

Fig. 8. Christen Købke, *Castel dell'Ovo in Naples*, 1839-40, infrared image. A fairly sketchy underdrawing (© Statens Museum for Kunst).

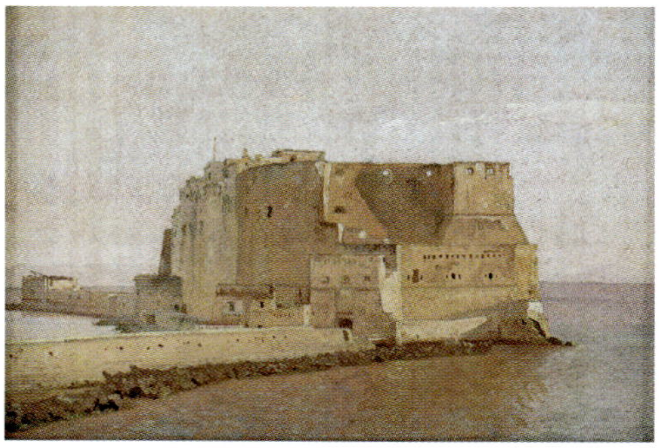

Fig. 9. Christen Købke, *Castel dell'Ovo in Naples*, 1839-40, oil on paper, adhered to canvas, 25.0 × 36.0 cm, Statens Museum for Kunst, inv. 3467 (© Statens Museum for Kunst).

the shorter dimensions of the oil sketch paper. Købke made a deliberate choice in deleting the bridge, and by doing so, strengthened the perspective line created by the pier. None of the boats were transferred to the underdrawing. Also, in the painting, all human activity on the ramparts and bridge is lacking. The purpose of the oil sketch was to capture light and atmosphere on the spot. Købke probably felt no need to render detail. Had he chosen to paint a larger version of the motif, he would most likely have included small details, as he did in other cases.

Købke precisely recorded the view of *Castel dell'Ovo in Naples* in his original drawing. He adjusted the subject in the painting by omitting significant architectural elements. He also reduced the level of precision during the transfer of the image from the drawing to the underdrawing/painting. Eckersberg would have drawn auxiliary lines and used a compass and ruler in order to reconstruct on canvas what he had already carefully constructed and recorded on paper. His pupils became less restricted in style, indicating a change in fashion towards the middle of the 19th century. Evidence of this trend can be seen by comparing Købke's careful drawing with his less restrained and sketchy underdrawing.

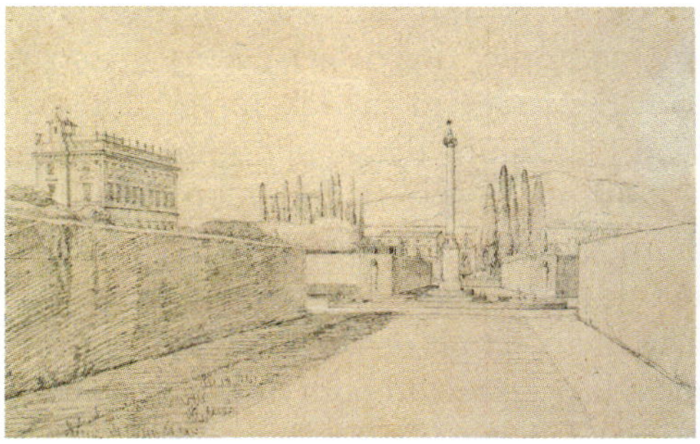

Fig. 10. Constantin Hansen, *In the Gardens of the Villa Albani*, 1841, pencil and white chalk on paper, 22.0 × 34.3 cm, Statens Museum for Kunst, inv. KKS10540 (© Statens Museum for Kunst).

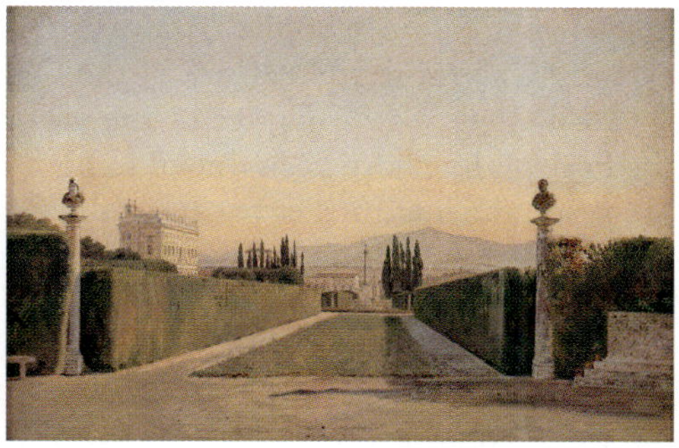

Fig. 11. Constantin Hansen, *In the Gardens of the Villa Albani*, 1841, oil on canvas, 34.0 × 50.0 cm, Statens Museum for Kunst, inv. KMS3843a (© Statens Museum for Kunst).

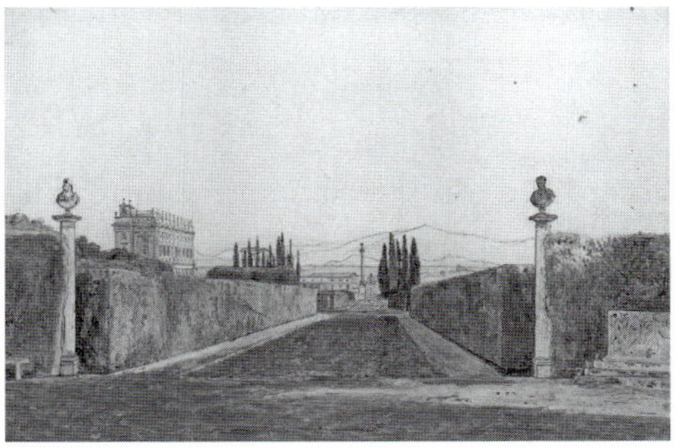

Fig. 12. Constantin Hansen, *In the Gardens of the Villa Albani*, 1841, infrared image, underdrawing (© Statens Museum for Kunst).

The Gardens of the Villa Albani

One of Købke's fellow students was Constantin Hansen (1804-80). He went to Italy in 1835 and stayed until 1844.

Hansen executed several drawings and paintings of *The Gardens of the Villa Albani* in 1841. Here we concentrate on the first drawing (fig. 10) and the second painting (fig. 11).[8] The most obvious difference between drawing and painting is the addition of tall, slender columns with busts to the painting in order to 'frame' the image. When the drawing is compared to the painting's underdrawing (fig. 12), the various elements

[8] Constantin HANSEN, *The Gardens of the Villa Albani, Rome* (1841), oil on canvas, 34,0 × 50,0 cm (Statens Museum for Kunst, inv. KMS3843a); Kasper MONRAD ed., *Turner and Romantic nature*, Exhibition catalogue, Statens Museum for Kunst, Copenhagen, 2004, cat. 58. Drawing in pencil and white chalk, 220 × 343 mm (Statens Museum for Kunst, inv. KKS10540). A more elaborate study by the same authors of this painting, the two drawings as well as an oil sketch published in "Hidden Drawings from the Danish Golden Age. Drawing and underdrawing in Danish Golden Age views from Italy", in *SMK Art Journal 2006*, Copenhagen, 2007, p. 111-119.

appear meticulously copied: the mansion in the underdrawing shows vertical lines extending above the actual size of the building, and other drawn perspective elements are easily detected; a quarter of the way down, the hedge to the left foreground and the horizon line have been drawn in both drawings, forming a base in both renderings. However, moving further backwards into the scene we observe that Hansen has accentuated depth by letting the hedge run past the small house surrounded by cypresses on the hill in the middle ground diminishing not only the herm of the second hedge but also the hedge itself in the underdrawing. The architecture of the buildings and the church or arch in the centre has, however, almost kept its size in both drawings whereas the ionic column in the distant centre is dramatically smaller in the underdrawing. In the preparatory drawing, the column towers over the park and is much taller than the group of cypresses to its right.

Even though Hansen deliberately zoomed out in the final version of *The Gardens of the Villa Albani*, several elements were changed in the interrelationship of the architectural elements as recorded in the preparatory sketch. By adding columns with busts in the foreground of the painting, the monumentality of the *veduta* has been strongly accentuated.

Conclusion

This preliminary study of drawings, underdrawings, oil sketches and paintings by Eckersberg and two of his pupils reveals a number of important aspects of artistic practice in Danish painting during the period 1800-1850. Preparatory drawings made from life were often carefully constructed and detailed. Architecture, however, became the most important part of the exercise; vegetation, random rocks or stones and human figures were treated as secondary elements. Drawings from life would in many cases be squared for transfer to a painting of a slightly larger scale. During the transfer process some elements could be changed, but the general outline was in most cases precisely retained. However, architectural elements were deliberately changed in some cases in order to enhance forms and harmony in the final image. From evidence revealed in paintings examined so far, we can state that, although at first sight the representations of Rome in Danish 19th century painting appear realistic, they were in many cases manipulated by the artist.

Jørgen WADUM
Statens Museum for Kunst
Sølvgade 48-50
DK – 1307 Copenhague
j.wadum@smk.dk

Paintings conservator and art historian, currently Keeper of Conservation at Statens Museum for Kunst, Copenhagen. Since 1984, specialised in the painting techniques of 16th and 17th century Dutch and Flemish artists. He has published and lectured internationally on a variety of subjects related to this and other issues of importance for the understanding and safeguarding of our cultural heritage; he was chair of the ICOM Committee for Conservation (ICOM-CC) Directory Board from 2002 to 2005.

Kasper MONRAD
Statens Museum for Kunst, Copenhagen

Chief Curator at the Statens Museum for Kunst, Copenhagen. He has curated (or co-curated) a number of exhibitions of Danish painting from the early 19th century, notably *Danish Painting: The Golden Age*, London 1984, *Caspar David Friedrich und Dänemark*, Copenhagen 1991, *The Golden Age of Danish Painting*, Los Angeles and New York 1993-94, *Christen Købke*, Copenhagen 1996, *Baltic Light. Early Open-Air Painting in Denmark and North Germany*, Ottawa, Hamburg and Copenhagen, 1999-2000, and *Turner and Romantic Nature*, Copenhagen 2004. He was a member of the Scientific Committee of the exhibition *Paysages d'Italie*, Paris and Mantova 2001.

Mikkel SCHARFF
The Royal Danish Academy of Fine Arts, School of Conservation, Copenhagen
ms@kons.dk

He is a paintings conservator and lecturer and is Head of the Departments of Paintings and Monumental Art at the Royal Danish Academy of Fine Arts, School of Conservation. He is specialized in the structural treatment of canvas paintings, analytical photography, technical art history and history of conservation. He has conducted research and written publications in these fields. He was previously coordinator for the ICOM Committee for Conservation's (ICOM-CC) Working Group on Painting Conservation and treasurer on the ICOM-CC Directory Board from 1996-2002.

THE UNDERDRAWING AND PAINTING TECHNIQUES OF A PORTRAIT OF *HUGO ERFURTH WITH DOG* (1926) BY OTTO DIX (1891-1969)

Susana Pérez Pérez, Andrés Sánchez Ledesma and Ubaldo Sedano Espín

Abstract: Otto Dix was one of the leading artists of the *Neue Sachlichkeit* (New Objectivity), the classicist movement which brought together a group of German artists around 1925 with the aim of breaking away from Expressionism through a new objective figuration. Dix gradually recovered from the horrors of the war and, impelled by the pessimistic belief that art no longer could change society, modified the critical tone and revolutionary beliefs of his Dadaist period to the taste of the new bourgeoisie he had previously mordantly criticized. The renewed emphasis on the human image brought with it the revival of portraiture, the genre in which Dix produced his most significant work during these years. Furthermore, as Peter Vergo explained (1992), Dix's return to portraiture was accompanied by the intention to revive the technique of the great German Renaissance painters (Dürer, Cranach and Baldung Grien), described in Max Doerner's book *The Materials of Painting* published in 1921. For this reason Grosz wittily dubbed Dix "Hans Baldung Dix".[1]

Dix has produced carefully executed paintings, with complex techniques inspired by the techniques of the German Renaissance Masters described by Max Doerner in 1921.

In his portrait of the photographer Hugo Erfurth[2] (fig. 1) (1926, tempera and oil on panel, 80 × 100 cm)[3], the artist represents the sitter in three quarter profile turned to the right with his German shepherd rather than with one of the instruments of his trade, thus indicating the high status he had attained. The left arm of the sitter is resting over a wooden armchair, in a hieratical yet tense posture. The exterior, Dix said, is the expression of the interior. That is why the artist sometimes exaggerates forms to intensify the personality of his model. The background is treated in a flat, neutral manner, inorder to avoid distracting the spectator. Dix brings the figure closer to the viewer, whilst at the same time he makes the model appear distant through the seriousness of his attitude and expression. The colour palette is limited and cold, the colours are flat with limited variations in tonality. The texture of each object is elaborate, achieving a physical feeling of realism. The smooth finish of the flesh areas contrasts with the printed drapery which has a slight relief. The flesh tones are shaded in a subtle manner and the eyebrows and hair are meticulously executed. The round and slightly raised glasses frame Erfurth's fixed expression.

The work is signed and dated '19 D 26' in the upper left corner of the painting. The inscription looks more like an old copyright mark than a signature.

[1] P. Alarcó, *Masterworks Museo Thyssen-Bornemisza*, Madrid, 2000, p. 182.

[2] A. Strobl, "Otto Dix und Hugo Erfurth. Der Maler im Zeitalter de Photographie", in *Münchner Jahrbuch der bildenden Kunst*, Munich, 1993.

[3] B. Miller, "Otto Dix and his oil-tempera technique", in *The Bulletin of the Cleveland Museum of Art*, oct. 1987, Vol. 74, n°8, p. 332-351.

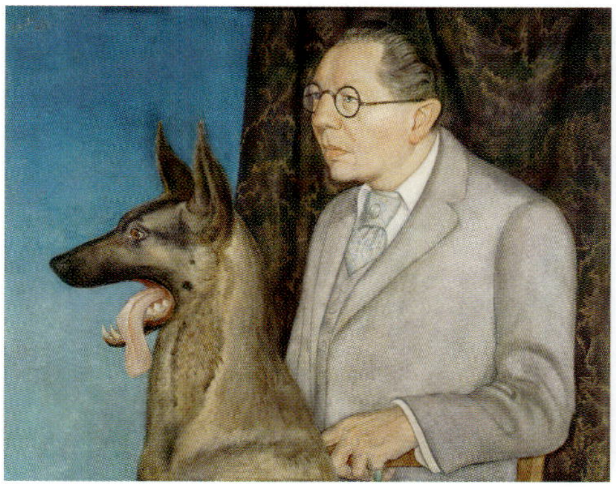

Fig. 1. Otto Dix, *Hugo Erfurth with Dog*, 1926, tempera and oil on panel, 80 × 100 cm (with frame: 97 × 117.5 cm), Madrid, Museum Thyssen-Bornemisza (© Hélène Desplechin).

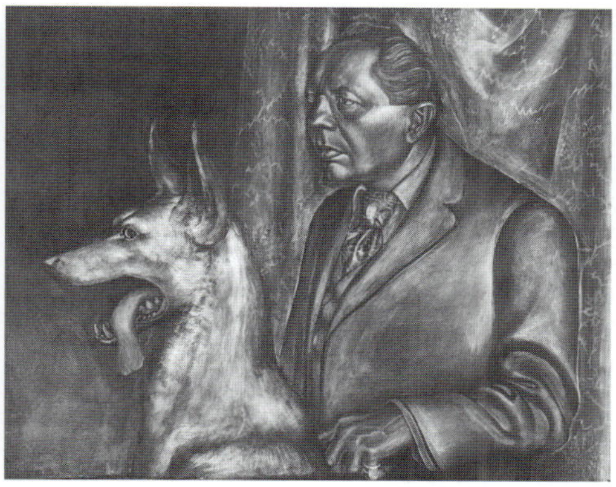

Fig. 2. Otto Dix, *Hugo Erfurth with Dog*, X-ray (© Rayxart).

On the reverse of the panel, an inscription with Dix' signature reads: *Mischtechnik/ Letzter Schicht Ölf. Leinöl/ Darf von Sept 27. Nicht/Verschiebt werden* DIX (Mixed technique/ Last layer oil. Linseed oil/ It should not be moved/ before Sept. 27 DI). In the upper right corner of this inscription we read: *LUKAS MATT/ MIT WACHS/50/50* (Lukas (commercial brand) matt/ mixed with wax/50/50). In both upper corners, there are pin holes, probably used to hold the preparatory drawing during transfer or possibly to protect the work during transport whilst the grisaille was still fresh.

There are two artists' boards with 'Erfurth and his dog', probably preparatory drawings, one in the Städtische Galerie Albstadt and the other in the Kupferstichkabinett in Berlin.[4] The size of the painting is 80 × 100 cm, the board in Albstadt 83.5 × 100 cm and the one in Berlin 84 × 100 cm. The three are almost identical in size.

The X-ray image shows that Dix executed a grisaille with lead white in which he establishes the lights and shadows prior to applying the colour layers. It also shows that Erfurth's glasses were reserved in the face. Lead white is suggested by light strokes in the X-ray in the figures of the photographer and the dog (fig. 2).

The underdrawing (figs. 3, 4) (not a common feature in a 20th century painting) is perceptible to the eye in some areas and makes Dix'preparatory work better understandable.[5]

The infrared reflectograms[6] reveal an elaborate preparatory drawing with few lines perfectly drawn with ink and brush. The profile of the face is elegantly drawn. The

[4] A. DÜCKERS, *Das Berliner Kupferstichkabinett. Ein Handbuch zur Sammlung* (Staatliche Museen zu Berlin Preussicher Kulturbesitz). Berlin, 1994, p. 463.

[5] U. LORENZ, *Otto Dix. Das Werkverzeichnis der Zeichnungen und Pastelle*, Weinmar, 2003, Vol. 2, p. 815. *Halbwelt auf Papier: Otto Dix und Conrad Felixmüller aus einer deutchen Privatsammlung*, Ostifildern-Ruit: Hatje Cantz; [Leipzig]: Stiftung Schleswig-Holsteinische Landesmuseen; Schloss Gottorf; Museum der bildenden Künste, Leipzig, 2000, Exhibition Catalogue.

[6] We used a the digital camera FujiFinePix UVIR.

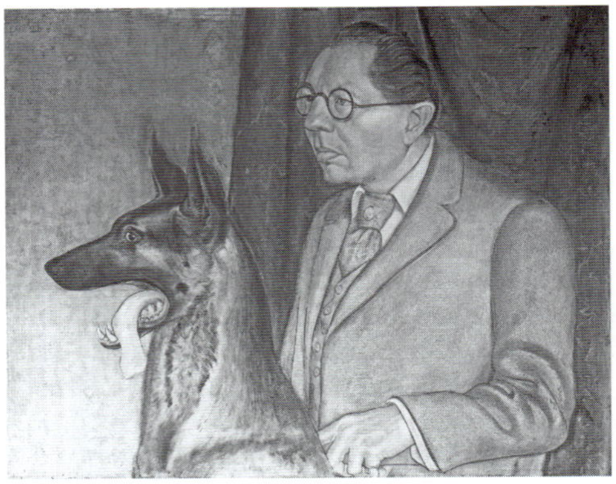

Fig. 3. Otto Dix, *Hugo Erfurth with Dog*, IRR assembly (© Hélène Desplechin).

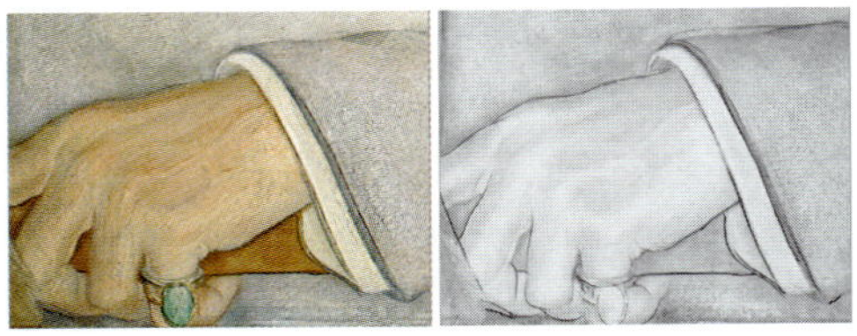

Fig. 4. Otto Dix, *Hugo Erfurth with Dog*, detail, colour photography and IRR (© Hélène Desplechin).

painting is very faithful to the drawing. It is possible that a large part of the drawing was transferred by sepia tracing and then traced over with black ink and brush. In a cross section the thin layer of sepia colour is visible under bone black pigment particles. Both drawings correspond to the draft of the composition on the artist's preparatory board.

The brush strokes, mainly continuous, seem to have been done with intensity, resulting in a slight underrelief in the surface.

Dix delimits the contour of the figure with great precision. He stops before some details like the hair. The strokes are confident. In some areas the artist reinforces a first line. The dog's hair and teeth are perfectly outlined in the underdrawing and meticulously painted. A continuous and energetic line delimits the suit of the sitter, his coat as well as the sleeves and collar of his shirt. Had the arm of the chair not been visible, the photographer could have been standing. The distance from the right shoulder to the neck is impossible, as is disproportionate the full size of the dog and its neck. This is typical in Dix: details are not placed at random, but correspond with the desire to bring out the essence of his models.

There are no important changes to the composition. Worth mentioning are the *pentimenti* in the placing of the glasses, slightly falling to the right in a first stage. They do not appear in the X-ray image because of the materials used, but they do in the IRR. The treatment of light and shade in the figures is shallow, with short parallel lines. Instead, the folds of the cloth are traced with long lines, without modelling. The most elaborated areas are the face and hand of the sitter and the head of the dog. The group ends up very similar to that on the preliminary drawing on board. Some parts of the drawing are almost visible on the surface, some other lines are visible through the thin paint layers, for example the dog's teeth, the folds in the sleeve and the photographer's chin. Summing up, Dix's working procedures consists in transfering a highly detailed, scaled preliminary drawing to the final support, making no significant changes during painting.

Dix might be one of the few artists in history whose pictorial technique had political repercussions. His painting was criticized for being barbaric. Since he was inspired by the German Old Masters, it is surprising that he failed to gain the favour of the socialist German rulers between 1933 and 1945, who promoted everything that was 'genuinely German'. On the contrary, he had to quit his position as art professor in the Dresden Academy for political reasons. He was labelled a 'degenerate artist' and he was said to offend the German moral and its spirit of fight. According to William Robinson, although Dix rarely used his art to support a political party or idea, he sometimes assumed the role of 'moralizer', following the tradition of Pieter Bruegel the Elder and Bosch.

Dix worked with a wide variety of techniques such as watercolour, print and sculpture but it is in painting that he developed a technique uncommon at the time. His method consisted in applying thin glazes of oil and tempera on wooden supports primed on both sides. These thin glazes were usually applied over a white ground that reflected the light and gave luminosity to the work. Dix worked out a carefull technique intended to create long lasting object. Dix used the technique of glazes in other paintings at the time.

In the study of the micro-samples obtained from different areas of the painting, the layer structure matches in essence the descriptions the artist made of his own technique.

The support consists of a wooden panel, 1 cm thick without reinforcements, stretcher or canvas on the front or reverse. It is primed on both sides, presumably to prevent warping. The thickness and materials of the ground layer are not the same on front and reverse. In the front we find (from lower to upper layer) (fig. 5):

- gypsum bound in animal glue (ground)
- a sanguine underdrawing composed of earth pigments (fig. 6)
- an underdrawing with bone black and earths (water-based technique), over the sanguine underdrawing
- a grey priming consisting of mastic varnish and linseed oil (almost imperceptible). Risaille executed with egg tempera. Oil and tempera paint layers. The paint layers are transparent, diluted in large quantities of resin or varnish. Details like hair and

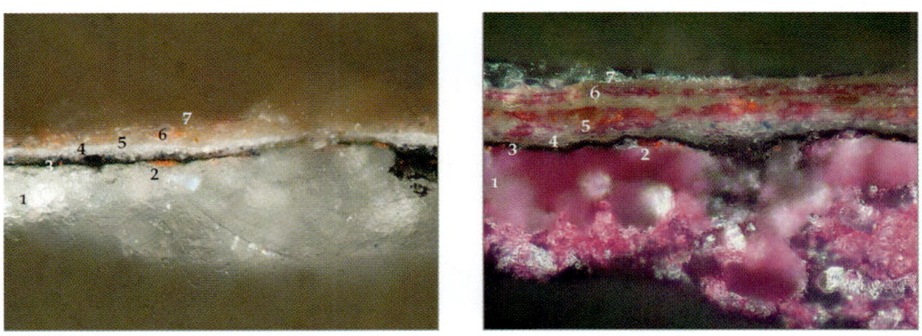

Fig. 5. Otto Dix, *Hugo Erfurth with Dog*, cross section with the different paint layers. 1. Gypsum and animal glue ground. 2. Sanguine from earth pigments. 3. Drawing, with bone black and earths (water technique). 4. Grey priming, with mastic varnish and linen oil. 5. Grisaille executed with egg tempera. 6. Oil and tempera paint layers. 7. Mastic resin and bees wax final varnish (© Artelab).

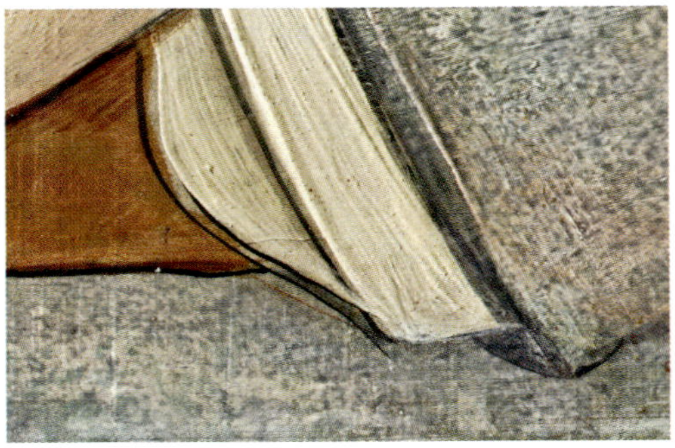

Fig. 6. Otto Dix, *Hugo Erfurth with Dog*, detail, sanguine under the black layer of the underdrawing (© Artelab).

shadows seem to have been applied over a water-based layer.Mastic resin and beeswax form the final varnish. It is not certain whether the varnish was applied by Dix or a restorer at a later date. It is worth noting that mastic varnish yellows considerably over time and that the varnish in this particular painting does not show the degree of yellowing that would be expected if the varnish had been applied by Dix.

On the reverse there is a gesso priming bound in animal glue, followed by a varnish layer that corresponds with the commercial brand Lukas, described by Dix' inscription, and mixed with wax at 50%.

Although Dix wanted to draw inspiration in Renaissance portraits, his execution of Hugo Erfurth's portrait is very different from, for example, the *Self portrait* by Dürer (1506) or *The merchant Georg Gisze* by Hans Holbein the Younger (1532). In his *Self Portrait*, Dürer looks at the spectator. An imaginary landscape in the background is the symbol of the artist's superiority over nature. Holbein also represents a man looking straight to the viewer and surrounded by the allegories of human knowledge. On the contrary, Dix' model does not look to the spectator. The artist gives equal importance to the living as to the objects, and creates a psychological tension between them. The presence of an elaborate and detailed underdrawing that corresponds almost exactly to the painting, supports the hypothesis of preliminary drawings similar to what is sometimes suspected in Old Masters.

Susana PÉREZ PÉREZ
Departamento de Restauración
Museo Thyssen-Bornemisza
Paseo del Prado 8
E – Madrid 28014
Tel: +34.91.420.39.44
Fax: +34.91.420.27.80
sperez@museothyssen.org

Conservator-restorer with a degree in Fine Arts, specialized in the conservation of paintings and sculpture. Employed in the conservation department of the Queen Sofia Museum of Modern Art in Madrid and currently conservator at the Thyssen Bornemisza Museum, Madrid.

Andrés SÁNCHEZ LEDESMA

Degree in biochemistry, head of the laboratory of the Department of Conservation at the Thyssen Bornemisza museum. Founder of the ArteLab laboratory for the scientific analysis of works of art.

Ubaldo SEDANO ESPÍN

Director of the Conservation department of the Thyssen Bornemisza museum. General coordinator of conservation and restoration projects in Latin America.

MAKING A PAIR OF *BIBLES MORALISÉES* IN THIRTEENTH-CENTURY PARIS: THE ROLE OF UNDERDRAWING

John Lowden

Abstract: Two illuminated picture books of the Bible of the type known as *Bibles moralisées*, one in the treasury of Toledo cathedral and the other divided between libraries in Oxford, Paris, and London, were long thought to be related as model and copy. First-hand examination of the blank backs of the pages in both shows that they are in fact based on a common workshop model, from which the main lines of the images were traced with a stylus. The resulting images were then gilded and painted by a large team of craftsmen. This use of pressure transfer for thousands of images on hundreds of folios is a technique without parallel in medieval manuscript production. It was probably adopted in order to enable two more or less identical books, each with an unprecedented number of images, to be produced to a short deadline. They may have been 'his and hers' volumes for King Louis IX of France and Queen Marguerite (married 1234), commissioned by his mother, Queen Blanche, widow of Louis VIII (died 1226).

The *Bibles moralisées* are picture books of the Bible, made for the queens and kings of France and their closest relatives between around 1220 and the first years of the fifteenth century.[1] In this paper I shall consider in detail evidence from two of the four earliest exemplars: the three-volume copy in the treasury of Toledo cathedral (fig. 1), documented in the will of King Alfonso X *el Sabio* as a gift to him from King Louis IX of France, and hence now termed the Bible of St Louis (or 'Toledo'),[2] and its three-volume companion (fig. 2), now divided between the Bodleian Library at Oxford (MS Bodley 270b), the Bibliothèque nationale de France at Paris (MS Lat. 11560), and the British Library at London (MSS Harley 1526-1527), and hence known as the 'Oxford/Paris/London' *Bible moralisée* (or 'OPL').[3] These two books are very similar. In his publication of Oxford/Paris/London at the beginning of the twentieth century, Laborde argued that Toledo was derived in some way from OPL.[4] In his publications of the 1960s to 1980s, Haussherr reversed the position, and argued that OPL was a more or less hasty

[1] I first considered this material in detail in J. LOWDEN, *The Making of the Bibles Moralisées. I. The Manuscripts, II. The Book of Ruth*, University Park, 2000. For discussion of the underdrawing see especially vol. I, p. 119-121 and 167-180. For the later *Bibles moralisées* see J. LOWDEN, 'The *Bible moralisée* in the fifteenth century and the challenge of the *Bible historiale*', in *Journal of the Warburg and Courtauld Institutes*, 2005, n. 68. In general see *Bibles moralisées* Electronic Bibliography: http://www.courtauld.ac.uk/people/lowden_john/bibliography.html

[2] *La Biblia de San Luis/Bible of Saint Louis*, 3 vol. facsimile and 2 vol. commentary, R. GONZÁLVEZ RUÍZ ed., Barcelona, 2001-4. J. LOWDEN, *Making*, I, p. 95-137.

[3] A. DE LABORDE, *La Bible moralisée illustrée conservée à Oxford, Paris et Londres. Reproduction intégrale du manuscrit du XIIIe siècle accompagnée de planches tirées de Bibles similaires et d'une notice*, Société française de reproductions de manuscrits à peintures, 5 vols., Paris, 1911-27; vol. 5 entitled *Étude sur la Bible moralisée illustrée*. J. LOWDEN, *Making*, I, p. 139-187.

[4] A. DE LABORDE, *Bible moralisée*, 'Notice' accompanying vol. III, p. 3 and 6.

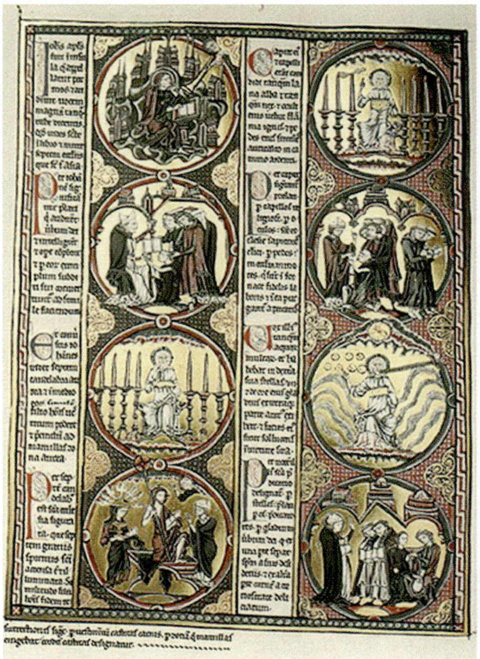

Fig. 1. Bible of Saint Louis, vol. III, fol. 159v, Toledo, Cathedral Treasury. Opening of Apocalypse.

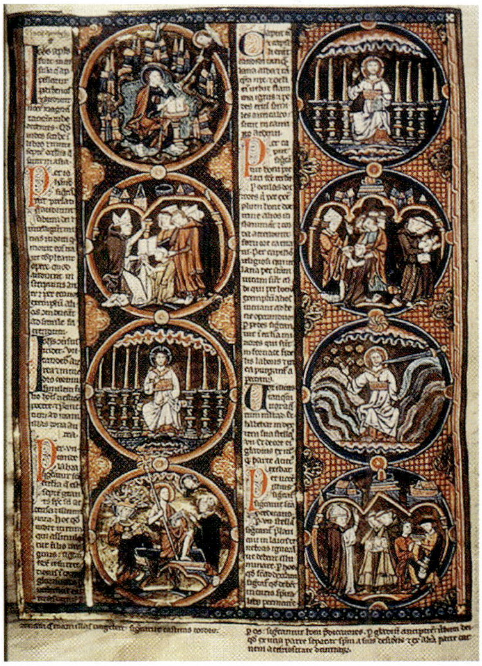

Fig. 2. MS Harley 1527, fol. 116v, London, British Library. Opening of Apocalypse.

and incompetent copy of Toledo.[5] The true situation, however, is unexpected, and much more interesting than either of these proposals.

Toledo and Oxford/Paris/London are both very large books, their pages more than 40 cm tall. For most of the three volumes of both *Bibles moralisées* the two proceed strictly in parallel. For example, the book of Ruth begins in the top right column of a verso page in both, and in both this page is (or in the case of Toledo was) the third folio of the 16th quire of volume I (many leaves have been cut from volume I of Toledo).[6] When you compare the eight medallions of this page in the two manuscripts it is at once apparent that the same complex and often unusual images appear in the same roundels in both. Only the choice of colours is conspicuously different. If we were to look more carefully we could observe how the texts that accompany the miniatures are often not quite the same, and in some instances, as at the top of the left column on

[5] E.g., R. HAUSSHERR, "Über die Auswahl des Bibeltextes in der Bible moralisée", in *Zeitschrift für Kunstgeschichte*, 1988, n. 17, p. 129. Crucial to this change was the article by A. HEIMANN, "Jeremiah and his Girdle", in *Journal of the Warburg and Courtauld Institutes*, 1962, n. 25, p. 1-8. The argument was accepted by R. BRANNER, *Manuscript Painting in Paris during the Reign of Saint Louis*, California Studies in the History of Art 18, Berkeley and Los Angeles, 1977, p. 49: 'It is generally agreed that the Oxford Bible was a fairly literal copy of Toledo, done in a quick and rather careless manner'.

[6] J. LOWDEN. *Making*, II, col. pls. VI, VIII. See also diagrams of quires in vol. I, p. 288-298.

this page, they are quite different.[7] Nonetheless, the sort of similarity that you see in this comparison, multiplied by the many hundreds of folios of the two three-volume books, is absolutely characteristic.

It is also characteristic that the two three-volume *Bibles moralisées*, exactly like the two one-volume examples, both now in the Österreichische Nationalbibliothek in Vienna (cod. 1179 and 2554), and both of which predate Toledo and Oxford/Paris/London by a few years, were worked entirely on only one side of the parchment sheets from which they were assembled.[8] This was a regular procedure in high quality illustrated books at the time, as for example in the frontispiece cycle of the Psalter of Queen Ingeborg, now at Chantilly, for it ensured that any bleeding through of the pigments from one side of the leaf to the other would not damage the painted surface. Obviously this was a costly decision to make, as it doubled the number of sheets of parchment that were required, and I think the *Bibles moralisées* are unique in using just one side of the parchment for the entire volume (the entire three volumes in the case of Toledo and OPL). In some cases the blank pages of such books were glued together at some date (as in volumes I and III of Toledo, but not volume II),[9] but this was not the case for most of the leaves of most of the *Bibles moralisées*.

The blank backs of some of the pages in Oxford/Paris/London and Toledo are slightly transparent, depending on the thickness and preparation of the particular sheet of parchment (fig. 4). Some are also more or less distorted by the pigments and the gilding with its coloured ground that was applied on the front of the leaf. These features are to be expected. What is more surprising is that the outlines of the compositions in the medallions on the front of the leaves are quite often clearly legible as grey lines on the blank backs of the leaves. What is more, in the most conspicuous cases these lines appear under normal viewing conditions to be in relief, as though impressed into the parchment by heavy pressure from the painted side of the leaf.

Raking light makes the relief much more clearly visible, although the result is not easy to photograph for the surfaces of the leaves are undulating and hence create deep shadows. But because we have thousands of medallions to look at there is plenty of evidence from which to choose.

In order to understand what has happened, and what it might imply, we need to look carefully at some details. The examples I will present are from the beginning of the Apocalypse in volume III of Toledo and Oxford/Paris/London (figs. 1, 2).[10] The comparable leaves are the first of quire 21 in Toledo (folio 159) and the first of quire 17 in Harley 1527 (folio 116) (the disparity of the quire numbering is considered below). We will look briefly at the first image in the left column (which accompanies a paraphrase

[7] J. LOWDEN, *Making*, II, *passim*.

[8] R. HAUSSHERR, *Bible moralisée. Faksimile-Ausgabe im Originalformat des Codex Vindobonensis 2554 der Österreichischen Nationalbibliothek*, Codices Selecti, XL-XL*, Graz, 1973; G.B. GUEST, *Bible Moralisée. Codex Vindobonensis 2554. Vienna, Österreichische Nationalbibliothek*, London, 1995; J. LOWDEN, *Making*, I, p. 11-54 and 55-94.

[9] J. LOWDEN, *Making*, I, p. 119.

[10] More generally see J. LOWDEN, "The Apocalypse in the Early-Thirteenth-Century *Bibles moralisées*: A Re-Assessment", in *Prophecy, Apocalypse and the Day of Doom*, ed. N. MORGAN, Proceedings of the 2000 Harlaxton Symposium, Harlaxton Medieval Studies 12, Donington, 2004, p. 195-219, and pls. 18-31.

of Apoc. 1:9-11), and then more extensively at the third. Here I have to say that the photographs of Toledo are taken from the recent facsimile publication as I was not able to get permission to make photographs directly from the manuscript. The reproduction in the facsimile of the gilding of the original is accurate only in its location: the texture, finish, and general appearance of the gold are all misleading. But if we look at the Toledo version of the first medallion of the Apocalypse, and compare it with the back of the same medallion in Oxford/Paris/London (figs. 3, 4) it appears that they are extraordinarily similar in outline and design. In fact it is difficult at first to see any detail in which the sketch visible on the back of the OPL image differs from the painted image on the front of the page in Toledo.

Let us move to the third medallion on this page in OPL and examine it more closely (fig. 6). The passage that is illustrated is from Apocalypse 1:13-14. The figure in the midst of seven golden candlesticks, is the 'one like unto the son of man, clothed with

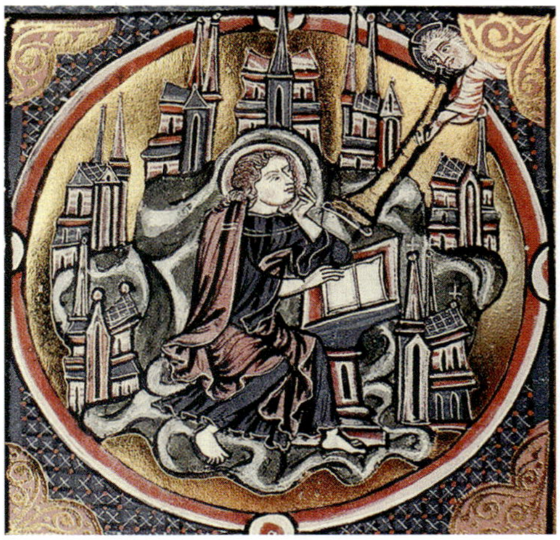

Fig. 3. Toledo III, fol. 159v. Detail of medallion 1.

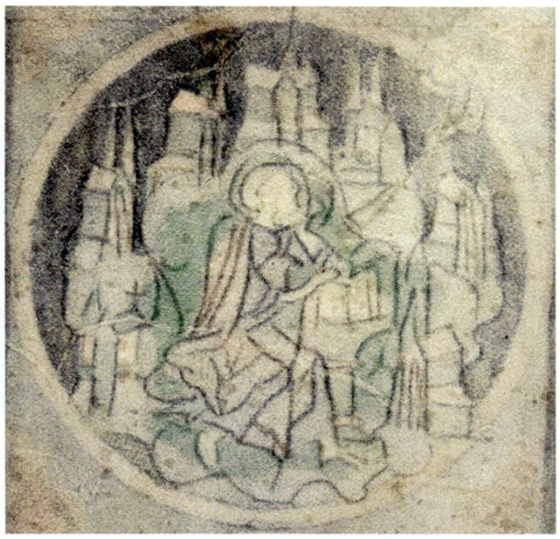

Fig. 4. Harley 1527, fol. 116r. Detail of back of medallion 1 (image reversed).

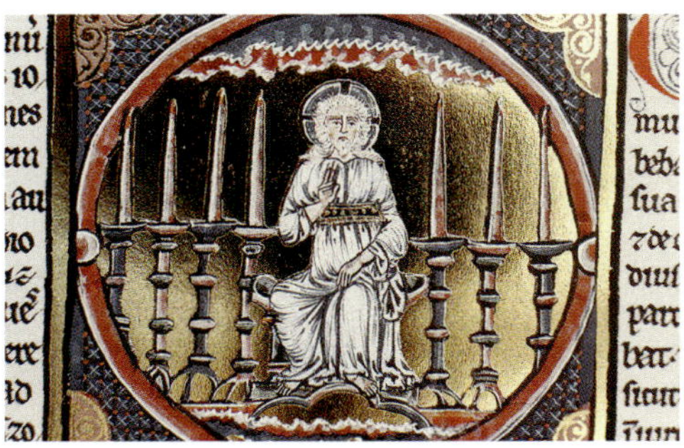

Fig. 5. Toledo III fol. 159v. Detail of medallion 3.

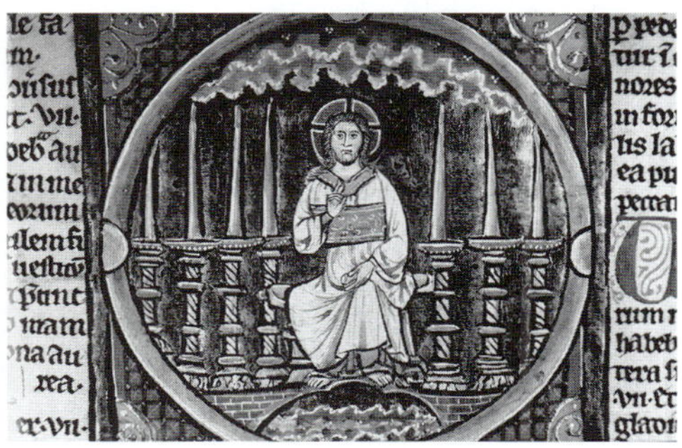

Fig. 6. Harley 1527, fol. 116v. Detail of medallion 3.

a garment down to the foot (*vestitum podere*), and girt about the breasts with a golden girdle (*precinctum ad mamillas zona aurea*). His head and his hairs were white like wool, as white as snow (*tamquam lana alba et tamquam nix*)'. We should note that in OPL the figure's hair is not white like wool or snow, the girdle about the breasts is not gold, and the garment does not come down to the figure's feet.

If we now look at the same figure under a strongly raking light (fig. 7) it can be seen that the main lines of the composition, including the folds of the drapery have been impressed into the parchment. Obviously this must have happened before the image was painted for to impress lines so deeply would have required the use of a stylus and heavy pressure, pressure that would surely have marked the paint surface if it had been applied over that surface.

Looking at the back of the image (fig. 9) (here shown in reverse to simplify the comparison) it can now be seen that the drawing visible on the blank parchment is the preliminary *underdrawing* for the image, executed on the front of the leaf by means of a stylus and heavy pressure. Although it is difficult to see, all the dark greyish lines visible on the reverse of this and other leaves are the result of compression of the parchment by the stylus, combined with rubbing of the resulting ridged lines which have picked up additional dust and dirt over the centuries.

Let us look at a few details in OPL (figs. 6, 7, 9). The Son of Man has a decorated hem at the v-shaped neck of his garment, which is clearly visible in the underdrawing. When we look at the belt around his breasts, however, we can see on the back of the leaf that the underdrawing was for a much narrower belt (note that because the raking light, views of the front of the leaf are lit from the left). Vertical furrow-like lines show up clearly as dark shadows, but horizontal furrowed lines are less clear, and horizontal painted lines (as for example the dark lines of the broad painted belt), are clearly visible. This is why the furrows that define the narrow belt in the raking light image appear

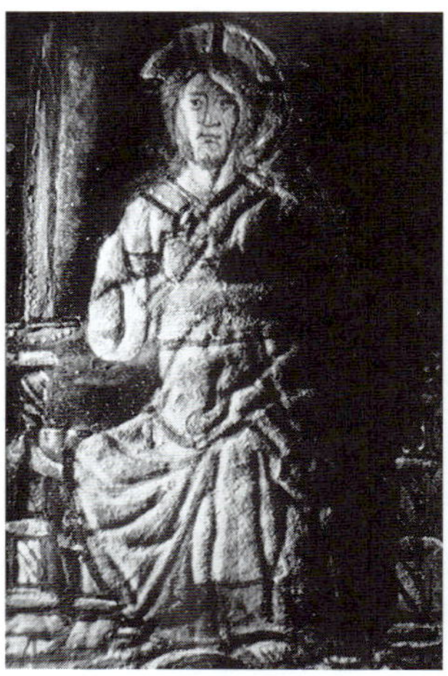

Fig. 7. Harley 1527, fol. 116v. Detail of medallion 3 under raking light.

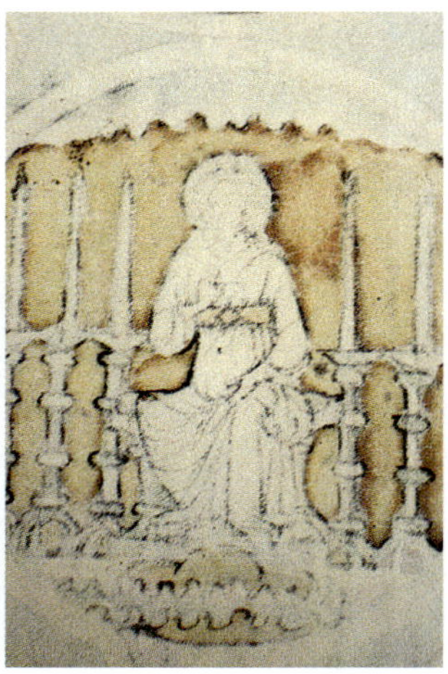

Fig. 8. Toledo III, fol. 159r. Detail of back of medallion 3 (image reversed).

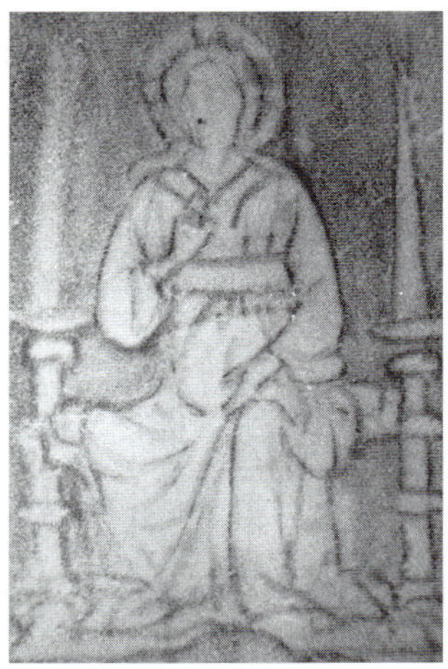

Fig. 9. Harley 1527, fol. 116r. Detail of back of medallion 3 (image reversed).

as pale lines, rather than dark ones). Moving down we see that the figure's left hand is turned palm upwards. The treatment of the leg below is also characteristic. In the painted image we see a broad triangular fold spreading out from the knee. But with raking light and a glance at the reverse we see that the underdrawing calls for parallel vertical folds, defining the lower leg as though with clinging drapery. Finally, we see that the zig-zag lower edge of the white garment, firmly defined by a black painted line, is not present in the underdrawing (as seen most clearly on the reverse). In the underdrawing the garment extends, as the text specifies, 'to the feet'.

Turning to comparison with the same figure in the same medallion in Toledo (figs. 5, 6) we notice at once that this image is far more carefully painted than in OPL (it is necessary to overlook the ugly gilding of the facsimile). In Toledo, also, the Son of Man's hair is white as wool or snow, as the text specifies, and the garment extends right down to his feet. We can note other differences too (differences that are undoubtedly minor, but which are yet crucial to the present discussion). Toledo lacks the v-shaped decorated hem at the figure's neck, as seen in OPL. The *zona aurea* is narrow in Toledo, not broad (as it is in OPL), and it is at least partially gold. The figure's left hand in Toledo is turned palm downwards, not upwards. And the leg below in Toledo is defined by parallel vertical lines, rather than being hidden by a broad, triangular fall of drapery.

Fortunately for my purposes, the facsimile of the Toledo manuscript reproduces all the blank pages of the original as well as all the painted pages. Although the result is not as clear as in the original, enough can be seen to establish what is going on, even without the use of raking light (figs. 7, 9). In the underdrawing of the image in Toledo, as seen on the reverse of the leaf (and again shown here in reverse for clarity), the v-shaped decorated hem at the neck is present. The underdrawing in Toledo also shows the narrow (rather than broad) belt around the breasts, and further we can see that in Toledo this was at one stage intended to be entirely gilded for the artist applied the same yellowy brown ground for the belt that underlies the gilding elsewhere in the image. Moving down the image we see that the underdrawing for the left hand in Toledo shows it turned palm upwards, not palm downwards as it is painted on the front of the leaf. Finally, in the underdrawing of Toledo the left leg is defined by parallel vertical folds in the drapery, as it is in the painted image, and the drapery extends as far as the feet without a horizontal division between under and over garments above the left ankle.

Turning to a comparison of the underdrawings of Toledo and Oxford/Paris/London, as these are revealed on the 'blank' backs of the leaves (figs. 8, 9), we can see that the two versions of the image are in fact close to identical at the underdrawing stage. Both have the v-shaped decorated hem at the figure's neck (although this is a little hard to see in Toledo). Both have a narrow belt round the chest. In Toledo this has a diamond shape in the centre — I am not sure whether or not this is in relief —, for there are heavy black lines that have bled through around the edges of the gilding in both manuscripts, and I think the diamond may be another case of this phenomenon). Clearer is the fact that both underdrawings show the figure with the left hand turned palm upward. Both also have the parallel vertical folds defining the leg below. Both have the garment extending to the feet, with no sign of a horizontal hem above the ankle, as in the painted version in OPL.

Returning to the front of the leaf (figs. 5, 6) we can see that although the two artists worked up the same underdrawing they achieved surprisingly different results. Oxford/Paris/London reproduced the v-shaped decorated hem at the neck whereas Toledo omitted it. Toledo reproduced the narrow belt, whereas OPL chose to substitute a much wider band of material. OPL reproduced the gesture of the left hand that the underdrawing called for, whereas Toledo changed it. As for the leg below, here Toledo followed the guidance of the underdrawing, whereas OPL painted the garment as rising in a zig-zag hem line to a triangular shape. The most conspicuous difference between the two however, the colour and treatment of the figure's hair and beard, was defined not by the underdrawing (which emphasized outlines and had no indications as to colour) but rather by the Toledo artist's closer adherence to the Apocalypse text (*tamquam lana alba et tamquam nix* 'white like wool, as white as snow').

Despite the losses of leaves from both Toledo and Oxford/Paris/London, together they preserve a total of 9880 medallions.[11] The kind of analysis performed here on one pair of medallions could thus be tried, in theory, many thousands of times. In practice, the very variable extent of the visibility of the pressure-traced underdrawing from one folio to the next means that many potential comparisons would be hard to demonstrate in a convincing manner, especially without photography under raking light. Yet assuming the results of such comparisons, where they could be made, would be similar (as I am confident is the case), it remains to consider more broadly how and why the working process that has been observed was employed.

To sum up: Toledo and Oxford/Paris/London are image for image, page for page, and quire for quire the same (with a crucial exception to be mentioned below). That is to say a bifolio or quire could be exchanged between the two, and its content would be the same. In detail, there would be differences, it is true. The work of the two artists of the Apocalypse medallion we examined was very different in style (figs. 5, 6). The Toledo artist was clearly working more slowly with much finer and more complex brushwork. The texts of the two books, as mentioned before, are also not precisely the same. Yet the overall consistency of the two books is remarkable. The exception to this pattern is found in the first quires of volume III in both Toledo and Oxford/Paris/London. The pressure-traced underdrawings appear for the first time in quire 13 of Toledo III (folio 97) and in quire 10 of Harley 1527 (folio 66). From this point onwards the two run in parallel with the exception of quire 20 of Toledo III (folios 151-158), which shows no signs of pressure-traced underdrawings and which has no equivalent in OPL.[12] Before this point the history of the two books must be different. I have argued elsewhere that work on the Toledo manuscript began with volume III, and it was only after some period, possibly years, that the decision was made to produce another 'twin' manuscript (OPL) alongside Toledo.[13] Because the early quires of Toledo III were already in

[11] J. LOWDEN, *Making*, I, p. 306, n. 29-30.

[12] J. LOWDEN, "The Artists of the Bible of Saint Louis", in *The Bible of Saint Louis*, commentary vol. II, p. 331.

[13] J. LOWDEN, *Making*, I, p. 118-127. F. BOESPFLUG and Y. ZALUSKA, "The New Testament (Except the Revelation) in the *Bible of Saint Louis*", in *The Bible of Saint Louis*, commentary vol. II, p. 347-447.

existence they could not be traced for inclusion in OPL. The early quires of Harley 1527 thus differ markedly from Toledo III.

So far as I know, no other medieval manuscript was produced in the same way as these two *Bibles moralisées*, with pressure traced underdrawings for all the images. I looked carefully at both so as to try to establish whether the drawings had been sketched with ink into one first, and then transferred (before they were painted) by tracing over with a stylus into the other. I am persuaded that this cannot have happened, for in many cases the pressure tracings are not sufficiently clear for the page in question to have been the model or donor (it must rather have been the recipient). Rather than the underdrawings of one being traced into the other we have instead to imagine another possibility: that there was originally a complete set of detailed sketches for all the images in such a *Bible moralisée*, sketches that were arranged on bifolios and quires exactly like Toledo and OPL (the accompanying texts must have been kept on separate schedules). These sketches would have been traced a bifolio at a time onto sheets that would go to make either Toledo or OPL (it is even conceivable, where the impression in Toledo or OPL is very clear, that the underdrawing was traced onto the bifolios of both Toledo and OPL at the same time — the so-called blind ruling of parchment by scribes regularly saw the transfer of the pattern through four sheets at once). To make an entire workshop model of a three-volume *Bible moralisée* was of course an amazingly extravagant plan. But it had a supremely important advantage: once the master set of drawings had been executed, the bifolios that formed the model could be distributed among as many artists as were available for the straightforward and doubtless rapid process of tracing them with a stylus onto the sheets that would go to make Toledo or OPL. Conceivably the workshop model could then be retained against the possibility that one or more further *Bibles moralisées* might be made in the future. Alternatively, and possibly in addition, it can be imagined that the workshop model, if no longer required, might even have been completed and have circulated as a 'normal' *Bible moralisée*. This seems to have happened with BL Add. MS 18719, a copy of OPL made around 1280, and subsequently used as the workshop model for BnF fr. 167, the Bible of Jean II le Bon, after which it appears to have been given to Saint-Louis de Poissy.[14]

Once the bifolios for Toledo or OPL had been traced they could be given to a team of craftsman, who could work concurrently, gilding and painting as rapidly as possible. They needed only to work up the sketches, following what was a simple overall plan for what the book was going to look like, for example in the choice and balance of colours.[15]

The demand that suggested the utilisation of this working practice must, I think, have been the need to produce not one, but two *Bibles moralisées,* and both to a short deadline. I have proposed elsewhere that the deadline was the marriage of Louis IX to Marguerite de Provence in 1234, and that the project was overseen by Louis's mother, Blanche of Castille.[16] Just as Toledo and OPL were, I believe, 'his and hers' *Bibles moralisées* for the king and queen of France, the previous generation of such books, Vienna 2554 and 1179, were, I have argued, made for Blanche and her husband, Louis VIII.

[14] J. LOWDEN, *Making*, I, p. 189-250.
[15] J. LOWDEN, *Artists*, p. 323-346.
[16] J. LOWDEN, *Making*, I, p. 127-132.

The making of Toledo and Oxford/Paris/London was indeed a daunting challenge, for even one such *Bible moralisée* represented by far the most ambitious project in manuscript illumination ever attempted. But by employing a workshop model and tracing its drawings into the bifolios of two manuscripts the requirements of the commission could be met. *Underdrawing* is indeed the key to *understanding* the making of these two extraordinary books.

John LOWDEN
Courtauld Institute of Art
Somerset House, Strand
UK – London WC2RR 0RN
john.lowden@courtauld.ac.uk
Tel: +44.20.7848.2668
Fax: +44.20.7848.2410

John Lowden has taught at the Courtauld Institute since 1982. He is the author of numerous articles, chapters in books, contributions to exhibition catalogues, encyclopaedias, etc., as well as *Illuminated Prophet Books* (1988), *The Octateuchs* (1992), *Early Christian and Byzantine Art* (1997-4th reprint 2003, translated into French, Greek, Japanese, and Korean), and *The Making of the Bibles moralisées* (2 vols., 2000), winner of the Gründler Prize (2002). He has long championed an applied codicological method for the study of illuminated manuscripts of all periods.

THE CONSERVATION ASSESSMENT OF THE PHILIPS OF CLÈVES BOOK OF HOURS (BRUSSELS, ROYAL LIBRARY, MS IV 40)

Lieve Watteeuw

Abstract: The *Hours of Philips of Cleve* (Brussels, Royal Library, MS IV 40), is a small but richly illuminated book of hours, created in Ghent and Bruges around 1485-1495. Although several masters were involved in its production, only the name of the Master of the Older Prayer Book of Maximilian is mentioned in literature. This paper will discuss the codicology of the complex manuscript and compare the underdrawing, materials and painting techniques of different illuminations, as recorded in the current conservation research. These aspects will be compared to materials and techniques used by other Flemish illuminators active in the second half of the fifteenth century.

Introduction

The technological study of medieval miniatures, is not as self-evident as that of larger scale paintings. Because of their small size and fragile characteristics, they do not leave the library store rooms as easily. In focus of this research is the 15th century Book of Hours of Philips of Clèves, Brussels, Royal Library, MS IV 40 (fig. 1). For ethical reasons, the manuscript curator and the conservation-research team decided not to take any micro-samples in the documentation phase, but to rely on a broad spectrum of documentation and analyses. In the research preliminary (2006-2007) to the conservation treatment, a collaborative interdisciplinary project was developed with art historians, conservation-scientists and book-conservators in the field of illuminated manuscripts.[1] Methods used were archival documentation, book-archaeology, interviews, photo macrographs, digital imaging, data management of typologies of illumination techniques and damage assessment of paint layers, micro-XRF and micro-Raman.

Illuminated manuscripts are usually the 'hidden' treasures in the contemporary library. Their status as medieval symbolic, royal or devotional objects has shifted in Western collections since the 16th century towards esteemed museum artefacts of high artistic value, and extremely vulnerable exhibition objects. This 'high-cult' value is taken into account by the growing academic curiosity for illuminator craftsmanship. Since 1990, the enhancing focus on conservation-science and annex conservation treatments became both an excellent vehicle to these strictly art-theoretical research fields. Insight in traditional methods of illumination was documented in a range of treatises on pigment preparation.[2]

[1] For generous support and discussion I thank: Jan Van der Stock (KULeuven, Research Center for Illuminated Manuscripts), Marina Van Bos and Steven Saverweys (Royal Institute for Cultural Heritage, KIKIRPA, Brussels), Bernard Bousmanne (Curator, Manuscript Department, Royal Library of Belgium), Nancy Turner (Senior paper conservator, J.P. Getty Museum, Manuscript Department) and Jan Wouters (KIKIRPA).

[2] CLARKE, 2001.

library of the Trappist Cistercians from Orval and the Monastery of Saint-Martin of Mondaye.[5] He said he never retouched any medieval miniature. The source of some modern interventions on the miniatures had to be sought in an earlier period. The Marchoul-Gilissen binding style, he used during his entire carrier (more than thirty years) was named by the binder *a l'antique*. The paradigm of the neo pastiche rebinding policy of an important generation of post-war bookbinders-restorers in Belgium became more intelligible.

It was imperative for further conservation of the pictorial layers, that the modern binding of the manuscript was removed again. Detailed codicological and archeological information was recorded (1961). The documentation was based upon a detailed description of technical features of the parchment and binding structure, in order to simulate the original conditions that were altered during the last five centuries. The basic principle is straightforward: every workshop or binder leaves technical traces behind in the course of composing, completing or rebinding a codex. Pricking or cutting the holes for the sewing of the manuscript — a preparatory action — make the stitching smooth, quick, and efficient. In order to bind a manuscript, new sewing holes are made or the previous ones are reused. In this case, one must recall that a manuscript is usually restitched according to practical considerations. Aesthetic and bibliophile arguments might determine later interventions. The succession of cuts and needle pricks in the Cleves Book of Hours could be analyzed as follows (folio 21, gathering 5, fig. 3). With the exception of bi-folio 33-34, which lacks the sewing perforations (cuts) from five threads of the first 15th century and original binding, twenty-two perforations of diverse form and size from knifes or needles were registered in the gathering fold. This gave evidence that, soon after the first compilation of the manuscript ca. 1485, one bi-folio was added (in the early sixteenth century). This folio had a distinct sub-division of sewing perforations and its miniature of St. Anna was clearly executed by a less experienced illuminator. The manuscript was sewn a second time, on four supports. It presumably remained in this state long after Joris Hoefnagel's additions at the end of the 16th century. Later, in the eighteenth century, the manuscript was re-sewed again on four bands according to the usual procedures, re-cut significantly for almost 10 mm on the three edges. There is more. In the nineteenth century, Charles de Brou, a notoriously passionate connoisseur, purchased the manuscript, which was then bound in fashionable green velvet with silver fittings, for the Museum of the Princes of Arenberg in the center of Brussels. During the First World War the codex disappeared, returning in public through the American dealer in 1958. In 1961 it was rebound, re-using the previous sewing pattern. In 500 years, the manuscript was adapted and rebound at least five times.

The research in the Royal Library (Manuscript Department) and the Brussels Royal Institute for Cultural Heritage on the Clèves Book of Hours continued with visual examination under binocular,[6] the development of an extensive visual database with macro photographs[7] documenting the techniques and the damages (Cumulus 5.5©/digital

[5] Leon Gilissen became in 1955 employed at the Royal Library in Brussels, first in the Chalcography Department, later in the manuscript Department.

[6] Zeiss, Stemi 2000, JVC color video camera JVC, Axio Vision archiving system.

[7] Nikon D 70, AF micro-Nikkor 60 mm.

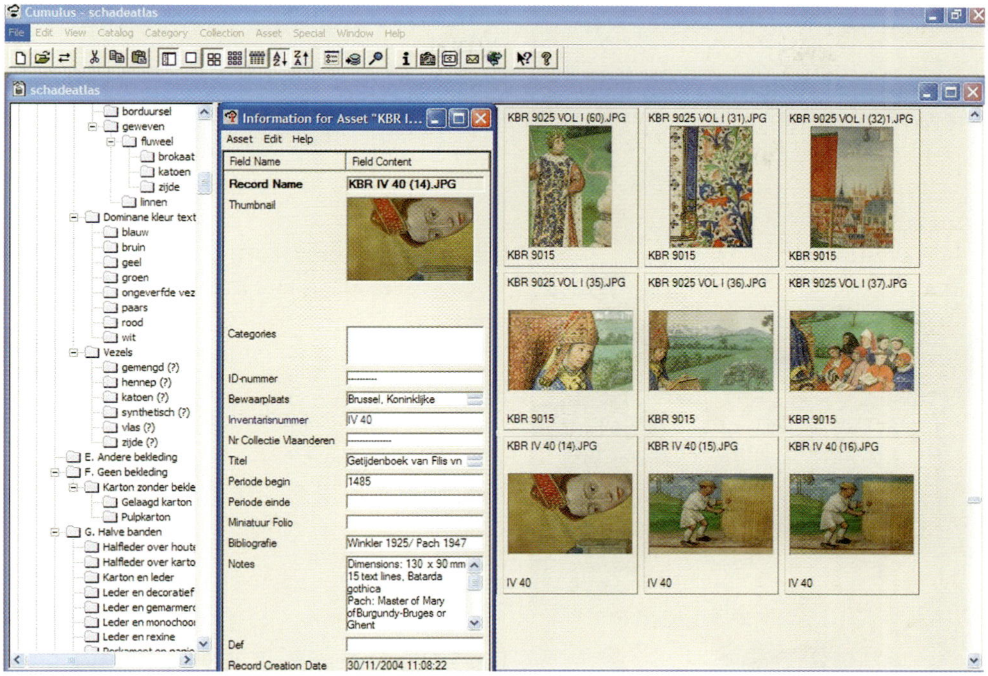

Fig. 4. Example showing the database with extensive visual documentation (Cumulus 5.5. © digital asset management software).

asset management software) (fig. 4).[8] Further infra-red photography and infrared reflectography[9] (camera) was combined with digital detail photographs with direct, transmitted and racking light.[10] After the examination of the visual documentation and data, further analytical non- and micro- destructive research was carried on.

A selection of the illuminations was further investigated with micro-XRF analyses and micro-Raman Spectrometry, providing more information about the materials used. These visual investigations on the paint layer are combined with micro x-ray fluorescence[11] (fig. 5).[12] The decision was made to work with micro-XRF in the first stage and

[8] The Cumulus database contains images and information in the fields of illustrating techniques, materials, condition, bookbinding techniques and materials from 112 illuminated manuscripts belonging to collections of monasteries, churches, libraries and archives in present Belgium. In total 800 information fields were developed, with reference information on 3500 images.

[9] InfraCAM SWIR — short-wave infrared — video.

[10] Digital camera MAMIYA RZ 67, digital back Light Phase I / Nikon D 70 and EOS I D-Mark II, 100 mm macro lens.

[11] Bruker Optics ArtTAX micro-XRF, equipped with a colour CCD camera for sample visualisation and fine positioning of the measuring head. (Experimental conditions, Rh X-ray tube, 50kV, 500µA, 50 keV, Mo25µm filter, in air and in helium flow).

[12] Until 2005 XRF spectra were taken wilt a Tracor X ray DuBois 404 Object Analyzer, equipped wilt a Mo X-ray tube (max. power 50 kV and current 1 mA) and a liquid nitrogen cooled Si(Li) detector. The spot size was approximately 1-2 cm, which is large for the fine brushstroke in an illumination. Moreover,

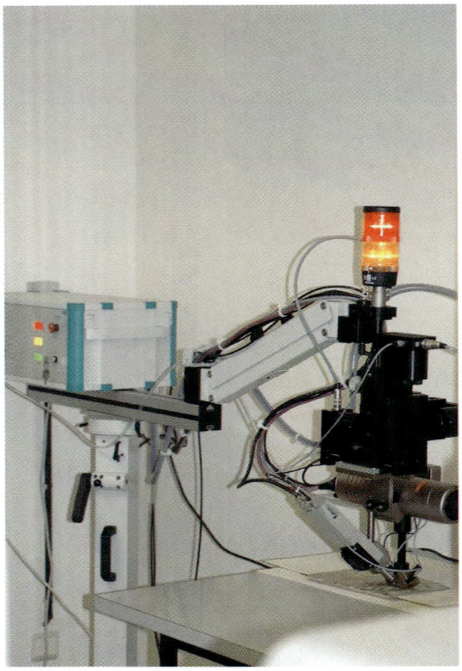

Fig. 5. Equipment for X-Ray fluorescence.

not destructively with micro samples. As a result, only a range of inorganic elements could be analyzed. To investigate the stratification of the paint layer on the point of measurement, the area was documented with highly detailed digital photographs taken with strong transmitted light, penetrating easily the thin parchment leaves. Brushstrokes, underdrawing, the thickness of paint layers, flaking and the writing material (like text lines in black iron gall ink on the verso of the folio) were made visible. Micro sampling was not an option, since the sampling technique is not accepted in miniature painting. The micro-destructive Q-tip method with rubbing of the paint layer was in this case evaluated as inappropriate by the curator and the conservation team. In 1997 this method was used on the Mayer Van den Bergh Breviary. The abrasion for sampling areas had resulted in a limited local damage, especially on the tactile features of the shiny paint layers. This damage is visible under magnification and weakens the pictorial integrity.

After the broad overview of the pigments in each illumination (15 to 25 spots in 37 miniatures recorded in the Cumulus database), a small selection of pigments was further analysed in-situ with the Renishaw InVia Raman spectrometer. In cooperation with the constructer Renishaw, a special optical gantry was developed for easy positioning

the movement and focusing of the system was difficult, and the positioning of the parchment folio's had to be vertical, a complex and delicate manipulating for bound codices and unbound leaves. The obtained spectra could not be exported to modern equipment because of technical reasons. Since 2005 a portable ArtTAX micro-XRF with a Rh X-ray tube (max. power 50 kV and current 700 /lA), a polycapillary lens for a micro-excitation spot (higher relative intensity field; lateral resolution up to 70/lm), a thermo-electrically cooled Si-drift detector, helium flushing for monitoring low Z elements and a CCD camera for positioning the X-ray beam on the object is in use.

and focusing of the laser beam on mid-sized objects. A big drawback of both non-destructive techniques is their inability to study the underlying paint layers. The poor depth resolution of the XRF is both an advantage and a disadvantage: while it hampers the interpretation, it does give insight about the exact underlying paint structure.

Conclusion

The illumination of the portrait of Philip (folio 121 verso) could be considererd as a *pars pro toto*. It was examined under the microscope with zenithal, transmitted and striking light. The examination revealed the typical stratification of the flesh shades on the velum, executed in five or six layers, and completed with shading (called in illuminating treatises *puncteeren* or *pointiller*) in concentrations of lead white and vermilion, to create the physiognomy of the portrait. A diagonal pattern for the heightening of the cheeks in the flesh tone was applied on top of this. In the almond-shaped eyes with their gray iris, a dot of azurite and one of lead white were used for the light reflection incoming from the left. The eye sockets are suggested with transparent diluted carbon black. The hard brushwork adds in relief shadows of thicker deep rose (fig. 6). On folio

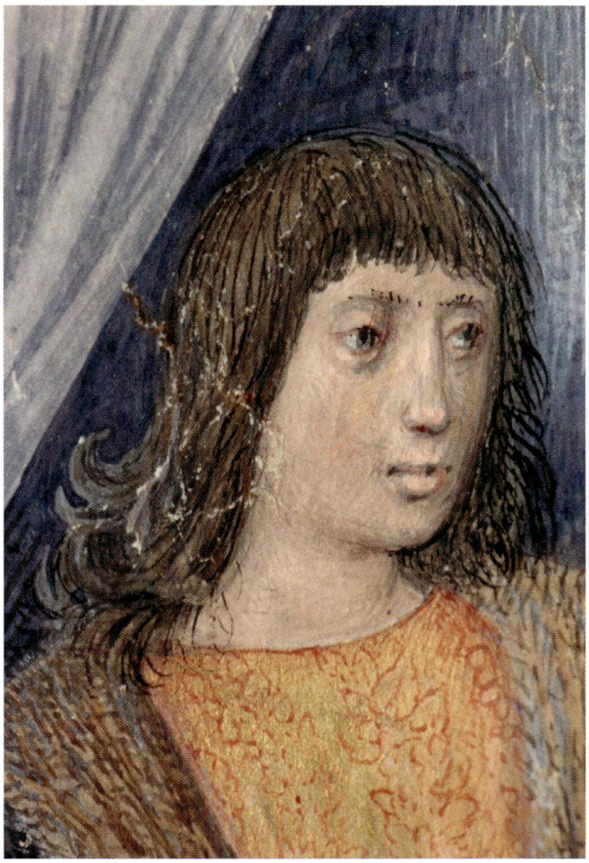

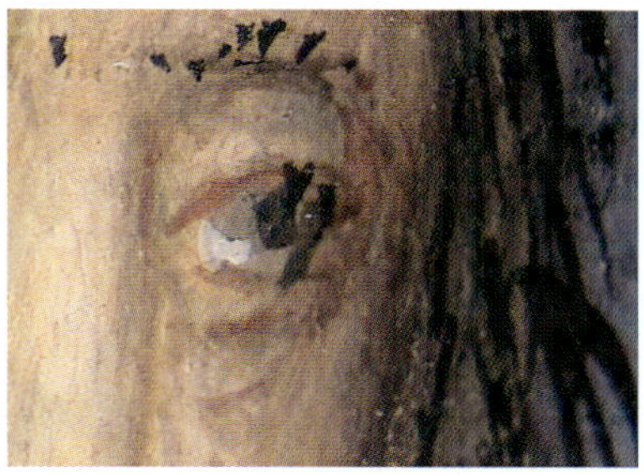

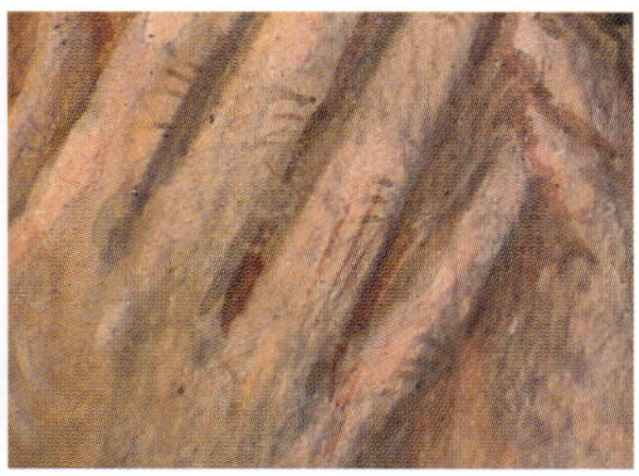

Fig. 6. The Book of Hours of Philips of Clèves, Brussels, Royal Library, MS IV 40, folio 127 verso: details of the face, eye and hands.

121 verso, a brushstroke of black ink tracing the outline of the eye pupil was measured. It contained the element calcium and iron (proper to the parchment-preparation), copper, mercury and considerable amounts of lead, due to preparation layer. A few millimetres further a vertical fold in the vellum resulted in losses of pigment in the blue canopy, and a modern intervention was detected, with elements of manganese and titanium. The New York antiquarian bookseller who dealed with the manuscript in 1958 has perhaps retouched the manuscript.

In the lacunas of the paint layer, a lightly painted line of the preparatory underdrawing was visible in the green cloak of the patron saint. An examination of the miniature with infra-red reflectography revealed a geometrical preliminary composition, but also a small modification of the position of the hands of Philips by the artist. Digital shots with transmitted light — which have been converted to black and white images contrasted in Adobe Photoshop© — revealed the original composition for the contour of the patron saint on the right. This is a remarquable feature. Only this illumination in the manuscript incorporates this type of underdrawing in a liquid medium containing small amounts of iron and copper.

The technological study of miniature painting is not as self-evident as that of paintings, due to their small scale and fragility. An interdisciplinary approach is essential in order to satisfy scholarly curiosity from the rare-book librarian, art historian, conservation-scientist and the conservator-restorer. The combination of the documentation and the analytical-data allows to gain knowledge on technique and degradation. A relevant range of basic documentation methods, as the observation and photography with the Stemi Zeiss binocular, the high resolution digital imaging and the data management with Cumulus gave relevant information to evaluate the condition. It allowed us to formulate a consequent conservation-restoration protocol and to perform consolidation and rebinding of the codex in respect of old and new skills on research and craftsmanship.

Lieve WATTEEUW
KULeuven, Illuminare, Research Centre for Illuminated manuscripts
The Royal Institute for the Study and Conservation of Belgium's Artistic Heritage
lieve.watteeuw@pandora.be

She studied conservation and restoration of graphic documents in Ghent (HICOREB 1991) and art history at KULeuven (2002). She is collaborator at the conservation department, graphic documents KIKIRPA and member of *Illuminare, Centre for the Study of the Illuminated Manuscript*, KULeuven. She works as an external conservator for the Royal Library in Brussels, Manuscript Department and for several city libraries and archives in Belgium. Her Ph.D. (april 2008) focused on the history of conservation/restoration of illuminated manuscripts from the Southern Netherlands.

LA CRITIQUE D'AUTHENTICITÉ DES VITRAUX ANCIENS

Isabelle Lecocq*

Résumé: Souvent inclus dans la clôture monumentale, les vitraux sont particulièrement exposés à des altérations qui les anéantissent, les transforment ou les dénaturent: la grêle, le vent, les incendies, l'altération du support et le vieillissement des matériaux ou encore les dégâts dus aux hommes comme le vandalisme, les changements de mode et la négligence. Il n'en existe guère d'indemnes de toute restauration. La question de leur authenticité est donc cruciale: avant de définir le style et les caractères techniques, il est nécessaire de faire la part des pièces originales ou, du moins, anciennes. L'exposé montrera à partir d'exemples précis, tirés de vitraux des anciens Pays-Bas, quels critères peuvent être dégagés pour effectuer cette critique (nature des verres et des matériaux de peinture, modalités d'utilisation de celle-ci) et quelles en sont les difficultés (repeints et recuissons qui perturbent irrémédiablement la lecture).

Après le rappel de quelques pratiques en matière de conservation et de restauration des vitraux monumentaux, la présente contribution discutera de la question de l'authenticité dans le domaine du vitrail. Seront ensuite examinés à partir d'exemples des anciens Pays-Bas quelques critères permettant d'effectuer une 'critique d'authenticité' et les difficultés inhérentes à celle-ci.

Pratiques d'antan en matière de conservation et de restauration

Les vitraux anciens belges ont tous été abondamment restaurés. Certaines archives témoignent d'un entretien régulier. On remarque que chaque édifice a généralement eu ses verriers attitrés, *grosso modo* jusqu'à la Seconde Guerre mondiale. On peut donc parler d'une véritable 'maintenance' des vitraux. Ne disposant plus des ressources techniques de leurs aînés, les verriers des périodes plus récentes utilisaient fréquemment en bouche-trou des pièces prélevées de vitraux démantelés[1]. Ces interventions, peu soucieuses de préserver la lisibilité et l'esthétique des vitraux, ont motivé des restaurations plus importantes, particulièrement au XIXe siècle. Des maîtres verriers comme le bruxellois Jean-Baptiste Capronnier[2]

* Je remercie les personnes qui m'ont aidée dans la préparation du présent article: Yvette Vanden Bemden et Dominique Vanwijnsberghe ont relu mon texte et m'ont fait bénéficier de leurs observations.

[1] Voir F. Perrot, «De l'utilité des bouche-trous pour l'histoire du vitrail», dans H. Scholz, I. Rauch, D. Hess, *Glas – Malerei – Forschung: internationale Studien zu Ehren von Rüdiger Becksmann*, Berlin, 2004, p. 73-76.

[2] Jean-Baptiste Capronnier (Bruxelles, 1814-1891) est une des personnalités majeures de l'art du vitrail belge du XIXe siècle. En 1839, il prit la direction de l'entreprise familiale fondée par son père, François (1779-1853), peintre doreur sur porcelaine d'origine française, établi depuis les années vingt à Bruxelles et gratifié en 1828 par le gouvernement hollandais d'un brevet de quinze années «pour l'invention d'un nouveau moyen pour peindre, dorer et argenter sur verre, cristal, faïence et porcelaine». À l'exposition de Paris de 1855, la seule médaille octroyée pour le domaine de la peinture sur verre lui fut décernée. Il a œuvré sur d'innombrables chantiers pour la création et la restauration de vitraux, tant en Belgique qu'à l'étranger. Voir principalement *Catalogue de la collection de cartons de vitraux peints de la succession de J.-B. Capronnier,*

(1814-1891) ou le liégeois Joseph Osterrath[3] (1845-1898) sont intervenus systématiquement sur de nombreux vitraux anciens en éliminant les bouche-trous, en renouvelant les plombs et toutes les pièces corrodées ou brisées et en refaisant les parties manquantes.

En définitive, quand on considère un vitrail ancien, on est confondu devant l'ampleur des restaurations. Il n'est pas rare que plus de la moitié des calibres (c'est-à-dire des pièces de verre) aient été remplacés et parfois bien davantage, si l'on considère par exemple le vitrail de François Buisseret de la collégiale Sainte-Waudru à Mons (fig. 1).

Il ne faut néanmoins pas conclure hâtivement à l'insignifiance de ces vitraux drastiquement restaurés. Ceux de l'abside du chœur de l'église Saint-Martin à Scry (v. 1559-1561) sont exemplaires à cet égard[4]. Récemment encore, ils étaient considérés comme totalement refaits et signalés à titre indicatif seulement, en raison de la reprise, dans la baie axiale, du motif du Christ en croix du vitrail de Jean Stouten de la cathédrale Saint-Paul à Liège (1557). Or, un examen approfondi a permis de déterminer que certains calibres, parmi les plus significatifs, sont anciens; le Christ du vitrail de la Crucifixion en conserve même davantage que son homologue liégeois. Par ailleurs, il importe de remarquer que des restaurateurs comme Capronnier et Osterrath ont généralement respecté l'image originale lorsqu'ils remplaçaient des calibres; on le constate en comparant les nouvelles pièces avec les originaux quand ceux-ci sont conservés (fig. 2). Les vitraux du chœur de la basilique de Tongres (1548-1550) ont été pratiquement renouvelés; les modèles utilisés pour leur conception peuvent néanmoins être aisément identifiés[5].

La situation peut parfois être vraiment complexe. Une gravure de Pierre Génard du milieu du XIXe siècle donne une idée de l'état de conservation du vitrail d'Engelbert II de Nassau représentant la Dernière Cène et conservé dans la chapelle du Saint Sacrement de la cathédrale d'Anvers[6]. Au même moment, l'historien Léon de Burbure fait

à Bruxelles. Vente publique le samedi 18 juin 1892 et le lundi suivant, s'il y a lieu, à 10 heures du matin, en la Salle Sainte-Gudule, 9, rue du Gentilhomme, par le ministère de MM. les Notaires De Doncker et Milcamps, Bruxelles, 1892; J. HELBIG, «Capronnier (François)», dans *Biographie nationale*, 31, Bruxelles, 1961, col. 168-169; J. HELBIG, «Capronnier (Jean-Baptiste)», dans *Biographie nationale*, 31, Bruxelles, 1961, col. 170-176; P. PAUWELS, «Capronnier, François», dans *Nationaal Biografisch Woordenboek*, 16, 2002, col. 242-243; P. PAUWELS, «Capronnier, Jean-Baptiste-François», dans *Nationaal Biografisch Woordenboek*, 16, 2002, col. 243-253.

[3] S. LAGNEAUX et M. PIROTTE, «Les ateliers Osterrath et leur production de vitraux religieux», dans *Art, technique et science: la création du vitrail de 1830 à 1930, Colloque international, Liège, Le Vertbois, 11-13 mai 2000*, Dossier de la Commission royale des Monuments, Sites et Fouilles, 7, 2000, p. 117-128.

[4] Voir I. LECOCQ, «Les vitraux du chœur de l'église Saint-Martin à Scry. Un éclairage inédit sur l'art du vitrail monumental dans la Principauté de Liège pendant la seconde moitié du XVIe siècle», dans *Annales du Cercle hutois des Sciences et des Beaux-Arts*, 57, 2004-2005, p. 49-100.

[5] Voir ID., «Copie, variation et invention dans l'art de la seconde moitié du XVIe siècle», dans *VIIe Congrès de l'Association des Cercles francophones d'Histoire et Archéologie de Belgique & LIVe Congrès de la Fédération des Cercles d'Archéologie et d'Histoire de Belgique, Louvain-la-Neuve, 26-28 août 2004*, Actes.

[6] Voir J. HELBIG, *Les vitraux de la première moitié du XVIe siècle conservés en Belgique. Province d'Anvers et Flandres*, Corpus Vitrearum Belgique, 2, Bruxelles, 1968, p. 36-51; S. GRIETEN, M. MEES, J. VAN DAMME et L. VAN LANGENDONCK, *Een venster op de hemel. De Glasramen van de Onze-Lieve-Vrouwekathedraal van Antwerpen*, cat. d'exposition (cathédrale d'Anvers, 7 sept.-31 octobre 1996), Anvers, 1996, p. 56-59; Z. VAN RUYVEN-ZEMAN, «Het Laatste Avondmaal-glas in de Onze-Lieve-Vrouwkerk te Antwerpen. Een oude toeschrijving herzien», dans *Belgisch Tijdschrift voor Oudheidkunde en Kunstgeschiedenis*, 57, 1988, p. 51-66.

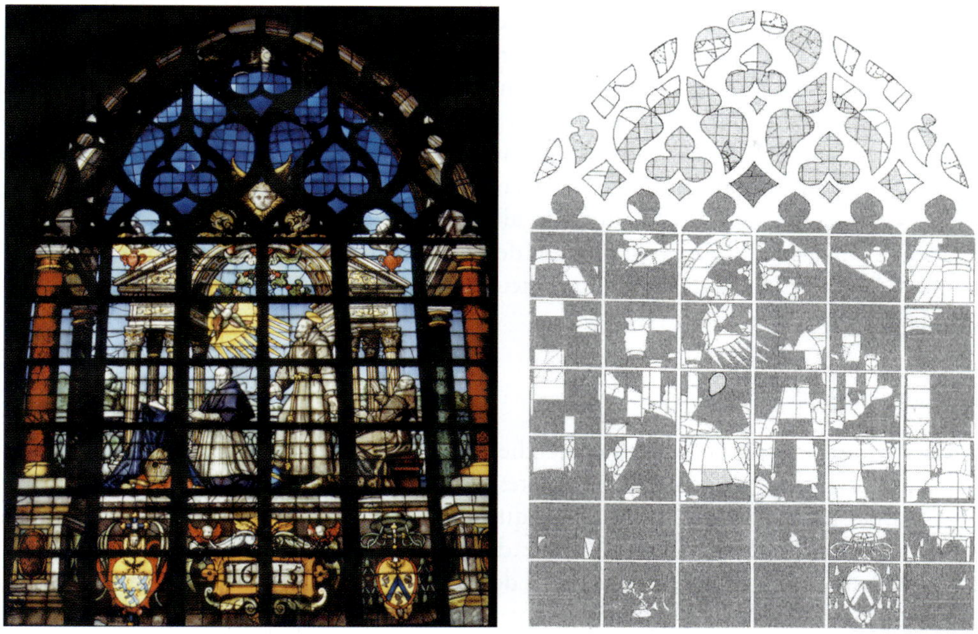

Fig. 1. Vitrail de François Buisseret, 1615, Mons, collégiale Sainte-Waudru (© IRPA). Les parties anciennes sont celles qui ne sont pas recouvertes par des grisés sur le schéma.

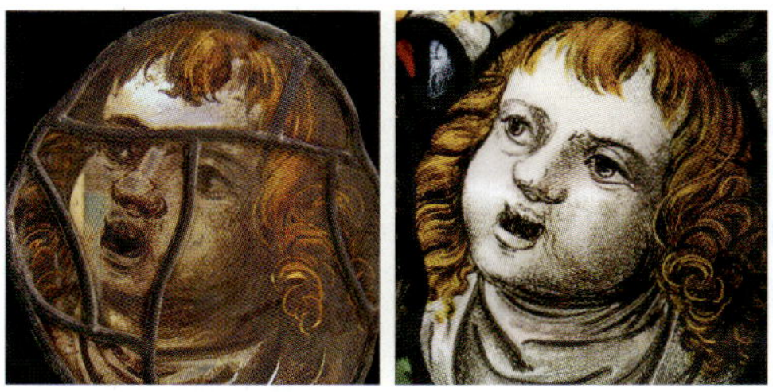

Fig. 2. Vitrail de la cathédrale Saint-Paul à Liège, 1530, détail avec une tête, original conservé dans une collection privée et copie réalisée par l'atelier Osterrath à la fin du XIXe siècle (© I. Lecocq).

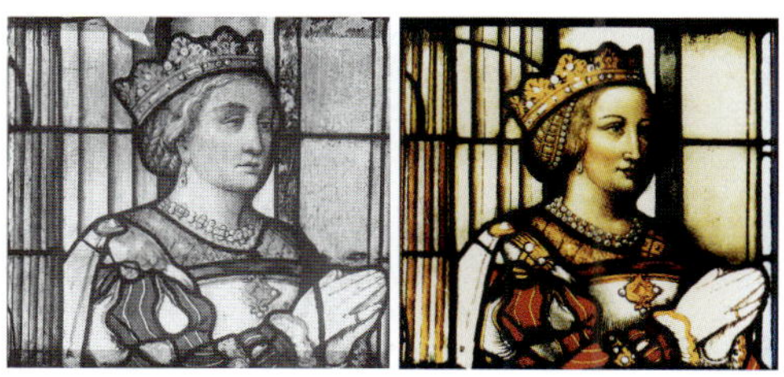

Fig. 3. Vitrail de Jean III de Portugal dans la chapelle du Saint Sacrement de la cathédrale Saints-Michel-et-Gudule, 1542. Détail avec la tête de Catherine d'Aragon, avant la restauration de 1950 et actuellement (© IRPA et Glaswerken Mortelmans).

des recherches dans les archives de la cathédrale et retrouve un texte qui mentionne la rétribution en 1503 du célèbre verrier bruxellois Nicolas Rombouts pour un vitrail placé dans la chapelle du Saint Sacrement. De là à en déduire qu'il s'agit du vitrail d'Engelbert II de Nassau, il n'y avait qu'un pas, que l'on a franchi allègrement. Le vitrail est restauré à partir de 1870 par Capronnier. Celui-ci introduit un fond derrière la Cène, de son invention semble-t-il. Il recompose un décor architectural à partir d'autres vitraux de Nicolas Rombouts. Par ailleurs, S. Van Ruyven-Zeman a montré en 1988 que la Dernière Cène reprenait un dessin de Lambert van Noort. Elle en a déduit que la Cène centrale fut détruite au milieu du XVIe siècle, lors des conflits religieux, et qu'elle dut alors être restaurée.

Une authenticité plurielle

En définitive, où faut-il donc rechercher l'authenticité d'un vitrail? En fait, il n'y a pas une authenticité, mais des authenticités. Chercher 'l'authenticité' d'un vitrail ancien est absurde: aucun vitrail n'est 'authentique' dans le sens plénier du terme, c'est-à-dire «transmis sans subir d'altération, même dans ses nuances»[7]. Les vitraux subissent au cours de leur existence des altérations et des restaurations de nature diverse (cf. *supra*).

On peut distinguer une 'authenticité formelle', une 'authenticité des matériaux' et une 'authenticité historique'[8].

L'authenticité formelle d'un vitrail concerne celui-ci en tant qu'œuvre matérielle élaborée par un concepteur et un réalisateur à un moment donné. Aucun vitrail n'est formellement pleinement authentique puisque aucun n'est resté en son état initial. Même lorsqu'ils travaillent à «l'identique», avec la volonté de reproduire le plus fidèlement possible les originaux qu'ils remplacent, les maîtres verriers altèrent le caractère des formes initiales. Lorsque Jean-Baptiste Capronnier s'efforce de copier une série de têtes provenant de vitraux de la cathédrale Saints-Michel-et-Gudule de Bruxelles, l'aspect est globalement satisfaisant, mais, vue de près, la facture n'a pas le même brio[9]. Parfois les restaurateurs modifient les détails iconographiques des parties qu'ils renouvellent. Après la Seconde Guerre mondiale, dans le vitrail de Jean III de Portugal de la même cathédrale (1542), la coiffe de Catherine d'Aragon a été enrichie d'un bandeau perlé (fig. 3).

L'authenticité des matériaux n'est jamais pleinement acquise. Par exemple, en Belgique, on ne possède plus aucun vitrail ancien avec son réseau de plombs d'origine. Et quand un verre est ancien, cela ne veut pas dire que la peinture l'est forcément: les restaurateurs n'hésitaient parfois pas à repeindre et à recuire des pièces de verre. Pour mémoire, la peinture la plus couramment utilisée par les maîtres verriers (la 'grisaille'), à base de poudre de verre et d'oxyde de fer, ne peut être durablement fixée que par

[7] Extrait de la définition de R. LEMAIRE, «Authenticité et patrimoine monumental», dans le *Dossier de la Commission royale des Monuments. Sites et Fouilles*, 5, 1999, p. 175.

[8] Voir le point de vue de R. LEMAIRE qui remarque judicieusement que «l'emploi du mot 'authenticité' non assorti d'une spécification appropriée est vidé de toute signification valable» (R. LEMAIRE, 1999, p. 184).

[9] I. LECOCQ et Y. VANDEN BEMDEN, «L'examen et la critique d'authenticité des vitraux des XVIe et XVIIe siècles», dans I. Lecocq (dir.), *Les vitraux de la cathédrale Saints-Michel-et-Gudule de Bruxelles. Histoire, conservation et restauration*, Institut royal du Patrimoine artistique, Scientia Artis, 2, Bruxelles, 2005, fig. 182, p. 229.

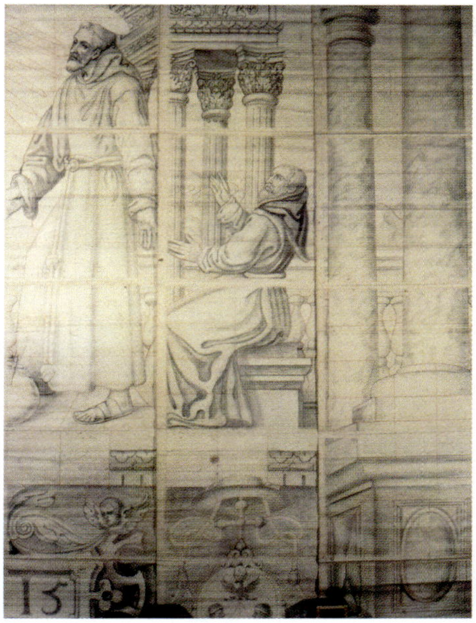

Fig. 4. Cartons pour la restauration du vitrail de François Buisseret, 1845-1848, Bruxelles, Musées royaux d'Art et d'Histoire, fonds Capronnier, boîtier n°54 (© I. Lecocq).

cuisson. Les grisailles recuites s'identifient parfois aisément par leur tonalité rougeâtre caractéristique.

L'authenticité historique prend en compte tous les événements (accidents et interventions volontaires) qui modifient un vitrail donné au cours du temps. C'est cette authenticité qui souvent a tendance à prévaloir: l'œuvre telle que nous la voyons résulte de toute une série d'interventions qu'il importe d'identifier et de documenter. Les historiens d'art du vitrail s'efforcent d'ailleurs, avant toute chose, de reconstituer ce qu'ils appellent «l'histoire matérielle». Celle-ci mène tout naturellement à la critique d'authenticité des vitraux, c'est-à-dire l'identification et la caractérisation des différentes interventions qui les ont modifiés et transformés. Cette critique d'authenticité doit idéalement être réalisée lorsque le vitrail peut être approché de près: sur un échafaudage ou en atelier, au moment de sa restauration.

En Belgique, la critique d'authenticité de maints vitraux anciens a été facilitée grâce aux cartons établis par l'atelier Capronnier pour ses restaurations et conservés notamment aux Musées royaux d'Art et d'Histoire[10] (fig. 4). Ils indiquent au moyen de chiffres la localisation des pièces remplacées. Ces cartons sont des témoins d'autant plus précieux que, dans certains cas, ils documentent précisément des œuvres aujourd'hui disparues, comme les vitraux de l'église Saint-Servais à Liège, détruits par un incendie le 21 août 1981. Le relevé des pièces remplacées effectué par Capronnier n'est malheureusement pas toujours complet. Pour le vitrail de Ferdinand Ier et Anne de Bohème de la cathédrale Saints-Michel-et-Gudule (1546), l'examen sur un échafaudage a amené à

[10] Une autre partie importante de ce fonds, dernièrement achetée par la communauté flamande, est déposée à Louvain au KADOC (Centre de Documentation et de Recherche pour la Religion, la Culture et la Société).

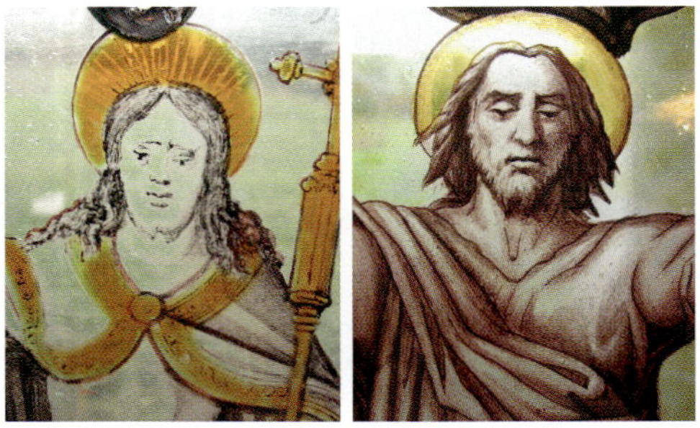

Fig. 5. Vitraux de la chapelle du château de la Follie (Écaussines), 1520-1530 et XIXe siècle (© I. Lecocq).

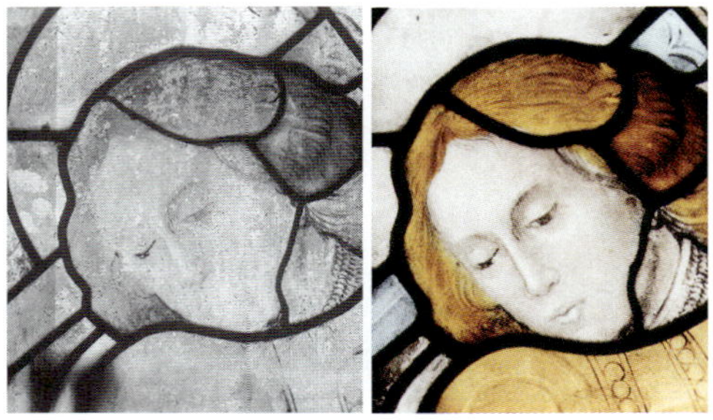

Fig. 6. Vitrail de Philippe le Beau et Jeanne de Castille dans le chœur de la cathédrale Saints-Michel-et-Gudule, 1520-1530. Détail avec la tête de saint Michel, avant la restauration de 1950 et actuellement (© IRPA et C. Van den Wijngaert).

revoir la critique d'authenticité proposée par Jean Helbig sur la base des cartons du restaurateur: beaucoup de calibres renseignés comme anciens sur ces cartons portent un numéro gravé par Capronnier, ce qui signifie qu'ils ont été remplacés. En définitive, moins d'un quart de la superficie de ce vitrail peut être considéré comme antérieur au XIXe siècle[11].

En l'absence de documents d'archives, toute une série de critères peuvent aider l'auteur de la critique d'authenticité à estimer l'ancienneté des pièces de verre[12], notamment leur homogénéité à l'intérieur du panneau, que ce soit par le style, la teinte de la grisaille, la continuité des lignes du dessin; l'état de conservation du verre, le verre ancien étant parfois corrodé superficiellement (cette corrosion prend des formes variées qui peuvent aller jusqu'à des cratères); l'aspect du verre, des irrégularités de surface caractérisant souvent les verres anciens; l'aspect de la peinture. La grisaille ancienne, grise et noire, a été appliquée, diluée et modulée par épargnes et grattages qui mettent le verre à nu (fig. 5, à gauche). La peinture récente est souvent rouge et plus épaisse. La technique est différente: la grisaille, appliquée en grandes quantités, est plus couvrante et moins enlevée (fig. 5, à droite).

Difficultés inhérentes à la critique d'authenticité

La distinction entre les parties anciennes et les parties remplacées est parfois délicate: certains calibres anciens peuvent avoir été repeints et recuits et, dans certains cas, il est

[11] I. Lecocq et Y. Vanden Bemden, 2005, fig. 177, p. 223
[12] I. Lecocq et Y. Vanden Bemden, 2005, p. 206-209.

difficile de savoir si on a affaire à un nouveau calibre ou à un calibre ancien mais repeint. Parfois aussi, l'ancienneté d'un calibre est assurée, sans que l'on puisse affirmer qu'il s'agisse du calibre original. Il est donc souvent préférable de parler de 'calibres anciens' plutôt que de 'calibres originaux'.

Par ailleurs, certains calibres anciens ne comportent parfois pratiquement plus de peinture, effacée par usure ou par manque d'adhérence. Des compléments peuvent avoir été peints sur un verre de doublage pour leur restituer une lisibilité et améliorer la perception globale du vitrail (fig. 6). Dans ce cas, il importe de ne pas se laisser abuser par la qualification de 'calibre ancien'. Le verre est ancien, certes, mais on ne conserve de la matière picturale que des traces ou des 'fantômes' qui ne rendent plus compte du savoir-faire des maîtres verriers.

La distinction de différentes époques parmi les calibres remplacés est parfois très hasardeuse. Dans ce contexte, on parlera de pièces remplacées à une date indéterminée. Il faut enfin signaler que beaucoup de calibres sont généralement douteux. Si une tête du XVIe ou du XVIIe siècle remplacée au XIXe ou au XXe siècle se reconnaît aisément, ce n'est pas le cas pour un segment de pilastre, uniquement peint de quelques lignes verticales, un fond simplement putoisé ou une partie de nuage. Il faut donc parfois accepter certaines approximations.

Conclusion

La notion d'authenticité, communément utilisée, doit être maniée avec prudence: elle est complexe et ambiguë. L'authenticité: oui,... mais quelle authenticité? Une interprétation hâtive peut mener à une politique de sauvegarde inadaptée ou dogmatique. Cette complexité ne dispense pas de réaliser une critique d'authenticité aussi complète que possible afin de solliciter pertinemment les rares témoins qui nous sont parvenus.

Isabelle LECOCQ
Institut royal du patrimoine artistique
Parc du Cinquantenaire, 1
B – 1000 Bruxelles
Tel: +32.(0)2.739.68.09
Fax: +32.(0)2.732.01.05
isabelle.lecocq@kikirpa.be

Née à Mons le 29 novembre 1972. 1995: licence en Histoire de l'Art et Archéologie, orientation Histoire de l'Art et Archéologie de l'Ère moderne et agrégation de l'enseignement secondaire supérieur en Initiation esthétique et histoire de l'art à l'Université de Liège. Janvier 1996 – juillet 1996: stage de spécialisation en Allemagne (Linnich) dans l'atelier de création et de restauration de vitraux OIDTMANN. Depuis le 17 mars 1997, attachée puis première assistante KIKIRPA, au sein du Département Conservation & Restauration (études et conseils pour le domaine du vitrail, suivi des études préalables et des interventions de conservation et restauration). 2001: doctorat en Philosophie et Lettres aux Facultés universitaires Notre-Dame de la Paix à Namur (sujet de thèse: *Le vitrail monumental en Wallonie pendant la seconde moitié du XVIe siècle*).

DIPTYQUES, TRIPTYQUES, QUADRIPTYQUES… INSTRUCTIONS POUR UN BON USAGE

Hélène Verougstraete

Résumé: Il y avait des diptyques très petits, certains tenaient dans la paume d'une main, d'autres étaient destinés à être suspendus aux courtines d'un lit. Leur usage ne pose pas de grand problème. Plus difficile à évaluer correctement est l'usage qui était fait des diptyques plus grands, ceux que l'on posait sur un meuble, sur le plan incliné d'un prie-Dieu, sur un coussin, ouverts à la contemplation, puis que l'on refermait et rangeait dans un sac ou 'couverte'[1]. La plupart des diptyques étaient mobiles. Le fait que les revers des deux feuillets[2] d'un diptyque soient le plus souvent recouverts de peinture atteste que toutes les faces étaient occasionnellement exposées au regard. Très souvent les deux feuillets sont hiérarchisés. La partie subordonnée pivotait au titre de volet pour recouvrir la partie principale. On ouvrait certains diptyques de gauche à droite et d'autres de droite à gauche, selon les cas. Partant de cette constatation, il est logique de penser qu'on puisse reconnaître sur le diptyque des 'instructions pour un bon usage'. Ces instructions relèvent parfois de la menuiserie, mais elles peuvent aussi être inhérentes à la représentation elle-même: la perspective, la répartition de la lumière… qui l'une et l'autre peuvent guider le regard, anticiper visuellement le mouvement de rotation du volet, hiérarchiser les espaces… Les conventions adoptées pour les diptyques s'appliquent également à certains triptyques et quadriptyques.

L'article sur les cadres des diptyques flamands des 15e et 16e siècles[3] dans le volume d'essais qui accompagnait le catalogue de l'exposition *Prayers and Portraits. Unfolding the Netherlandish Diptych* en 2006-2007[4] est à l'origine du présent travail. Les organisateurs de l'exposition nous avaient demandé d'écrire un article sur les cadres des diptyques. La tâche n'était pas aisée, étant donné que la plupart de ces diptyques sont aujourd'hui démembrés, et qu'ils sont rares à conserver leur cadre d'origine. J'aurais pu tenter d'en établir le catalogue et de décrire ce qui subsiste de la menuiserie originale. Mais je me suis rendu compte que la menuiserie ne peut se comprendre isolément.

[1] Jeffrey C. SMITH, «The Practical Logistics of Art: Thoughts on the Commissioning, Displaying and Storing of Art at the Burgondian Court», dans *In Detail, New Studies of Northern Renaissance. Art in Honor of Walter S. Gibson*, ed. Laurinda S. Dixon, 1998, 27-48. Rares sont les diptyques encore complets avec leur sac, comme le *Diptyque Matheron*, attribué à Nicolas Froment, représentant René d'Anjou et Jeanne de Laval, c. 1476 (Paris, Louvre, inv. R.F.665).

[2] Le terme 'volet' s'applique correctement au seul feuillet qui a la fonction de fermer, de protéger, l'autre partie constituant alors, en termes de menuiserie, la 'partie dormante'. Nous préférons pour désigner les deux éléments du diptyque le terme 'feuillet' utilisé dans les textes de l'époque (voir Lorne CAMPBELL, «Diptychs with Portraits», dans *Essays in Context. Unfolding the Netherlandish Diptych*, eds. John Oliver Hand et Ron Spronk, Cambridge, New Haven-Londres, 2006, p. 33-45).

[3] Hélène VEROUGSTRAETE, «Diptychs with instructions for use», dans *Essays in context. Unfolding the Netherlandish Diptych*, eds. John Oliver Hand et Ron Spronk, Cambridge, New Haven-Londres, 2006, p. 156-171.

[4] John Oliver HAND, Catherine METZGER et Ron SPRONK, *Prayers and Portraits. Unfolding the Netherlandish Diptych*, catalogue d'exposition, National Gallery of Art Washington, Koninklijk Museum voor Schone Kunsten Antwerpen, novembre 2006-mars 2007, en association avec les Harvard University Art Museums, Cambridge and Yale University Press, New Haven, 2006.

Dans la peinture des Primitifs flamands, où la frontière entre le monde matériel et le monde spirituel est estompée, la menuiserie est au service du sens que l'artiste donne à son œuvre, comme on le verra plus loin. Seule la compréhension globale de l'objet, de la hiérarchie des représentations sur chaque élément du diptyque, triptyque... permet le juste maniement, la bonne ouverture ou fermeture. Inversement, seule une correcte ouverture (et fermeture) rend justice au sens global de l'objet. La menuiserie est au service d'une peinture qui adopte des conventions et un langage codé. Dans l'art raffiné des Primitifs flamands, les conventions sont nombreuses, elles peuvent être ambivalentes, ou incompatibles. Un donateur, subordonné à la Vierge et Enfant qu'il implore, sera le plus souvent relégué sur le volet, qui pivotera. Mais un volet a aussi fonction de fermer et protéger, et on verra à l'occasion un saint patron, ou même une Vierge et Enfant assurer cette fonction au sens littéral. Ils sont représentés sur un volet qui pivote sur le donateur pour en assurer la protection, au sens littéral et figuré.

La plupart des diptyques, triptyques... sont aujourd'hui figés dans une ouverture muséale dont le caractère irréversible découle de la destruction d'éléments incompris, jugés secondaires. Tout au plus soupçonne-t-on confusément que certains diptyques devaient s'ouvrir de gauche à droite, et d'autres de droite à gauche, sans toujours comprendre les raisons de cette diversité. J'ai donc pris comme point de départ cette exigence d'une ouverture qui n'était pas indifférente, pour me dire que l'usager de l'époque devait se trouver en face de consignes, d'instructions 'pour bien ouvrir' et 'pour bien fermer'. Il est impensable qu'il ait procédé 'par essais et erreurs'. C'est à rechercher ces consignes que je me suis appliquée. J'ai réalisé des maquettes des diptyques quand cela était possible[5]. Mes conclusions dépassent la stricte menuiserie des cadres mais concernent aussi la représentation elle-même, la perspective et l'éclairage par exemple, avec des implications lourdes de sens non seulement pour les diptyques mais aussi pour les triptyques, quadriptyques... L'usager savait comment ouvrir et fermer, il avait sous les yeux des consignes.

La conclusion la plus originale à laquelle je suis arrivée est peut-être celle qui concerne l'usage que les Primitifs flamands ont fait de la perspective. Une juste représentation de l'espace ne semble pas avoir été leur préoccupation. Les peintres ont utilisé la perspective pour guider le regard du spectateur, l'instruire sur la hiérarchie des représentations, des espaces... Ce système est appliqué dans le registre supérieur de l'*Agneau mystique*. A ma connaissance, on n'avait jamais relevé que le spectateur ne pouvait pas lire ce registre de manière linéaire, il ne peut pas y voir une séquence de grands personnages alignés à un même niveau. Au contraire, guidés par la perspective, il doit voir des personnages hiérarchisés (fig. 1). Le langage codifié de la perspective s'organise autour de l'axe vertical central occupé par la divinité, placée sur une marche. Un seul et même carrelage orne la marche et le sol sous Marie et saint Jean. Ces trois personnages se situent dans un même espace, hiérarchisé par la marche. De part et d'autre de l'axe central, les lignes fuyantes des carrelages s'étalent en éventail. Le carrelage sous les anges est décoré différemment et ses lignes fuyantes se déploient en lignes obliques qui sont en discontinuité avec celles du carrelage central. Nous sommes invités à élever la partie

[5] Les revers des peintures sont rarement représentés dans les livres. Fait exception le livre de Angelica DÜLBERG, *Privatporträts. Geschichte und Ikonologie einer Gattung im 15. und 16. Jahrhundert*, Berlin, 1990.

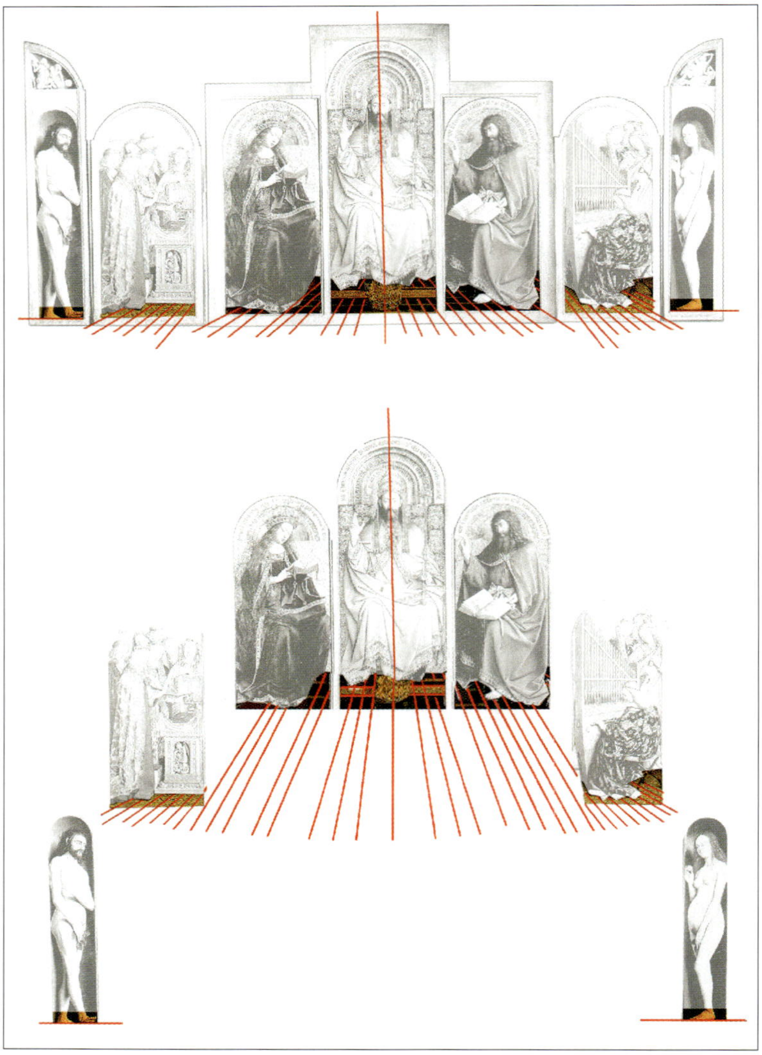

Fig. 1. Schéma de la perspective de Van Eyck, *Agneau Mystique*, 1432, registre supérieur, dimensions de la partie centrale: 230 × 255,8 cm (Gand, cathédrale Saint-Bavon). La perspective discontinue d'un panneau à l'autre invite à la lecture d'un monde hiérarchisé.

centrale (avec la divinité, Marie et saint Jean), pour harmoniser l'éventail des lignes des deux carrelages. Adam et Eve ont les pieds sur terre. L'inversion de la perspective du pied d'Adam, vu en contre-plongée, est tellement appuyée que nous ne pouvons ignorer le message: Adam et Eve n'appartiennent pas aux sphères célestes. Chacun se situe à son niveau, sur une marche, sur un premier carrelage, sur un second carrelage, sur terre…

La perspective, dans le registre supérieur de l'*Agneau mystique*, reflète la hiérarchie du monde. Lorsque les historiens de l'art ont souligné le manque de savoir-faire des peintres flamands en matière de représentation de la perspective et leur méconnaissance du point de fuite unique (pourtant bien maîtrisé à l'époque par les italiens), ils ignoraient que les Primitifs flamands avaient d'autres ambitions que de réaliser un ensemble cohérent de lignes fuyantes. Le passage à une plus juste représentation de la perspective — à l'italienne — a dû être pour les flamands un bouleversement, l'abandon d'un système codifié, d'un langage symbolique. Ce passage impliquait peut-être pour certains un renoncement autant qu'une conquête.

Revenons à la manière correcte d'ouvrir ou de fermer le diptyque. La perspective joue en la matière un rôle important. La frontalité d'une représentation, d'une Vierge à l'Enfant par exemple, sur un des deux feuillets invitait le spectateur à se situer face à cette représentation, à en privilégier la contemplation immédiate (fig. 2). Les lignes fuyantes, en demi éventail, des carrelages et/ou plafonds sur le volet, avec un donateur par exemple, rendaient cet espace subordonné (généralement le donateur est représenté sur le volet droit, celui de la gauche héraldique), et anticipaient le mouvement du volet. Dans le cas du donateur sur le volet droit, le diptyque s'ouvre de gauche à droite. Les conventions picturales qui régissent la préséance sont connues, mais imparfaitement puisqu'on ne comprend pas pourquoi les artistes y dérogent parfois. Certaines conventions étaient peut-être incompatibles entre elles. Une séquence historique (se déroulant de gauche à droite) pouvait entrer en conflit avec l'attribution des dextres et sénestres héraldiques. Quoi qu'il en soit, dans des espaces hiérarchisés, les liens entre les personnages sont indiqués d'un geste, d'une attitude, d'un doigt pointé, d'un regard... Dans le *Diptyque de Martin Van Nieuwenhove* auquel on songe immanquablement, les personnages sont hiérarchisés par la perspective. La Vierge est vue frontalement tandis que Martin est placé devant des lignes fuyantes. Le miroir derrière Marie unit des personnages situés dans l'espace unique d'une même chambre. La perspective souligne l'humilité de Martin (exprimée par les lignes fuyantes) qui implore Marie avec intensité puisqu'il se trouve dans le même espace qu'elle (ce qu'atteste le miroir).

Mais les représentations ne sont pas toujours hiérarchisées. Certaines se succèdent dans une séquence narrative ou historique, de gauche à droite. La perspective frontale de chaque épisode invite à la contemplation successive de chaque événement comme dans le *Retable de la vie du Christ* de Roger Van der Weyden (fig. 3). Dans ce cas, il s'agit d'un triptyque fixe, mais le même usage d'épisodes qui se succèdent frontalement vaut pour certains diptyques et triptyques articulés.

Les instructions pour un bon usage se manifestent sur le diptyque fermé autant que sur le diptyque ouvert. Elles peuvent relever de la menuiserie: de la mouluration du cadre, de la disposition du crochet de fermeture, de l'emplacement des charnières... Elles peuvent aussi être peintes: le décor, la perspective, l'angle d'incidence de la lumière... On a donné à certains diptyques la forme d'un livre, avec les feuillets taillés en trompe-l'œil sur les chants des planches. La forme de livre était à elle seule une instruction sur la juste ouverture, qui devait se faire, comme pour tous les livres, de droite à gauche.

Certains objets complexes se déploient par étapes successives, formant un diptyque à un certain stade de leur déploiement, pour devenir ensuite triptyque et/ou quadriptyque. Le déploiement par étapes pouvait exposer des contenus successifs. Il existait

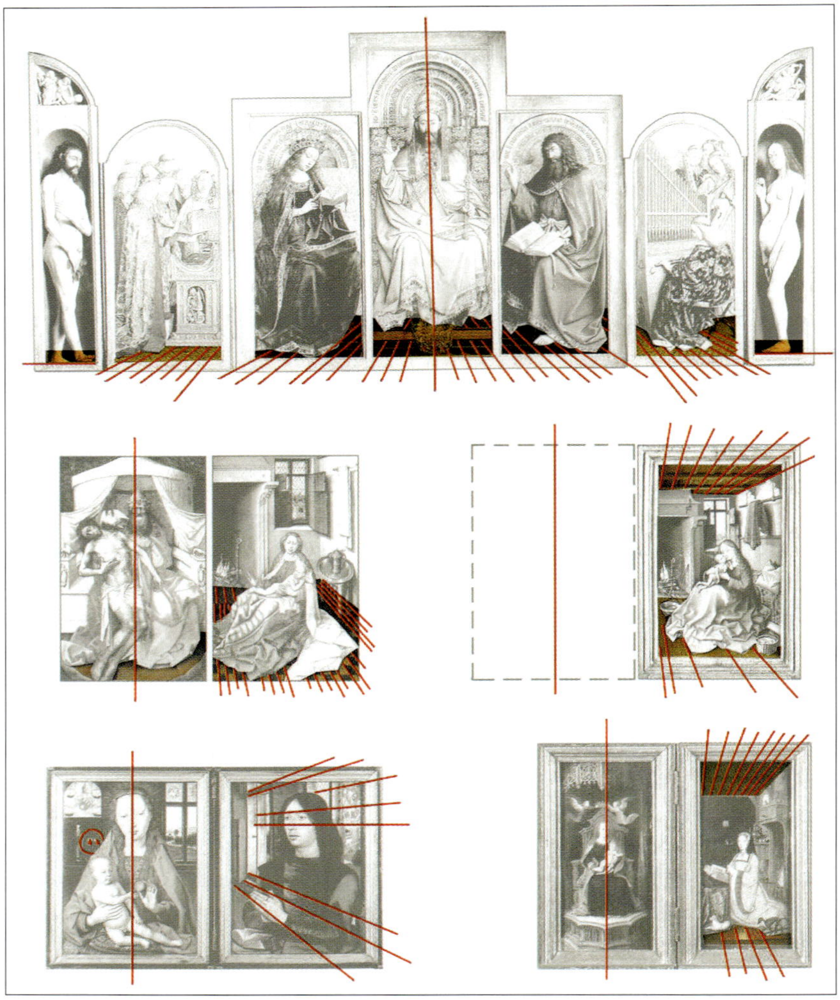

Fig. 2. Exemples de représentations hiérarchisées au moyen de la perspective. La frontalité d'une représentation appelle à sa contemplation directe. Dans le cas d'un diptyque, un demi éventail de lignes fuyantes (carrelages, plafonds, voûtes…) marquent l'espace secondaire du volet). Van Eyck, *Agneau Mystique*. Attribué à Robert Campin, *Diptyque avec Trinité et Madone devant l'âtre*, c. 1420/1425, bois, 34,3 × 24,3 cm (Saint-Pétersbourg, Ermitage, inv. 447, 448). Atelier de Robert Campin (Jacques Daret?), *Madone devant l'âtre*, 1428/1432? (Londres, National Gallery, inv. 6514). Hans Memling, *Diptyque de Martin van Nieuwenhove*, 1487, bois, chaque élément 52,5 × 41,5 cm, cadre original inclus (Bruges, Hôpital Saint-Jean, inv. O.SJ178.I). Attribué au Maître de 1499, *Diptyque de Marguerite d'Autriche*, après 1501, bois 30,6 × 14,6 cm (Gand, Musée des Beaux-Arts, inv. 1973-A).

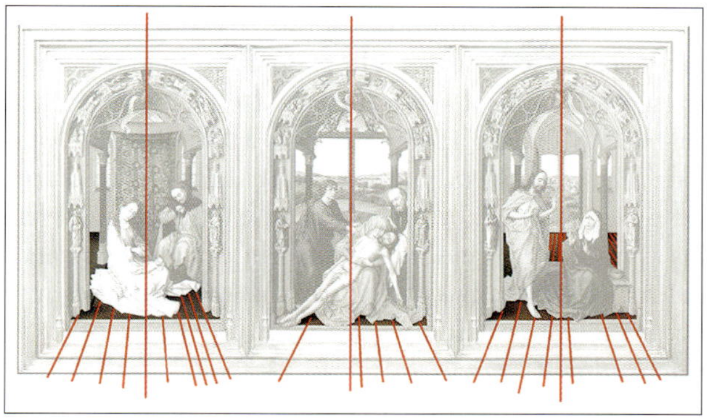

Fig. 3. Exemple de trois espaces successifs: la disposition frontale répétée dans chaque composition correspond au mode narratif ou historique. Rogier van der Weyden, *Retable de Miraflores*, chaque panneau: 71 × 43 cm, cadres originaux perdus (Berlin, Gemäldegalerie, cat. 534 A).

même des diptyques réversibles, véritables 'objets à transformation'. On constate que la réalité matérielle de l'objet et le contenu spirituel sont indissociables.

'Pour bien ouvrir' un diptyque: les instructions sur le diptyque fermé

On conserve au Louvre un diptyque eyckien sur chêne taillé en forme de livre: le *Diptyque de la Vierge et Enfant et de saint Jean Baptiste*, c. 1440 (Paris, Louvre, inv. RF.38-22). Saint Jean-Baptiste occupe le feuillet gauche. Une reconstitution sur base des restes de la polychromie originale (fig. 4) (marbrure peinte verte du côté de saint Jean-Baptiste, et rouge du côté de la Vierge et Enfant) montre que le diptyque fermé, avec ses ais 'jaspés', ressemble beaucoup à bon nombre des livres des 17e et 18e siècles, avec leurs ais couverts de papier marbré[6]. L'adoption de couleurs différentes pour chaque ais peut surprendre. Mais ces couleurs pourraient avoir un sens symbolique: le vert, couleur de l'attente, convenait au précurseur et le rouge celle de l'accomplissement, celle du cœur, de la charité, du sang versé convenait à la Vierge et Enfant, selon un code de couleurs rarement identifié dans les peintures flamandes anciennes. J'ai observé jadis au Louvre, à l'aide d'un microscope, que sous la dorure actuelle du cadre intérieur du diptyque, subsistait une polychromie rouge. La dorure sur la tranche avec ses feuillets taillés en trompe-l'œil est originale. Ni la dorure au dos du livre, ni les traces de charnières ne paraissent originales. Il est possible que ces éléments remplacent un dos en cuir, comme on en observe dans d'autres diptyques 'en forme de livre'. Les stucs au centre des ais ne sont pas originaux non plus. Il devait y avoir à cet emplacement une pierre semi-précieuse tenue par quatre griffes dont les emplacements sont bien visibles autour du stuc. Quatre traces de forte densité (éléments métalliques?) se voient sur la radiographie.

[6] Les sources d'inspiration de cette marbrure eyckienne et sa relation avec les manuscrits et les papiers marbrés ont été évoquées dans Hélène VEROUGSTRAETE et Roger VAN SCHOUTE, «The Significance of Marblings and Other Decorative Paintings on Reverses in Early Netherlandish Painting», dans *Painting Techniques. History, Materials and Studio Practice. IIC. Contributions to the Dublin Congress, 7-11 September 1998*, Dublin, 1998, p. 98-100. Hélène VEROUGSTRAETE et Roger VAN SCHOUTE, «Frames and supports of some Eyckian Paintings», dans *Investigating Jan van Eyck. Proceedings of a Symposium, National Gallery, London*, eds. Susan Foister, Sue Jones et Delphine Cool, Londres-Turnhout, 2000, p. 107-117.

Les diptyques en forme de livre restent une exception. De nombreux diptyques comportent, lorsqu'ils sont fermés, deux revers asymétriques. On exposait au regard la face la plus élaborée. Parmi les asymétries, il y a la mouluration des cadres. Dans le *Diptyque de Martin Van Nieuwenhove* de Hans Memling, 1487 (Bruges, Hôpital Saint-Jean, inv. O.SJ178.I), le cadre au revers de Martin descend en biseau vers le panneau[7], alors que le cadre au revers de la Madone présente une arête vive. Le cadre à biseau au revers de Martin doit se trouver par-dessus avant ouverture du diptyque. Il devait évidemment toujours y avoir moyen de reconnaître le haut du bas. Le portrait de Martin constitue le volet et devra pivoter. Le feuillet principal (partie dormante du diptyque) porte la Madone. Notons que le *Diptyque de Jean de Carondelet* de Jean Gossaert, 1517 (Paris, Louvre, inv. 1442-1443) présente la situation inverse: le biseau se situe au revers de la Madone qui constituait le volet; au revers du feuillet avec Jean de Carondelet, l'arête vive du cadre correspond à la partie dormante. On peut se demander si par là l'artiste a souhaité évoquer la fonction de protection de la Madone, celle que remplit au sens littéral tout volet. Cette fonction de protection sera reconnaissable dans certains triptyques à trois volets égaux, dont la fermeture finale peut être assurée par la représentation d'un saint patron protecteur, comme on le verra plus loin.

D'autres revers de diptyques étaient hiérarchisés au moyen de décors asymétriques, par exemple une grisaille ou un autre motif élaboré d'un côté et, de l'autre côté, une couleur décorative ou une marbrure. Pour préparer l'ouverture du diptyque de Jan Provoost, *Le Christ portant la croix- Donateur de 54 ans*, 1522 (Bruges, Hôpital Saint-Jean, inv. O.SJ191) le crâne dans la niche doit se trouver par-dessus (fig. 5), comme l'ont fait remarquer les auteurs[8]. Doit se trouver également au-dessus, la grisaille qui orne le revers du volet du *Diptyque de la Vierge à l'Enfant avec donateur* de Hans Memling (Chicago, The Art Institute, Ryerson Collection et don d'Arthur Sachs, n° 1933.1050 et 1953.467).

La perspective et l'incidence de la lumière sur le diptyque fermé jouaient également un rôle. Pour les revers des volets de diptyques, les artistes flamands des 15e et 16e siècles ont opté pour une disposition frontale du motif, souvent une niche avec un éclairage latéral très accentué (fig. 7). Dans les triptyques, par contre, on observe généralement sur chacun des volets une perspective oblique, vers la gauche et vers la droite, à partir des montants centraux des cadres (d'où part le regard frontal du spectateur), le tout baigné dans une lumière latérale généralement douce, projetée par une source unique. L'éclairage latéral dans le cas des diptyques paraît répondre à une convention. Une niche se prête bien à un éclairage latéral qui éclaire l'une de ses parois et plonge l'autre dans l'ombre. Cet éclairage anticipe de manière illusionniste la rotation du volet: en pivotant, la perspective de la niche est progressivement modifiée, de frontale à oblique. Peu à peu, seule la paroi éclairée reste visible (fig. 7). La paroi proche de l'usager, ombrée

[7] Hélène VEROUGSTRAETE et Roger VAN SCHOUTE, «Cadres et supports chez Memling», dans *Memling Studies. Proceedings of a Symposium*, éd. Hélène Verougstraete, Roger Van Schoute et Maurits Smeyers, Leuven, 1997, p. 269-286. Hélène VEROUGSTRAETE et Roger VAN SCHOUTE, *Cadres et supports dans la peinture flamande aux 15e et 16e siècles*, Heure-le-Romain, 1989.

[8] Ron SPRONK, «Diptyque d'un frère mineur», dans *Bruges et la Renaissance. De Memling à Pourbus*, catalogue d'exposition, Groeningemuseum, Bruges, 1998, p. 36-38.

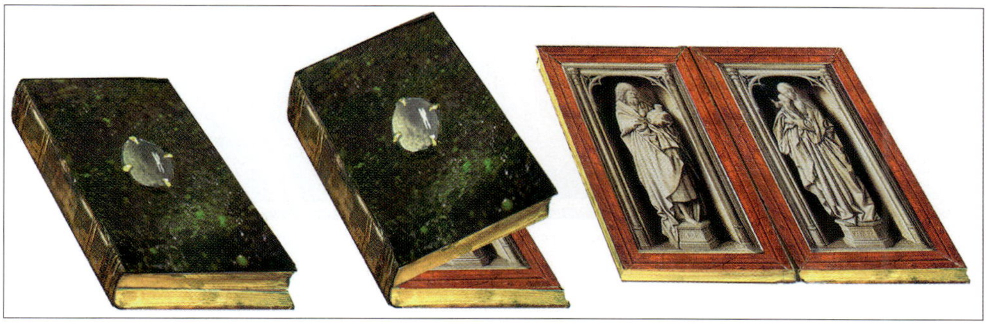

Fig. 4. Reconstitution de l'aspect original: atelier de Van Eyck, *Diptyque avec Vierge et Enfant et saint Jean-Baptiste*, c. 1440, 38,2 × 23,4 cm (Paris, Louvre). La forme de livre invite à l'ouverture de droite à gauche.

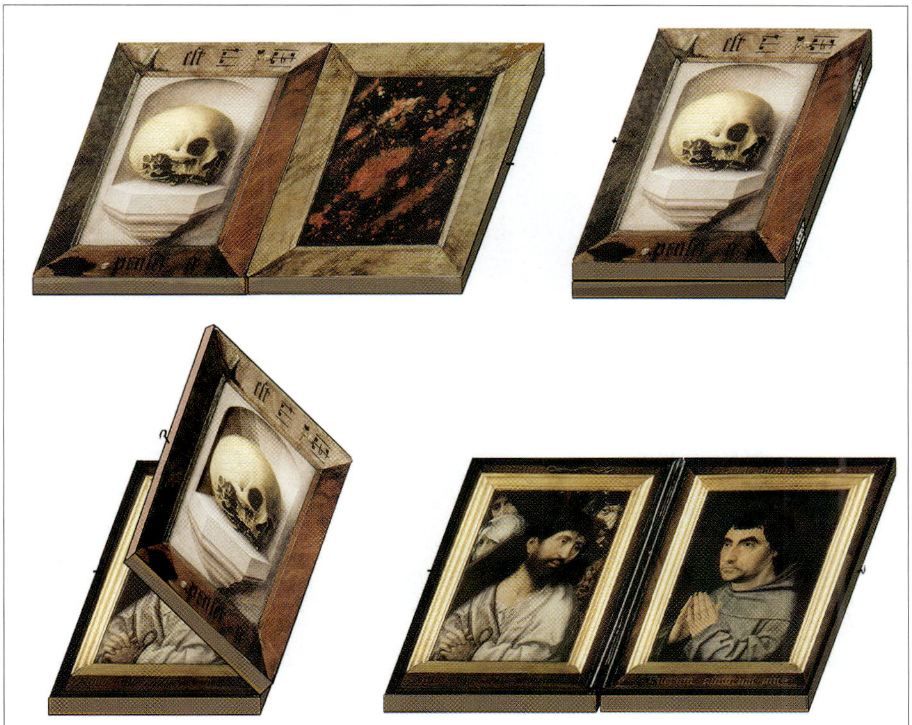

Fig. 5. Disposition d'un diptyque à revers asymétriques. Pour préparer à la juste ouverture, le diptyque doit être posé avec la représentation la plus élaborée au-dessus. Jan Provoost, *Christ portant la croix et donateur de 54 ans*, 1522, bois, chaque élément: 50 × 40,2 cm y compris le cadre original (Bruges, Hôpital Saint-Jean, inv. O.SJ191.1).

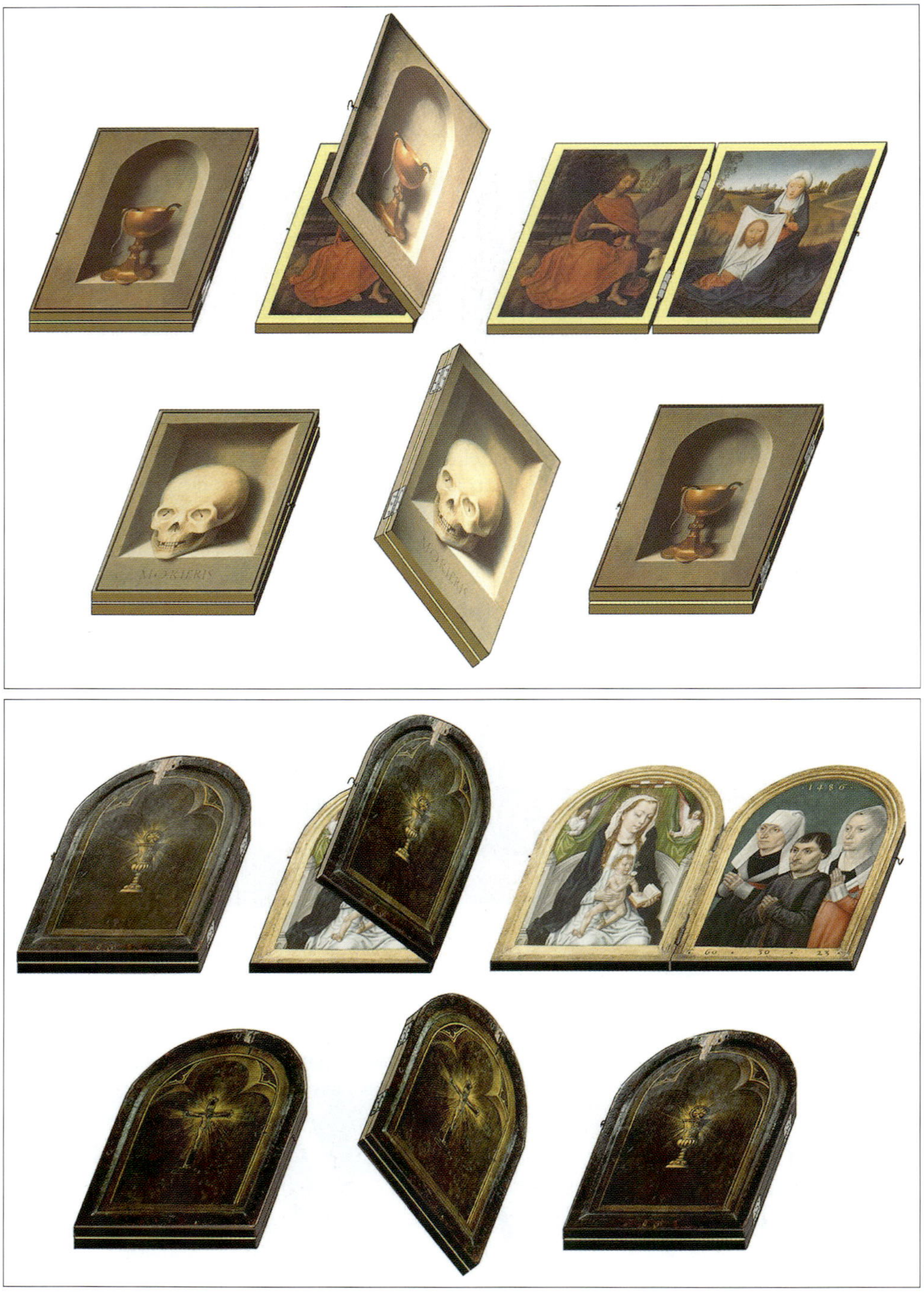

Fig. 6. Diptyque avec des revers similaires: des niches vues de front et éclairées latéralement. Le schéma évoque le maniement pour une juste ouverture. L'éclairage latéral permet de positionner correctement le diptyque en vue de son ouverture: il faut saisir le côté ombré de la niche pour ouvrir ou retourner. Hans Memling, *Diptyque Bembo* (saint Jean-Baptiste; revers: crâne; sainte Véronique; revers: calice de saint Jean l'Évangéliste), c. 1470/1475? c. 1480/1483?, chêne, cadres originaux perdus, 31,6 × 24,4 × c. 0,5 cm et 31,2 × 24,4 cm (Munich, Alte Pinakothek, inv. 652; Washington, National Gallery of Art, Samuel H. Kress Collection, inv. 1952.5.46). Maître de la Légende de sainte Ursule, *Diptyque de la Vierge à l'Enfant avec trois donateurs*, 1486, bois, chaque volet: 32 × 24,5 cm cadres originaux inclus (Anvers, Musées royaux des Beaux Arts, inv. 5004-5004 bis).

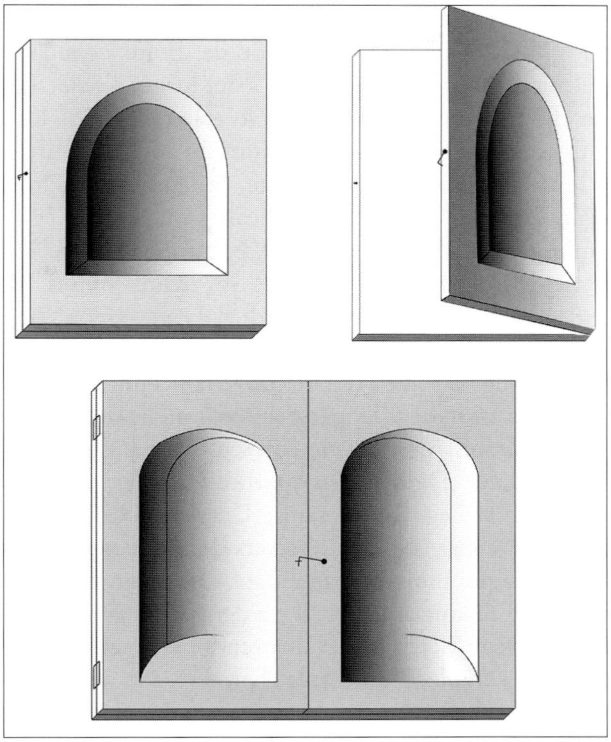

Fig. 7. Schéma avec niches de diptyque et de triptyque. Dans le cas du diptyque, la perspective est frontale et l'éclairage latéral est appuyé. Le diptyque s'ouvrira en prenant en main le côté ombré de la niche. Dans les triptyques, les niches sont mises en perspective vers la gauche et vers la droite, ce qui correspond à la place centrale qu'occupe le spectateur (au centre, en face des montants médians du cadre). Une seule source d'éclairage baigne l'ensemble des niches.

par l'artiste, s'enfonce rapidement dans l'ombre. Pour respecter l'intention illusionniste du peintre, nous devons saisir en main le côté de l'ombre pour ouvrir le diptyque. Ce qui importe ensuite, c'est l'emplacement à gauche ou à droite des charnières, parce que notre geste (celui de saisir le côté ombré de la niche) tantôt ouvrira l'objet, tantôt le retournera afin de le mettre en bonne position pour une juste ouverture. Ceci est illustré dans les schémas de la figure 6. Dans l'ouvrage d'Angelica Dülberg déjà cité, de nombreux diptyques allemands du 16e siècle sont reproduits avec leur cadre et leur revers, ce qui est rare. Le plus souvent, les cadres et revers sont souvent omis, jugés sans intérêt! L'existence d'une convention en matière d'éclairage nous a été inspirée par l'observation des diptyques reproduits dans le livre de Dülberg. Cette convention y est appliquée dans la majorité des cas. A l'intérieur est représenté un couple: l'homme à gauche, la femme, à droite. Cette dernière est représentée sur le volet, orné au revers d'un crâne dans une niche. Le système d'éclairage est presque toujours le même: la paroi gauche de la niche est dans l'ombre, la droite est vivement éclairée, pour une ouverture qui se fait de gauche à droite, selon la convention décrite plus haut. Des niches au revers de nombreux diptyques flamands sont éclairées dans le même esprit. Le *Diptyque Bembo* de Hans Memling, c. 1470/1475?-1480/1483?, cadres originaux perdus (Munich, Alte Pinakothek, inv. 652; Washington, National Gallery of Art, Samuel H. Kress Collection, inv. 1952.5.46) présente aux revers deux niches peintes en trompe-l'oeil, l'une cintrée, l'autre carrée (fig. 6). L'asymétrie des deux revers est peu prononcée. Dès lors, l'usager pourrait hésiter à disposer le diptyque d'un côté ou de l'autre. Nous pensions jadis que deux ouvertures étaient possibles, de gauche à droite ou de droite à gauche[9]. C'était une erreur. L'éclairage de la niche avec le crâne invite à retourner l'objet (vu l'emplacement des charnières) afin de présenter la niche avec le calice de saint Jean l'Évangéliste. L'éclairage de cette niche invitait alors à l'ouverture du diptyque. Il en découle que Véronique constitue le volet et saint Jean-Baptiste, le feuillet principal. Nous sommes donc en présence d'un diptyque des deux saint Jean.

Le *Jeune homme en prière* de Memling, c. 1485/1494 (Madrid, Fundacion Colleccion Thyssen-Bornemisza, inv. 1938.1) (fig. 8), tourné vers la droite, les mains jointes, est le volet gauche d'un diptyque dont la partie principale est perdue. Le revers présente un bouquet de fleurs dans une niche vue de manière frontale, avec un éclairage latéral accentué. Friedländer et Pächt[10] avaient suggéré que l'œuvre provenait d'un diptyque, mais selon d'autres, il s'agissait plutôt d'un triptyque. La perspective frontale de la niche évoque effectivement un diptyque. La manipulation s'avère correcte si on saisit le côté ombré de la niche pour faire pivoter le volet. Dans le *Diptyque de la Vierge et Enfant avec trois donateurs*, attribué au Maître de la Légende de sainte Ursule, 1486 (Anvers, Musée royal des Beaux-Arts, n° 5004-5004 bis) (fig. 6), les revers des feuillets portent l'un, un crucifix, et l'autre, un calice. Chaque motif est situé dans une arcature peinte en clair sur fond sombre. Au côté droit de chaque arcature, il y a deux filets jaunes évoquant la lumière et, au côté gauche, un seul filet. On devait prendre en main le côté du filet unique (côté dans l'ombre). Selon l'emplacement de la charnière, soit on retournait l'objet, soit on l'ouvrait. Cette manipulation conduit à une juste ouverture, celle qui faisait pivoter le volet avec les donateurs. On a pensé que ces peintures au revers étaient tardives, mais le respect d'une convention qui est sans doute tombée assez vite dans l'oubli me fait plaider en faveur de l'ancienneté de ces simples arcatures.

Les systèmes de fermeture et les charnières pouvaient également guider l'usager. La fermeture se fait souvent au moyen d'un crochet métallique, formé d'une tige terminée en boucle ou angle droit d'un côté, cloué au cadre (les vis qu'on observe parfois ne sont jamais originales). Dans un diptyque, les éléments de fermeture sont ancrés dans la tranche des cadres: le crochet, fixé au volet, tombe en place, à la fermeture, dans l'anneau (ou sur le clou) du feuillet principal (fig. 7). Ce crochet n'est pas fixé au feuillet principal ou 'dormant', parce qu'il retomberait et n'assurerait plus la fermeture. Dans le cas d'un volet de triptyque, le crochet de fermeture est souvent cloué sur le plat du cadre du volet droit tandis que l'anneau (ou le clou) est fixé au côté gauche. S'il y a deux crochets, leur sens est parfois alterné, comme c'est le cas pour le *Triptyque d'Adriaan Reins* de Hans Memling, 1480 (Bruges, Hôpital Saint-Jean, inv. O.SJ177.I). L'emplacement de crochets peut éventuellement, dans le cas de diptyques démembrés, donner de précieuses informations.

Notons encore que certains diptyques semblent avoir comporté un système de fermeture analogue à celui des manuscrits, peut-être avec des lanières en cuir ancrées dans

[9] VEROUGSTRAETE et VAN SCHOUTE, 1997, p. 272-274.

[10] Max J. FRIEDLÄNDER, *Hans Memling and Gerard David.* Commentaires et notes Nicole Veronee-Verhaegen, trad. Heinz Norden, vol. VI. *Early Netherlandish Painting*, II, New York-Washington, [1937] 1971, p. 110; Otto PÄCHT, *The Master of Mary of Burgundy*, Londres, 1948, 54. Cité dans Dirk DE VOS, *Hans Memling. L'œuvre complet*, Anvers, 1994, p. 262-263.

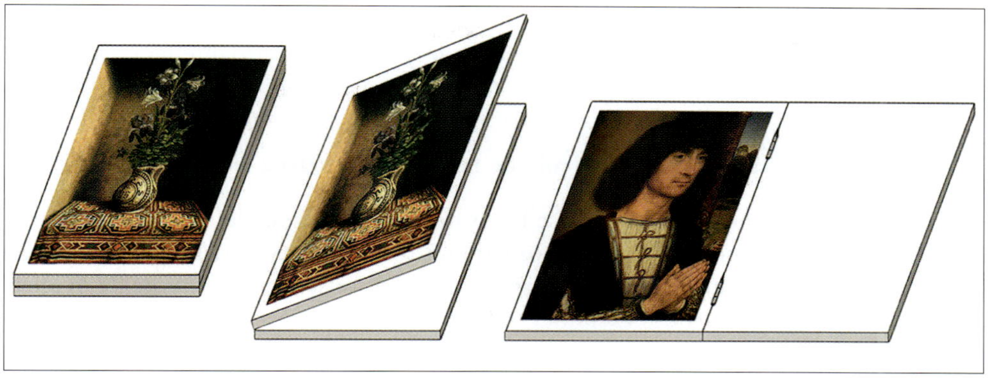

Fig. 8. Reconstitution évoquant le maniement d'un revers avec niche vue frontalement et éclairée latéralement. Hans Memling, *Jeune homme en prière*, c. 1485/1494, cadre original perdu, chêne, 29,2 × 22,5 (Madrid, Fondation Thyssen-Bornemisza, inv. 1938.1). Saisissant le côté ombré de la niche, on pratique correctement l'ouverture.

les bords des cadres. Le petit diptyque en forme de livre du Maître des Portraits Benson (Anvers, Musée Mayer Van den Bergh, inv. 254) présente un trou au centre de chaque bord extérieur de ses feuillets, trou qui devait probablement permettre l'insertion d'une cheville pour fixer une lanière. Le *Diptyque de Martin Van Nieuwenhove* présente quelque chose de semblable: la radiographie met en évidence l'existence, dans les bords extérieurs latéraux des volets, de deux groupes, l'un dans le haut, et l'autre dans le bas, de deux trous réguliers qui pourraient avoir servi à un usage analogue.

L'articulation des feuillets se faisait la plupart du temps avec des charnières. Elle pouvait aussi, dans le cas de petits diptyques, s'opérer au moyen de deux anneaux métalliques entremêlés, fichés en biais dans les montants des cadres. Parfois, et c'est le cas dans les diptyques en forme de livres, les deux feuillets étaient solidarisés au moyen de bandelettes de cuir, de parchemin ou de tissu[11]. On l'a évoqué déjà pour le diptyque eyckien conservé au Louvre (fig. 4). Dans les triptyques, la charnière de la partie centrale, celle qui porte le poids des volets, comporte la pièce la plus forte, avec le plus grand nombre de charnons. Dans le cas des diptyques, cette fonction portante n'a plus cours puisque l'objet est déposé, et cette distribution inégale des charnons ne semble pas avoir été appliquée. Certaines peintures représentent des petits diptyques suspendus aux courtines d'un lit. Le *Diptyque avec la Madone dans une église et l'abbé Chrétien de Hondt*, du Maître de 1499, 1499 (Anvers, Musée royal des beaux-Arts, n° 255-256, 530-531) ne présente pas de traces de charnières anciennes. Sans doute les deux feuillets devaient-ils être solidarisés dans un même plan en vue d'une suspension, qui aurait été instable si le diptyque avait été articulé. L'artiste a représenté sur le feuillet droit un petit diptyque suspendu aux courtines du lit qui pourrait évoquer le *Diptyque de Hondt* lui-même. Un diptyque dont les deux feuillets étaient solidarisés dans un même plan pouvait être retourné, s'il était peint au revers. S'il était dépourvu de peinture au revers, on ne le

[11] DÜLBERG, 1990, p. 66-67, note 370.

retournait pas. Les pratiques ont évolué au cours des temps. Il existe dans la seconde moitié du 16e siècle des diptyques à suspendre dont la partie dormante ne porte pas de peinture au revers.

'Pour bien fermer' un diptyque: le rôle de la perspective dans le diptyque ouvert

La question de la perspective dépasse le cas des diptyques, où elle est cependant particulièrement importante. La perspective joue un rôle dans l'éventuelle hiérarchisation des feuillets. Elle présente des espaces unifiés ou discontinus; elle les présente de manière frontale ou oblique. Quelques exemples ont été choisis pour clarifier les intentions qui ont été, je pense, celles des Primitifs flamands en la matière. La précocité de certains exemples semble indiquer qu'il y avait déjà, au début du 15e siècle, une tradition dans le domaine. Certaines lignes servent à guider notre regard. Dans le *Diptyque de Hondt*, la voûte de l'église n'aligne pas les éléments frontaux dans un seul axe vertical central mais en trois axes verticaux décalés qui dirigent le regard l'un vers la fenêtre haute, l'autre vers une croix et le troisième vers le retable sur l'autel (cette multiplication des axes verticaux est un peu rectifiée dans la *Madone dans une église* de Jean Van Eyck, c. 1425, Berlin, Staatliche Museen, Gemäldegalerie, cat. 525C).

Certaines perspectives visent à représenter un espace unique ou unifié (fig. 9). C'est le cas de nombreuses *Annonciations*, avec un plancher ou carrelage qui déploie sa perspective en éventail continu sur les deux feuillets, à partir d'une ligne verticale à leur jonction (pour rappel, les lignes horizontales restent parallèles à l'horizon). Un demi éventail correspond, selon l'inclinaison des lignes, à un feuillet gauche ou droit du diptyque. Dans le *Diptyque de Hondt* les espaces diffèrent d'un feuillet à l'autre: le Vierge est représentée dans une église et Chrétien De Hondt, dans un intérieur. Cependant l'éventail des lignes du sol (et même celui de la voûte et du plafond) se poursuit d'un volet à l'autre, manifestant la volonté de l'artiste d'unifier l'espace, de souligner l'unité spirituelle de la représentation.

D'autres œuvres présentent des espaces hiérarchisés (fig. 2). Nous avons évoqué déjà le registre supérieur de l'*Agneau mystique* de Van Eyck. Dans le *Diptyque de la Trinité et de la Vierge à l'Enfant devant l'âtre* attribué à Robert Campin, c. 1420/1425 (Saint-Pétersbourg, Ermitage, inv. 447-448), la Trinité constitue le feuillet principal que la perspective frontale nous invite à contempler préférentiellement. La Vierge à l'Enfant lui sont subordonnés dans une perspective en demi éventail vers la droite, typique d'un volet droit. Par sa forte similitude avec ce volet, le tableau avec la *Vierge et Enfant* récemment attribué à l'atelier de Robert Campin (Jacques Daret?) 1428/avant 1432? (Londres, National Gallery, inv. 6514)[12] nous paraît être un volet droit qui faisait pendant à une scène frontale, aujourd'hui disparue. Vu l'absence de traces de charnières dans l'œuvre de Londres et la présence de traces de suspension, et vu aussi sa petite taille, ce fut peut-être un diptyque non articulé, à suspendre.

[12] Lorne CAMPBELL, «*Review*, Hans Memling. The complete Works, by Dirk De Vos», dans *The Burlington Magazine,* CXXXVII, 1995, p. 253.

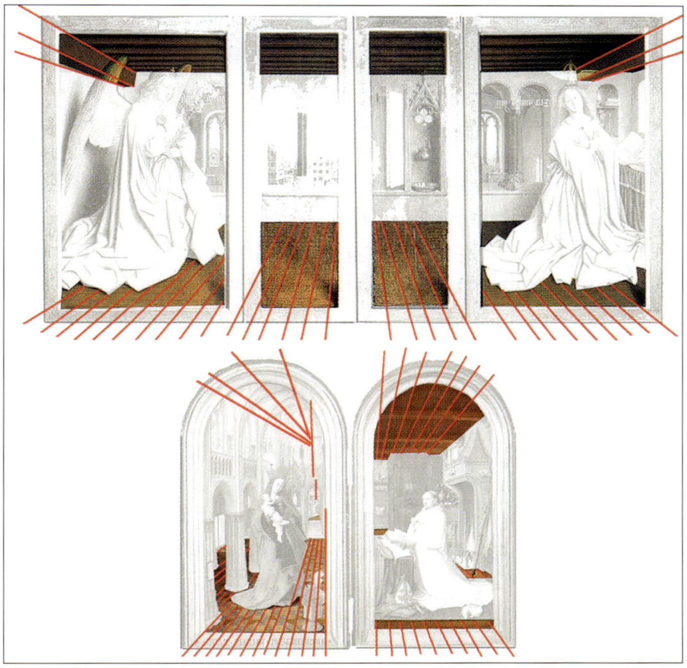

Fig. 9. Exemples de perspectives adoptées pour représenter un espace unique ou unifié. Van Eyck, *Agneau Mystique*, 1432, détail du registre supérieur à l'extérieur du retable. Chêne, partie centrale: 230 × 255,8 cm. *Diptyque avec la Vierge dans une église et l'abbé Chrétien de Hondt*, 1499, bois, chaque élément: 37 × 22 cm y compris le cadre original (Anvers, Musées royaux des Beaux-Arts, inv. 255-256, 530-531). La perspective se prolonge d'un panneau à l'autre. La perspective peut se poursuivre harmonieusement dans un seul intérieur, mais aussi dans des espaces différents unis dans une même spiritualité.

Un exemple plus tardif d'espaces différents hiérarchisés est constitué par le *Diptyque de Marguerite d'Autriche*, attribué au Maître de 1499, après 1501 (Gand, Musée des Beaux-Arts) (fig. 2).

Dans un même espace, une perspective peut être discontinue comme dans le *Diptyque de Martin van Nieuwenhove*. Marie, présentée de front, focalise notre attention. Martin est représenté devant une paroi de la même pièce, mise en perspective, dans un éventail de lignes obliques (cette fois organisées autour d'un axe horizontal) qui remplit la même fonction que les éventails de lignes fuyantes des carrelages ou des plafonds. Ces obliques hiérarchisent les feuillets et suggèrent la rotation du volet (fig. 2). Le reflet dans le miroir nous rappelle que les personnages se trouvent réunis dans un même espace, malgré la discontinuité de la perspective. Memling fut, peut-être, de tous les Primitifs, celui qui a porté l'art de la perspective (non réaliste) à un très haut degré de raffinement et de complexité.

Parfois chaque feuillet présente une vue frontale qui correspond à un espace indépendant (fig. 3), invitant à la contemplation successive de chaque feuillet isolément, ce qui répond aux exigences d'une narration ou d'une succession historique des scènes.

Entre diptyque, triptyque et quadriptyque: des usages subtils et des objets 'à transformation'

Il a existé des objets complexes dont l'usage était sophistiqué. L'ouverture ou la fermeture de certains retables se faisaient en étapes successives, passant par le stade de diptyque pour continuer à se déployer en triptyque ou quadriptyque. Le célèbre *Quadriptyque Anvers-Baltimore*, anonyme, Gelderland (?) (Baltimore, The Walters Art Gallery, inv. 37.1683 et Anvers, Musée Mayer Van den Bergh, inv. I)[13] ne devait probablement pas s'ouvrir en accordéon, comme il a parfois été proposé. Nous suggérons que les faces rouges étaient exposées à l'extérieur (fig. 10). Un cadre à Anvers se profile en arrondi sur un de ses chants. On reconnaîtra cet arrondi sur le schéma du retable fermé et on comprendra sa raison d'être: faciliter l'ouverture. Le cadre correspondant de Baltimore, aujourd'hui perdu, présentait probablement le même arrondi. Une première ouverture exposait un diptyque avec saint Christophe et le baptême du Christ, chaque fois avec un ciel rouge. Ensuite, selon le volet que l'on ouvrait, on était en présence d'un triptyque relatif à l'Enfance de Jésus ou d'un autre relatif à la vie adulte de Jésus. L'ouverture finale en quadriptyque déployait l'histoire complète de la Rédemption. Le quadriptyque fermé était peint en rouge, couleur souvent adoptée pour les revers d'œuvres pré-eyckiennes. A une première ouverture en diptyque, les cieux étaient rouges, couleur inhabituelle pour les cieux, mais qui signifiait bien un état entre ouverture et fermeture, sans doute au sens littéral et spirituel. La distribution actuelle du quadriptyque résulte d'un désir de couper en 'deux parts égales' (un élément biface, et un élément avec revers rouge) un objet dont on ne comprenait plus les subtilités.

Cette ouverture en étapes qui exposent des contenus successifs pourrait également avoir été le fait des triptyques à trois feuillets d'égales dimensions (fig. 11). Dans ces triptyques, une hiérarchisation des représentations réglait les modalités de la fermeture. On repliait d'abord un diptyque premier. On repliait ensuite le second volet. Ces triptyques fort intéressants sont aujourd'hui presque tous démembrés ou sans cadres, ils ont des revers parquetés... Un exemple rare, encore dans son état primitif, est un triptyque italien de l'entourage de Paolo Veneziano, c. 1300, avec au centre un Homme des douleurs, à gauche la Vierge et à droite saint Jean. Au revers du feuillet gauche: deux dominicains, un oiseau, et le Christ trônant (Dordrecht, Musée Simon Van Gijn, prêté par une collection privée)[14]. Les visages du Christ et de Marie y sont inclinés l'un vers l'autre, instruction pour refermer prioritairement ce diptyque, avant le volet avec saint Jean. Le revers avec les dominicains a bien les allures de revers de diptyque. Matériellement, on ne pouvait d'ailleurs pas faire autrement. Pour superposer à la fermeture trois (ou parfois quatre) feuillets, l'adjonction de petites lattes, et de hautes charnières était nécessaire. Ces lattes sont toujours en place à Dordrecht. Dans ma proposition de reconstitution du *Quadriptyque Anvers-Baltimore*, j'en ai mis de semblables.

[13] Hans NIEUWDORP, Régine GUISLAIN-WITTERMAN et Léopold Kockaert, «Het pre-Eyckiaans vierluik Antwerpen-Baltimore», dans *Bulletin IRPA-KIK* 20, Bruxelles, 1984, p. 70-98. Hans NIEUWDORP, «The Antwerp-Baltimore polyptych: A portable altarpiece belonging to Philip The Bold, Duke of Burgundy» dans Henk VAN OS et al., *The Art of Devotion in the Late Middle Ages in Europe* 1300-1500, cat. d'exposition, Rijksmuseum Amsterdam, Amsterdam, 1994, p. 137-150. Hans NIEUWDORP et Nicole GOETGHEBEUR, «Anonymous Master. Two panels of the Antwerp-Baltimore Quadriptych», dans Hélène MUND, Cyriel STROO, Nicole GOETGHEBEUR avec la collaboration de Hans NIEUWDORP, *The Mayer Van den Bergh Museum Antwerp*, vol. 20 du *Corpus of the Fifteenth Century Painting in the Southern Netherlands and the Principality of Liège*, Bruxelles, 2003, p. 282. E. Melanie GIFFORD, «A pre-Eyckian Altarpiece in the Context of European Painting Materials and Techniques Around 1400», dans *Corpus of Illuminated Manuscripts. Flanders in a European Perspective. Manuscript Illumination Around 1400 in Flanders and Abroad*, Louvain, 1995, p. 357-370.

[14] DÜLBERG, 1990, 83, fig. 9-10. VAN OS, 1994, p. 106-109.

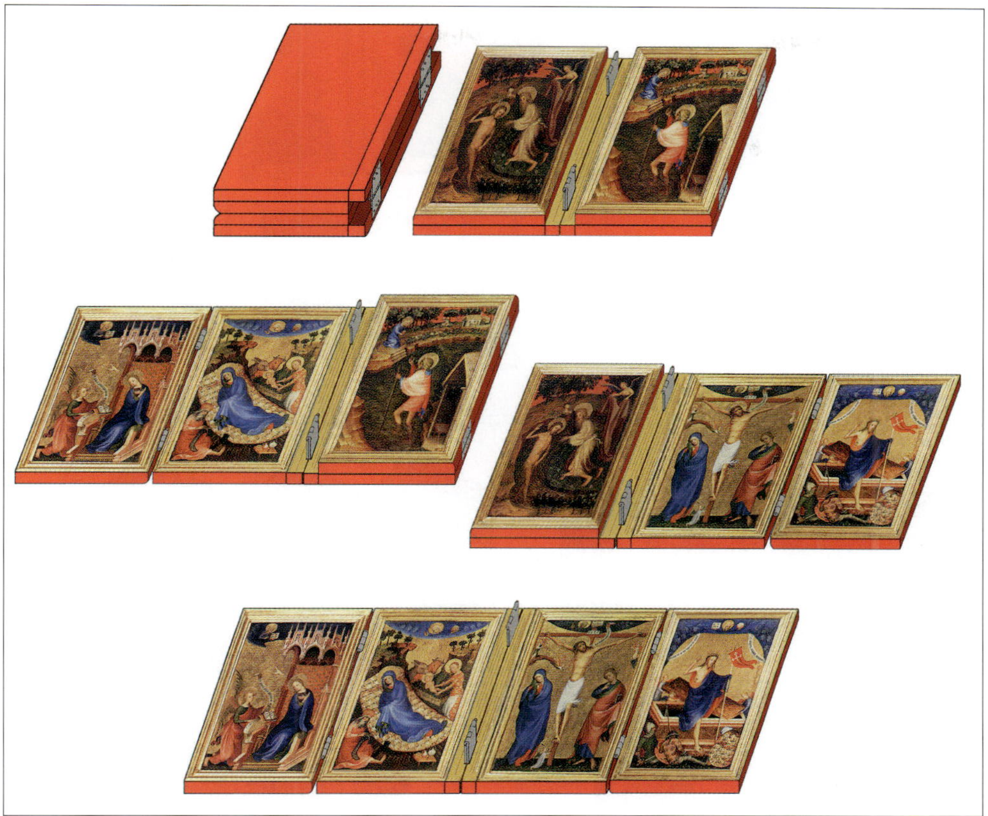

Fig. 10. Reconstitution d'un retable aujourd'hui démembré, et de son ouverture progressive. Maître anonyme, Gelderland (?), *Quadriptyque Anvers-Baltimore*, Annonciation, Nativité, Crucifixion, Résurrection, Baptême du Christ (revers de a) et saint Christophe (revers de d); c. 1400, chaque élément 37,9 × 26,5 cm, cadres originaux à Anvers seulement (Baltimore, The Walters Art Gallery; Anvers, Musée Mayer Van den Bergh). Les faces rouges étaient à l'extérieur. Les cieux rouges de saint Christophe et du Baptême du Christ correspondent à une ouverture intermédiaire, incomplète, invitant aux ouvertures suivantes qui exposent des contenus successifs.

La présence de ces lattes et charnières protubérantes n'empêche pas le peintre de visualiser, par son langage pictural, la hiérarchisation des feuillets. Dans le *Triptyque de Benedetto Portinari* de Hans Memling, 1487 (panneau central: Staatliche Museen zu Berlin, Preussischer Kulturbesitz, Gemäldegalerie, inv. 528B et volets: Florence, Uffizi, inv. 1100 et 1090) la hiérarchisation est indiquée par l'Enfant qui pointe du doigt vers Benedetto, liant la Vierge et Enfant au portrait, pour former le diptyque premier sur lequel se refermait ensuite le volet avec l'effigie du saint qui, plongé dans sa lecture, se soustrayait au diptyque premier. Dans l'état fermé, la hiérarchie subsistait matériellement plus fort encore. Benedetto était placé contre Marie, et il était recouvert et protégé par son saint patron. Dans le *Triptyque de Jean De Witte* attribué au Maître de

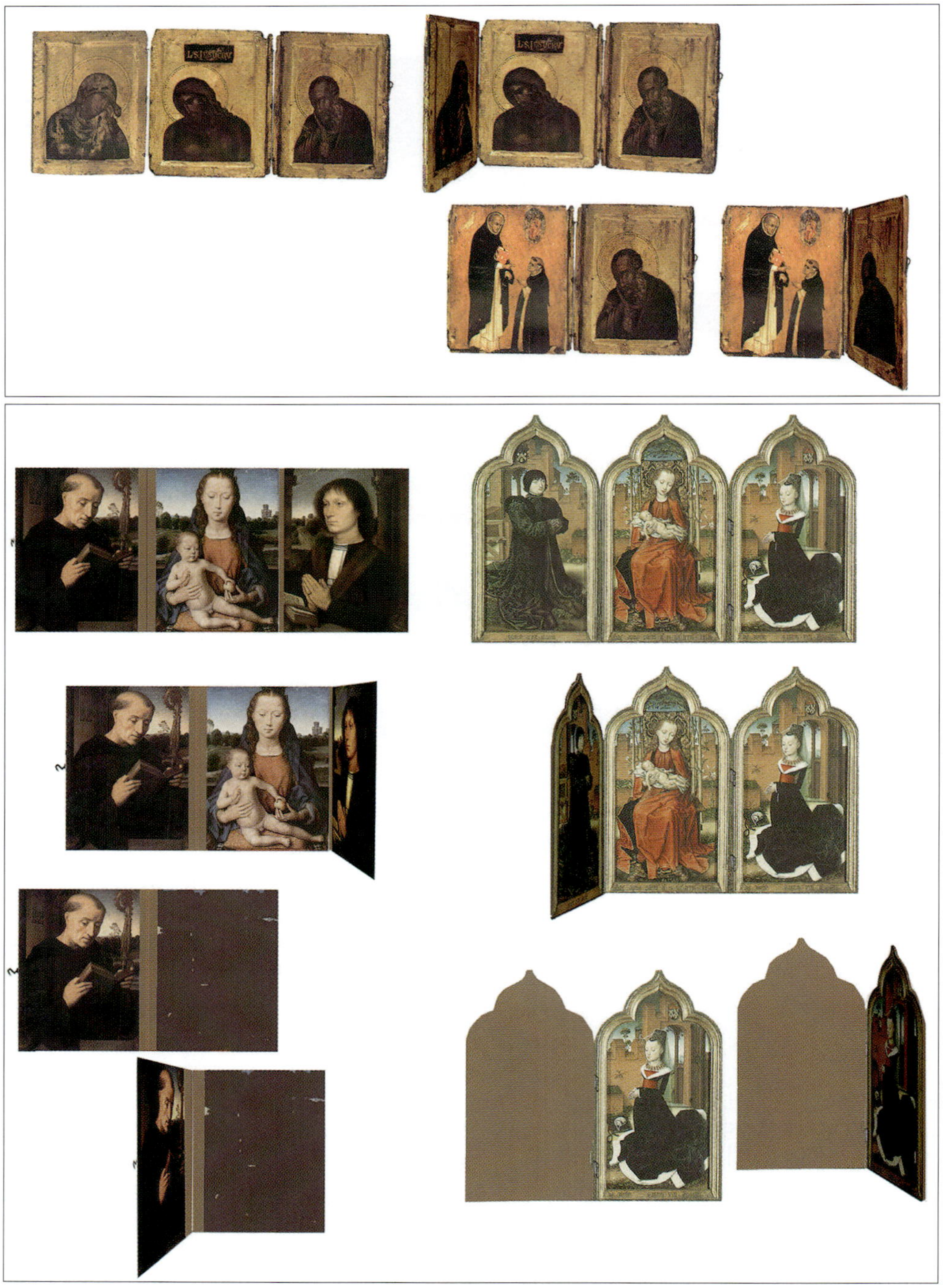

Fig. 11. Diptyques à trois volets de mêmes dimensions. La première fermeture se fait avec un volet privilégié. Entourage de Paolo Veneziano, c. 1300, *Homme des Douleurs, Vierge et saint Jean l'Évangéliste* (Dordrecht, Musée Van Gijn); Hans Memling, *Triptyque de Benedetto Portinari*, 1487, bois (Berlin, Staatliche Museen Gemäldegalerie: partie centrale) et Florence (Galleria degli Uffizi: volets). Maître de 1473, *Triptyque de Jean De Witte* (Bruxelles, Musées Royaux des Beaux-Arts, inv. 7007).

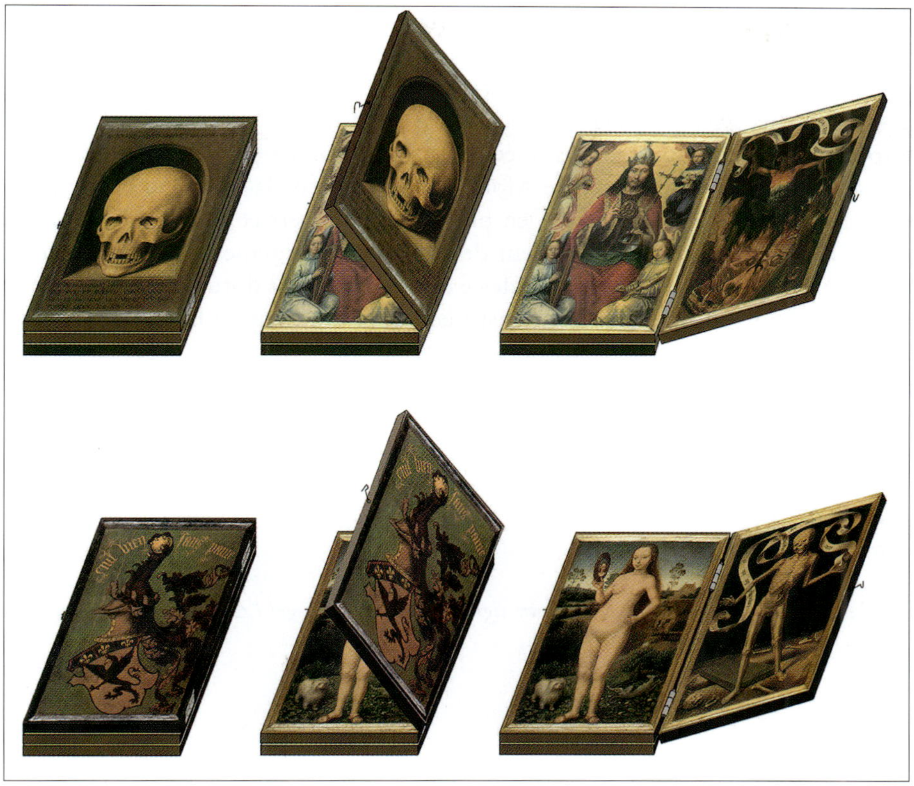

Fig. 12. Reconstitution d'un retable, qui se présente aujourd'hui en six éléments séparés et qui était à l'origine réversible (diptyque double). Memling, *Retable de la Rédemption céleste et de la Vanité terrestre*, chaque panneau: c. 22,1 × 13,1 cm (Strasbourg, Musée des Beaux-Arts, inv. 51). Les deux diptyques sont hiérarchisés par les écritures, en capitales sur le diptyque majeur, en cursives gothiques sur le diptyque plus intime.

1473 (Musée des Beaux-Arts, Bruxelles, inv. 7007)[15], les grosses charnières protubérantes existaient encore en 1960, comme en atteste une photographie datée. Le donateur regarde la Vierge et Enfant; il est placé à la droite héraldique, et c'est lui qui occupe le premier volet, celui qui se repliait en premier lieu. Le regard baissé de la donatrice placée à la gauche héraldique, correspond à la modestie de sa position secondaire. Ses mains sont également baissées en prières, quoique initialement le peintre les avait dessinées vers le haut[16]. Elle se repliera en second lieu. Dans ces deux cas, les yeux baissés sont propres au personnage qui se soustrait au dialogue principal et qui se trouve sur le second volet.

[15] Bart FRANSEN et Pascale SYFER-D'OLNE, «Le Maître de 1473», dans Pascal SYFER-D'OLNE, Roel SLACHMUYLDERS, Anne DUBOIS, Bart FRANSEN et Famke PEETERS, *The Flemish Primitives. Masters with Provisional Names. Catalogue Royal Museums of Fine Arts of Belgium,* vol. IV, Bruxelles, 2006, p. 391-413.

[16] FRANSEN et SYFER-D'OLNE, 2006, p. 398.

Il a existé des diptyques réversibles. Le petit retable de Memling avec la *Rédemption céleste* et la *Vanité terrestre*, c. 1486 (Strasbourg, Musée des Beaux-Arts, inv. 51) est un bon exemple (fig. 12). Les six éléments actuels avec cadres originaux collés proviennent de trois panneaux sciés dans l'épaisseur. On sait aujourd'hui, grâce à l'observation de la correspondance des fissures ou des traces de sciages, que les panneaux suivants étaient disposés dos à dos: les armoiries/le Mort; le Rédempteur/la Vanité; l'Enfer/le crâne. Quatre cadres sont dorés, deux sont en pleine couleur (vert et brun). Ces deux cadres en pleine couleur constituent l'extérieur de l'objet, conformément à un usage qui était de mettre à l'extérieur une couleur modeste et de réserver les dorures plus riches et plus fragiles pour l'intérieur. L'ensemble constituait deux diptyques. On ouvrait le premier diptyque, celui de la Rédemption; on le refermait, on retournait l'objet et on ouvrait l'autre diptyque, plus secret et familier. La hiérarchie entre les deux diptyques était concrétisée par l'usage de capitales romaines du côté de la *Rédemption céleste* et de lettres gothiques du côté de la *Vanité terrestre*[17].

Hélène VEROUGSTRAETE

Voir Hélène VEROUGSTRAETE, *Les couleurs signifiantes chez Bruegel l'Ancien et chez ses successeurs.*

[17] VEROUGSTRAETE et VAN SCHOUTE, 1997, p. 272-274.

L'ENCADREMENT DES PEINTURES DANS LES PAYS-BAS MÉRIDIONAUX AU XVIIe SIÈCLE

Elisabeth Bruyns

Résumé: De nombreuses informations concernant l'encadrement des peintures au XVIIe siècle figurent dans les documents d'archives. La délimitation des tâches entre menuisiers, fabricants de cadres, ébénistes, sculpteurs et doreurs y est notamment évoquée. L'étude d'encadrements dans divers musées et églises a permis d'identifier plusieurs types de cadres. Outre des cadres simplement moulurés noirs et/ou dorés, des encadrements ornés de motifs peints ou sculptés (guirlandes de feuilles et oves) et quelques cadres entièrement sculptés ont été étudiés.

Cette contribution s'inscrit dans le contexte de notre thèse de doctorat sur l'encadrement des peintures dans les anciens Pays-Bas du sud au XVIIe siècle[1]. Notre objectif est de déterminer dans quelle mesure les œuvres conservées possèdent encore des encadrements anciens et quels types d'encadrements étaient en usage à cette époque[2].

1. L'apport des archives anversoises

La première partie de notre travail a consisté en l'étude de divers documents d'archives publiés et inédits[3]. Ces documents apportent des informations sur les matériaux utilisés dans la fabrication d'encadrements mais aussi sur leur décoration, leurs prix, leurs formats et les accessoires qui pouvaient y être ajoutés. On y trouve également de multiples indications sur les métiers qui intervenaient dans leur fabrication ainsi que sur la répartition des tâches entre ces différents métiers.

Les encadrements de peintures étaient réalisés par des menuisiers (*schrijnwerkers*) ou par des fabricants de cadres (*lijstmakers*), de panneaux (*paneelmakers*) ou des *tafereelmakers* (fabricants de 'tableaux'[4]). Tandis que les menuisiers formaient un métier indépendant, les fabricants de panneaux et de cadres appartenaient à la guilde de Saint-Luc. Les cadres en ébène étaient fabriqués par les ébénistes (*ebbenhoutwerkers*), qui adhéraient

[1] Ces recherches sont menées au Laboratoire d'étude des œuvres d'art de l'Université catholique de Louvain sous la direction du Prof. H. Verougstraete.

[2] Peu de publications existent à ce sujet. Une étude approfondie a été publiée sur les cadres hollandais du XVIIe siècle (P.J.J. VAN THIEL et C.J. DE BRUYN KOPS, *Prijst de Lijst. De Hollandse schilderijlijst in de zeventiende eeuw*, La Haye, 1984).

[3] Nous avons consulté les archives d'Anvers, Bruges, Bruxelles, Gand, Louvain et Malines. Les documents examinés à Anvers proviennent principalement du fonds des guildes et métiers (GA) et des dossiers de procès des archives de la ville (SAA). Quelques documents sont issus des archives de la guilde de Saint-Luc conservées à l'Académie des Beaux-Arts (AAA).

[4] La traduction du terme *tafereelmaker* par fabricant de tableaux peut prêter à confusion. Les *tafereelmakers* étaient bien des artisans qui fabriquaient des supports de bois (et des cadres) pour des peintures. Parmi les quelques deux cents artisans que nous avons identifiés, certains se voient attribuer une profession différente selon les documents. Quoi qu'il en soit, ces trois dénominations (*lijstmakers, paneelmakers, tafereelmakers*) désignent des artisans qui réalisaient des cadres et/ou des panneaux pour des peintures.

soit à la guilde, soit au métier des menuisiers. Les cadres devaient être sculptés par des sculpteurs (*beeldhouwers*), tandis que les doreurs (*vergulders*) et 'étoffeurs' (*stoffeerders*[5]) s'occupaient de la dorure et de la peinture. Les fabricants de cadres ont aussi sculpté et doré eux-mêmes leurs encadrements.

a. L'ordonnance de 1617 et son application

Le 13 novembre 1617, les *tafereelmakers* anversois demandèrent une ordonnance aux magistrats de leur ville[6]. Le même jour, les menuisiers en demandèrent une également[7]. Deux ordonnances semblables furent donc rendues le 11 décembre suivant[8]. L'ordonnance des *tafereelmakers* devait poser les fondements de la réglementation de leur métier et prévenir les querelles entre fabricants de 'tableaux' (*tafereelmakers*), de cadres de miroirs (*spiegelcasmakers*) et de cadres en bois tendre (*weeckelijstmakers*). Selon cette ordonnance, personne ne pouvait être affranchi dans le métier des *tafereelmakers* et des *weeckelijstmakers* avant d'avoir appris le métier durant deux ans et passé l'épreuve. Les artisans qui souhaitaient travailler les bois durs réalisaient à titre d'épreuve un panneau de 6 pieds[9] de longueur (±172 cm) et 5 de largeur (±143 cm), bien conçu et collé, et un cadre de mêmes dimensions, avec des vis[10] (art. 1). Ceux qui ne voulaient fabriquer que des cadres de miroirs ou des cadres en bois tendre ne suivaient qu'un an d'apprentissage et ne passaient pas d'épreuve (art. 2). Cette ordonnance imposait également aux artisans de marquer leurs panneaux et leurs cadres d'une marque personnelle[11] (art. 7). Différents formats de panneaux y sont évoqués: 26 sous (*stuyvers*), florins[12], 8 sous, *stooters* et demi-*stooters* (art. 10). Ces formats s'appliquaient évidemment aussi aux encadrements.

Les menuisiers pouvaient, comme auparavant, fabriquer des cadres et des panneaux (art. 12). Ils devaient également marquer leurs réalisations d'une marque personnelle et utiliser les formats cités. Les menuisiers qui voulaient réaliser des travaux concernant les *tafereelmakers* devaient passer la grande épreuve des menuisiers (qui consistait à réaliser un buffet ou un dressoir) (art. 13). Les *tafereelmakers* ne pouvaient pas utiliser le hêtre «sauf pour le nez[13] (*neuse*) et les cadres intérieurs (*binnen lijsten*)» (art. 15).

[5] *Stofferen* signifie décorer, peindre, dorer, polychromer (J. VERDAM, E. VERWIJS, *Middelnederlandsch woordenboek*, La Haye, 1912, vol. 7, col. 2173-5).

[6] SAA, GA4346. Les références de cette requête sont données dans J. VAN DAMME, «De Antwerpse tafereelmakers en hun merken. Identificatie en betekenis», dans *Jaarboek van het Koninklijk Museum voor Schone Kunsten Antwerpen*, 1990, p. 194.

[7] La requête des menuisiers figure dans SAA, GA4578.

[8] Une copie de l'ordonnance des *tafereelmakers* figure dans SAA, GA4575, *Tafereelmakers en Lijstmakers*, n°5, 11/12/1617. L'ordonnance des menuisiers a été publiée par J. VAN DAMME, 1990, p. 235-6 (SAA, GA4335, f°78v°-81v°).

[9] Un pied d'Anvers = 286,8 mm (H. DOURSTHER, *Dictionnaire universel des poids et mesures anciens et modernes contenant des tables des monnaies de tous les pays*, Bruxelles, 1840, p. 403).

[10] *Vijsen* peut être traduit par vis (J. VERDAM, E. VERWIJS, 1929, vol. 9, col. 541).

[11] De telles marques ont été observées par J. Wadum sur les cadres des peintures de la chambre d'hiver du château de Rosenborg. J. WADUM, «The Winter Room at Rosenborg Castle. A Unique Survival of Antwerp Mass-Production», dans *Apollo*, août 1988, p. 82-87.

[12] Un florin valait 20 sous (*stuyvers*) (J. VERDAM, E. VERWIJS, 1889, vol. 2, col. 2209-10). Un *stooter* valait 2,5 sous (J. VERDAM, E. VERWIJS, 1912, vol. 7, col. 2245-6).

[13] Le terme *neuse* désigne sans doute la battée du cadre.

L'ordonnance de 1617 permettait aux artisans qui ne voulaient réaliser que des cadres en bois tendre et des cadres de miroirs de travailler après un apprentissage d'un an et sans avoir passé d'épreuve. Cette décision permit à de nombreux artisans peu qualifiés de devenir maîtres, ce qui engendra régulièrement des différends entre les artisans qui travaillaient le bois dur et ceux qui travaillaient le bois tendre. Pour mettre fin à ces différends, il fut décidé le 8 octobre 1646[14] que ceux qui voudraient fabriquer des cadres en bois tendre et des cadres de miroirs devraient apprendre le métier chez un franc maître pendant deux ans et réaliser comme épreuve un cadre en bois tendre, avec vis, d'une longueur de 5 pieds (±143 cm) et d'une largeur de 4 pieds (±115 cm) (art. 1). Les artisans qui deviendraient maîtres après la publication de cette ordonnance pourraient travailler le bois dur et le bois tendre après deux années d'apprentissage et après avoir passé les deux épreuves, l'une pour le bois dur (l'épreuve mentionnée dans l'ordonnance de 1617) et l'autre pour le bois tendre (art. 2).

En 1649, les fabricants de cadres déplorent que des étrangers importent à Anvers de nombreux panneaux, cadres et cadres de miroirs[15]. Ces panneaux et cadres étaient réalisés et vendus par des non maîtres, qui ne respectaient pas la réglementation concernant le marquage. Les fabricants de cadres demandent donc que les amendes prévues soient étendues à tous les cadres et panneaux apportés ou vendus par des non maîtres. Le 15 décembre 1649[16], il fut décidé que dorénavant, il serait interdit d'importer à Anvers des panneaux, cadres et cadres de miroirs pour les vendre «en secret ou publiquement».

b. L'origine des lijstmakers

La réalisation de manteaux de cheminée et d'encadrements de pièces (*lijsten voor camers*)[17] fut le sujet de conflits réguliers entre menuisiers et fabricants de cadres. Dans les archives de ces procès, est évoquée l'origine du métier des fabricants de cadres. En 1663, les menuisiers affirment qu'il leur a toujours appartenu de fabriquer des encadrements, des panneaux, des cadres de miroirs et autres travaux de ce genre, et que les *tafereelmakers* et les fabricants de cadres faisaient autrefois partie de leur métier, dont ils s'étaient détachés pour entrer dans la guilde de Saint-Luc[18]. De tout temps, les fabricants de cadres auraient été des 'ouvriers invités' (*werkgasten*) des menuisiers. En 1617 ou peu avant[19], les *lystmakers* auraient abandonné les menuisiers et auraient constitué

[14] Ordonnance transcrite dans J.B. VAN DER STRAELEN, *Jaerboek der vermaerde en kunstryke gilde van Sint Lucas binnen de stad Antwerpen*, Anvers, 1855, p. 105-8.

[15] SAA, GA4575, *Tafereelmakers en Lijstmakers*, n°14, 20/11/1649 et n°21, 29/11/1649.

[16] SAA, GA4575, *Tafereelmakers en Lijstmakers*, 15/12/1649.

[17] Ces encadrements de pièces étaient placés sur les murs à des hauteurs variées et servaient à suspendre des tapisseries ou des cuirs dorés. On pouvait aussi placer des peintures au-dessus de ces cadres. Voir J. VAN DAMME, «17de-eeuwse kamerlijsten. Een weinig bekend interieur onderdeel», dans *Gentse bijdragen tot de interieurgeschiedenis*, vol. 33, 2004, p. 1-7.

[18] SAA, GA4575, *Tafereelmakers en Lijstmakers*, n°19, 7/5/1663.

[19] Une gravure d'Adriaen Collaert d'après un dessin de Sebastiaen Vrancx, représentant saint Luc peignant la Vierge (1614), est bordée des attributs des métiers appartenant à la guilde de Saint-Luc d'Anvers. La présence des attributs des fabricants de panneaux et de cadres montre qu'à ce moment des cadres étaient déjà fabriqués par des artisans appartenant à la guilde (gravure citée dans P.J.J. VAN THIEL & C.J. DE BRUYN KOPS, 1984, p. 28 et illustrée dans A. KEERSMAEKERS, «De schilder Sebastiaen Vrancx (1573-1647) als rederijker», dans *Jaarboek van het Koninklijk Museum voor Schone Kunsten Antwerpen*, 1982, p. 170).

un métier à part entière dans la guilde de Saint-Luc. Cette même année, ils auraient demandé une ordonnance, car ils n'en disposaient pas jusque-là[20]. D'après les menuisiers, le métier des fabricants de cadres n'aurait jamais été cité avant l'ordonnance du 11 décembre 1617, où ces derniers sont mentionnés avec l'apposition du mot *weecke*, ce qui signifie qu'il s'agissait de fabricants de cadres en bois tendre[21].

Lorsque les fabricants de cadres entrèrent dans la guilde de Saint-Luc, il leur fut permis de réaliser des encadrements et des supports de bois en dehors du métier des menuisiers[22]. Toutefois, leurs tâches étaient limitées à la réalisation de supports et de cadres. Par leur entrée dans la guilde, ils ont obtenu de pouvoir réaliser certains travaux nécessaires aux peintres et communs aux menuisiers, mais rien d'autre. Ils souhaitaient toutefois étendre leurs compétences à toutes sortes de 'tableaux' et d'encadrements.

c. *Le travail de l'ébène*

Avant 1620, tant les menuisiers que les membres de la guilde de Saint-Luc se disputaient le droit de rattacher les ébénistes à leur corporation[23]. En 1620, un procès opposa les menuisiers aux ébénistes[24]. Dans l'une des pièces de ce procès, il est expliqué que les fabricants de cadres et de panneaux seraient restés dans la guilde de Saint-Luc sous prétexte que leurs travaux étaient directement liés aux peintures réalisées par les membres de la guilde et qu'ils devaient enduire ceux-ci au moyen de pinceaux, ce qui revenait uniquement aux peintres[25]. Quelques artisans avaient toutefois souhaité faire partie du métier des menuisiers, pour pouvoir réaliser toutes sortes de travaux en ébène. Selon les menuisiers, les membres de la guilde ne disposaient d'aucun argument permettant d'établir que les ébénistes devaient faire partie de leur métier, si ce n'est que les *tafereelmakers* devaient en faire partie. De ce fait, les ébénistes souhaitant réaliser des cadres devraient aussi adhérer à la guilde.

Au cours d'un litige survenu en 1589, les doyens de la guilde avaient requis de pouvoir fabriquer les 'bacs' (*backen*), les 'tableaux' (*tafereelen*) et les cadres (*lysten*) des tableaux d'autels (*aultaertafelen*[26]), sans plus. Ils n'auraient donc pas prétendu réaliser d'autres cadres que ceux utilisés pour les autels. Or, les cadres en ébène n'étaient pas utilisés pour les autels. Les ébénistes ne devaient donc pas dépendre des *tafereelmakers* de la guilde. Aucun ébéniste n'a du adhérer à la guilde pour pouvoir réaliser des cadres en ébène. Les ébénistes les réalisaient en faisant partie du métier des menuisiers. Seuls quelques ébénistes ont adhéré à la guilde, non pas pour fabriquer des cadres en ébène mais pour pouvoir vendre des peintures.

[20] Il s'agit de la requête du 13 novembre 1617 et de l'ordonnance du 11 décembre de cette année.

[21] SAA, GA4575, *Tafereelmakers en Lijstmakers*, n°19, 7/5/1663, p. 10.

[22] SAA, Procès G7122, 1660.

[23] R. FABRI, «De 17de-eeuwse Antwerpse kunstkast. Typologische en historische aspecten», dans *Verhandelingen van de Koninklijke Academie voor Wetenschappen, Letteren en Schone Kunsten van België*, 53, 1991, p. 113. SAA, GA4337, AAA, n°98 (72*), 27/02/1621, SAA, Procès A929, 18/7/1675, Procès A648, 27/2/1621.

[24] SAA, Procès D4868, 1620.

[25] SAA, Procès D4868, 5/05 et 6/08/1620.

[26] *Woordenboek der Nederlandsche Taal*, La Haye-Leyde, 1898, vol. II, col. 290, 3).

Certains ébénistes étaient en effet employés par des peintres (peut-être pour réaliser des cadres), et recevaient comme salaire des peintures[27]. Tant qu'ils ne faisaient pas partie de la guilde, ces ébénistes ne pouvaient pas vendre ces peintures eux-mêmes. Ils seraient donc entrés dans la guilde pour vendre des peintures et en même temps continuer le travail de l'ébène et la menuiserie, sans plus faire partie des menuisiers. Les doyens de la guilde fournirent trois extraits de leurs registres, datant de 1595, 1600 et 1612[28] et montrant l'adhésion de trois fabricants de cadres en ébène. Les menuisiers soulignent que ces trois artisans ont bien été reçus dans la guilde comme fabricants de cadres en ébène, et non comme ébénistes en général[29]. Il n'existerait cependant à Anvers aucun ébéniste se limitant à la fabrication de cadres. Les doyens de la guilde ne pourraient donc pas tirer prétexte des cadres que les *tafereelmakers* réalisent en d'autres bois pour attirer dans la guilde tous les ébénistes. Les ébénistes appartenant au métier des menuisiers et ayant payé le 'demi-métier' pour pouvoir travailler l'ébène pouvaient réaliser toutes sortes de travaux en ébène, tandis que les ébénistes inscrits parmi les *tafereelmakers* devaient se limiter aux encadrements.

Les membres de la guilde se basent également sur leur ordonnance de 1617 interdisant aux menuisiers de réaliser des travaux concernant les *tafereelmakers* sans avoir subi la grande épreuve des menuisiers[30]. Les ébénistes qui, pour pouvoir travailler l'ébène, n'avaient adhéré qu'au 'demi-métier', ne pourraient donc réaliser aucun cadre. Les menuisiers nient toutefois que le travail de l'ébène puisse concerner les *tafereelmakers* et affirment en détenir le monopole. Ce point de l'ordonnance de 1617 n'aurait été prévu que pour empêcher les menuisiers d'accueillir des artisans payant le 'demi-métier' et réalisant uniquement des cadres ou des panneaux, ce qui aurait préjudicié le métier des *tafereelmakers*. Selon les menuisiers, cette ordonnance ne concerne pas le travail de l'ébène, notamment dans les formats qui y sont cités: ces formats ne s'appliquent pas aux cadres ou panneaux en ébène, «puisqu'on ne fabrique pas de panneaux en ébène», ni des cadres en ébène à ces formats.

Il fut enfin décidé que les ébénistes ayant accompli un apprentissage de trois ans et acquis le 'plein' métier, devaient faire partie des menuisiers[31]. Les ébénistes qui avaient une compétence limitée, comme la fabrication de cadres, adhéreraient à la guilde de Saint-Luc. Ces candidats ébénistes devaient avoir accompli un apprentissage de seulement deux ans chez un menuisier ou un ébéniste. Leur épreuve consistait à réaliser un pied de crucifix. Ils devaient en outre payer 30 florins comme droits d'entrée. Ceux qui avaient versé leur redevance à la guilde de Saint-Luc auparavant ne devaient pas s'acquitter de ce droit. Selon R. Fabri, beaucoup d'ébénistes s'inscrivaient donc d'abord dans la guilde de Saint-Luc comme fabricants de cadres, car le prix de leur entrée était moins élevé, puis adhéraient à la corporation des menuisiers.

[27] SAA, Procès D4868, 20/08/1620.
[28] AAA, n°98 (72*).
[29] SAA, Procès D4868, 20/08/1620, art. 8.
[30] SAA, Procès D4868, 20/08/1620.
[31] R. FABRI, 1991, p. 113. AAA, n°98 (72*), 27/02/1621, SAA, GA4337, Procès A929, 18/7/1675, Procès A648, 27/2/1621.

d. *La sculpture des cadres*

Bien que réservé aux sculpteurs, ce travail était également réalisé par les fabricants de cadres. Selon les *lystmakers*, tous les maîtres de la guilde de Saint-Luc pouvaient pratiquer tant la sculpture que la fabrication de cadres[32]. En 1689[33], les sculpteurs expliquent qu'ils sont gravement lésés par les fabricants de cadres, car ces derniers sculptent leurs travaux et détournent leur clientèle. Ils demandent donc d'interdire à quiconque d'exercer la sculpture, à moins d'avoir suivi un apprentissage de quatre ans et d'être inscrit dans la guilde de Saint-Luc. Les fabricants de cadres ne pourraient plus sculpter eux-mêmes leurs travaux, mais devraient faire appel à un maître sculpteur. Ils devraient se limiter à des travaux sans éléments sculptés, conformément au modèle de leur épreuve. Les fabricants de cadres demandent cependant de pouvoir continuer à tailler des feuillages (*loofwerck*), «ainsi qu'ils l'ont fait depuis de nombreuses années»[34]. Le 25 octobre, les magistrats se prononcèrent en faveur des sculpteurs[35]: les fabricants de cadres ne pourront plus sculpter leurs travaux et devront faire appel à un sculpteur. Cette interdiction mènerait, d'après les fabricants de cadres, à la ruine de leur métier[36]. Ils demandent donc de pouvoir continuer à sculpter leurs cadres comme ils l'ont appris de leurs maîtres et promettent de ne pas s'occuper d'autre chose que de leurs cadres.

e. *La dorure et la peinture des cadres*

L'ordonnance du 25 octobre 1689 protégeait également les doreurs et 'étoffeurs'[37]. Au terme d'un apprentissage de quatre années, leur épreuve consistait à 'étoffer' une sculpture de Marie et à dorer un cadre sculpté (art. 6). Ce document interdit en outre les mélanges d'or fin et de 'faux or' (ainsi que d'argent fin et de 'faux argent') et impose le marquage des travaux de la marque personnelle du maître et de la lettre F s'il s'agit d'or ou d'argent fin (*fyn silver oft goudt*) et de la lettre V (*valsch*) s'il s'agit de faux or ou argent (art. 8).

En 1696, les fabricants de cadres affirment qu'ils ne pratiquent la dorure que sur leurs cadres, comme cela leur est permis depuis toujours[38]. Ils sollicitent de pouvoir fabriquer et dorer au moyen de faux or ou argent les cadres nommés *kip kap* ou travaux à la douzaine (*dozijn werck*)[39]. Les fabricants de cadres demandent aussi d'être autorisés à fabriquer des cadres lisses avec un bord doré à l'or fin ou argenté, laissant les autres travaux, notamment les cadres sculptés et les cadres entièrement dorés à l'or fin ou argentés aux doreurs. Un marchand déclare avoir envoyé en Espagne plus de huit mille cadres réalisés, dorés et argentés par les fabricants de cadres. Ce à quoi les doreurs répondent que

[32] SAA, Procès G7122, 14/09/1662.

[33] SAA, Pk778, f°140 r°-v°, 25/08/1689. Cité dans A.K.L. THIJS, «De nijverheid», dans *Antwerpen in de XVII^{de} eeuw*, Anvers, 1989, note 78, p. 142.

[34] SAA, GA4575, *Tafereelmakers en Lijstmakers*, n°2, 2/9/1689.

[35] Ordonnance publiée par J.B. VAN DER STRAELEN, 1855, p. 149-51 (art. 4).

[36] AAA, n°150 (153*), 6, 13 et 27/02/1696.

[37] J.B. VAN DER STRAELEN, 1855, p. 149-151.

[38] AAA, n°150 (153*), 6, 13 et 27 février 1696.

[39] SAA, GA4573, 8/02/1696.

la dorure et l'argenture leur reviennent exclusivement et qu'ils les exercent «au même prix et aussi rapidement» que les fabricants de cadres. On peut donc supposer que l'on faisait appel aux fabricants de cadres pour éviter de passer par l'intermédiaire des doreurs, ce qui devait être plus long et plus coûteux. Il fut décidé que les fabricants de cadres pourraient continuer à fabriquer, cuivrer ou dorer au moyen de faux or leurs encadrements dénommés *kip kap* et travaux à la douzaine, à la condition que ceux-ci ne soient pas plus larges que deux pouces et demi de la ville (6,5 cm)[40], ni plus épais qu'un pouce de la ville (2,6 cm), sans pouvoir utiliser pour cela de l'or fin ou de l'argent[41].

Quelques mois plus tard, les étoffeurs, les fabricants de cadres et les marchands de peintures s'exprimèrent à nouveau[42]. Pour les étoffeurs, les fabricants de cadres doivent se limiter à leur ordonnance ou leur épreuve, qui ne prévoit pas la dorure de leurs travaux. Plusieurs fabricants de cadres déclarent ne jamais avoir eu l'habitude de cuivrer leurs encadrements. D'autres affirment devoir et pouvoir cuivrer leurs cadres, sous peine de ne pouvoir subsister. Un représentant des marchands de peintures déclare enfin que les fabricants de cadres ont toujours doré leurs cadres et que si cela leur était interdit, cela serait très préjudiciable au commerce de peintures.

Le 24 juillet 1696, le fabricant de cadres Christiaan van Reynevelt affirma que l'épouse de son père, ancien du métier des fabricants de 'tableaux' et de cadres, dorait, décorait, ébénait (?) (*morissen*[43]) et peignait des encadrements[44]. Il a également vu réaliser cela par la plupart des ouvriers de son père, sans le moindre désagrément. Lui-même a d'ailleurs fabriqué certains encadrements qui ont été dorés chez son père. Plusieurs fabricants de cadres déclarent la même année[45] ne pas s'occuper de dorure: ils ne l'ont pas apprise, et ne l'ont pas vue pratiquer par leurs maîtres.

Un document non daté se base sur l'origine du métier des *tafereelmakers* et des fabricants de cadres pour expliquer le fait que ceux-ci dorent et peignent leurs travaux[46]. Les *tafereelmakers* et les fabricants de cadres seraient issus du métier des menuisiers. «Au temps du défunt monsieur Petro Paulo Rubens», il devint impossible pour les menuisiers de fournir en supports tous les peintres. Dès lors, quelques menuisiers se sont consacrés exclusivement à la réalisation de supports et de cadres pour satisfaire cette demande. Ces maîtres ont pris en apprentissage des élèves, qui ont appris la fabrication de cadres, 'tableaux' et cadres de miroirs, mais sont devenus incapables de passer la grande épreuve des menuisiers. Ces artisans ont donc constitué un petit métier parmi les menui-

[40] Un pied anversois (286,8 mm) = 11 pouces. Un pouce anversois = 26,07 mm. (H. DOURSTHER, *op. cit.*, p. 403).

[41] SAA, GA4575, *Vergulders en Stoffeerders*, 13 et 27/2/1696.
AAA, n°150 (153*), 6, 13 et 27/02/1696.

[42] SAA, GA4575, *Vergulders en Stoffeerders*, 22-23/7/1696.

[43] Ce terme pourrait à notre avis être en rapport avec le *morissenhout*, qui désignait un bois utilisé pour imiter l'ébène à un prix moins élevé (R. FABRI, «Aan- en afwezigheid van merktekens op 17de-eeuwse antwerpse kunstkasten», dans *Merken opmerken. Merk- en meestertekens op kunstwerken in de Zuidelijke Nederlanden en het Prinsbisdom Luik. Typologie en methode*, Louvain, 1990, p. 202).

[44] R. FABRI, «De 17de-eeuwse Antwerpse kunstkast. Typologische en historische aspecten», dans *Verhandelingen van de Koninklijke Academie voor Wetenschappen, Letteren en Schone Kunsten van België*, 53, 1991, p. 151. SAA, GA 4575, *Vergulders en Stoffeerders*.

[45] E. DUVERGER, *Antwerpse Kunstinventarissen uit de zeventiende eeuw*, vol. XII, Bruxelles, 2002, p. 398-9, 25/08, 12 et 26/11/1696.

[46] SAA, GA4575, *Vergulders en Stoffeerders*.

siers. Ils auraient demandé à se faire admettre dans la guilde pour être reconnus et pouvoir peindre leurs cadres, les dorer, les argenter ou les recouvrir d'autres métaux.

Les doreurs et étoffeurs affirment que la dorure et la polychromie n'ont jamais été pratiquées par les fabricants de cadres. Ces derniers prétendent cependant avoir pratiqué la dorure depuis plus de cinquante ans. Les anciens fabricants de cadres ont non seulement doré et étoffé leurs travaux avec de l'or, de l'argent ou du cuivre, mais ont aussi réalisé et étoffé des encadrements de cheminées et de pièces avec du marbre rouge, blanc ou autrement. Ils donnent pour argument qu'un tonnelier ou un tailleur ne travaille pas sur base de son épreuve. Ils ne doivent donc pas se limiter à la fabrication de cadres et il devrait leur rester permis d'étoffer et de dorer leurs travaux, car ils ont toujours pu agir ainsi, sous la protection de la guilde.

Un autre document non daté, postérieur à 1646[47], interdit aux fabricants de cadres de dorer, décorer (*cireren*) ou étoffer leurs cadres autrement qu'en les peignant 'communément' en noir (*swerten*). Les fabricants de cadres doivent s'adresser aux étoffeurs et doreurs pour tout travail de peinture ou de dorure. Les menuisiers ne peuvent pas non plus étoffer leurs cadres autrement qu'en les peignant au moyen de 'noir de colle' (*lijmswert*), comme leurs autres travaux, «sans pouvoir dorer, peindre à l'huile ni étoffer autrement».

2. Typologie

Le deuxième volet de nos recherches consiste à examiner *in situ* (dans les musées et institutions religieuses) des peintures du XVIIe siècle[48] possédant un encadrement d'époque et à les documenter au moyen de photographies et de relevés des profils des cadres et de leurs assemblages. À ce jour, nous avons constitué un répertoire de plus de 150 cadres. Plusieurs types d'encadrements utilisés au XVIIe siècle ont pu être identifiés. Nous ne présentons ici que les catégories qui peuvent être définies avec certitude grâce au nombre d'exemplaires observés. D'autres catégories verront sans doute le jour dans la suite de notre étude. Nous évoquerons également la fixation des toiles dans les cadres, la suspension des encadrements et les marques que nous avons relevées.

a. Cadres simplement moulurés, peints en noir et dorés

Une première catégorie rassemble des cadres simplement moulurés, peints en noir et ornés d'un ou plusieurs bords dorés. Certains possèdent un profil très simple, constitué d'un large champ plat et d'une doucine (fig. 1a). Ces cadres simples étaient peut-être complétés par un second cadre ou insérés dans des lambris. Les documents d'archives mentionnent d'ailleurs des encadrements composés d'un cadre intérieur (*binnenlyst*) et d'un cadre extérieur (*buytenlyst*). La majorité des cadres appartenant à cette première catégorie présente toutefois un profil plus élaboré (fig. 1b).

[47] SAA, GA4573, après le 8/10/1646.

[48] Nous ne nous sommes pas limitée de façon stricte au XVIIe siècle: nous avons pris en compte des peintures de la fin du XVIe siècle et du début du XVIIe siècle. Dans notre répertoire, figurent également des cadres qui ont été modifiés pour permettre leur adaptation aux dimensions d'œuvres pour lesquelles ils n'étaient pas prévus.

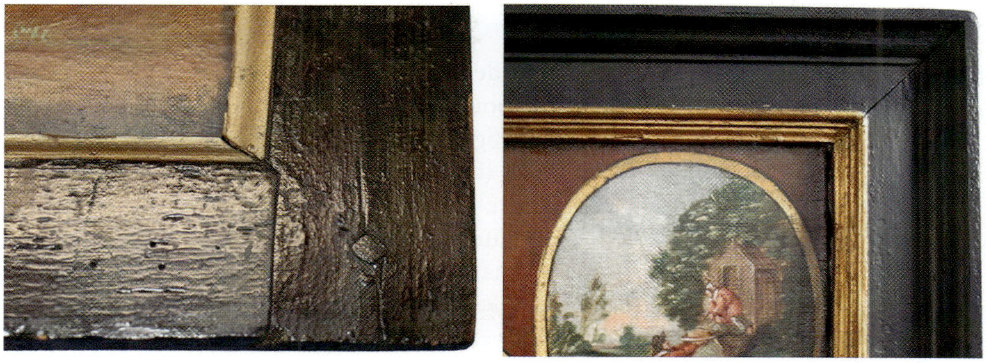

Fig. 1. Cadres moulurés, peints en noir et dorés, détails. a. Anonyme, *Retour du troupeau*, XVIIe siècle, collection privée. b. Anonyme, *Vie de sainte Jacqueline*, 1635, Lessines, Hôpital Notre-Dame à la rose (clichés de l'auteur).

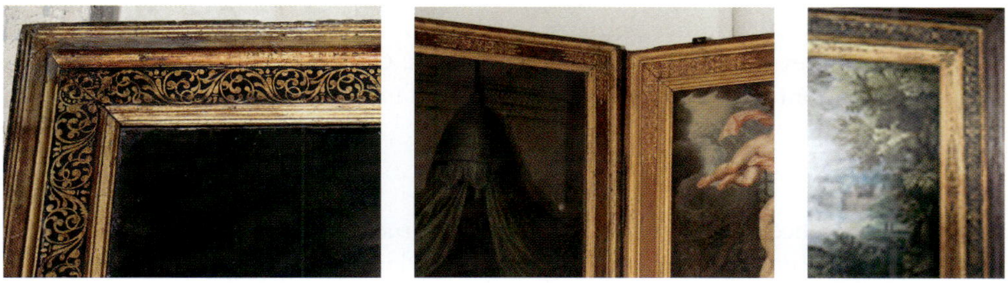

Fig. 2. Cadres aux champs plats ornés de motifs dorés, détails. a. Anonyme, *Glorification du nom de Jésus*, début XVIIe siècle, Malines, église des Saints-Pierre-et-Paul. b. H. De Clerck, *Triptyque du martyre de saint Sébastien*, vers 1617, Asse, église Saint-Martin. c. Détail d'un cadre de la chambre d'hiver du Château de Rosenborg (clichés de l'auteur).

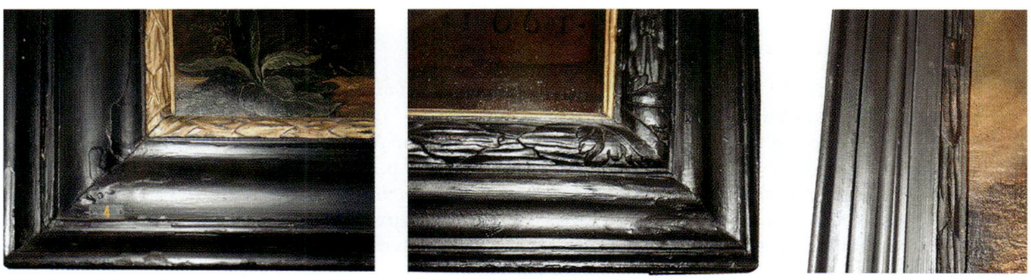

Fig. 3. Cadres ornés de feuillages sculptés, détails. a. Entourage de Pieter Lastman, *La multiplication des troupeaux de Jacob*, Anvers, Museum Maagdenhuis, cat. 4. b. Anonyme, *Portrait de sœur Barbara Happaert*, 1661, Bruges, Musée de la Poterie, inv. 0.P83.I. c. Anonyme, *Sainte Cécile*, Turnhout, Musée du Béguinage (clichés de l'auteur).

b. Cadres ornés de rinceaux dorés

Déjà en usage au XVIe siècle[49], ce décor de rinceaux ou de motifs plus géométriques dorés sur fond sombre (noir ou brun) se poursuit au début du XVIIe siècle. Ces décors peints couvrent tout le périmètre du cadre (fig. 2a), ou se limitent aux angles et éventuellement au centre des montants et traverses (fig. 2b). Les cadres du *triptyque du martyre de saint Sébastien* d'Hendrik De Clerck (vers 1617, Asse, église Saint-Martin) sont ornés de décors de ce type[50] (fig. 2b). Les moulures des cadres sont dorées et les plats sont peints en brun. Ces champs plats sont rehaussés de motifs dorés, peints au milieu de chaque montant ou traverse et dans les angles. Des motifs peints figurent également sur les cadres des tableaux de la Chambre d'hiver du Château de Rosenborg[51] (Danemark) (fig. 2c): chacun de ces tableaux possède trois cadres: un cadre intérieur doré et orné de motifs décoratifs peints en noir; un second cadre peint en noir enrichi d'ornements dorés et enfin un cadre extérieur en chêne uni, qui maintient la peinture et les deux autres cadres dans les lambris.

c. Cadres ornés de motifs sculptés

De nombreux cadres ornés de décors sculptés ont pu être examinés. Deux motifs sont récurrents: les guirlandes de feuilles et les oves. Ces motifs peuvent occuper toute la largeur des montants et traverses, ou seulement l'intérieur du cadre. L'espace restant est alors mouluré.

c.1. Feuillages

Nous avons rencontré fréquemment des cadres ornés de feuillages sculptés, alternant dans certains cas une feuille simple et une feuille double. Ces feuillages sont sculptés tantôt de façon très schématique, tantôt avec plus de détails (fig. 3a, 3b, 3c). Dans les exemples les plus détaillés, ils s'apparentent à des feuilles de laurier. Ces feuilles peuvent être dirigées vers le milieu ou vers les extrémités des montants et traverses, souvent symétriquement par rapport à un motif central. Autour de ces feuillages s'enroulent des rubans et les motifs sculptés peuvent être recouverts d'un autre type de feuille aux angles du cadre (fig. 3b). Ces guirlandes de feuillages sont dorées ou noires. En 1689[52], les fabricants de cadres anversois demandent de pouvoir continuer à sculpter des rinceaux

[49] H. VEROUGSTRAETE, R. VAN SCHOUTE, *Cadres et supports dans la peinture flamande aux 15e et 16e siècles*, Heure-le-Romain, 1989, p. 63-64.

[50] Ce triptyque a été restauré à l'Institut Royal du Patrimoine Artistique. D. ALLARD (dir.), *SOS Peintures anciennes. Sauvegarde de 20 œuvres sur panneau*, Bruxelles, p. 158-166. Une autre œuvre d'Hendrik De Clerck, le *Triptyque du martyre des saints Chrysante et Darie* (Bruxelles, église Notre-Dame de la Chapelle, 1619), possédait aussi un encadrement orné de rinceaux dorés. Ces décors, qui se trouvent sur les champs plats des cadres des volets du triptyque, ne sont plus visibles qu'en relief sous une couche de peinture noire.

[51] Ces tableaux, ainsi que leurs cadres, firent l'objet d'une commande unique à des ateliers anversois. Voir à leur sujet: J. WADUM, «The Winter Room at Rosenborg Castle. A Unique Survival of Antwerp Mass-Production», dans *Apollo*, août 1988, p. 82-87. Nous remercions J. Wadum pour les renseignements qu'il nous a transmis.

[52] SAA, GA4575, *Tafereelmakers en Lijstmakers*, n°2, 2/9/1689.

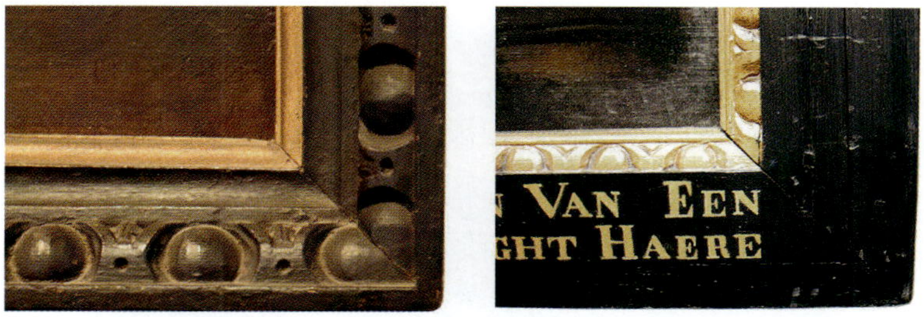

Fig. 4. Cadres ornés d'oves sculptés, détails. a. Theodoor Van Loon, *Les pélerins d'Emmaüs*, Anvers, Museum Maagdenhuis, cat. 38. b. Jacques Van Horen, *Guérison d'une jeune fille de 's Hertogenbosch*, Malines, Couvent des Jésuites (clichés de l'auteur).

(*loofwerck*), «comme ils l'ont fait depuis de nombreuses années». Il s'agissait peut-être de ce type de décor.

c.2. Oves

Les motifs d'oves sculptés sont traités de diverses façons: la forme des oves varie, ainsi que leurs dimensions et les éléments qui les séparent (fig. 4a, 4b). Plusieurs tableaux du XVIe siècle ont été munis de tels encadrements au XVIIe siècle: par exemple le *Triptyque Colibrant* de Goossen Van der Weyden (1516, Lierre, église Saint-Gommaire)[53]. Ce type de cadre semble avoir été utilisé fréquemment pour des tableaux d'autels. Ces motifs d'oves servent aussi d'encadrement dans les bordures de tapisseries[54].

d. Cadres entièrement sculptés

À ce stade de nos recherches, nous n'avons rencontré que peu d'exemplaires à classer dans cette catégorie. Citons par exemple le cadre d'un *Bouquet de fleurs* de Gaspar Pieter Verbruggen le Jeune, sculpté par Laurent Van der Meulen[55] (fig. 5). Ce cadre est composé d'une étroite guirlande de fleurs dorée qui jouxte la toile, d'une bordure d'acanthes et d'un décor composé de fleurs, de feuilles et de chérubins. Peu de cadres sculptés sont cités dans les inventaires après décès publiés par E. Duverger. La sculpture des encadrements de peintures fut toutefois le sujet de litiges entre les fabricants de cadres et les sculpteurs. Il devait donc s'agir d'un marché important pour les deux métiers. Les cadres évoqués dans ces documents n'étaient peut-être pas entièrement sculptés, et pourraient correspondre plutôt aux cadres partiellement sculptés de la catégorie précédente.

[53] H. LEEMANS, *De Sint-Gummaruskerk te Lier*, Anvers-Utrecht, 1972, p. 297.

[54] Trois tapisseries de la tenture du Triomphe de l'Eucharistie de Rubens sont encadrées d'oves. N. De Poorter souligne d'ailleurs que ces motifs d'oves doivent être considérés comme des cadres de peintures. N. DE POORTER, *The Eucharist Series* (*Corpus Rubenianum Ludwig Burchard*, II), Bruxelles-Londres-New York, 1978, t. I, p. 65, 210, fig. 209,214,216.

[55] Bruges, Groeningemuseum, inv.0.231.I. Voir S. VANDENBERGHE, «Een gebeeldhouwde lijst van de Mechelse beeldhouwer Laurent Van der Meulen (1645-1719) en enige aan hem toegeschreven werken», dans *Jaarboek van de Brugse Stedelijke Musea 1983-4*, Bruges, 1985, p. 307-313.

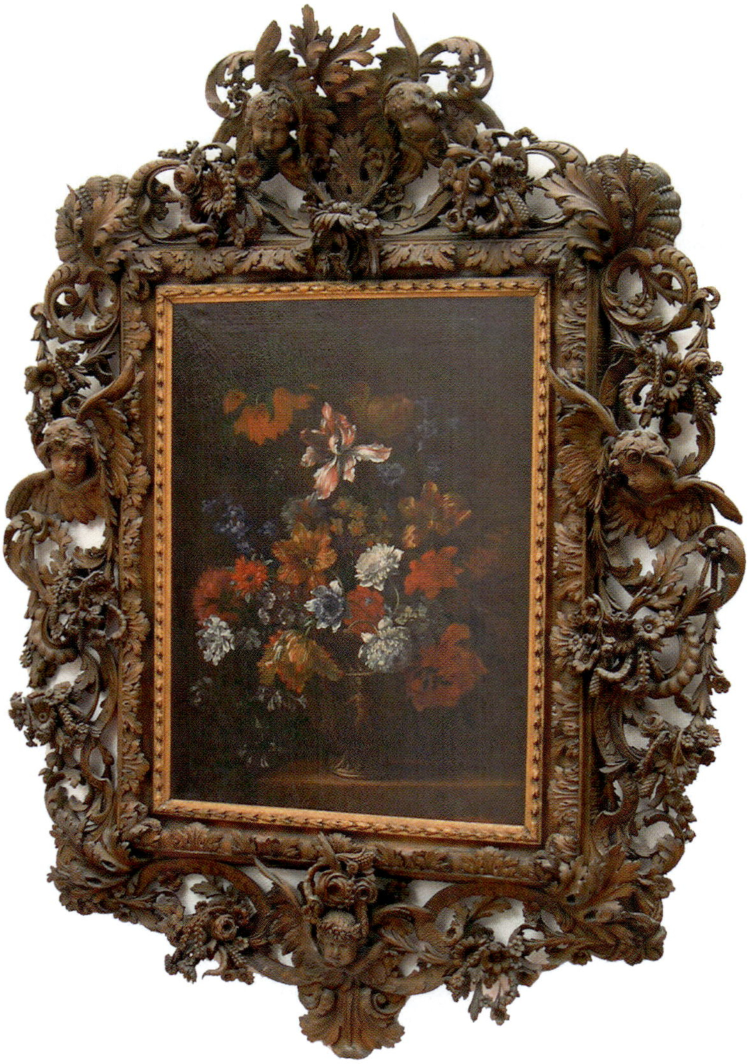

Fig. 5. Cadre entièrement sculpté. Gaspar Pieter Verbruggen le Jeune, *Bouquet de fleurs*, Bruges, Groeningemuseum, inv. 0.231.I (cliché de l'auteur).

L'image qui ressort des documents d'archives doit donc être nuancée et complétée par les observations effectuées *in situ*. Elle est également discordante en ce qui concerne les cadres sculptés d'oves: nous n'avons pas trouvé la moindre mention d'encadrements de ce type dans les documents d'archives, alors que nous en avons rencontré beaucoup. À l'inverse, nous n'avons pas, à ce jour, pu examiner un grand nombre de cadres en ébène, alors qu'il s'agit d'un type récurrent dans les inventaires après décès et que la fabrication de ces cadres en ébène fut également à l'origine de conflits entre métiers.

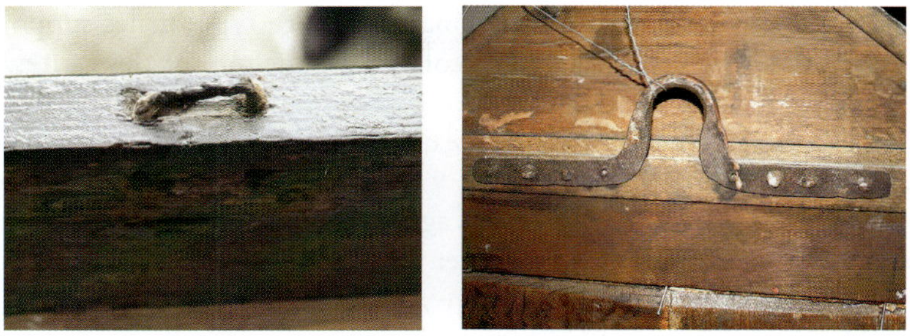

Fig. 6. Modes de suspension. a. Trous de suspension. Jacques Van Horen, *Guérison de Ioannes De Regere*, Malines, Couvent des Jésuites. Trous de suspension. b. Ferrure. Anonyme, *Triptyque de la Crucifixion*, 1611, Lessines, Hôpital Notre-Dame à la rose (clichés de l'auteur).

e. Fixation des toiles — suspension — marques

Plusieurs cadres observés ne comportent pas de battée. Ils étaient destinés à recevoir une peinture sur toile qui était directement clouée sur le cadre, sans l'intermédiaire d'un châssis. Un tel montage permettait de réduire le poids de l'ensemble, mais avait pour inconvénient que la toile était directement en contact avec la paroi. Au cours des restaurations, ces toiles ont souvent été montées sur des châssis et des lattes ont été ajoutées au revers du cadre pour constituer une fausse battée.

Deux méthodes de suspension peuvent être distinguées: les trous de suspension[56] (fig. 6a) et les ferrures (fig. 6b). Dans ces trous de suspension était passé un ruban, qui permettait de suspendre la peinture à une fixation dans le mur. Dans certains cas, ces trous de suspension sont abîmés ou même brisés, sans doute à cause du poids excessif du cadre et de la peinture. Ces trous de suspension pouvaient toutefois suffire à supporter le poids d'une toile dans un cadre simple, du type par exemple des cadres plats que nous avons observés. L'absence de châssis allégeait également le poids de l'ensemble. Les cadres plus lourds étaient suspendus au moyen de ferrures clouées. Des ferrures peuvent aussi remplacer des trous de suspension abîmés. Des représentations d'intérieurs[57] montrent que les peintures pouvaient être suspendues au moyen d'anneaux dans lesquels était passé un ruban noué. Nous n'avons rencontré ce type d'anneaux que dans deux cas.

L'ordonnance du 11 décembre 1617 obligeait les fabricants de cadres et les menuisiers anversois à marquer leurs panneaux et leurs cadres d'une marque personnelle[58].

[56] Trous de suspension ou de maintien: trous percés en biais depuis l'arrière du cadre vers la tranche de la traverse supérieure. H. VEROUGSTRAETE, R. VAN SCHOUTE, 1989, p. 75.

[57] Citons par exemple deux œuvres de David Teniers le Jeune, *La galerie de peintures de l'archiduc Léopold-Guillaume*, Vienne, Kunsthistorisches Museum, inv. 739 et *Le singe peintre*, Madrid, Musée du Prado, inv. 1.805.

[58] Concernant ce type de marques, voir J. VAN DAMME, «De Antwerpse tafereelmakers en hun merken. Identificatie en betekenis», dans *Jaarboek van het Koninklijk Museum voor Schone Kunsten Antwerpen*, Anvers, 1990, p. 193-236 et plusieurs articles publiés par J. WADUM.

Tandis que nous avons pu relever diverses marques de ce type sur les supports de bois des œuvres examinées, nous n'en avons encore relevé aucune sur des cadres. De nombreux cadres portent cependant des *tracés du menuisier*, c'est-à-dire de fines entailles servant à situer la battée ou l'arasement à réaliser. Nous avons également relevé sur un cadre de fins tracés comportant les signes X, V et I[59]. La signification précise de ces marques n'a pas encore pu être déterminée. Il s'agissait sans doute de mesures ou de signes permettant d'identifier les pièces à assembler.

À ce stade de nos recherches, plusieurs types d'encadrements en usage au XVIIe siècle ont pu être définis. Si certains cadres présentent des décors qui existaient déjà au XVIe siècle, d'autres sont ornés de motifs nouveaux (guirlandes de feuillages et oves sculptés), que l'on retrouve dans d'autres formes d'art de l'époque. La suite de nos recherches aura pour objectif d'affiner les catégories établies et, nous l'espérons, d'en constituer d'autres. Nous examinerons également les nombreuses sources iconographiques disponibles (par exemple les cabinets d'amateurs).

Elisabeth BRUYNS
Université catholique de Louvain
Laboratoire d'étude des œuvres d'art
par les méthodes scientifiques
Rue Cardinal Mercier 53
B – 1348 Louvain-la-Neuve
Tel: +32.10.47.48.97
Fax: +32.10.47.48.95

2004: licence en Histoire de l'Art à l'Université Catholique de Louvain. 2004-2005: assistante au Laboratoire d'étude des œuvres d'art de Louvain-la-Neuve. Depuis juillet 2005: bénéficiaire d'une bourse des Fonds Spéciaux de Recherche de l'Université Catholique de Louvain pour la préparation d'une thèse de doctorat sur *L'encadrement des peintures dans les Pays-Bas du sud au XVIIe siècle.*

[59] Anonyme, *Vanitas*, 1606, Lessines, Hôpital Notre-Dame à la rose, inv. TA 0018.

IRR UPDATE: PRACTICAL AND INTERPRETATIVE ISSUES

Molly Faries

Abstract: Infrared reflectography (IRR) has become a more complex technology in the last decade. The technique has been transformed by the transition from analog to solid state detectors, and the choice in imaging devices has greatly increased, ranging from low- to high-end cameras operating at different wavelengths. Patterns of use need to develop that can provide guidelines not only for practitioners in the field but also for those who otherwise use and interpret infrared documents. In initiating discussion of topics that will help define such guidelines, this paper will focus on an aspect that is fundamental to IRR technology — wavelength response (with and without filters and/or using different cameras in combination). The effects of different wavelengths will be presented visually, by imaging details of paintings, rather than by using test panels or graphs.

This paper derives from a session at the Bruges conference that was organized to bring out some of the inter-related factors making infrared reflectography, as it is practiced today, a more complex technique. One factor, unquestionably, is the diverse array of devices available that image in the infrared. The list this author compiled for the 2006 conference of IR cameras in current use by symposium participants included twenty-five different devices, with six different sensor types. This list, which includes a selected bibliography of recent publications and applications, is now available on-line at the Louvain-la-Neuve website.[1] Other factors are new developments in camera/detector positioning systems,[2] and improvements in system optics. At the Bruges conference, Rachel Billinge, research associate at the London National Gallery, presented a newly-designed camera with optimized resolution, and the camera has since been published by her and others in *Studies in Conservation*.[3] Additionally, IRR is now often done in conjunction with high-resolution digital scanning in visible light. Ron Spronk presented images at the Bruges conference produced by a camera of this type used at Harvard University Art Museums, the Phase One LightPhase CCD digital camera back on a Hasselblad camera body, which is also slightly sensitive in the infrared. This camera

[1] See the list at: http://labart.fltr.ucl.ac.be/eng/menu/recherche/colloque.html

[2] The prototype camera developed at London National Gallery involves a positioning system for the detector (see note 3 below); a positioning system (for the lens) is also part of the set-up devised by Duilio Bertani (see D. BERTANI, "La Reflectografía Infrarroja", in *El Trazo Oculto*, edited by Gabriele Finaldi and Carmen Garrido, exh. cat. Museo Nacional del Prado, Madrid, 2006, p. 54-63 and D. BERTANI and L. CONSOLANDI, "High Resolution Imaging in the Near Infrared", in *Digital Heritage*, edited by Lindsay MacDonald, Oxford, 2006, p. 211-38); another positioning method has been developed in Spain by T. Antelo, J. Alejandre, M. del Egido, A. Gabaldón, J.M. Menéndez, J. Torres, C. Vega: "VARIM Project" (Computer Vision applied to Mechanized Infrared Reflectography, corresponding authors: araceli.galdaldon@iphe.mcu.es; jta@gatv.ssr.upm.es); and Art Innovation b.v., Hengelo, The Netherlands, offers several tripod systems, one of which is computer controlled, capable of automating the making of mosaics, see: www.art-innovation.nl

[3] D. SAUNDERS, R. BILLINGE, J. CUPITT, N. ATKINSON and H. LIANG, "A New Camera for High-Resolution Infrared Imaging of Works of Art", in *Studies in Conservation*, 51 (2006), p. 277-290.

efficiency of the devices, particularly, indium gallium arsenide's, which is the highest (the implications of which are discussed below). The high sensitivity of indium gallium arsenide versus the lead sulfide of the vidicon has been plotted elsewhere.[11] Its high sensitivity as opposed to two other detector materials, indium antimonide and mercury cadmium telluride, has also been published.[12]

In order to gauge the penetration of infrared light at different wavelengths, almost all researchers in this field have made test panels with pigments in layers of varying thickness applied over different underdrawing materials; and these have served as the basis for IRR theory. It is useful to keep these panels, since they can help predict the performance of newly-developed detectors and provide comparisons with known results. Still, reconstructions and test panels cannot replicate all the complexities of historical painting techniques, nor can they approximate the effects of aging or the varied thicknesses and mixtures in paint. In this article, comparisons will be taken from several fifteenth- and sixteenth-century works, since it is necessary to provide examples from actual paintings in order to supplement the theoretical data. These can establish visual reference points and help refine our expectations of IRR technique.

The following examples present a few basic pigments: a blue and a green, whose penetration is of course critical to infrared study, and lead white. As more focal plane array cameras come into common use, these comparisons can hopefully be verified and expanded.

Azurite blue. The first example is a detail of the figure of Peter in an anonymous early sixteenth-century Antwerp painting depicting the *Arrest of Christ* (Cambridge, Massachusetts, Busch-Reisinger Museum, Harvard University Art Museums). Azurite has been identified in this painting, and although cross-sections were not taken in Peter's blue garment, the blue is most likely azurite.[13] Figure 2 shows the figure of Peter at different wavelengths — accomplished in this case by different cameras as well as different filters.[14] The images speak for themselves. In the image in the upper left, next to the color detail, the blue remains almost completely opaque. The image was produced by Harvard's Phase One camera in infrared mode, i.e. in a wavelength range from only 0.9 to 1.1 microns. As one would expect, the result is comparable to an infrared photograph.[15] The images in the middle row were both produced by the same Hamamatsu infrared vidicon, but with different filters, one cutting on at c. 850 nm, on the left, and one at 900 nm (Kodak 87A), on the right. It seems surprising that a small difference in wavelength could have such a marked effect: the image on the left approximates the

[11] The graph can be found at: http://www.corebyindigo.com

[12] See D. SAUNDERS et al., 2006, p. 279, fig. 1.

[13] For the painting and cross-sections, see R. SPRONK and C. VAN DAALEN, "Two Scenes from the Passion at the Harvard Art Museums: a Tale of Two Antwerp Workshops", in *Making and Marketing: Studies of the Painting Process in Fifteenth- and Sixteenth-Century Netherlandish Workshops*, Molly Faries ed., Turnhout, 2006, p. 43-53.

[14] I am grateful to Henry Lie of the Straus Center for Conservation, Harvard University Art Museums, for discussing these images with me and for allowing me to reproduce them.

[15] The result is similar to a published detail produced by Art Innovation's MuSIS camera, which is also responsive up to 1.1 microns, and is actually slightly better than an infrared photograph (at 0.7-0.9 μm); for these infrared images of this painting, see the article by H. LIE, 2003.

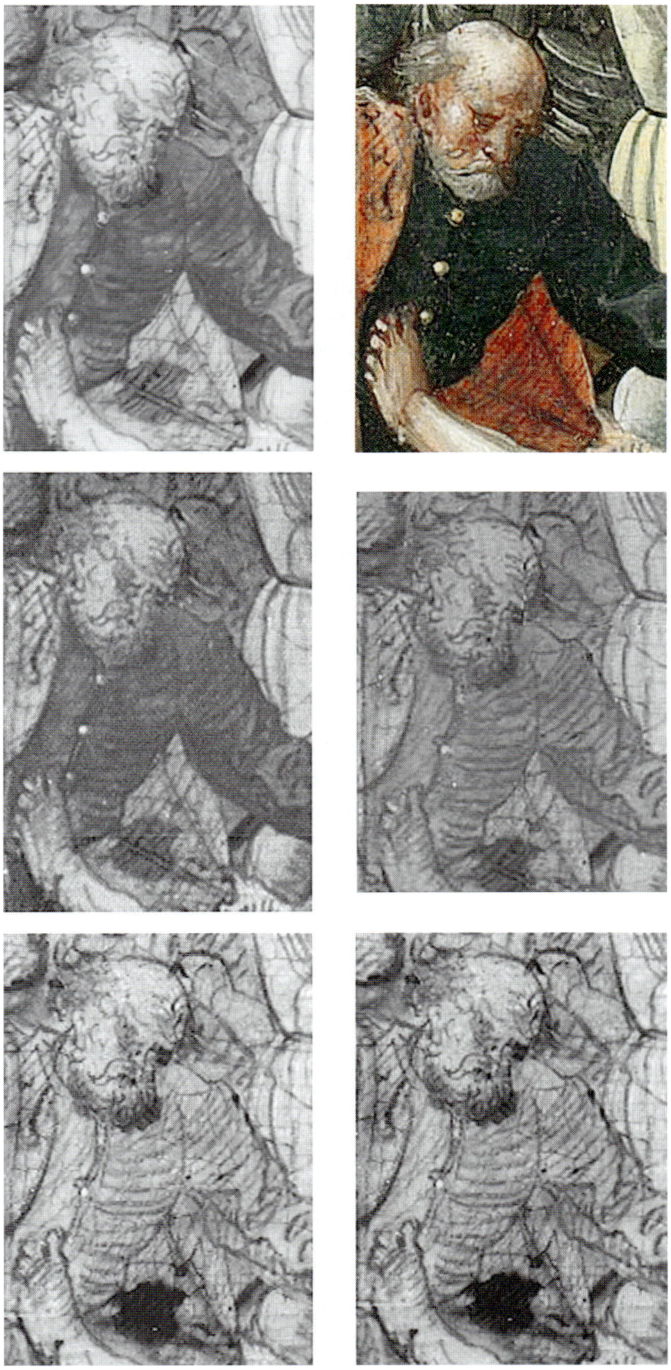

Fig. 2. Comparative images of blue (azurite) at different wavelengths: upper row, left: 0.9-1.1 μm (CCD), right: VIS; middle row, left: from c. 0.85 μm (vidicon), right: from 0.9 μm (vidicon); lower row, left: to 1.7 μm (InGaAs), right: 1.5-1.8 μm (PtSi) (© Images courtesy of the Straus Center for Conservation, Harvard University Art Museums; composite by Molly Faries).

digital infrared photograph, while the one on the right comes closer to the images produced by the indium gallium arsenide and platinum silicide cameras. This illustrates why it is so important to use the proper filter — one cutting on at 900 nm or above — with the infrared vidicon. In the infrared images in the bottom row, the effect of paint almost completely vanishes. These images were produced by cameras working up to and beyond 1.7 microns: an indium gallium arsenide camera on the left, whose wavelength response drops off sharply at 1.7 microns, and a platinum silicide camera on the right, whose wavelength response continues beyond 1.7 microns, potentially to 2 microns and further.[16] The performance of the two cameras is quite similar (note, for instance, that the retouch just above the foot turns much darker in these images). This appears to illustrate a recent point published by J.R.J. van Asperen de Boer: that, because of the extremely high signal-to-noise ratio of indium gallium arsenide focal plane arrays, they can be expected to perform as well as devices sensitive further in the infrared, even though they are working at shorter wavelengths.[17]

Malachite green. This example is included because it was published previously in the papers of this colloque. In 1995, Catherine Metzger, John Delaney, Colin Fletcher and Elizabeth Walmsley compared the IRR results of a Hamamatsu infrared vidicon with those obtained by the National Gallery's prototype platinum silicide camera of an angel in Francesco di Giorgio Martini's *God the Father Surrounded by Angels* (Washington, D.C., National Gallery of Art, Kress Collection). The angel wears a green robe identified as synthetic malachite (fig. 3). According to these authors, the vidicon was incapable of penetrating the green while the platinum silicide camera, working in the range of 1.5-2.0 microns, easily revealed the underlying "brushy undermodeling".[18] This result accords well with what researchers had already determined by imaging test panels. Using those made for their important article on IRR theory in 1994, the same authors predicted that all the underdrawing materials they studied but one would be best detected under malachite using a device that could image beyond 2 microns.[19] J.R.J. van Asperen de Boer had made a similar observation earlier: the theoretical thickness of malachite necessary to continue to hide underdrawing was greatest from approximately 1.7 to 2.1 microns.[20] I myself have used the malachite test panel made by J.R.J. van Asperen de Boer for his 1970 dissertation and found, using six different cameras, that platinum silicide cameras penetrated the thickest layers of malachite paint whereas the vidicon

[16] Again, these images derive from cameras used at Harvard University's Straus Center for Conservation, a Sensors Unlimited indium gallium arsenide camera, and an Inframetrics platinum silicide camera; they are described in the list mentioned in note 1.

[17] J.R.J. VAN ASPEREN DE BOER, "Slowly Towards Improved IRR Equipment", in *Recent Developments in the Technical Examination of Early Netherlandish Panel Painting. Methodology, Limitations & Perspectives*, M. Faries and R. Spronk eds., Turnhout and Cambridge, MA, 2003, p. 62: "An experiment by the author in 1995 with a Sensors Unlimited 128 × 128 InGaAs camera was very promising. In the region 0.8-1.7 microns, the penetration of malachite and azurite paints was as good as that around 2 microns. This can be explained by the high signal-to-noise ratio".

[18] C. METZGER, J. DELANEY, C. FLETCHER, and E. WALMSLEY, "A Platinum Silicide Camera Applied in the Study of Underdrawings", in *Le dessin sous-jacent dans la peinture, Colloque X. Le dessin sous-jacent dans le processus de création*, H. Verougstraete and R. Van Schoute eds., Louvain-la-Neuve, 1995, p. 197.

[19] E. WALMSLEY et al., 1994, p. 224, fig. 5.

[20] J.R.J. VAN ASPEREN DE BOER, 1970, p. 37, fig. 8a, and p. 38, fig. 9.

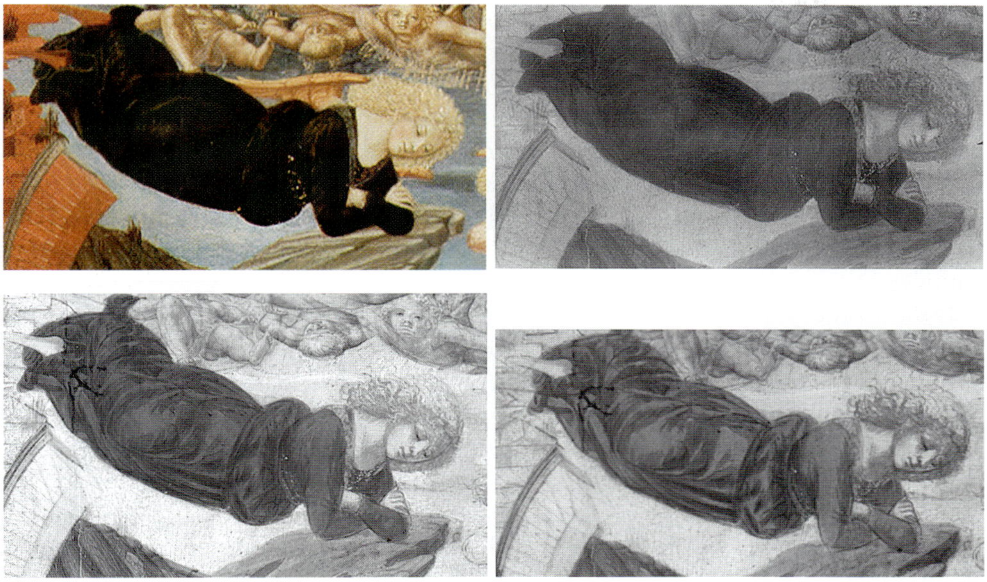

Fig. 3. Comparative images of green (synthetic malachite) at different wavelengths: upper row, left: VIS, right, 0.9-c.1.4+ μm (vidicon); lower row, left: 0.4-1.7 μm (InGaAs), right: 1.5-2.0 μm (PtSi) (© Images courtesy of Paintings Conservation, National Gallery of Art, Washington, DC, composite by Molly Faries).

did not.[21] More recent imaging tests, however, show that the situation with malachite may not be this straightforward.

In figure 3, the image in the upper row to the right of the color detail was made with a Hamamatsu vidicon. The malachite green appears to be entirely opaque, although in this case, the scanning of the IRR negatives to create the digital composite exaggerated this effect. Still, the penetration of this pigment has clearly been accomplished more fully in the images in the lower row. At my request, Elizabeth Walmsley, Paintings Conservator, Washington National Gallery, very kindly re-documented this area of the painting using an indium gallium arsenide camera responsive up to 1.7 microns (fig. 3, lower left). The image in the lower right was taken with the Washington National Gallery's Kodak platinum silicide camera configured to 1.5-2.0 microns. The visibility of the underdrawing in both of the composites in the bottom row is quite comparable, while in the vidicon image, the underdrawn washes can only be seen easily in the neck of the angel and the foot. The wash-like strokes are actually present in the green robe but can barely be discerned,[22] and they would probably be overlooked entirely without knowledge of what the solid state cameras revealed.

[21] The cameras included the Hamamatsu infrared vidicon, various indium gallium arsenide cameras, and several platinum silicide cameras.

[22] In several e-mails (27 February 2007 and 01 March 2007), Elizabeth Walmsley confirmed that the washy underdrawing strokes are actually visible in the vidicon images. In addition, in the documentation of this detail with the National Gallery's InSb camera, the strokes were visible in the image obtained with a J filter (1.1-1.4 μm), i.e. comparable to vidicon range.

In May of 2007, I was able to carry out another test on malachite green with the help of Dianne Modestini, Paintings Conservator, Kress Program in Painting Conservation, Conservation Institute, New York University, who made test panels (with both chalk and gesso grounds) with varying layers of malachite (Kremer pigments) over a number of different types of underdrawing materials. We compared three different IRR cameras: a Hamamatsu vidicon, a Sensors Unlimited indium gallium arsenide camera, and an Inframetrics platinum silicide camera.[23] The vidicon did not penetrate the green, while the InGaAs and PtSi cameras both penetrated it completely.

Up until this point, it has been thought that malachite was one of the few pigments requiring IRR equipment responsive to wavelengths longer than 2 microns. Since the use of malachite green has been reported much more frequently for Italian than for fifteenth- and sixteenth-century northern paintings, researchers in this field should remain watchful for cases that would help further define the pattern of use for this pigment.

Lead white. Due to the prevalence of lead white, it will be important to collect more comparative IR images of this pigment. I include the present example because it has also been published previously, in J.R.J. van Asperen de Boer's 1988 inaugural lecture upon his appointment as professor at the University of Groningen.[24] Depicting *The Visitation*, the painting is attributed to the Master of the Life of Mary (Rotterdam, Museum Boijmans-van Beuningen).[25] The critical area in this painting illustrates how overlying paint can affect the appearance of underdrawing, especially where the white inner lining of Elizabeth's mantle extends beyond the fold lines indicated in the underdrawn layout. At the point where the thicker edge of the lining crosses the underdrawing, the paint interferes with the registration of the underdrawn contour (fig. 4). It is only in the image made by the platinum silicide camera that the line continues uninterrupted for its full length (fig. 4, lower right). In this case, the indium gallium arsenide results seem to lie somewhere between those of the PtSi extended range camera and the vidicon. Given the extent to which lead white occurs in underpainting, priming, and other mixtures, this is a significant finding which should encourage the gathering of additional comparative images.

The details in figure 5 from an anonymous sixteenth-century Utrecht painting illustrate another phenomenon that cannot be ignored. They derive from a long panel painted on both sides depicting the *Works of Charity for the Poor of St. James* and *Last*

[23] One of the cameras was New York University Conservation Institute's Hamamatsu n 214 vidicon. For the other two, I owe thanks to both Donald Pancza of Sensors Unlimited, Inc., part of Goodrich Corporation, Princeton, N.J. for providing a SU320M-1.7RT/RS170 indium gallium arsenide camera and to Jim Coddington, who provided the Museum of Modern Art's Inframetrics platinum silicide camera. Some of Sensors Unlimited's models obtain the highest sensitivity in the 0.9-1.7 μm range; see http://www.sensorsinc.com

[24] J.R.J. VAN ASPEREN DE BOER, *Over materieel onderzoek van schilderijen*, printed lecture, 19 January 1988, p. 6-7.

[25] See the IRR of this painting by M. Faries in the catalogue by J. CHAPUIS, *German and French paintings, Fifteenth and Sixteenth Centuries*, Rotterdam, Museum Boymans-van Beuningen, 1995, cat. n°3.

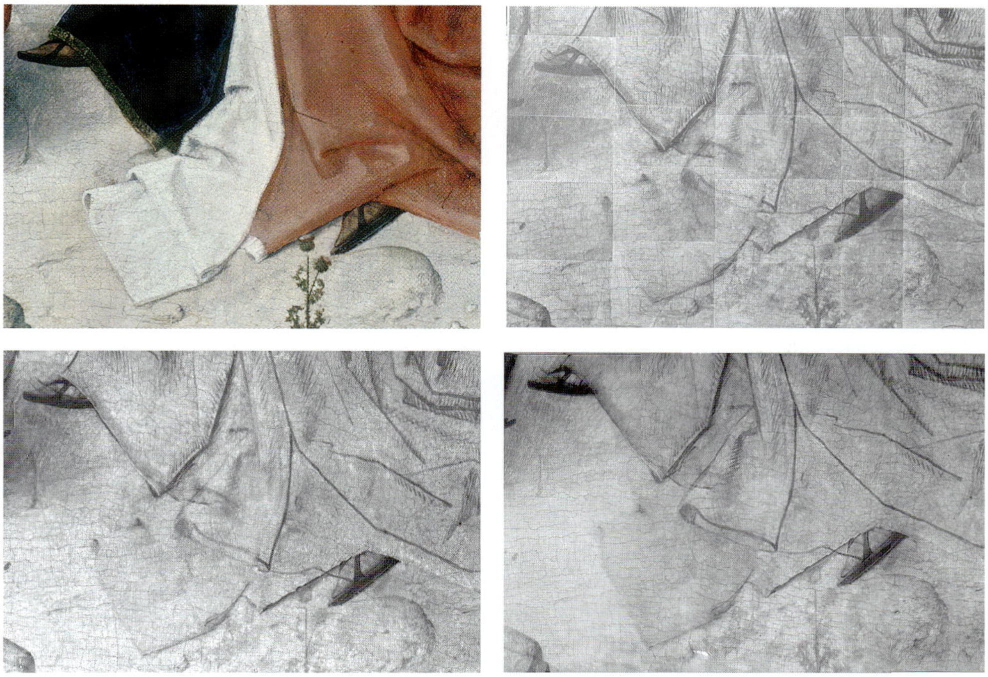

Fig. 4. Comparative images of white (lead white) at different wavelengths: upper row, left: VIS, right: 0.9-c.1.4+ μm (vidicon); lower row, left: 0.9-1.7 μm (InGaAs), right: 1.1-2.5 μm (PtSi) (© Images, upper left: courtesy M. Faries; upper right: IRR M. Faries; lower left: IRR and digital composite: M. Faries; lower right: IRR and digital composite: M. Faries).

Judgment (Utrecht, Centraal Museum).[26] The detail shows the head of the Virgin Mary in the Last Judgment scene. Just as underdrawing can become transparent in traditional vidicon reflectograms, it can also become transparent at longer wavelengths. In the platinum silicide image, the penetration of the paint is so strong that it even brings out the pattern of the wood grain, and no underdrawing appears (fig. 5, lower right). By contrast, the vidicon (fig. 5, upper left) is capable of detecting an underdrawn profile to the right of Mary's head. When the extended range camera is calibrated to 1.1 to 1.4 microns, it can reproduce the vidicon results (fig. 5, lower left). If this painting had been studied only in the range deemed the optimum for the penetration of paint, the underdrawing would have been missed.

The routine that developed out of experiences like this was the use of two basic filters with the University of Groningen's Mitsubishi M700 platinum silicide camera: one for IR imaging above 1.5 microns and one for the range from 1.1 to 1.4, a range comparable to the vidicon (these ranges, of course, can be further refined using other band pass filters). In my own experience using Groningen's camera, I encountered

[26] Inv. 2473, some of the IRR done by M. Faries in 2004 is illustrated in L.M. HELMUS, *Stichting Victor, Restauratie en natuurwetenschappelijk onderzoek in het Centraal Museum*, Utrecht, 2007, p. 64-65, 77.

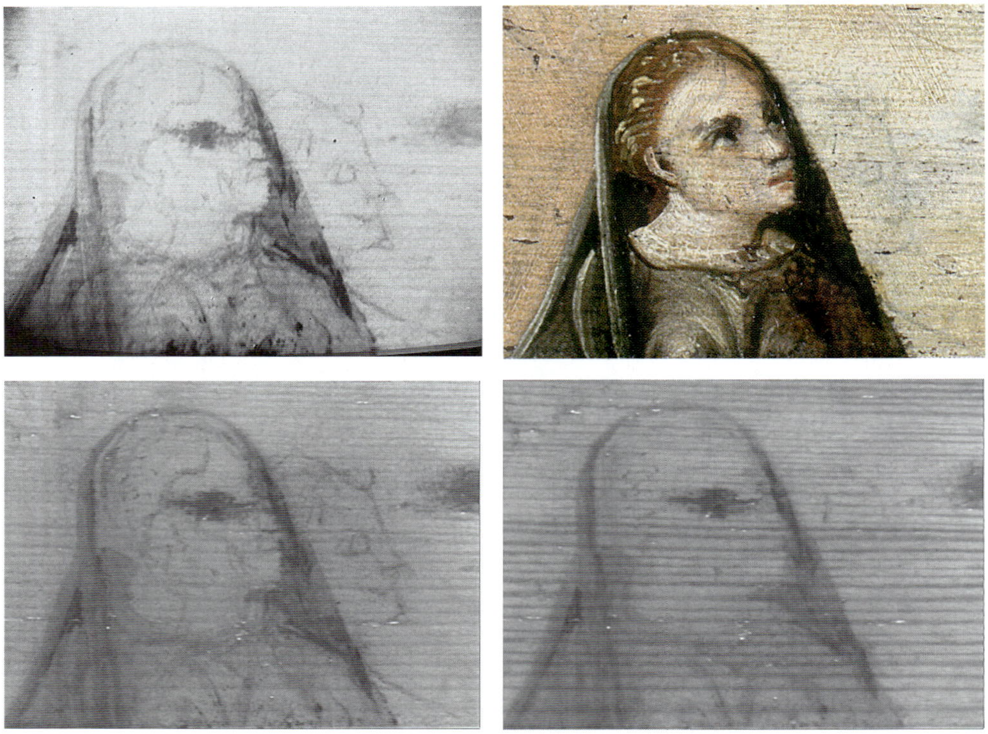

Fig. 5. Comparisons of a detail of the head of Mary in a *Last Judgment* scene (Utrecht Centraal Museum) by an anonymous sixteenth-century north Netherlandish master: upper row, left: 0.9 to 1.4+ μm (vidicon), right: VIS; lower row, left: 1.1 to 1.4 μm (PtSi), right: 1.5 to 2.5 μm (PtSi) (© Images, upper left: courtesy M. Faries; upper right: IRR: J.R.J. van Asperen de Boer, © Stichting RKD; lower left and right: IRR: M. Faries).

many more instances that required the vidicon range than the extended range around 2 microns.[27] This would suggest that those using extended range cameras cannot rely on longer wavelengths alone for IRR examinations but should double-check results using filters that can image "downwards" towards visible light, that is, at shorter and shorter wavelengths. In some cases, CCD cameras sensitive in the infrared between 0.9 to 1.1 microns can supplement the IRR study.

There are other benefits deriving from the use of cameras or filters that allow imaging closer to visible light. Paint changes can often be clarified beyond the gross comparison of the underdrawing with the final surface by gradually allowing more and more of the paint layers to register. A comparison that I published some years ago is still an excellent example of the usefulness of imaging at different wavelengths to elucidate

[27] Transparency can also occur in portions of the underdrawing. I noted this on at least once occasion in the black chalk underdrawing in a portrait by Jan van Scorel: at 1.1.-1.4 μm, all lines of the underdrawing could be detected, but at 1.5-2.5 μm, some lines disappeared. Some works may therefore require information derived from a combination of spectral ranges.

Fig. 6. IRR comparison of a painting by Vincent van Gogh at different wavelengths showing an underlying portrait of a woman: left: 0.9-1.4+ μm (vidicon), right: 1.1-2.5 μm (PtSi) (© IRR and digital composites: M. Faries).

compositional change. IRR revealed that a number of heads in Jan van Scorel's SS. Stephen and James polyptych (Douai, Musée de "La Chartreuse"), were underdrawn as profiles but painted in foreshortened views. Images of the heads in the range of infrared photography (0.7 to 0.9 microns) showed that they were first painted following the underdrawing and only changed late in the painting process to the more dramatic, foreshortened views. Some heads were completely painted in the intermediate stage, but one was only painted half-way before being changed.[28]

Some researchers use band filters and the resultant spectral response at specific wavelengths to help identify pigments.[29]

More cases may be found in modern painting where devices sensitive up to and beyond wavelengths of 2 microns are necessary. In a published study of one of Picasso's early paintings, it was only the extended range camera (PtSi) that revealed the sketches and caricatures of the first stage of the work that underwent further compositional change over a period of years, 1899 to 1903[30] (to my knowledge, the painting has not

[28] M. FARIES, "Underdrawings in the workshop production of Jan van Scorel, a study with infrared reflectography", in *Nederlands Kunsthistorisch Jaarboek*, 26 (1975), p. 122, figs. 21a-b; p. 128, fig. 25; p. 136, figs. 29a-b.

[29] See, for instance, J.K. DELANEY, E. WALMSLEY, B.H. BERRIE, and C.F. FLETCHER, "Multispectral Imaging of Paintings in the Infrared to Detect and Map Blue Pigments", in *Scientific Examination of Art: Modern Techniques in Conservation and Analysis* (Arthur M. Sackler Colloquia of the National Academy of Sciences, Washington, D.C.), Washington, D.C., National Academies Press, 2005, p. 120-35.

[30] A. HOENIGSWALD, "Works in Progress: Pablo Picasso's Hidden Images", in *Picasso, The Early Years 1892-1906*, ed. M. McCully, National Gallery of Art, Washington, D.C., 1997/98, p. 299-309.

yet been studied by an indium gallium arsenide camera). In my own work, I found significant differences in the results produced by the infrared vidicon as compared with an extended range camera (PtSi) in the study of a work by Vincent Van Gogh, *A Patch of Grass* (Otterlo, Kröller-Müller Museum).[31] As figure 6 illustrates, the platinum silicide camera better disclosed the darker and lighter shapes of an underlying portrait of a woman. The form is similar in appearance to portraits from the period of Van Gogh's well-known *Potato Eaters* (this painting has also not been studied by an indium gallium arsenide camera).

A Word about Resolution

High resolution has always been a requisite of infrared study. One aspect of traditional vidicon scanning involved imaging at a very small scale and training all of the image points of the vidicon on a section of a painting no larger than 1 to 1.5 cm wide to better observe the character and handling properties of underdrawing materials. Paintings were almost never documented in their entirety at this scale. Those IRR (indium gallium arsenide) cameras recently customized to improve resolution represent a very promising development, since they can obtain this degree of "sharpness" for the documentation of the complete painting. The London National Gallery's new SIRIS camera can operate at low, medium, and high resolutions at 2.3 pixels per mm, 4.6 pixels per mm, and 9.2 pixels per mm,[32] while that developed by D. Bertani at the University of Milan typically records images at 6 to 8 pixels per mm, or 36 to 64 pixels per square mm.[33] Other researchers refer to lines per mm or MTF (modular transfer function) to express resolution. For many of us used to working with images, it would be a help if these results could be visualized in some way. The chart published by J.R.J. van Asperen de Boer in the papers of this conference in 2001 may offer a possible solution, since it determines a camera's ability to image line pairs spaced increasingly closer together.[34]

Such a chart could serve in the same way as a test panel — as an objective reference for comparisons.

Concluding Remarks

In an effort to encourage more studies of this type, this paper has added several examples to the small group of comparative details of paintings imaged at different wavelengths in the infrared. These additions are key examples since they have figured in earlier discussions of IRR technology. They clearly illustrate that wavelength range is critical and that the sensitivity of the device at that range is equally important.

[31] M. FARIES, "Analytical Capabilities of Infrared Reflectography: An Art Historian's Perspective", in *Scientific Examination of Art: Modern Techniques in Conservation and Analysis* (Arthur M. Sackler Colloquia of the National Academy of Sciences, Washington, D.C.), Washington, D.C., National Academies Press, 2005, p. 87-104.

[32] D. SAUNDERS et al., 2006, p. 280.

[33] E-mail from D. Bertani, 15 September 2006.

[34] J. VAN DER WEERD, M.R.A. HEEREN, and J.R.J. VAN ASPEREN DE BOER, 2001, p. 237-239, fig. 4.

Such comparative images reveal that there is now more to negotiate during an infrared examination. Subtle manipulations are required as one determines the greatest penetration of most of the paint, the greatest visibility of all of the underdrawing, and the most informative registration of changes during the painting process. Ideally, practitioners should have as many options as possible at their disposal, and they must be aware of the capabilities and limitations of various types of IRR imaging devices. Using filters is clearly advisable, and researchers can also consider using different cameras in combination.

In my opinion, the traditional infrared vidicon has not been completely outmoded. This is implied in part by the number of times the University of Groningen's PtSi camera had to be configured to a wavelength range of 1.1 to 1.4 microns in order to maintain the visibility of the underdrawing. Nonetheless, with the increasing use of focal plane arrays, the vidicon's drawbacks will only become more and more evident, to be precise: its limited sensitivity at longer wavelengths (which relates to a limited ability to use filters) and its inherent geometric distortion. It should now also be evident that high-sensitivity cameras imaging up to 1.7 microns and extended range cameras imaging beyond 1.7 microns are coming to have preferred status. To date, indium gallium arsenide cameras perform surprisingly well when compared with extended range devices, and the fact that they have now been adapted to produce high resolution images only enhances their practicability. Extended range cameras may come to serve still another purpose as research instruments. Since their wavelength response is the broadest and can easily be calibrated by using filters, they may be the most appropriate cameras to use to set up wavelength comparisons for different pigments and different schools of painting. In any case, it is clear that practitioners will be defining ways of making IRR an even more versatile technique in the years ahead.

Molly FARIES
Hawthorne Drive 837
IN – Bloomington 47401
Tel: 001.812.330.9567
faries@indiana.edu

Ph.D. Bryn Mawr College, 1972. Directed two long-term IRR research projects, National Endowment for the Humanities Basic Research Grant (1984-87) and the Samuel H. Kress Foundation Grant for Art History Study (1990-1997) and co-directed (with M.P.J. Martens) *Painting in Antwerp before Iconoclasm, a socio-economic approach* (2000-2004, a project funded by NWO). Recent publications include: *Recent Developments in the Technical Examination of Early Netherlandish Painting* (co-edited with R. Spronk, Brepols, 2003) and *Making and Marketing: Studies of the Painting Process in Fifteenth- and Sixteenth-Century Netherlandish Workshops* (Brepols, 2006). Professor emerita, Departments of Art History at Indiana University/Bloomington and the University of Groningen.

ENQUÊTE SUR DES MARQUES D'AUTEUR

Elisabeth Martin*

Résumé: Après une introduction montrant l'intérêt des radiographies pour déterminer le statut des œuvres, prototype ou réplique, ce sont les marques d'auteur cachées, signature ou inscription nominative, qui retiennent l'attention. Suivant les cas, les découvertes résultent de l'établissement de dossiers scientifiques, photographies infrarouges et radiographies, ou des observations en cours de restauration, plus particulièrement lors du nettoyage ou du dérentoilage. Plus ou moins spectaculaires, elles modifient le regard que l'on porte sur les œuvres soit en confirmant son auteur, soit en modifiant l'attribution, soit en sortant l'œuvre de l'anonymat.

L'étude scientifique des œuvres d'art et les observations en cours de restauration sont concernées par la quête de l'original, thème choisi pour ce colloque. Toutefois, l'original doit-il être compris comme se référant en miroir au colloque précédent sur les copies? Doit-on différencier le prototype de la réplique ou interpréter le thème comme celui de l'attribution à un artiste donné?

Original ou réplique autographe: contribution de la radiographie

Le premier versant de la problématique sera évoqué par quelques exemples illustrant la contribution des radiographies.

Un couple d'œuvres, toutes deux au château de Fontainebleau, présente la même composition célébrant *Flore*, allégorie qu'Ambroise Dubois a peinte pour le château de Fontainebleau autour de 1600 pour le roi Henri IV[1]. Aucune n'est signée. Une des versions était entrée dans les collections nationales en 1964 et avait été considérée à l'époque comme l'original (fig. 1). L'autre tableau (fig. 2a) a été expertisé assez récemment avant son achat. Les deux exemplaires sont identiques à l'œil nu quant à la disposition des motifs picturaux. Seul le format en hauteur les distingue. La radiographie de l'exemplaire acquis récemment (fig. 2b) présente de nombreux repentirs, en particulier dans le vêtement de la figure féminine. Au stade initial le drapé était plus ample sur l'épaule et plus couvrant sur la poitrine qu'il ne l'est dans l'état final. La radiographie de l'exemplaire étudié en 1964 ne comporte aucun repentir et la composition reprend l'état définitif de la version qui avait nécessité une recherche dans la disposition du vêtement. Le tableau (fig. 1) est donc plus tardif, réplique ou copie de l'original de Dubois (fig. 2a); copie anonyme, semble-t-il, d'après l'aspect radiographique présentant une grande différence de subtilité dans le modelé des visages.

Toutefois, la présence de repentirs sur la radiographie ne constitue pas l'apanage des originaux. Des répliques peuvent avoir subi de petites améliorations à partir du prototype et

* Cet article est le résultat d'un travail collectif: photographes-radiologues et restaurateurs. Qu'ils soient tous remerciés de même que les conservateurs, responsables des œuvres citées.

[1] D. VÉRON-DENISE, «Acquisitions», dans *Revue du Louvre*, 2001, n°4, p. 92.

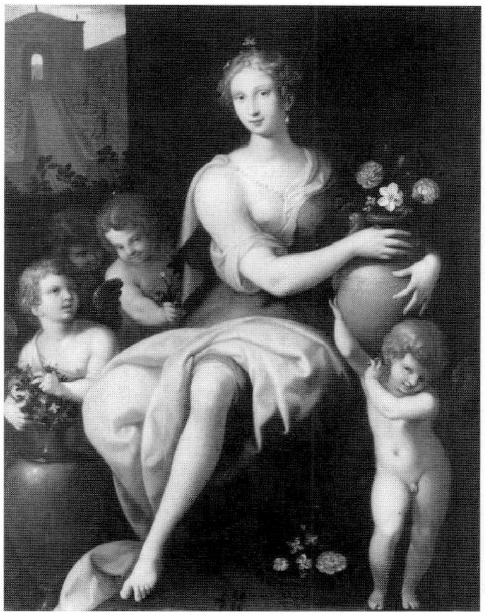

Fig. 1. Anonyme, *Flore*, Château de Fontainebleau (© C2RMF, J. Marsac).

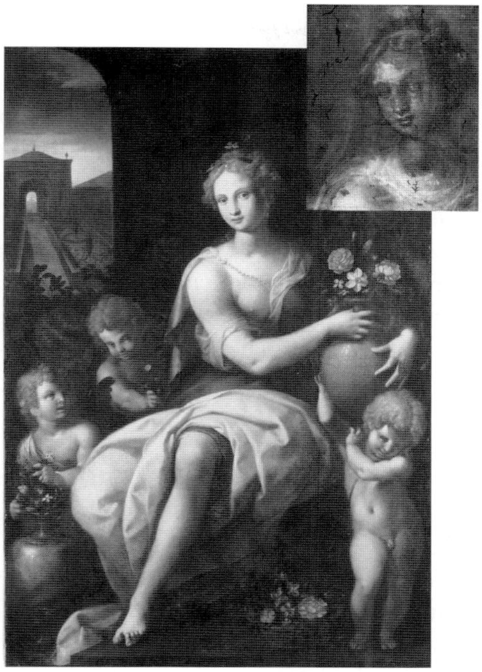

Fig. 2. Ambroise Dubois, *Flore*, Château de Fontainebleau. a. Ensemble (© C2RMF, J. Marsac). b. Détail radiographique de la figure féminine (© C2RMF, G. de Puniet).

manifester des repentirs en radiographie. On peut résumer cet apport des radiographies de la manière suivante: de façon générale, pour deux œuvres autographes, lorsqu'une zone identique à l'œil nu à la zone correspondante de l'autre tableau comporte des repentirs sur la radiographie, elle appartient au prototype. En revanche, lorsque les détails diffèrent à l'œil nu, c'est sur la radiographie de la réplique que les repentirs sont discernables.

Un cas singulier concerne deux *Madeleine repentante* à l'iconographie et aux dimensions très proches; l'une, au County Art Museum de Los Angeles, est signée G. de La Tour, l'autre, conservée au Louvre, est signée La Tour[2]. Dans la version de Los Angeles, il y a un petit repentir situé sur l'épaule de la blouse, large au stade initial, rétrécie par le peintre ensuite. Un repentir similaire est aussi présent sur la version du Louvre où le modèle paraît plus âgé. Comment expliquer ces deux modifications identiques au même endroit sur deux versions autographes, ou du moins faites dans l'atelier du peintre? D'après nous, Georges de La Tour a peint une première version de cette *Madeleine* avec une blouse large, exemplaire aujourd'hui disparu comme c'est le cas pour de nombreux tableaux du peintre[3]. Pour réaliser des séries, une spécialité du maître lorrain, de

[2] E. MARTIN, *A propos de l'œuvre de La Tour. Les repentirs et leur signification. 12ème Comité de l'ICOM pour la Conservation*, Lyon, 1999, p. 385-390. Les radiographies des deux tableaux sont publiées dans ce texte.

[3] J. THUILLIER, *Georges de La Tour*, Paris, 1992.

La Tour en a probablement fait le 'patron' qui a été utilisé, peut-être à quelques années d'intervalle, pour exécuter les scènes dont l'une est à Los Angeles et l'autre à Paris. Dans un souci d'amélioration et pour accentuer le dynamisme de la perspective, les deux versions auraient été modifiées en rétrécissant la blouse, déjà peinte selon les contours du patron. Elles différeraient donc à l'œil nu du prototype large si celui-ci était disponible et la confrontation possible. Si, en revanche, on considère la *Madeleine* de Los Angeles comme le prototype du thème en raison de son âge plus juvénile, le changement sur l'épaule serait assimilé à un repentir 'de recherche' pour atteindre la juste ampleur de la manche, mais logiquement la réplique postérieure n'aurait pas de repentir et aurait d'emblée intégré la phase finale du premier exemplaire; elle aurait été peinte à manche étroite; à moins qu'il ne s'agisse d'un 'lapsus pictural'.

Des attributions en question

Les marques d'auteur, dont la découverte constitue le corps de cet article, désignent à la fois des signatures apposées sur la face de l'œuvre et des noms calligraphiés sur le revers.

Toutes, sauf une, ont été découvertes depuis l'an 2000, date à laquelle ce type de bilan avait été entrepris dans un but méthodologique[4]. Suivant les cas, ce sont les photographies infrarouges, les radiographies ou les observations en cours de restauration, plus particulièrement lors du nettoyage ou du dérentoilage, qui ôtent le masque.

Plus ou moins spectaculaires, ces réapparitions modifient le regard que l'on porte sur les œuvres soit en confirmant son auteur, soit en modifiant l'attribution, soit en sortant l'œuvre de l'anonymat.

Confirmation de l'auteur présumé

Lors de l'établissement des dossiers préliminaires à la restauration, les photographies infrarouges facilitent la lecture d'une signature obscurcie par les encrassements de surface. Ainsi un tableau du Musée des beaux-arts de Valenciennes, *La Prédication de saint Jean-Baptiste*, était arrivé sous l'attribution de Pourbus l'Ancien, mais la signature était devenue illisible. L'examen infrarouge a permis la lecture du nom «FRANCISCUS POURBUS»; il a aussi révélé la date de 1574 qui n'avait pas été conservée dans les archives du tableau. Il s'agit donc d'une œuvre de la maturité du peintre, décédé en 1581.

La radiographie est aussi porteuse d'information quant à l'identité du peintre et récemment, en reprenant le dossier du *Combat de l'amour et de la chasteté*, destiné au *Studiolo* d'Isabelle d'Este, la signature «PETRUS PERUSINUS» a été localisée sur la radiographie dans l'angle inférieur gauche de la toile[5]. Cette signature usée était passée inaperçue jusque-là. Si l'œuvre était considérée, sinon comme un chef d'œuvre du Pérugin, du moins comme une œuvre autographe, la découverte de la signature constitue une

[4] E. MARTIN, J. MARSAC, M.A. BELCOUR, R. MALLET et F. MAURIER, «A la découverte de la signature du peintre», dans *Techné*, n°13-14, 2001, p 101-111.

[5] E. MARTIN et J.P. RIOUX, «Comments on the technique and the materials used by Perugino, through the study of a few paintings in French collections. The Painting technique of Pietro Vanucci, called Il Perugino», dans *Kermes*, Florence, 2004, p. 43-56.

information intéressante par rapport à la personnalité du peintre: quelles sont les œuvres qu'il signait et pourquoi, car à cette époque les artistes n'apposaient pas leur nom sur l'ensemble de leur production.

Changement d'attribution

La radiographie d'une œuvre quasiment de la même époque, aussi sur toile, a remis en question son auteur. Il s'agit de *Saint Sébastien* du musée de Rouen arrivé avec une attribution à Vincenzo Foppa. La radiographie a clairement permis de déchiffrer le nom de «LUDOVIC BREA». Pour la date il est difficile de se prononcer. Foppa et Brea ont travaillé tous deux vers 1490 pour le polyptyque Della Rovere de Santa Maria di Castello à Savone et les influences réciproques sont plausibles sans être flagrantes. Cependant, cette marque pose plusieurs problèmes. On distingue nettement que les contours ont été obtenus par l'utilisation d'un poncif avec dépose de blanc de plomb. Cet usage est pour le moins exceptionnel à deux titres. D'une part les ponctuations laissées par le poncif sont le plus souvent en noir de carbone, décelable en infrarouge et non en radiographie, et d'autre part il est peu fréquent que ce soit la signature qui fasse l'objet d'un tel traitement. La graphie en capitales romaines est cependant bien celle utilisée par le peintre sur d'autres œuvres, comme la *Déposition de Croix* du couvent niçois de Cimiez, datée de 1512. À Rouen, il y a un tiret au-dessus du C terminal pour l'abréviation de la forme latine *LUDOVICUS*[6]. Le mode d'abréviation du tableau niçois comporte pour sa part une sorte de boucle ouverte. On voit combien l'enquête est ici complexe. Il y a donc lieu de prendre au sérieux cette attribution tout en restant critique. Faut-il y voir une œuvre de l'atelier de Brea qui, d'après les travaux récents, semble avoir fait appel à plusieurs mains et qui a pu en partie sous-traiter une œuvre de catégorie inférieure dans sa production? Faut-il considérer que cette marque est apocryphe, comme le suggère C.L. Schwok[7]?

C'est souvent la vigilance et le long tête-à-tête des restaurateurs avec la matière qui permet de découvrir des signatures remettant en cause l'attribution traditionnelle des œuvres.

Le décadrage d'un *Combat des Amazones* conservé à Dunkerque a permis d'améliorer la visibilité de l'angle inférieur droit[8] et a rendu possible la lecture de «p. jordaens F». Il y a ainsi eu confirmation du patronyme Jordaens qui avait été transmis. Le prénom associé était J II. Le J correspond non à Jacob mais à Jean. Le dictionnaire Bénézit le cite sous la forme Hans II Jordaens (1581-1653) et attribue bien ce tableau à ce peintre actif dans la première moitié du XVIIème siècle dans les Pays-Bas du sud. Le dictionnaire Thieme-Becker mentionne, pour sa part, qu'il est fils d'un certain Petrus Jordaens. Il semble donc que la signature découverte en cours de restauration permette d'attribuer ce *Combat des Amazones* non au fils mais au père, peintre sans doute lui aussi, mais sans *corpus*.

[6] Je remercie Anna De Florani pour sa suggestion à propos de l'abréviation lors du colloque *L'arte dei Brea tra Francia e Italia*, Gênes, 31 octobre 2005 (cfr. note 4).

[7] M.L. SCHWOK, *Louis Brea*, Paris, Arthena, 2005. L'auteur mentionne pour ce tableau (p. 203) une terminaison erronée «Ludovico Brea».

[8] Travail réalisé par P. Mandron.

Le *Martyre de saint Sébastien*, conservé au Musée des beaux-arts de Nancy, illustrera un changement d'attribution plus spectaculaire. Donné à Jean Le Clerc, il est l'œuvre d'un autre lorrain, Jean Allemant, comme la signature a pu le mettre en évidence lors de la suppression d'un repeint qui la masquait[9] tel un voile de pudeur sur un nom peu glorieux (fig. 3a, 3b). Une certaine confusion avec le nancéien Georges Lallemant régnait jusqu'alors au sujet d'un portrait de François de Vaudémont et ses enfants exécuté en 1618[10]. Le doute est levé: Jean Allemant, natif de Plombières-les-Bains cité thermale des Vosges, existe bel et bien. Le chapeau de cardinal visible dans l'angle inférieur gauche pourrait, peut-être, mener au commanditaire et à la commande.

Emergence d'un auteur pour des œuvres anonymes

Un meuble présenté à l'exposition universelle de 1855 y fut acquis par l'impératrice Eugénie qui s'en défit dix ans après. Après diverses péripéties, ce bonheur-du-jour fait partie du mobilier du château de Compiègne. La structure en bois sculpté et les médaillons peints avec scènes de genre ou motifs floraux, ont été entièrement restaurés il y a peu. L'ébéniste, Alphonse Giroux, était connu, mais au cours de la restauration des médaillons qui faisaient partie intégrante de cet objet exubérant, le monogramme du peintre a été mis au jour sous un vernis obscurci[11]. Il s'agit de Charles Labbé, artiste spécialisé dans les fleurs et les paysages qui exposa au Salon de 1836 à 1876. La date de 1855 a de plus été apposée sur un des six médaillons floraux.

Pour terminer ce panorama restreint, il y a aussi l'apport des rentoileurs lors de l'élimination des toiles de rentoilage. Ainsi un dessin représentant un ange, marouflé sur toile, avait fait anciennement l'objet d'un rentoilage. Lors de la donation de cette œuvre au musée de Vic-sur-Seille en Lorraine, l'intervention sur le support[12] a permis de découvrir une inscription difficile à lire. C'est la photographie sous rayonnement UV qui a facilité la lecture, sans doute en raison de l'utilisation d'une encre ferro-gallique. On peut lire le prénom célèbre du peintre «DOMENICHINO». L'inscription comporte de plus le nom de Pietro Odescalchi, membre d'une célèbre famille de collectionneurs romains du XVIIe au XIXe siècle, dont fit partie également un prince Pietro (1789-1856). Faut-il considérer cette marque comme digne de foi et l'œuvre comme étant un ange du Dominiquin? Les avis sont partagés.

Le dernier exemple concerne une œuvre de médiocre facture, représentant des *putti* (fig. 4a) et intégrée dans le trumeau d'une cheminée du musée national de la coopération franco-américain de Blérancourt. Ces enfants semblent être une copie pudique d'après les angelots du *Martyre de Saint Liévin*[13] (Bruxelles, Musées royaux des beaux-arts) peint par Rubens. La radiographie (fig. 4b) indique qu'il s'agit d'une toile de réemploi et on

9 Restauration d'A. Jarry.

10 Dictionnaire Bénézit, 1972.

11 I. CABILLIC, B. DUCOS, R. FÉVRIER, F. MAISON, M-A. PAULIN, «La restauration d'un bonheur-du-jour d'Alphonse Giroux», dans *Techné* n°19, 2004, p. 82-88. La couche picturale des médaillons a été restaurée par C. Des Cloizeaux.

12 La restauration du support a été confiée à M. Cailleteau.

13 Un mémoire de l'Ecole du Louvre sur ce tableau a été soutenu par A. Collange en 2002, *Etude scientifique de la technique picturale de N. de Largillière, celui qui souriait derrière les anges.*

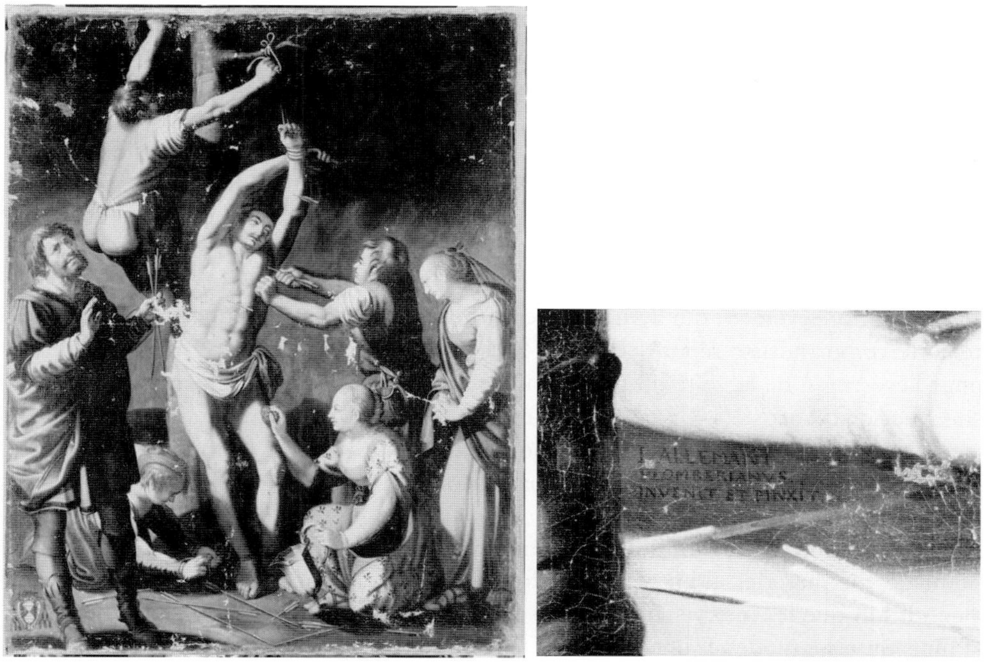

Fig. 3. J. Allemant, *Le martyre de saint Sébastien*, Nancy, Musée des beaux-arts. a. En cours de restauration (© C2RMF, J. Marsac). b. Détail de la signature après dégagement du repeint (© C2RMF, J. Marsac).

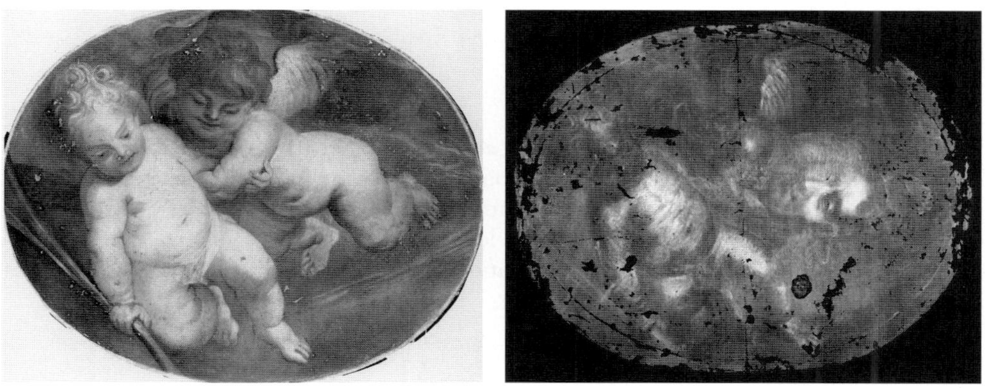

Fig. 4. Anonyme, *Deux putti jouant*, Château de Blérancourt. a. Ensemble (© C2RMF, G. de Puniet). b. Radiographie avec le portrait sous-jacent. Le côté droit devient le haut du portrait (© C2RMF, J. Marsac).

discerne sans ambiguïté le portrait d'un homme; l'ovale est, comme il est normal pour un portrait, avec son grand axe vertical. La perruque et l'habit évoquent le XVIIIe siècle français. Lors de l'élimination de la toile de rentoilage, rendue nécessaire par l'état de l'œuvre, l'inscription «peint par N. de/ Largillierre/1691» est apparue au revers de la toile originale. Deux autres portraits d'hommes, conservés l'un au musée Jacquemart-André à Paris[14], l'autre à Cherbourg au musée Thomas Henry, avaient au revers une marque d'auteur semblable découverte de la même façon. On note pour les trois tableaux la graphie du XVIIIe siècle *Largillierre* et non celle que l'on a coutume d'utiliser de nos jours *Largillière*. Il va de soi que si l'inscription doit être considérée comme une marque d'auteur c'est davantage en rapport avec le portrait qu'avec les *putti*. Un travail fructueux de comparaison des radiographies de N. de Largillière a été fait ainsi qu'un prélèvement qui a pu montrer l'utilisation d'une technique compatible avec celle de ce portraitiste réputé (1656-1746). La présence d'une laque rouge de qualité pour l'exécution de la veste du personnage peu apparente sur la radiographie, est aussi en faveur de cet artiste[15]. Ce tableau est encore dans les ateliers de restauration du Centre (C2RMF), et la décision de dégager le portrait d'un inconnu peint par un des plus grands portraitistes du XVIIIe siècle n'a pas encore été prise.

En conclusion, la restauration est un moment privilégié pour étudier les œuvres d'un point de vue matériel. Il en résulte parfois une remise en question de leur attribution à condition que les intervenants restent dans un état de vigilance permanent, même si la moisson reste modeste: une à deux découvertes par an dans les ateliers de restauration des musées de France[16].

Elisabeth MARTIN
Centre de Recherche et de Restauration des Musées de France
Quai François Mitterrand 14
F – 75001 Paris
Tel: +33.(0)1.40.20.57.86
elisabeth2.martin@culture.fr
yves.martin28@wanadoo.fr

Elle est entrée en 1973 comme chimiste au Laboratoire de recherche des musées de France. Après une formation complémentaire en Histoire de l'Art, elle a été responsable des études en peinture dans cette institution. Depuis la création du Centre de recherche et de restauration des musées de France en 1999, elle coordonne les activités scientifiques au service de la restauration. Elle est l'auteur de nombreuses publications dans la revue *Techné* et dans les colloques internationaux.

[14] Tableau précédemment attribué au peintre Louis Tocqué (cfr. note 4).

[15] La laque rouge est observable sur la coupe stratigraphique effectuée et dans une fenêtre de dégagement qui a été faite par A. Hochart-Giaccobi.

[16] D'après les données accumulées sur une trentaine d'années.

SIMULATION VISUELLE DE L'ÉTAT ORIGINAL D'UNE ŒUVRE PAR TRAITEMENT D'IMAGES

Elisabeth Ravaud et Myriam Eveno*

Résumé: De nombreuses œuvres d'art sont modifiées, mutilées ou découpées au cours des siècles. L'une des principales problématiques de leur étude scientifique est de rechercher leur état d'origine. Dans la peinture présentée ici, les examens de laboratoire ont révélé l'existence d'une composition sous-jacente complète. Les données obtenues par l'imagerie scientifique (photographies sous différents rayonnements, réflectographie infrarouge et radiographie), l'analyse des prélèvements ainsi que les observations faites en cours de restauration, ont permis la compréhension de la composition originale. Grâce au traitement infographique des images, une reconstitution virtuelle de la première composition a été menée en intégrant l'ensemble des données scientifiques. L'informatique et les logiciels de traitement d'images offrent aujourd'hui un important potentiel dans le domaine de la recherche sur les œuvres d'art. Ils permettent d'illustrer et d'expliquer en images les analyses et observations des scientifiques mais aussi de simuler des états perdus ou l'effet d'une restauration future.

La quête de l'original dans le domaine des œuvres d'art peut conduire à des situations où le retour à l'état d'origine n'est techniquement pas possible, ou reste trop aléatoire en terme de résultats ou encore n'est pas souhaitable car il ferait disparaître des étapes de l'histoire matérielle de l'œuvre. L'outil informatique peut alors permettre de restituer en partie cet état original de manière virtuelle. L'exemple choisi pour illustrer cette possibilité est un tableau du Musée des beaux-arts de Bordeaux représentant une *Vierge de Pitié entourée de saints* (fig. 1). La Vierge, assise devant une tenture en brocarts appliqués, tient son fils inanimé sur ses genoux. Elle est entourée à sa droite de sainte Barbe, saint Siméon, saint Sébastien et à sa gauche de saint André et sainte Catherine. La scène se situe devant un paysage et un ciel nuageux. Ce panneau de 242 cm sur 90 cm est constitué de quatre planches horizontales de chêne, débitées sur maille, dont l'assemblage était renforcé par deux montants verticaux, chevillés et situés aux tiers du panneau. Seules les traces de chevilles qui maintenaient ce montage subsistent actuellement. Le panneau a probablement été coupé latéralement puisqu'il est actuellement agrandi par deux planches étroites à fil vertical. Cette intervention fut sans doute consécutive à une altération importante du bois[1]. Le cadre original[2] est en partie conservé (le montant droit et une partie de la traverse inférieure sont refaits). Les traverses du

* Les auteurs remercient M. Lebihan et M. Acevedo du Musée des beaux-arts de Bordeaux, l'équipe du Centre de Recherche et de Restauration des Musées de France participant à l'étude de 1995: Jacqueline Bret, Frédéric Henriot, Carole Juillet, Patrick Mandron, Jean Perfettini, Elisabeth Martin, Gérard de Puniet et Patrick Le Chanu. Les auteurs remercient également N. Reynaud du Département des Peintures du Musée du Louvre ainsi que Jean-Louis Bellec (C2RMF) qui a réalisé la numérisation de la radiographie.

[1] Les archives du musée indiquent que le tableau était en très mauvais état quand il fut donné par l'église Sainte-Croix de Bordeaux.

[2] La largeur de la traverse supérieure du cadre est de 7,7 cm et de 6,8 cm pour les autres côtés.

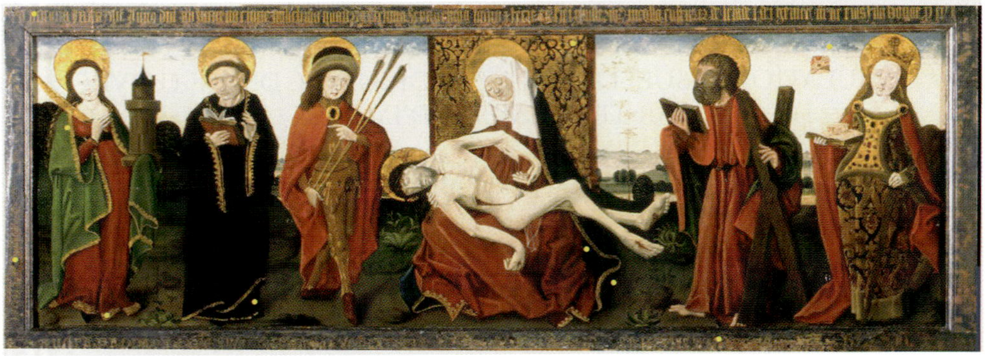

Fig. 1. *Vierge de Pitié entourée de saints*, photographie en couleurs après restauration avec emplacement des prélèvements (© C2RMF, Joël Requilé).

cadre sont appliquées et chevillées au panneau. Les montants (original à gauche et refait à droite[3]) sont posés sur les deux planches étroites agrandissant le panneau. Les assemblages du cadre sont à coupe d'onglet en haut et à coupe droite en bas, renforcés dans chaque cas d'une cheville. On observe que les deux montants ne gardent aucune autre trace de cheville.

Les noms des saints visibles sont inscrits sur la traverse inférieure avec une calligraphie identique à celle de l'inscription «Hans Clot M… fecit» lisible sur le montant gauche. Une calligraphie plus gothique[4] est observée sur la traverse supérieure où figurent le nom du commanditaire, un certain Guillaume Nicolas de Létade, et la date de 1469.

Nous savons peu de choses sur cette œuvre, si ce n'est qu'elle fut donnée en mauvais état en 1838 au musée par l'église Sainte-Croix de Bordeaux.

Une première restauration fut réalisée en 1942. Un cliché radiographique fait en 1941 avait déjà permis de déceler quelques modifications de la composition et des repeints. Deux fenêtres furent faites dans la couche picturale mettant en évidence les parties de deux personnages d'une composition sous-jacente. À cette époque, le restaurateur mentionne une reprise du parquetage, ce qui indique l'existence d'une intervention antérieure sur le support, actuellement aminci à 13 mm.

Une étude du tableau au Centre de Recherche et de Restauration des Musées de France (C2RMF) en 1995, en vue de sa restauration, mit en évidence l'existence d'une composition sous-jacente complète. Une première synthèse détaillée des résultats fut présentée à l'occasion d'une exposition-dossier à Bordeaux en novembre 1997.

Le dossier a été repris dans le but d'exploiter ces données avec des nouveaux moyens informatiques de traitement d'image et de proposer une reconstitution virtuelle de cette composition maintenant cachée. Les données ont été traitées avec le logiciel Adobe Photoshop® 7.0. Le matériel dont nous disposions pour ce travail était une radiographie argentique en sept bandes, des prélèvements (fig. 1) et des sondages (fig. 2) réalisés au

[3] Le montant droit du cadre avait disparu avant l'entrée de l'œuvre au Musée des beaux-arts de Bordeaux.
[4] Communication de Nicole Reynaud.

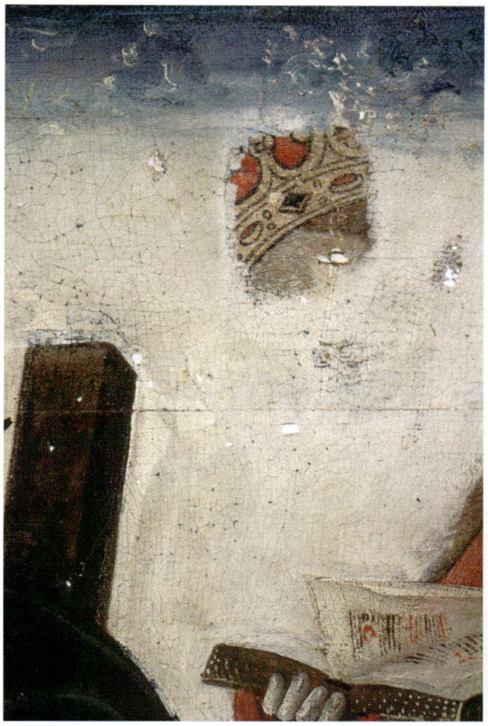

Fig. 2. Détail photographique en couleurs de l'ouverture réalisée en regard de la tête de sainte Catherine (© C2RMF, Joël Requilé).

Fig. 3. Coupe stratigraphique d'un prélèvement dans l'auréole de sainte Barbe. 1. Carbonate de calcium. 2. Impression orange à base de minium, d'oxydes de fer et de carbonates de calcium. 3. Assiette. 4. Feuille d'or. 5. Fine couche organique séparant les deux compositions. 6. Fine couche préparatoire à base de blanc de plomb. 7. Assiette. 8. Feuille d'or (© C2RMF, Myriam Eveno).

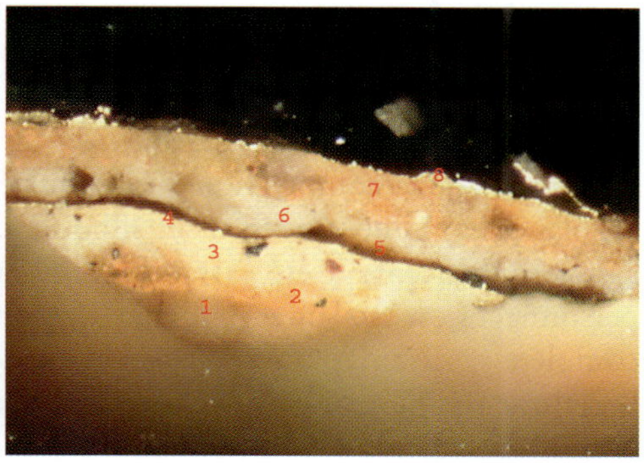

cours de la restauration. Les coupes stratigraphiques[5] réalisées à partir des prélèvements montrent la succession des deux compositions séparées par une fine couche organique (fig. 3). La première composition est peinte sur une préparation au carbonate de calcium et une impression orange à base de minium, d'oxydes de fer et de carbonates de calcium, que l'on retrouve sur le panneau comme sur le cadre (ce qui témoigne de l'originalité de ce dernier). La seconde composition est peinte sur une fine préparation au blanc de plomb.

Le traitement des images a consisté en une numérisation des radiographies et leur assemblage informatique. Le réseau du parquetage très présent sur l'image a été atténué numériquement. La composition sous-jacente a été relevée le plus complètement possible grâce à différents calques. La composition cachée représente également une *Vierge de Pitié* mais différente de la composition visible (fig. 4). Le groupe central du tableau rassemble la Vierge assise sur un banc de pierre (qui préfigure peut-être le tombeau du Christ?), maintenant son fils sous l'aisselle. Celui-ci a le bras droit retombant vers le sol. Saint Jean soutient la tête du Christ à gauche et saint Longin, identifiable à sa lance et à sa main essuyant ses yeux[6], est situé à droite. Latéralement les donateurs

[5] Les échantillons préparés en coupe stratigraphique ont été observés au microscope optique et étudiés au microscope électronique à balayage couplé à un système d'analyse dispersif en énergie (MEB-EDS).

[6] Ce geste rappelle que Longin était atteint d'un mal des yeux guéri miraculeusement par une goutte du sang du Christ: voir Louis RÉAU, *Iconographie de l'Art Chrétien*, tome III, PUF, 1958, p. 812.

sont présentés par leurs saints patrons. À gauche saint Jean-Baptiste, identifiable à ses attributs (l'agneau sur un livre et sa croix), présente un donateur agenouillé en prière. À droite, sainte Catherine, avec la roue de son martyre, pose ses mains protectrices sur la donatrice et sa fille, qui portent des vêtements de même tissu rayé et tiennent de longs chapelets de perles. À partir des donateurs en prière se déploient des phylactères encadrant la scène centrale. L'ensemble montre une disposition en triptyque avec une scène centrale liée à la passion du Christ et deux parties latérales avec les donateurs et leurs saints patrons, sans rapport avec cet événement. Le fond est formé d'un motif répétitif représentant un chien courant et une fleur stylisée avec des feuilles lancéolées, y compris derrière la Vierge. La plus grande partie du motif du fond apparaît opaque sur la radiographie, ce qui n'exclut pas que certains détails soient peints à l'aide de pigments radiotransparents. Ces motifs ne sont pas parfaitement superposables entre eux, ce qui indique qu'ils ont été réalisés séparément sans procédé global de reproduction. Le sol est parsemé de fleurs ou de touffes d'herbes et deux arbres grêles encadrent la composition.

À partir de ces données recueillies directement sur la radiographie, les éléments radiographiques de la composition visible ont été atténués, en particulier le ciel très chargé en blanc de plomb ainsi que les silhouettes des saints visibles. Le relevé de la composition sous jacente a été extrapolé (fig. 5) avec une certitude variable. La limite des personnages est facilement repérable grâce à l'arrêt des motifs du fond. Ce fond a été complété grâce aux parties localisées sur la radiographie, en respectant la disposition et le rythme des motifs. Le geste du bras et de la main droite de saint Jean-Baptiste relève d'une iconographie traditionnelle où le saint désigne l'agneau. La partie basse du donateur a été prolongée vers le bas pour reposer au sol comme les donatrices. Le bras gauche du Christ semble être situé plus près du corps. Si la racine d'une cuisse du Christ est bien visible sur la radiographie, l'emplacement des jambes n'a pas été déterminé. Celles-ci se situent probablement sous la jambe droite de la seconde composition, très opaque sur la radiographie.

La mise en couleur de la composition cachée (fig. 6) repose sur l'observation et l'analyse précise des stratigraphies et des sondages en cours de restauration. Sous la couche noire du cadre, les couleurs originales ont été retrouvées: la traverse supérieure et les deux montants étaient de couleur rouge avec des motifs ou des inscriptions dorés, la traverse inférieure était de couleur grise avec quelques traces rouges. Les auréoles sont à la feuille d'or sur mixtion beige, la robe de la Vierge est peinte à l'aide d'une laque rouge. Le fond et le sol ont été peints en noir, en réservant l'emplacement des personnages. Dans la partie supérieure, les motifs du chien sont obtenus par une couche de blanc de plomb sur le fond noir[7]. Dans la partie inférieure, les fleurs ou touffes d'herbes sont peintes avec du jaune de plomb et d'étain sans silicium. Enfin l'ensemble, fond et sol, a été recouvert d'un glacis de vert au cuivre. Les deux ouvertures en cours de restauration donnent également des informations précieuses. La couronne de sainte Catherine est dorée et ornée de pierreries, son voile est rouge et sa chevelure brune. Le vêtement de saint Longin est de couleur brune. Certaines parties restent non documentées (car les prélèvements réalisés en 1995 avaient été réalisés avant tout dans le cadre de la restauration).

[7] Le motif a été documenté grâce à un prélèvement ciblé à partir de l'image radiographique.

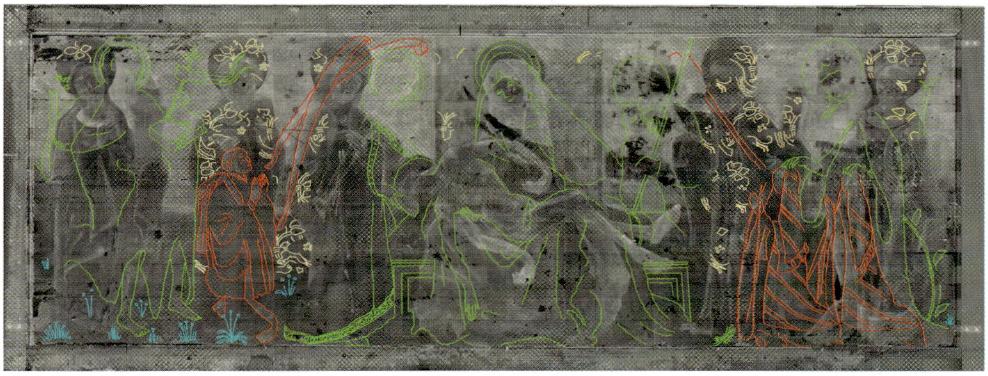

Fig. 4. Relevé de la composition sous-jacente à partir de la radiographie (© C2RMF, Elisabeth Ravaud).

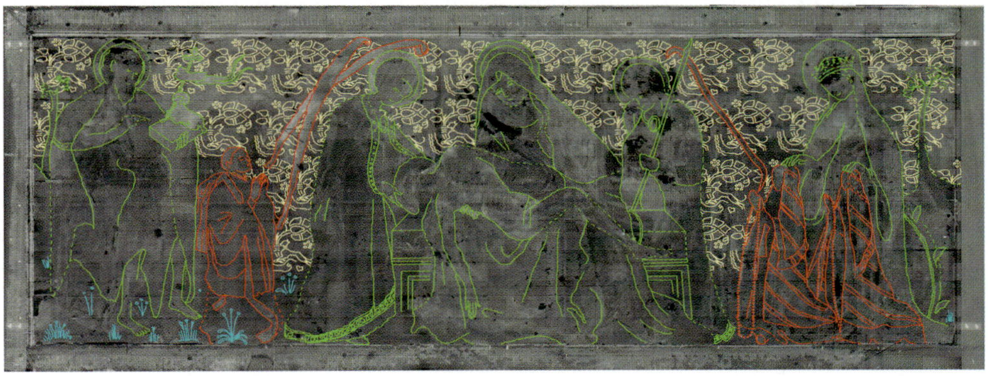

Fig. 5. Extrapolation de la composition (© C2RMF, Elisabeth Ravaud).

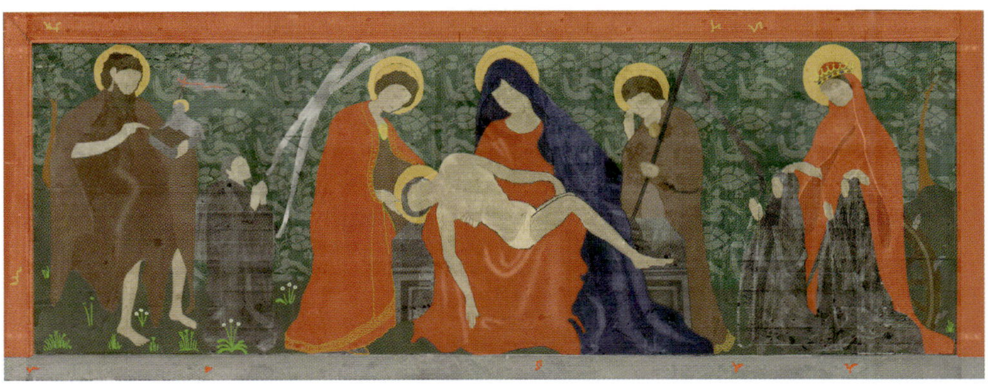

Fig. 6. Mise en couleur de la composition cachée (© C2RMF, Elisabeth Ravaud, Myriam Eveno).

Pour certains personnages, la couleur est connue par l'identification du saint: Jean-Baptiste, vêtu d'une peau de bête, est en brun. La couleur la plus probable du vêtement du saint Jean l'Evangéliste est rouge. Dans cette hypothèse, on note l'alternance des couleurs employées pour les différents saints. Pour le donateur et les donatrices, la radiographie nous indique seulement que les parties chargées en blanc de plomb sont l'écharpe retombant par terre chez l'homme et les rayures pour les vêtements des donatrices. N'ayant pas d'information supplémentaire, ces personnages ont été laissés avec le seul contraste radiographique.

La composition sous-jacente, en grande partie reconstituée, représente avant tout un document de travail, qui permet la synthèse en une seule image de l'ensemble des informations connues. Elle pourra être le point de départ d'une nouvelle recherche pour tenter de préciser la datation et la provenance du tableau. Ce travail s'appuiera sur les éléments stylistiques et iconographiques repérés, comme par exemple le motif du fond à thème floral et animalier. Il prendra également en compte les caractéristiques techniques du tableau comme le mode d'assemblage du panneau et du cadre et la nature de la double préparation[8].

Le support de chêne et la préparation au carbonate de calcium sont très évocateurs d'une provenance nordique (Flandres, France du nord, Allemagne du Nord ou Angleterre). Les assemblages du cadre, à coupe d'onglet en haut et droite en bas, sont fréquemment retrouvés dans la peinture flamande[9]. La couleur rouge du cadre est retrouvée dans des œuvres de la fin du XIVe siècle: les panneaux de Walcourt, conservés à Namur, présentent un cadre de couleur rouge, peint en trompe-l'œil[10].

À partir de ce constat, quelques comparaisons iconographiques ont été faites avec des œuvres flamandes. Le motif des arbres grêles avec des touffes de feuillages est retrouvé dans de nombreuses miniatures du XIVe siècle comme la miniature ms 72 de la Morgan Library à New York, mais également dans les *Heures du Maréchal de Boucicaut*[11] datées de 1405-1408. Les petites touffes aux feuilles retournées vers le bas sont relevées dans les *Heures de Gand*[12] datées de 1410-1425. Le donateur de petite taille avec les phylactères est souvent représenté, comme dans les *Heures de Daniel Rym*[13] de 1415-25 et encore dans le volet gauche du *Triptyque Seilern* du Maître de Flémalle[14] (1425-1430). Le même mouvement ondulé du voile de la Vierge est noté dans la *Présentation au Temple* de Melchior Broederlam daté de 1394-1399, conservé au Musée des beaux-arts de Dijon. Ces différentes analogies permettent de situer la première composition au tout début du XVe siècle.

La représentation virtuelle d'une composition cachée, illustrée par le tableau de Bordeaux, est un exemple d'application du traitement d'image informatique. La reconstitution

[8] La seconde couche préparatoire orange est rarement rencontrée.

[9] H. VEROUGSTRAETE-MARCQ et R. VAN SCHOUTE, *Cadres et supports dans la peinture flamande aux 15e et 16^{e}siècles*, Heure-le-Romain, 1989.

[10] H. VEROUGSTRAETE-MARCQ et R. VAN SCHOUTE, 1989, p. 317.

[11] *Les Heures du Maréchal de Boucicaut* sont conservées au Musée Jacquemart-André à Paris.

[12] *Enluminure de sainte Barbe*, livre d'heures, ms 170, fol. 172v, Walters Art Gallery, Baltimore.

[13] *Enluminure de Daniel dans la fosse aux lions, vénéré par Daniel Rym*, ms 166, fol. 168 v, Walters Art Gallery, Baltimore.

[14] Institut Courtauld de Londres.

de retables démembrés[15], la visualisation d'une restauration future sont d'autres applications de cet outil.

Elisabeth RAVAUD
Centre de Recherche et de Restauration des Musées de France
elisabeth.ravaud@culture.gouv.fr

Médecin, ancien interne des Hôpitaux de Paris, ancien chef de clinique, spécialisée en radiodiagnostic. Elle obtient un Diplôme d'Etudes approfondies en Histoire de l'Art en 1995. Elle rentre au Centre de Recherche et de Restauration des musées de France en 1993 où elle est chargée des études radiographiques. Ses recherches portent sur le développement d'une sémiologie fine et systématique en radiographie conventionnelle et sur le développement de nouvelles techniques d'imagerie.

Myriam EVENO

Physico-chimiste, diplômée du Conservatoire National des Arts et Métiers. Elle entre au Centre de Recherche et de Restauration des Musées de France en 1989 où elle se spécialise dans la microscopie électronique à balayage, la fluorescence X et la diffraction des rayons X appliquées aux matériaux des œuvres d'art. Elle a en charge les études de la matière picturale des peintures de chevalet.

[15] E. RAVAUD, «Contribution des examens radiologiques à la reconstitution de retables vénitiens du XIVe siècle», dans *Autour de Lorenzo Veneziano. Fragments de polyptyques vénitiens du XIVe siècle*, catalogue d'exposition, Musée des beaux-arts de Tours, 22 Octobre 2005-23 Janvier 2006, Milan, 2005, p. 45-55.

À LA RECHERCHE DE L'ORIGINAL

Patrick Le Chanu

Résumé: La notion d'originalité pourrait, au premier regard, sembler partagée par le marché de l'art, les musées et le monde de la restauration. Mais un examen des enjeux qu'elle recouvre dans chacun de ces univers et par voie de conséquence des définitions qui y sont associées devrait nous permettre de mieux situer l'univers des musées, de l'histoire de l'art et plus spécifiquement celui des laboratoires au sein de cet écheveau complexe.

En matière d'histoire de l'art et des techniques artistiques, le sens du mot original évoque bien une œuvre 'd'origine' émanant directement de son auteur, à la différence de certaines copies ou d'œuvres caractérisées comme 'dans le goût de', 'style', 'manière de', 'genre de', 'd'après', 'façon de'. Un texte juridique trop peu connu en France indique bien que ces dernières formules «ne confèrent aucune garantie particulière d'identité d'artiste, de date de l'œuvre, ou d'école»[1]. Il convient cependant de ne pas oublier le sens développé depuis le dix-septième siècle et devenu désormais dominant dans le langage courant. Celui-ci qualifie une œuvre, une personne, un comportement hors du commun, profondément individuel et nouveau[2]. Nous garderons en mémoire les deux sens de ce mot tout au long de cet article, quitte à passer de l'un à l'autre.

Bien loin d'un exercice scolastique, cette réflexion sur le sens des mots que nous employons est méthodologiquement salutaire et fondamentale. Le glossaire multilingue élaboré dans le cadre du programme Narcisse a permis de définir un vocabulaire technique commun[3]. Mais ce qui est en jeu avec des notions aussi fondamentales que l'originalité, ce sont les fondements mêmes de notre travail.

Notre discipline, l'histoire de l'art et plus singulièrement l'histoire de la peinture, côtoie d'autres activités très proches: la restauration et le marché de l'art, qui sont à la recherche d'une forme d'originalité en accord avec leurs objectifs propres. Envisager la recherche de l'originalité dans une perspective géographique et historique mais aussi en relation avec ces deux domaines peut nous aider à en élaborer une définition conforme aux valeurs et aux exigences de notre discipline. C'est important d'un point de vue scientifique mais aussi pour la restitution au public de la meilleure part de notre travail.

En effet, le but ultime de notre activité est la transmission d'un message au public sur le processus d'élaboration des œuvres et, au-delà, sur ce qui fonde le sens et l'importance de l'art ancien pour nous aujourdhui. Les vraies questions sous-jacentes que

[1] Décret n°81-255 du 3 mars 1981 sur la répression des fraudes en matière de transactions d'œuvres d'art et d'objets de collection.

[2] «qui a sa marque propre, unique» (av.1662, Pascal) «qui s'exprime d'une manière qui lui appartient en propre» (1691, un auteur original), dans *Dictionnaire historique de la langue française*, Robert, sous la direction d'Alain Rey, 1992, t. 2, p. 1384.

[3] CD-ROM Narcisse, *Glossaire multilangue*, Arquivos Nacionals/Torre do Tombo, Lisbonne, 1993.

pose la recherche de l'originalité sont donc intimement liées à notre activité au service des musées et notamment:

Pourquoi est-il important de se rendre au musée pour y voir de ses yeux des œuvres originales plutôt que de les voir en feuilletant un livre ou sur un écran d'ordinateur? Pourquoi est-il plus important de voir une œuvre originale qu'une copie? Est-il préférable de voir de ses yeux une mauvaise copie peinte plutôt que la reproduction mécanique d'une œuvre originale? Si nous acquiesçons à cette dernière idée, nous serons en accord avec Walter Benjamin lorsqu'il affirme que «la plus lamentable représentation de Faust dans un théâtre de province est supérieure à un film sur le même sujet»[4].

Histoire et géographie

L'originalité a une histoire, elle a aussi une géographie. Toutes deux nous donnent de bonnes indications sur le caractère relatif de leur définition, ne serait-ce que dans le cadre de la civilisation ancienne de l'Europe à laquelle nous nous limitons. Son histoire s'exprime de façon particulièrement significative grâce à l'apport de l'étude du dessin sous-jacent. Ce que nous démontre celle-ci, pour ce qui est des anciens Pays-Bas, c'est la remontée progressive vers la surface du tableau de l'expressivité individuelle longtemps contenue au stade du dessin sous-jacent comme la liberté de pensée et la reconnaissance de la valeur de l'individu avaient pu l'être, avant les conquêtes de l'humanisme, dans le tréfonds des consciences. Dans la peinture du quinzième siècle, ce que Paul Philippot[5] nomme le 'schéma liturgique', c'est à dire la distance existentielle entre les personnages du tableau, appartenant clairement au monde du sacré, même lorsqu'il s'agit de personnages profanes, et le spectateur; cette distance fait de l'artiste une sorte d'intercesseur religieux. Dans le même temps, la représentation picturale atteint de tels sommets de 'vérisme' que la vraisemblance des objets et des êtres du tableau paraît équivalente à celle qu'ils ont dans le monde réel. Dès lors, l'originalité suprême s'apparente à la négation de l'originalité, par l'effacement de toute trace de réalisation matérielle[6]. C'est au stade souterrain du dessin sous-jacent qu'est cantonnée alors l'expression libre, spontanée et donc pour nous originale, qui, avec le temps, ne cesse de s'émanciper jusqu'à 'faire surface' entre autres dans l'œuvre de maîtres tels que Rubens et Rembrandt.

Invention et exécution

Parallèlement au jaillissement de l'expressivité individuelle, aboutissant à la peinture 'graphique' de Rubens, de Rembrandt ou de Georges de La Tour[7] et de manière plus

[4] W. BENJAMIN, *L'œuvre d'Art à l'époque de sa reproductibilité technique*, traduit de l'allemand par Maurice de Gandillac, 2003, note 1, p. 16.

[5] P. PHILIPPOT, *La Peinture dans les Anciens Pays-Bas*, Paris, 1994.

[6] De là, la comparaison des peintures de van Eyck à des miroirs dans l'éloge de Lucas de Heere que cite van Mander: «Mais c'est en vain qu'on louerait une chose plutôt que l'autre, car entre eux luttent ici les plus nobles joyaux; tout s'anime et paraît sortir du cadre. Ce sont des miroirs, oui des miroirs, et non point des peintures!». Henri HYMANS, *Le Livre des Peintres de Carel van Mander*, réimpression 1979, APA – G.W. Hissink & C°, p. 35.

[7] Nous pensons ici tout particulièrement à la façon très graphique avec laquelle cet artiste peint les cheveux.

générale, à toute la peinture baroque, la signification certes d'abord artistique, mais immédiatement après, plus commerciale de l'original, au sens d'une œuvre intégralement inventée et exécutée par un seul et même artiste, a bien sûr certainement toujours eu son importance. Ce fut même peut-être particulièrement le cas dans la peinture des anciens Pays-Bas, avec son système d'organisation artisanal qui, dès le Moyen-Age, avait fait de l'œuvre, dans la plupart des cas, le fruit d'un processus d'élaboration collectif, un peu comme le film aujourd'hui. Cette organisation de la production artistique ayant persisté, dans l'atelier de Rubens par exemple, on conçoit le sens que lui-même et que ses clients les plus prestigieux accordent à des précisions telles que: *originale da mia mano e lacquila fatta dal Snyders (Prométhée)* ou bien encore: *Originale tutto da mia mano* (Daniel dans la fosse aux lions)[8].

La même échelle de valeur prévaut pour des œuvres moins prestigieuses comme l'indique là aussi le vocabulaire employé. Ainsi, Picart, le marchand établi à Paris, parle à propos des œuvres originales de Teniers de *Principalen*. Toujours à propos de Rubens, un client de Matthijs Musson parle quant à lui d'*origineele, principaele stucken*[9].

Le partage du travail dans les ateliers médiévaux a sans doute en partie été à la source de ce débat fondamental de l'histoire de l'art ancien: la distinction entre l'invention et l'exécution, dont découlera en partie celui du dessin et de la couleur. Elle est très bien illustrée par un passage du Journal de Voyage du Cavalier Bernin écrit par Paul Fréart de Chantelou[10]. Ce dernier était l'accompagnateur du Cavalier Bernin lors d'un séjour de celui-ci à Paris en 1665 avec comme projet une grandiose transformation du Palais du Louvre, transformation qui ne vît d'ailleurs pas le jour. L'embarras de Chantelou face à une réflexion du roi remarquant que le Bernin «ne loue pas grand-chose» amène à une discussion mettant bien en valeur, tout à la fois une possible cabale contre l'artiste étranger, une certaine conception de la création picturale et l'idée d'une absence d'originalité de l'art français[11]: «Le quatrième, l'abbé Butti étant venu à l'hôtel Mazarin, je lui ai conté ce que le Roi m'avait dit du Cavalier, le jour précédent, afin qu'il l'en avertit et qu'il se rendit plus facile à voir les choses et les louer. L'abbé a dit que c'était Le Brun[12] qui était auteur de ce bruit, à cause que le Cavalier ne louait pas ses ouvrages, qui en effet ne valent rien et ne le méritent pas, et qu'il peint comme un flamand. Je lui ai reparti qu'au moins ne pouvait-on lui ôter qu'il n'ait beaucoup d'invention. Cette invention, a-t-il répliqué, n'est tirée que des dessins de Jabac. Il y a à Rome un Carluccio[13] qui est bien un autre peintre»[14]. On trouve là l'expression d'une doctrine artistique sur ce qu'est la peinture, fondée sur deux moments essentiels qui caractérisent

[8] Cité par R. MARIJNISSEN, *Tableaux*, Elsevier, Bruxelles, 1985, p. 24.

[9] R. MARIJNISSEN, 1985, d'après Denucé, p. 18-19. Cette échelle de valeur fort ancienne s'exprime également dans un texte de Bernard Palissy: «Une pourtraiture qui aura esté contrefaite à l'exemple d'une autre pourtraiture, la contrefacture ou pourtraiture qui aura esté faite, ne sera jamais tant estimée comme l'original». *Le Littré, Le Dictionnaire de Référence de la langue française*, 2007, vol. 13, p. 631.

[10] P.F. DE CHANTELOU, *Journal de Voyage du Cavalier Bernin en France*, 1981. Ce texte avait été publié une première fois par Ludovic Lalanne à partir de 1877. Les éditions de la *Gazette des Beaux-arts* en publièrent une édition complète en tirage limité en 1885.

[11] Cette scène se déroule le 4 septembre 1665.

[12] Il s'agit, bien sûr, de Charles Le Brun (Paris, 1619 - id., 1690) confirmé depuis 1664 et malgré la disgrâce de Fouquet comme premier peintre du roi Louis XIV.

[13] Carlo Maratta (1625-1713).

[14] P.F. DE CHANTELOU, 1981, p. 155.

en même temps des 'écoles': l'invention, apanage des italiens, puis l'exécution, étape moins noble dans laquelle il n'est pas difficile tantôt de reconnaître la supériorité 'technique' des artistes des anciens Pays-Bas, tantôt de la stigmatiser: «Il peint comme un flamand». Le style pictural 'français' quant à lui sera le fruit d'une lente maturation. En effet, la France aura à s'émanciper de ces deux modèles entre lesquels elle était tout à la fois géographiquement et stylistiquement prise en étau: l'Italie, lieu de la peinture 'mentale' et les anciens Pays-Bas, patrie de l'exécution soignée et de la technique artisanale comme l'avait, en son temps, très malicieusement signalé Vasari saluant l'invention de van Eyck.

Pour une vision 'néo-nationaliste' de l'histoire de l'art au cœur de laquelle se situe notamment la question des influences, prises non pas comme thème fondamental de l'histoire de l'art, mais comme ingrédients des styles 'nationaux', on conçoit que ce ne soit pas avant l'impressionnisme que l'on puisse parler d'un art original français.

On remarquera par ailleurs qu'au-delà de la répartition géographique des tâches, est exprimée dans le texte de Chantelou l'idée selon laquelle l'invention et l'originalité ressortent plus du dessin que de la peinture. Comme le dira beaucoup plus tard Max Friedländer, le caractère direct et la spontanéité sont indissolublement liés à l'originalité.

Originalité et marché de l'art

Aux yeux des juges, encore aujourd'hui, l'authenticité ne dépend pas nécessairement de l'exécution de l'œuvre par l'artiste en personne. Cela est vrai fréquemment pour les décisions des tribunaux rendues en matière d'art contemporain mais pas uniquement. On voit fréquemment alors apparaître la notion de co-auteur. Ceci finalement n'est qu'un nouvel épisode du débat entre l'invention et l'exécution. La différence étant ici que certains artistes contemporains jouent consciemment avec cette question, voire même en font le sujet principal de leur œuvre. Ainsi, aux antipodes de van Eyck dont plusieurs œuvres ne portent pas de signature puisqu'elles sont censées porter en elles-mêmes tout à la fois son humilité et son nom par leur seule perfection, seule la signature semble faire du Porte-Bouteille fabriqué par un autre une œuvre de Duchamp.

Si l'on examine donc ce que semblent être les critères de fixation des prix de l'œuvre d'art par le marché, on serait tenté de dire sans hésiter que les œuvres les plus chères y sont *a priori* celles dont l'invention et l'exécution sont le fait d'un maître incontesté. Mais le marché lui aussi a ses incohérences. Ainsi, on ne peut manquer de s'étonner des prix considérables atteints par les dérivations des compositions breugeliennes. Leurs auteurs furent des tâcherons d'atelier dont les appointements devaient être fort modestes. Ce qui se vend ici, c'est non pas l'exécution, souvent très médiocre et pour longtemps encore anonyme[15] mais c'est, au-delà de l'invention initiale d'un genre, une marque. Le succès et surtout la spéculation commerciale sur ces œuvres ont sans doute beaucoup cru depuis leur création et ce nonobstant le fait que les compositions, l'invention y sont souvent fort maladroites et très éloignées des visions grandioses de Pierre Bruegel l'ancien. On pourrait dire que le succès actuel de ces œuvres sur le marché donne raison à Walter Benjamin et à sa réflexion sur l'aura des œuvres anciennes.

[15] On sera bien en peine de retrouver les auteurs de ces innombrables tableaux, même si les enquêtes en cours aideront sans conteste à y mettre de l'ordre.

Originalité et reproduction

Si la recherche de l'original doit être entendue en première instance comme la distinction des œuvres authentiques d'avec les copies plus ou moins fidèles à leurs modèles, on ne peut s'empêcher d'évoquer aujourd'hui la question du développement exponentiel de la reproduction photographique des œuvres et de leur diffusion inégalée dans l'histoire. Les catalogues, les affiches, les images numériques accessibles sur des supports variés dont, bien sûr, la toile[16], ont singulièrement démocratisé l'accès à ces images et une certaine connaissance de l'histoire de l'art. Il importe que les historiens d'art prennent position sur cette question, il en va de l'avenir des musées. En effet, la diffusion démocratique de l'art par la duplication et la facilité d'accès aux reproductions mécaniques a pu, comme nous allons le voir, dans un premier temps, être l'objet d'une certaine réticence.

Dans sa définition de l'authenticité, intimement liée à la recherche de l'original, Walter Benjamin attache une importance cruciale à la présence physique de l'œuvre: «Le *hic et nunc* de l'original constitue ce qu'on appelle son authenticité»[17]. Mais pour lui, ce *hic et nunc* n'est qu'un résidu laïcisé du pouvoir de la relique dans la religion catholique[18]. Benjamin n'envisageant pas l'art comme un langage à part entière, et un langage fondé sur la mise en forme d'une matière colorée et des sensations qu'elle engendre, il ne peut que bâtir sa théorie de l'aura: «Ce qui fait l'authenticité d'une chose est tout ce qu'elle contient de transmissible de par son origine, de sa durée matérielle à son pouvoir de témoignage historique»[19]. «L'unicité de l'œuvre d'art et son intégration à la tradition ne sont qu'une seule et même chose»[20]. «Tous ces caractères se résument dans la notion d'aura, et on pourrait dire: à l'époque de la reproductibilité technique, ce qui dépérit dans l'œuvre d'art, c'est son aura»[21]. Mais celle-ci a des aspects très superficiels. L'aura de la tradition, c'est un peu le charme de la brocante, qui veut que tout objet ancien surpasse ses équivalents contemporains de par la charge de vécu qu'il semble véhiculer. Toute exigence artistique est absente d'une telle réflexion. Envisagés de ce point de vue en effet, les tableaux breugeliens offerts sur le marché, même si ils ne sont pas originaux, sont authentiques.

Dans l'art sacré, l'œuvre avait pour Benjamin une fonction rituelle, ce qui resterait d'ailleurs à prouver puisqu'au moins à partir du Concile de Trente, les images ont beaucoup plus un rôle pédagogique que rituel. Bossuet, bien après les Pères des Conciles[22]

[16] Mot français désignant *The web*.

[17] W. BENJAMIN, 2003, p. 13.

[18] «Or, c'est un fait de la plus haute importance que ce mode d'existence de l'œuvre d'art, lié à l'aura, ne se dissocie jamais absolument de sa fonction rituelle», W. BENJAMIN, 2003, p. 22.

[19] W. BENJAMIN, 2003, p. 16.

[20] W. BENJAMIN, 2003, p. 22.

[21] W. BENJAMIN, 2003, p. 16.

[22] On trouve cette affirmation dans le troisième Canon du Concile œcuménique de 869. Ce dernier est cité dans la *Revue de l'Art*, CNRS, Paris, 1973, n°21, p. 9: «...et il convient que, avec l'accord de la raison et d'une tradition très ancienne, les icônes soient honorées et adorées par extension en raison de l'honneur qui s'attache aux objets essentiels qu'elles désignent, au même titre que le livre sacré des Saints Evangiles et le modèle de la précieuse croix». Le Concile de Trente reprend effectivement cette affirmation: «Que l'on doit conserver, principalement dans les églises, les images de Jésus-Christ et des saints, pour

indiquait que «Bien loin de croire comme eux (les idolâtres) que quelque divinité habite dans les images, nous ne leur attribuons aucune vertu que celle d'exciter en nous le souvenir des originaux»[23].

L'une des articulations essentielles de la démonstration de Benjamin est «qu'avec la sécularisation de l'art, l'authenticité devient le substitut de la valeur cultuelle». Et «De plus en plus, à l'unicité de ce qui apparaît dans l'image cultuelle, le spectateur tend à substituer l'unicité empirique du créateur ou de son activité créatrice»[24]. Au-delà d'une définition de la notion d'aura avec les limites que nous lui avons vues, auxquelles s'ajoute comme un refus de reconnaître pour part dans le culte moderne de l'authenticité les conséquences de la spéculation financière, Benjamin analyse de manière très critique les effets du développement des reproductions d'œuvres d'art.

Selon lui, le problème posé par celles-ci est que «la technique de reproduction détache l'objet reproduit du domaine de la tradition. En multipliant les exemplaires, elle substitue à son occurrence unique son existence en série. Et en permettant à la reproduction de s'offrir au récepteur dans la situation où il se trouve, elle actualise l'objet reproduit»[25]. «Rendre les choses spatialement et humainement 'plus proches' de soi, c'est chez les masses d'aujourd'hui, un désir tout aussi passionné que leur tendance à déposséder tout phénomène de son unicité au moyen d'une réception de sa reproduction». Très naturellement la disparition de l'aura du fait de l'absence du *hic et nunc* de l'œuvre originale aboutit au rejet de toute forme de reproduction: «Tout ce qui relève de l'authenticité échappe à la reproduction — et bien entendu pas seulement à la reproduction technique»[26]. Il y a là l'expression d'un effroi devant 'l'horreur de la société de masse' présente également chez un Herbert Marcuse.

Cette horreur de la 'domination démocratique'[27] se comprend d'autant mieux que nous avons là affaire à des penseurs qui ont assisté à la tragédie de l'adhésion des masses au nazisme en Allemagne avant de le fuir pour sauver leur vie. Leur réponse a été l'élaboration d'une théorie néo-aristocratique de l'art.

Ainsi, Marcuse développe l'idée qu'en étant reproduits et diffusés, les chefs-d'œuvre de l'art perdent leur qualité subversive. Il dénonce les «défenseurs de la culture de masse» qui «insistent sur le fait que les classiques ont quitté le mausolée et sont revenus à la vie, par le fait qu'ainsi le public est éduqué. C'est vrai, mais s'ils reviennent à la vie comme classiques, ils revivent comme autres qu'eux-mêmes, ils sont privés de leur force antagonique, de leur étrangeté qui était la dimension même de leur vérité. Le but et la fonction de ces œuvres ont donc fondamentalement changé. Si, à l'origine,

leur rendre un juste tribut d'honneur et de vénération: non pas que l'on croie qu'il y ait en elles quelque divinité ou quelque vertu pour laquelle on leur doive rendre un culte, ou qu'il faille leur demander quelque chose, ou mettre en elles sa confiance, comme faisaient autrefois les adorateurs des idoles; mais parce que l'honneur qu'on leur rend se rapporte aux originaux qu'elles représentent». P. SFORZA PALLAVICINI, *Histoire du Concile de Trente*, Paris, 1845, t. 3, p. 631.

23 BOSSUET, *Exposé de la Doctrine Catholique*, cité dans *Le Littré, Le Dictionnaire de Référence de la langue française*, 2007, vol. 13, p. 630.

24 W. BENJAMIN, 2003, p. 23.

25 W. BENJAMIN, 2003, p. 17.

26 W. BENJAMIN, 2003, p. 14.

27 H. MARCUSE, *L'Homme unidimensionnel*, Paris, éditions de Minuit, 1968, p. 90.

elles étaient en contradiction avec le *status quo*, cette contradiction a maintenant disparu»[28] et plus loin: «Cette imagerie a perdu son efficacité. Le fait qu'elle prend place dans la cuisine, le bureau, le magasin, qu'elle est mise en circulation dans un but commercial, pour les loisirs, est en un sens, une désublimation — il remplace une satisfaction médiatisée par une satisfaction immédiate». La hantise de Marcuse est de ce fait le renforcement de la cohésion sociale et du contentement. Mais que dire du pouvoir subversif de ces grandes œuvres, dès lors que leur diffusion, en l'absence de moyens de reproduction mécanique, était limitée à un petit groupe de privilégiés qui n'avait par définition, d'autre intérêt que le maintien du *status quo*?

Il nous semble que l'examen de la pensée de ces deux auteurs nous aide à proposer une position des musées d'aujourd'hui: leur credo est la recherche de l'original, tant par l'enquête historique que la compréhension de ses techniques de fabrication, facilitant les classifications, voire les attributions. Au-delà d'un 'verdict' sur l'originalité d'une œuvre, qui demanderait un clair positionnement déontologique de l'activité des laboratoires par rapport au marché, c'est la démarche scientifique elle-même, en reconstruisant le scénario de la création, qui porte l'originalité. Mais les exigences de cette recherche ne peuvent en rien être aussi implacables que celle du marché, car les musées doivent faire leur la vision des grands classiques qui ne rechignaient pas à prendre en considération, voire à admirer les copies de qualité[29].

Les musées ont un devoir de démocratisation de la culture, par la diffusion de reproductions de qualité, qui dans leur domaine propre, peuvent elles aussi, et pourront de mieux en mieux, à l'instar des copies «fidèles et faciles» appréciées par Roger de Piles[30], apporter l'équivalent de l'œuvre originale. Mais ils doivent dans le même temps et de manière inlassable, transmettre l'idée que la rencontre d'œuvres originales est irremplaçable, car, pour paraphraser un titre célèbre, l'œuvre d'art n'est 'message' que parce qu'elle est matière, non pas avant, mais dans le même instant, celui de la sensation.

Patrick Le Chanu
patrick.lechanu@culture.gouv.fr

Né en 1959. Etudes d'Histoire de l'art à l'Université de Paris-IV Sorbonne et à l'Ecole du Louvre. Depuis 1987: conservateur des Musées de France puis du patrimoine. 1991: conservateur au Laboratoire de recherche des Musées de France. Ancien pensionnaire de l'Académie de France

[28] H. Marcuse, 1968, p. 89.

[29] C'est ce que nous avons tenté de montrer dans: «Tout ce qu'il y a de beau à Rome», Colloque XV, *Copies, répliques, pastiches*, 2006, p. 106-114.

[30] «Et de ces copies, j'en trouve de trois sortes. La première est faite fidèlement, mais servilement. La seconde, est légère, facile et non fidèle. Et la troisième est fidèle et facile. La première, qui est servile et fidèle rapporte à la vérité, le dessein, la couleur et les touches de l'original: mais la crainte de passer les bornes de la précision, et de manquer à la fidélité, appesantit la main du copiste, et la fait connaître de ce qu'elle est, pour peu qu'elle soit examinée. La seconde, serait plus capable d'imposer, à cause de la légèreté du pinceau, si l'infidélité des contours ne redressait des yeux habiles. Et la troisième, qui est fidèle et facile, et qui est faite par une main scavante et légère, et surtout dans le temps de l'Original, embarasse les plus grands connoisseurs, et les met souvent au hazars de prononcer contre la vérité, quoique selon la vraisemblance». Roger de Piles, «L'idée du peintre parfait», dans *Abrégé de la vie des Peintres*, Paris, 1699.

à Rome. 1997: chargé de cours à l'Ecole des Mines de Paris. 1999: responsable de la filière d'étude des peintures de chevalet, de la polychromie, des arts graphiques et de la couleur au Centre de recherche et de Restauration des Musées de France. Actuellement: conservateur en chef du patrimoine, conseiller pour les musées à la Direction des Affaires Culturelles de Champagne-Ardennes. Domaines d'intérêt: peinture des anciens Pays-Bas du XVe au XVIIe siècle, notamment par l'étude du dessin sous-jacent. Etudes comparatives entre l'art italien et l'art des anciens Pays-Bas.

UNDERDRAWINGS:
TRANSFER METHODS AND IDENTIFICATION USING IRR

Ingo Sandner

Abstract: Replica's and copies often enjoy little consideration. However, opinions on the status of 'original' of a painting may differ widely in some cases. The long discussion on Rembrandt's work is a spectacular example.[1] The analysis of the underdrawing plays a key role, as sketching the composition on the ground is one of the first steps in the realization of a painting, and indeed, often a critical step in elaborating a composition. In the case of the Nuremberg and Berlin Emperor's portraits discussed below, some evidence leads to the assumption that the panels in Nuremberg are originals while the portraits in Berlin were created afterwards.

Infrared reflectography revealing the underdrawing may provide information not only on the draughtsman and his working habits, but also on the method used for transferring the composition. Basically, the following methods can be described:

1) the composition is drafted freehand on the ground after a sketch or preliminary drawing. The search for form occurs during the painting process and becomes obvious from corrections (*pentimenti*). A typical example is offered by Lucas Cranach the Elder (fig. 1).[2]
2) the composition after a sketch or preliminary drawing is drafted freehand on the ground; corrections occur at the stage of the drawing. Large-scale corrections during the painting process are redundant and rare. This method may be observed in particular if crayons were used for underdrawing. A painting by Conrad von Soest from Dortmund (fig. 2)[3] or another by an anonymous master from Cologne around 1325 (fig. 3) illustrate this procedure.
3) the main elements of the composition are roughly drafted after a sketch or preliminary drawing with an erasable drawing tool, such as a charcoal stick. On this draft, the actual underdrawing is carried out with a brush or nib. Before painting, the lines of the rough sketch are removed. This effective method may well have been applied by many painters. In cases of a comparatively free underdrawing with few corrections, the painter may always have thus proceeded. The painting of *Venus and Amor as a honey thief* at the Kröller-Müller-Museum in Otterlo might be seen as a rare example of painting on a rough sketch[4]; the actual underdrawing is hardly visible

[1] Rembrandt Research Project (RRP).

[2] *Unsichtbare Meisterzeichnungen auf dem Malgrund – Cranach und seine Zeitgenossen*, catalog of the exhibition and colloquium in Eisenach 1998.

[3] Ingo SANDNER, "Unterzeichnungen auf Gemälden des Conrad von Soest", in *Conrad von Soest, Neue Forschungen über den Maler und die Kulturgeschichte der Zeit um 1400*, Bielefeld, 2004, p. 37-59; Uta REINHOLD and Peter WELLER-PLATE, "Le *Kreuzaltar* de Conrad von Soest", in *Le dessin sous-jacent dans la peinture, Colloque XI*, Louvain-la-Neuve, 1997, p. 117-120.

[4] Published in *Ästhetik und Wissenschaft, Beiträge zur Restaurierung und Denkmalpflege*, Arbeitsheft 8, Landesamt für Denkmalpflege Sachsen, 2006, p. 21.

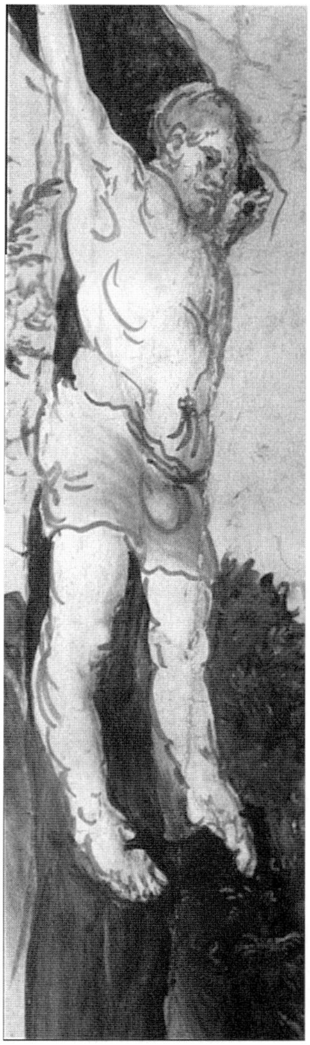

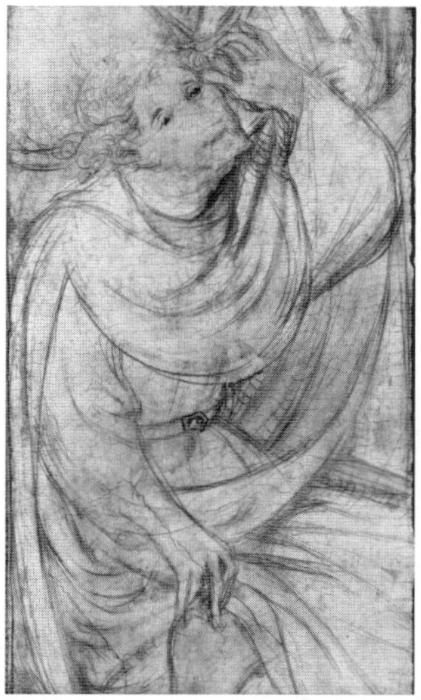

Fig. 2. Conrad von Soest, *Kreuz Altar*, Wildungen, detail. The composition is drafted after an idea sketch freehand on the ground, obvious from a correction of groups of lines.

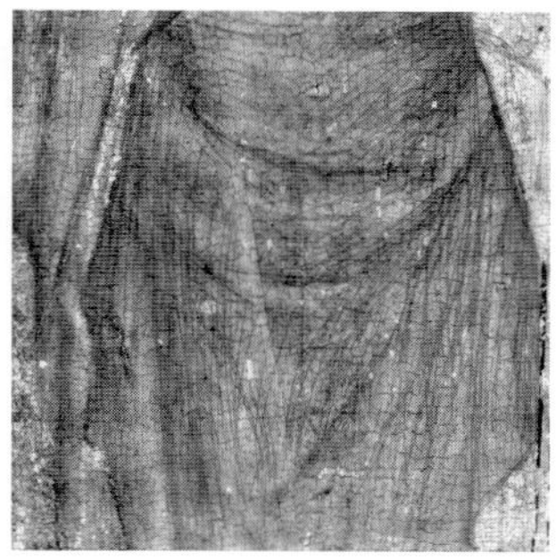

Fig. 3. A painter from Cologne, around 1325, detail. The composition is drafted freehand on the ground after a sketch, obvious from a correction of groups of lines.

Fig. 1. Lucas Cranach the Elder, *Crucifixion*, around 1500, Vienna, Kunsthistorisches Museum, detail. The composition is drafted freehand on the ground after a sketch.

(figs. 4, 5). But in a number of paintings by the Munich master Jan Polack (fig. 6), around 1500,[5] or the master of the St. Bartholomew altar in Cologne, the rough sketch was removed (fig. 7).[6]

[5] Ingo SANDNER, "Unterzeichnungstypen auf Bildtafeln der Werkstatt Jan Polacks", in *Jan Polack – von der Zeichnung zum Bild, Malerei und Maltechnik in München um 1500*, Augsburg 2004, p. 77-93.

[6] Ingo SANDNER, "Die Infrarot-Reflektographie als Untersuchungsmethode. Ergebnisse an den Kölner Tafelbildern des Meisters des Bartholomäus-Altars", in *Genie ohne Namen. Der Meister des Bartholomäus-Altars*, Köln 2001, p. 162-173; Ulrike NÜRNBERGER, "Die Unterzeichnungen in den Gemälden des Meisters des Bartholomäus-Altars. Zur Arbeitsmethode der Werkstatt", in *Genie ohne Namen. Der Meister des Bartholomäus-Altars*, Köln 2001, p. 151-161.

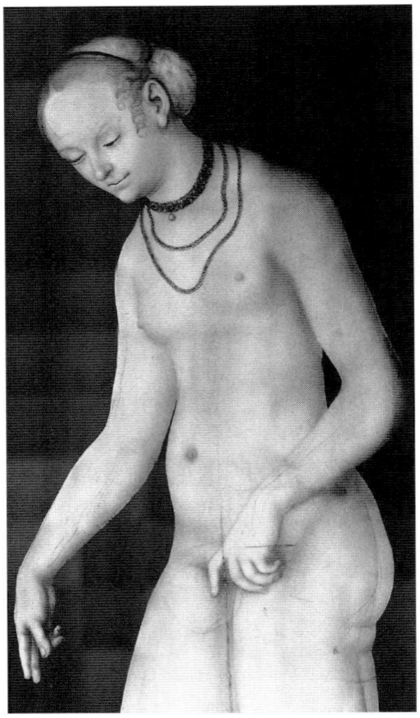

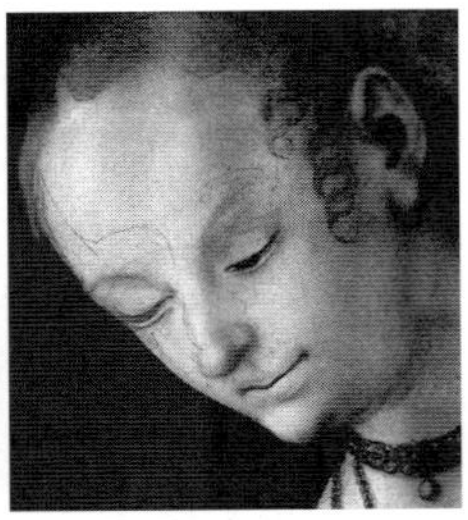

Fig. 5. Lucas Cranach the Elder, *Venus and Amor*, detail from figure 4.

Fig. 4. Lucas Cranach the Elder, *Venus and Amor*, Otterlo, Kröller-Müller-Museum. The composition is drafted after a sketch. The rough sketch is not removed and still detectable.

4) the composition is well elaborated in a preliminary drawing; the search for form takes place outside the painting. The preliminary drawing is transferred to the ground of the painting using aids, such as enlargement through a square grid. There is evidence that Lucas Cranach and Altdorfer used this method.[7] If the copies were 1:1 (cartoon), tracing methods were helpful. Apart from pouncing (*spolvero*), known since around 1500 but probably rarely used, tracing was done mainly by colouring the back of the preliminary drawing and then pressing the main lines through with a stylus.[8] The actual preliminary drawing was then carried out on the support. Using this method, either the main lines were traced, as often seen in portraits, or the transfer was executed to the last detail. As no further corrections were introduced, it was possible to model highlights and shadows with hatching. The underdrawing becomes part of the painting, as the shadows contribute to the surface impression. The use of tracing methods is no indication of a copy. Rafael often used tracing methods for transferring his sketches.[9] But many other painters also drew the sketch 1:1 on paper and traced the composition or parts of it on the ground. Tracing lines

[7] Martin SCHAWE, "Albrecht Altdorfers *Susanna im Bade* von 1526. Überlegungen zum Werkprozeß", in *Niederdeutsche Beiträge zur Kunstgeschichte*, Bd. 38, München/Berlin 1999, p. 93-103.

[8] Andreas SIEJEK and Katrin KIRSCH, *Die Unterzeichnung auf dem Malgrund. Graphische Mittel und Übertragungsverfahren im 15.-17. Jahrhundert, Kölner Beiträge zur Restaurierung und Konservierung von Kunst und Kulturgut*, Band 11, Siegl München, 2004.

[9] The Princeton Raphael Symposium (1983), Princeton, New Jersey.

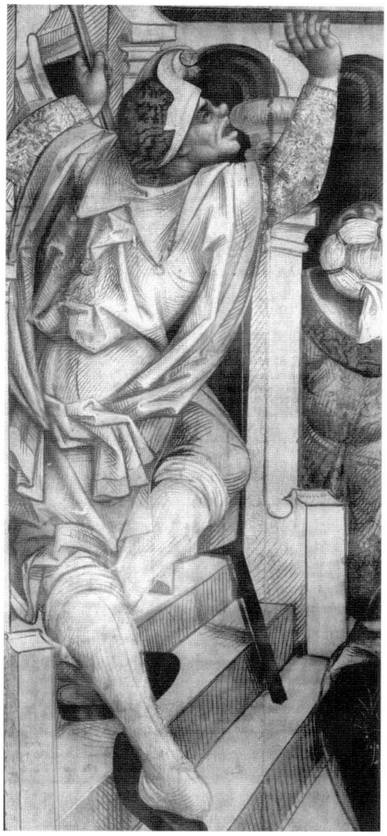

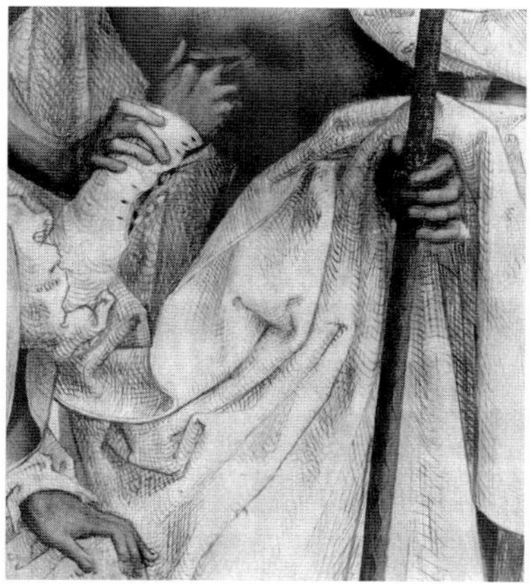

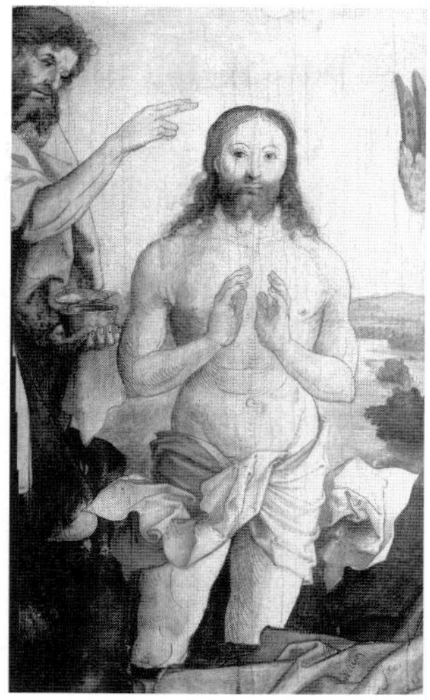

Fig. 6. Jan Polack workshop, *Ecce homo*, detail, Munich, Bayerische National Museum (BNM). The composition is drafted after a sketch with an erasable drawing medium, such as charcoal sticks. The main forms are roughly sketched.

Fig. 7. Master of the St. Bartholomew altar, detail, Cologne, Wallraf-Richartz-Museum. The composition is drafted after a sketch with an erasable drawing medium, such as charcoal sticks. The main forms are roughly sketched.

Fig. 8. Wolf Traut, *Christening*, Nuremberg, Germanisches National Museum (GNM), detail. The composition is drafted in the form of a well-elaborated preliminary drawing. The search for form takes place outside the painting area. The preliminary drawing is transferred to the painting area using aids.

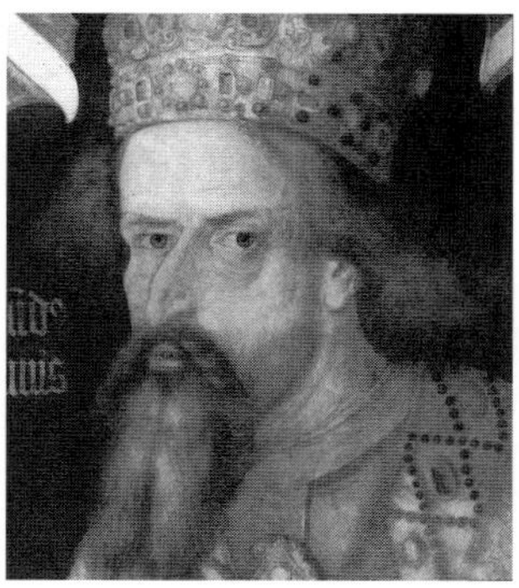

Fig. 9. Albrecht Dürer, *Emperor Sigismund*, Nuremberg, GNM, detail. The underdrawing shows no search of form and reveals the use of a tracing method for the main lines.

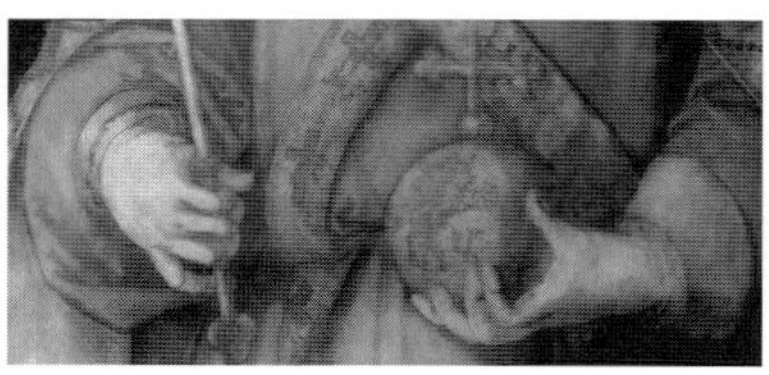

Fig. 10. Albrecht Dürer, *Emperor Sigismund*, Nuremberg, GNM, detail. Under the actual underdrawing, which was carried out very sparingly and almost without hatches, thin tracing lines can be detected.

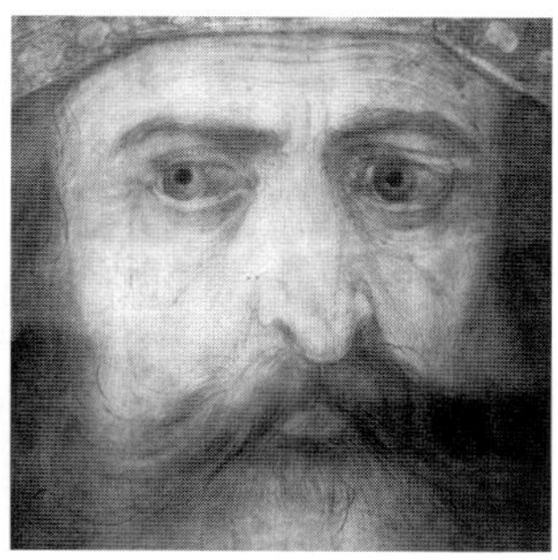

Fig. 11. Albrecht Dürer, *Emperor Carl the Great*, Nuremberg, GNM, detail. The Emporer's head is drafted directly on the ground with hatches and fine brush lines.

can be made visible to some extent in paintings by Nuremberg painter Wolf Traut (fig. 8).[10]

As nearly all painters applied tracing methods in addition to free sketches on the ground, infrared reflectography does not identify easily copies. However, it becomes easier when the personal drawing style of a master can be identified. Even though tracing methods were used, the sketches on the ground often retain characteristic features. This will be illustrated by Dürer's portraits of Emperors.

The knee-length portraits of Emperors Carl the Great and Sigismund at the Germanisches Nationalmuseum in Nuremberg were underdrawn in a very different way. According to a document, Dürer received on February 16, 1513 from the City Council of Nuremberg 85 guldens, 1 pound of pfennigs, and 10 shillings *fur zwo groß Tafel die pildnuß kayser Carls und kayser Sigmunds gestalt zu malen in die heilthums kammer*[11] (for two large panels to paint the picture of the figure of emperor Carl and emperor Sigmund in the holiness chamber). None of the Nuremberg paintings was signed. It is interesting to compare them to the half-length portraits of both Emperors in the State Gallery of Art (Gemäldegalerie, Old Masters Paintings) in Berlin. Thomas and Adelheit Brachert have attempted to clarify the question with infrared photography.[12] Whether the Berlin panels are preliminary stages, replicas or copies, to the Nuremberg portraits remains a debated subject. The underdrawings might provide substantial help to the debate since infrared reflectographs now exist for the Nuremberg portraits.

Dürer obviously made a preliminary drawing of the Emperor Sigismund's characteristic head in the necessary dimensions. The underdrawing on the Nuremberg panel reveals no search of form but it does reveal the use of a tracing method for the main lines (fig. 9). Under the actual underdrawing, carried out very sparingly and almost without any hatches, thin tracing lines can easily be detected. Search of form took place outside the painting (fig. 10). Considering these findings, the question whether it is a copy or a replica cannot be settled at this stage.

Dürer's approach to the Nuremberg portrait of Emperor Carl the Great is different. He invented some idealistic picture. It would be fair to assume that the master also made a cartoon. But even in the macro area, no tracing lines, tracing points or rough sketch are detectable. Evidently Dürer drafted the Emperor's head immediately on the ground with hatches and fine brush lines, like an autonomous drawing (fig. 11). The method reminds of his idealized self-portrait of 1500, where he also carefully modelled the front view of the face with hatches, even though finer and more detailed. Tracing lines cannot be detected in that painting either.[13] The 'invented' appearance of Carolus was carried out immediately on the ground. There is proof enough for the fact that this is not a duplication but a first formulation. Brachert detected impressed lines in patches of

[10] Ingo SANDNER, *Unterzeichnungen auf Gemälden Nürnberger Meister. Dürer und sein Kreis*, in: reference 2, p. 277-302, ill. p. 284-285.

[11] Quoted from: *Die Gemälde des 16. Jahrhunderts, Germanisches Nationalmuseum*, Nürnberg, Stuttgart, 1997, p. 206.

[12] Thomas and Adelheit BRACHERT, "Neues zu Dürers Kaiserbildern", in *Restauro 1*, 1989, p. 22-39.

[13] Gisela GOLDBERG, Bruno HEIMBERG and Martin SCHAWE, *Albrecht Dürer. Die Gemälde der Alten Pinakothek*, München, 1998, p. 323.

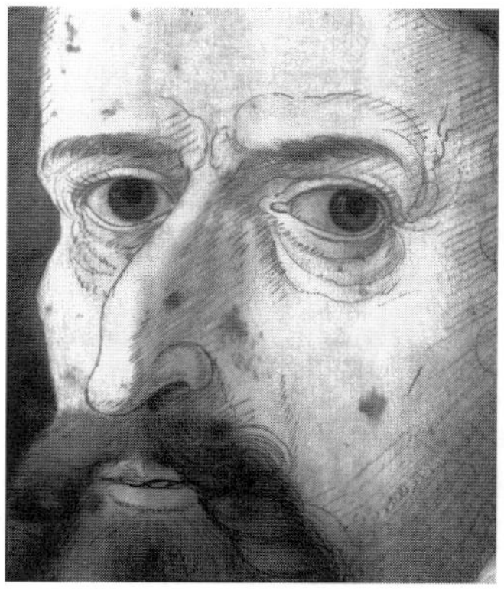

Fig. 12. Albrecht Dürer, *Emperor Sigismund*, Berlin, Gemäldegalerie, detail. The portrait was underdrawn with a crayon, probably dark natural chalk. Thin tracing lines can be detected.

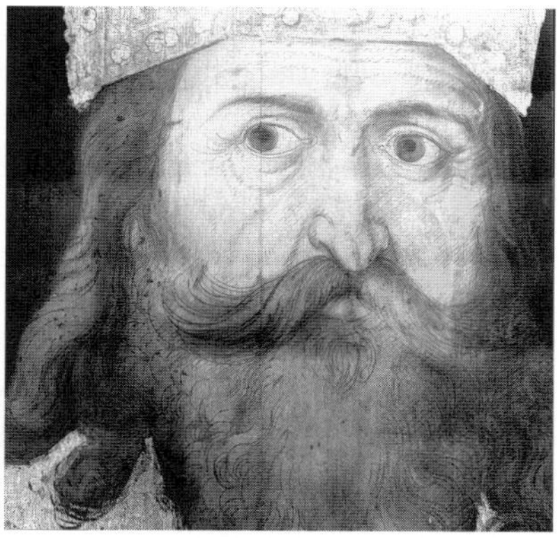

Fig. 13. Albrecht Dürer, *Emperor Carolus*, Berlin, Gemäldegalerie, detail. The portrait was underdrawn with a crayon, probably of dark natural chalk. Thin tracing lines can be detected.

the guilded parts in both the Nuremberg and Berlin portraits of Carolus.[14] He concluded that in both cases the same cartoon had been used. This seems possible for the crown, of which the appearance was known. Dürer might have done a preliminary drawing in 1:1 size. A freehand drawing of the crown by Dürer has survived in the Germanisches Nationalmuseum in Nuremberg. The findings establish that the Nuremberg painting of Carl the Great is the original creation. The findings with regard to the portrait of Emperor Sigismund may not be as convincing, yet it may be assumed that both Nuremberg paintings are original creations.

In addition to later copies of Dürer's Emperor's paintings, two much disputed portraits of both Emperors with exactly the same head size survive in Berlin (figs. 12, 13).[15] Both bear Dürer's monogram and are dated 1514. Winzinger considers the Berlin paintings as preliminaries to the Nuremberg portraits.[16] Brachert, who did not have digital IR reflectographs at hand in 1989, supports Winzinger's thesis. He draws his conclusion from interpretations of the X-ray.[17] On the Carolus panel, the initial inscription, mentioning the date of Emperor Carl IV's coronation in 1346, appears under the visible text. From this the author concludes that at first Emperor Carl IV was to be depicted. Such change of concept would date from before the Nuremberg portraits. Moreover, the frontal view and the parted beard were in analogy with images of Emperor Carl IV.[18] Brachert's considerations are logical but they contrast with the evidence provided by the underdrawing in the Nuremberg Carolus. The thesis of the Berlin Emperors being a

[14] T. and A. BRACHERT, 1989, p. 23.
[15] Earlier in the Wellesney collection and in a private collection Zürich.
[16] F. WINZINGER, "Umstrittene Werke Albrecht Dürers: Die Kaiserbrustbilder", in *Zeitschrift des deutschen Vereins für Kunstwissenschaft XXXI*, H. 1/4, Berlin, 1977.
[17] T. and A. BRACHERT, 1989, p. 32.

'preliminary stage' to the Nuremberg paintings is also opposed by the fact that their underdrawings are positioned over tracing lines. The interpretation of the X-ray would thus need some reconsideration. The date of 1514 on the Berlin panels is also one year after the documented payment for the Nuremberg knee-length portraits. The 85 guldens cover the overall performance and cannot be a partial payment, as Brachert suggests.

Incidentally, those supporting the attribution to Dürer equal in numbers those opposing the attribution. The underdrawings of the Berlin panel paintings will add some arguments to the debate.

Both Emperor's portraits in Berlin were underdrawn, not with a fine pointed brush as in Nuremberg, but with a crayon, probably dark natural chalk. The higher resistance against wiping of natural chalk versus charcoal sticks, and the granular consistence of rubbed-off parts, allow to assume that it was a crayon. Brachert suggests a lead stylus was used. Recent tests carried out by Andreas Siejek with different drawing instruments on dummies demonstrated a rather 'bound' stroke for metal styluses.[19] Crayon lines, however, corresponded with the present case. The utilisation of this instrument for underdrawing is remarkable, as it has so far not been noticed in a similar way in other paintings by Dürer. On the other hand, a fair number of freehand drawings in dark crayon by Dürer have been preserved. It has also been mentioned above that in Cranach's workshop crayons were repeatedly used for underdrawings or rough sketches on the ground. The use of a crayon for the underdrawing may not be typical of Dürer but can not be excluded either. It may also be taken for certain that the visible underdrawing of both portraits lies upon tracing lines which are still faintly perceptible in a few places. While this might apply to head and crown in the case of Emperor Sigismund, Brachert's observations of imprinted lines from a transfer method in the areas of guilded parts have already been mentioned. But also in the face there was no search of form as there was in the Nuremberg Carolus. Again, the main lines might have been transferred through tracing after the Nuremberg model. The actual underdrawing lies upon the tracing lines. It is clearly duplicating, yet rather freely executed. The faces were laid out in firm strokes and regular hatches. Emperor Sigismund's head in particular, in its superior and accomplished style, indeed reminds of Dürer. Compared to this, the underdrawing in the Carolus painting seems flatter and less accomplished. Brachert too expresses doubt whether this is Dürer. An assistant's contribution may be assumed.

To sum up, the Nuremberg Emperor paintings may be considered as original creations despite counterarguments resulting from the X-ray. Their underdrawings incline to attribute them to Dürer himself. The case is clearer regarding the Berlin paintings. Judging from their underdrawings it is inconceivable to consider them to be preliminary stages to the Nuremberg portraits. The accomplished underdrawing of the portrait of Emperor Sigismund suggests Dürer's hand in the free duplication via tracing lines. Even if there is no proof, it will probably have to be categorised as a replica. The case is more open in the portrait of Emperor Carolus. While Dürer's hand can definitely be identified on the Nuremberg panel, the Berlin painting features a fairly simple duplication via tracing lines without any characteristic particularities. An assistant's contribution

[18] T. and A. BRACHERT, 1989, p. 33-37.
[19] A. SIEJEK and K. KIRSCH, 2004, p. 120-121.

seems possible. Whether it must be considered a copy rather than a replica is questionable. As far as the Berlin portraits are concerned, the analysis and judgement of the surface must play a major part.

Ingo Sandner
Schwenkstraße 5
D – 01326 Dresden
Tel: 0049/351/2684016
ingsandner@aol.com

1968: thesis in conservation and restoration of paintings and sculptures. 1977: graduation in history of art (University of Leipzig). 1968-1974: chief restorer at the Institute for Preservation of Monuments in Dresden (*Institut für Denkmalpflege, Dresden*). 1974-1993: university lecturer and professor for conservation and restoration at the Academy of Fine Arts Dresden (*Hochschule für Bildende Künste, Dresden*). 1994-2006: professor for conservation and restoration at the University of Applied Sciences Cologne (FH Köln). 2006: retired.

DATATION DU DÉCOR PEINT DES PORCELAINES ET DÉTECTION DES FAUX. L'EXEMPLE DE TOURNAI

Jacqueline Couvert, Laurence Lenne et Ludovic Recchia[1]

Résumé: Dans le domaine de la porcelaine, la détection des faux ou des imitations de pièces d'époque est parfois ardue tout comme l'est l'identification des surdécorations. L'idée de Jacqueline Couvert a été d'appliquer aux pièces de Tournai — présentées à des fins d'examen par Laurence Lenne, spécialiste en céramiques anciennes — une analyse par microfluorescence X, utilisée couramment dans l'étude de peintures de chevalet pour l'identification des pigments minéraux. Le but était de vérifier si, dans la peinture sur porcelaine comme dans la peinture sur chevalet, le zinc apparaissait dans la palette vers 1830. Il résulte des examens que seules les couleurs utilisées à partir d'environ 1835-1840 comprennent du zinc en quantité significative dans leur composition. Cette découverte permet de situer le décor peint des porcelaines par rapport à cette période pivot. Il importera de vérifier l'application de cette observation aux productions d'autres manufactures belges et étrangères. Ludovic Recchia — doctorand et collaborateur scientifique au Musée de Mariemont — a complété cette étude en évoquant le contexte de création de ces objets.

1. Contexte historique, technique et rappel des données connues sur la fabrication des couleurs pour pâtes tendres

L'étude et la datation des décors des porcelaines de Tournai reposent actuellement sur des critères croisés tels: l'évolution stylistique des formes et des décors à partir notamment de jalons (pièces signées et/ou datées, commandes documentées...), les innovations technologiques (pose de l'or, coulage), la comparaison avec des productions en pâtes tendres anglaises et françaises (emprunts, imitations...), etc. Il existe ainsi une classification de la production en trois périodes durant le XVIIIe siècle: la première entre 1750 et 1763, la deuxième entre 1763 et 1775, la troisième entre 1775 et 1800. La production de porcelaines polychromes est attestée jusqu'en 1830 mais la manufacture a continué ses activités jusqu'en 1890.

Au XVIIIe siècle, le savoir-faire se transmet de génération d'artisans en génération d'artisans. Au-delà des gestes et des méthodes découvertes empiriquement et transmises d'abord oralement, la connaissance scientifique de la composition des pâtes et des émaux était détenue par une catégorie d'individus appelés 'arcanistes'. Tantôt artistes, tantôt artisans, tantôt hommes d'affaires, ces érudits poursuivaient la quête de secrets dont l'exploitation commerciale n'était pas évidente. François-Joseph Peterinck, au départ étranger au monde de la porcelaine, exerça des activités très contrastées. On a dit à son sujet: «En même temps [que sa fonction de directeur de la manufacture], il enseigne [l'architecture] à l'académie, il bâtit des maisons, il construit des ponts et des routes, il

[1] J. Couvert (Laboratoire d'étude des œuvres d'art par les méthodes scientifiques, UCL). L. Lenne, antiquaire, a sélectionné les pièces provenant de collections privées. L. Recchia (Musée de Mariemont) a mis à disposition pour analyse quelques pièces de ce musée.

fait sauter des fortifications, il fouille la terre pour y chercher des minéraux, et dépense, à ces occupations diverses, un temps que les travaux de sa fabrique semblaient devoir absorber tout entier»[2]. Cette description révèle le contexte de la porcelaine au siècle des Lumières: aventureux, expérimental, vecteur de prestige social. Pour lancer son affaire, Peterinck débauche des ouvriers français, les frères Dubois, dont l'itinéraire est bien connu. Consignant, entre 1751 et 1756, des recettes probablement empruntées à ces deux transfuges ou d'autres aujourd'hui inconnus, il devient durant un demi-siècle le détenteur d'une science qui fera de Tournai un haut lieu de la porcelaine.

Les pâtes qu'utilisait Peterinck, pourtant proches de celles utilisées en France dès la première moitié du XVIIIe siècle, devinrent rapidement désuètes suite à la découverte de kaolin qui permettait la fabrication de vraie porcelaine (dure). On en oublia même le secret au point qu'Alexandre Brongniart (1770-1847), chimiste, directeur de la manufacture nationale de porcelaines de Sèvres de 1800 à 1847, tenta d'en comprendre le principe lorsqu'il visita la manufacture belge entre 1835 et 1844[3]. Les contemporains savaient que les pâtes utilisées à Tournai ne différaient que peu de celles produites dans les manufactures françaises mais les spécificités de chaque pâte comptaient énormément. À Tournai, il s'agissait d'un mélange de marne argileuse et d'argile figuline ayant une fritte pour fondant.

Cette marne (ou *marle*) était une terre de Bruyelles, village tout proche de Tournai, où s'approvisionnaient également les arcanistes de Delft, de Lille et d'autres fabriques du nord de la France. Cette terre constituait l'ingrédient de base de la pâte mise au point par Peterinck. On faisait aussi venir des terres blanches des environs de Belœil et Chimay[4], du plomb et de l'étain d'Angleterre[5].

Par nostalgie romantique sans doute, la pâte tendre fut à la mode au milieu du XIXe siècle. Le rapporteur du jury mixte international de l'exposition universelle de Paris de 1855 nous en dit ceci: «Cette porcelaine, qui a précédé la porcelaine dure en France et dans d'autres pays, est peu convenable pour les usages domestiques, parce qu'elle se raye facilement, sa fabrication est difficile et coûteuse. Telles sont les anciennes porcelaines de Tournai, qui ont presque totalement disparu; et les pâtes tendres vieux Sèvres, qui ont acquis une si grande réputation par leur décoration, à laquelle cette porcelaine se prête éminemment, à cause de la fusibilité de son émail». Il n'est pas question dans ce rapport des productions en porcelaine tendre de Tournai qui existent pourtant toujours à ce moment, mais bien de celles de la fabrique de Maximilien de Bettignies (1845-1907) à Saint-Amand (1817-1880). Selon ce rapport, ne possédant pas d'atelier de décoration, de Bettignies faisait décorer ses blancs par des artistes parisiens indépendants, fort nombreux à l'époque. Outre la qualité de l'émail, cet amandinois était parvenu à fabriquer des vases de grande taille en utilisant la technique du coulage mise au point à Tournai en 1790[6].

[2] E. SOIL DE MORIAMÉ, *Recherches sur les anciennes porcelaines de Tournay*, Tournai, 1883, p. 76-80.

[3] SOIL DE MORIAMÉ, 1883, p. 132. Les archives de Brongniart mentionnent une visite à Tournai en 1835.

[4] Archives Générales du Royaume (AGR), Conseil des Finances, farde 2029.

[5] AGR, Conseil des Finances, fardes 2029, fol. 1768; farde 2030, fol. 1774 à 1787.

[6] E. GREBER, *Traité de céramique. Terres cuites, produits réfractaires, faïences, grès, porcelaines*, Paris, Encyclopédie Roret, Larousse, édition revue et corrigée, 1950 (première édition s.l., s.d.), p. 16.

Le rapporteur officiel belge à l'exposition parisienne de 1855 dresse le même constat curieux de la disparition de la porcelaine tendre de Tournai: «Quand à la porcelaine tendre, elle ne se fabrique plus qu'à Saint-Amand, où elle a été transportée de Tournay par Maximilien de Bettignies. Elle est exclusivement employée par les décorateurs de Paris pour citer ou falsifier le vieux Sèvres: son débit est très restreint et l'on conçoit qu'on ait cherché à éluder les lois prohibitives de nos voisins en s'installant chez eux. La manufacture impériale de Sèvres a repris la fabrication de la porcelaine tendre pour les grandes pièces décoratives, après avoir perdu, pendant un demi-siècle, une pâte qui avait fait son ancienne réputation. Quoiqu'il en soit, et de Sèvres et de Saint-Amand, nous ne devons pas regretter la perte, pour la Belgique, d'une fabrication difficile, coûteuse et fort peu propre aux besoins habituels de la consommation. L'avenir appartient décidément à la porcelaine dure...».

Si Saint-Amand permit la fabrication de faux Sèvres, il est probable que des blancs de Tournai du XIXe siècle aient reçu des décors (tournaisiens notamment) de style XVIIIe siècle par des décorateurs indépendants. La connaissance de la composition chimique des couleurs peut donc apporter une réponse à de nombreuses interrogations, notamment en ce qui concerne des pièces jugées aberrantes parce que falsifiées ou simplement inédites (et uniques?). C'est ce qu'annoncent les résultats des analyses réalisées récemment au Laboratoire d'étude des œuvres d'art de l'UCL. Rappelons dans les grandes lignes la composition des couleurs à Tournai à l'époque citée.

La fabrication des couleurs était une science en soi. On fondait les minerais métalliques dans des creusets en y ajoutant d'autres ingrédients. On obtenait un «culot» qui était ensuite broyé et mêlé aux fondants. Pour éviter toute confusion, rappelons que les oxydes métalliques responsables de la coloration pouvaient composer tant des couleurs de petit que de grand feu. Le bleu de cobalt est à la fois une couleur de grand et de petit feu. Pour le bleu et les divers ingrédients de ses couvertes, Peterinck se fournissait à Rotterdam[7]. Le pourpre de Cassius et le carmin se préparaient en précipitant l'or dans l'oxyde d'étain, en y ajoutant, pour le carmin, un peu d'oxyde d'argent; le rouge peut avoir diverses tonalités avec le fer quelques fois mêlé au cuivre et à l'étain. Pour les autres couleurs: le vert vient des oxydes de chrome et de cuivre; le jaune de l'étain, de l'antimoine et du plomb; le blanc de l'étain; le noir du fer, du cuivre et du manganèse; le *chatiron* (cerne brun foncé du dessin de décors orientaux ou strasbourgeois) du manganèse et du fer; les différents bruns ont tous pour base l'oxyde de fer. Il est parfois évoqué que les oxydes d'iridium et de platine peuvent donner du gris et même du noir pour le second[8]. Au XIXe siècle, de nombreuses combinaisons d'oxydes permettent des nuances méconnues jusqu'alors: des nuances de bleu peuvent être modifiées en l'utilisant avec l'oxyde de zinc ou d'alumine, les proportions d'oxydes de zinc et d'étain éclaircissent les nuances des ocres pâles ou foncés obtenus à partir du fer, etc[9].

La cuisson de la porcelaine avait lieu en deux temps: la cuisson au grand feu ou feu de biscuit, ensuite la cuisson du biscuit recouvert d'émail à température moins élevée.

[7] AGR, Conseil des Finances, fardes 2029, 31/12/1756; 04/1757.

[8] M.A. SALVÉTAT, *Leçons de céramique professées à l'Ecole Centrale des Arts et Manufactures ou Technologie céramique comprenant les notions de chimie, de technologie et de pyrotechnie, applicables à la fabrication, à la synthèse, à l'analyse, à la décoration des poteries*, Paris, Mallet-Bachelier, tome 2, 1857, p. 446.

[9] SALVÉTAT, 1857, p. 476.

Parfois le même four était utilisé pour ces deux cuissons. Certaines manufactures utilisaient des fours à deux étages (biscuit en dessous, vernis au dessus). D'autres manufactures disposaient de plus petits fours pour cuire l'émail. À Sèvres, les cuissons exigeaient 75 à 100 heures pour le biscuit, 30 heures pour l'émail. Le combustible utilisé était le bois. Peterinck fit néanmoins des essais infructueux au charbon en 1768.

Les décors polychromes engendraient des étapes extrêmement délicates qui expliquent le coût important d'une porcelaine polychrome. Les décors composés d'oxydes résistant au grand feu étaient appliqués sur le biscuit, c'est-à-dire sous l'émail tandis que les couleurs dites de «petit feu» étaient appliquées au-dessus de celui-ci. L'émail était composé de sable, de minium, de carbonate de soude et de potasse[10]. Les couleurs, broyées à l'état de poudre, étaient délayées dans l'eau et mêlée de céruse (ou blanc de plomb) pour les rendre plus fusibles. Elles s'appliquaient sur l'émail comme toute peinture à l'eau. Lors de la cuisson, elles faisaient corps avec l'émail. Comme chaque couleur exigeait une température de cuisson appropriée, chaque pièce passait au four plusieurs fois si nécessaire.

Sans détailler ces procédés, on imagine sans peine la difficulté de créer une porcelaine au XVIIIe siècle. Où étaient consignées ces connaissances techniques? Dans des cahiers d'atelier ou de laboratoire au XVIIIe siècle; ensuite, à l'aube du XIXe siècle, dans des manuels techniques imprimés qui laissent penser à tort que la cuisine devint plus aisée. Dans tous les domaines techniques, la publication de manuels techniques devint très populaire durant le siècle de l'industrie. La plus célèbre collection de manuels est sans aucun doute celle de la *Librairie Encyclopédique de Roret*. Le premier manuel se rapportant à la céramique relève d'un certain M.P. Magnier et date de 1827: *Nouveau manuel complet du Porcelainier, Faïencier, Potier de Terre*. Il fut plusieurs fois réédité jusqu'au XXe siècle. En 1844, parut aussi un *Nouveau Manuel Complet de la Peinture sur Verre, sur porcelaine et sur émail*, témoignage de l'importance de la peinture dans les décors peints.

Le plus célèbre de tous les ouvrages techniques sur la porcelaine reste le *Traité des Arts céramiques ou des poteries* d'Alexandre Brongniart (Paris, 3 volumes et un atlas, en 1844). Ce savant, qui dirigea la manufacture de porcelaine de Sèvres interrogea les préfets de départements français sur les techniques en usage dans l'industrie faïencière de son territoire. L'ouvrage fut complété par son successeur, Alphonse Salvétat, et une première fois réédité en 1857 puis une deuxième en 1877[11].

Au XIXe siècle, les entreprises se livrèrent un combat sans merci, innovant sans cesse en matière de produits et de coûts de production. Il y eut de nombreuses innovations en matière de chromie. Citons à titre d'exemple l'emploi du chrome de grand feu (1400° C pour les pâtes dures), isolé en 1797 et employé comme couleur en 1802[12]. Comme nous venons de l'évoquer plus haut, de nombreuses autres couleurs naquirent au XIXe siècle. À Tournai, l'âge d'or de la polychromie se termine sans doute avant le milieu du XIXe siècle. La manufacture avait conservé ses antiques recettes tout en s'ouvrant à l'innovation.

Une fois que les analyses qui nous intéressent aujourd'hui auront été généralisées à un ensemble plus grand de spécimens, issus d'autres manufactures notamment, nous pourrons

[10] SALVÉTAT, 1847, p. 473.

[11] 4e édition, 1977 (éd. Dessain et Tolra).

[12] A. FAŸT-HALLÉ et B. MUNDT, *La porcelaine européenne au XIXe siècle*, Fribourg, Office du Livre, 1983, p. 14.

en confronter les résultats aux informations contenues dans certains manuscrits ainsi, bien entendu, qu'aux objets eux-mêmes. Un nouvel éclairage s'en dégagera certainement.

2. Analyses par microfluorescence X et datation

La microfluorescence X est une méthode d'analyse couramment utilisée par les historiens de l'art pour l'étude des pigments minéraux utilisés dans la peinture de chevalet. Parmi les pigments blancs, les blancs de plomb (2PbCO3. Pb(OH)2), de zinc (ZnO) et de titane (TiO2) peuvent être considérés comme les trois pigments majeurs[13]. Le blanc de plomb, employé depuis l'Antiquité, fut le seul pigment blanc important utilisé par les peintres jusqu'au milieu du XIXe siècle. À partir de 1834, il fut remplacé progressivement par un pigment non toxique, le blanc de zinc (ZnO ou oxyde de zinc). Enfin, le blanc de titane s'imposa vers 1920.

Dans le domaine de la porcelaine, l'oxyde de zinc se retrouve dans la composition des colorants[14] mais agit également comme fondant en mélange avec d'autres fondants. Il permet d'augmenter notamment la brillance, la dureté et la résistance chimique de la couverte[15]. Il fut employé dès 1838 à Bristol (Angleterre) comme substitut de l'oxyde de plomb.

L'hypothèse de départ était la suivante: si les colorants employés avant l'abandon des décors polychromes à la manufacture de Tournai vers 1830 ne présentaient pas de zinc en quantité significative dans leur composition, les colorants utilisés pour le décor de pièces fausses ou sur-décorées ultérieurement pourraient au contraire en contenir. Une méthode de datation de la porcelaine (avant ou après cette date pivot) pouvait dès lors être adoptée.

A cet effet, la couverte a été analysée de même que les couleurs du décor peint de trente objets réalisés entièrement à la Manufacture de Tournai de toutes périodes. L'identification des composants chimiques de la couverte est nécessaire pour mettre en évidence les éléments chimiques propres aux couleurs elles-mêmes. Ces pièces de références proviennent des collections du Musée royal de Mariemont (notamment une palette de couleurs) ou de collectionneurs privés. Les couleurs analysées sont le bleu, le jaune, le pourpre, le vert, le noir, le brun et le rouge. La présence de zinc n'a été décelée dans aucune couleur sauf parfois de façon résiduelle ou en faible quantité dans certains verts où il pourrait jouer un rôle d'opacifiant.

D'autre part, nous avons sélectionné un échantillonnage de quinze pièces anciennes en porcelaine de Tournai dont les décors polychromés sont postérieurs (sur-décor tardif) ou

[13] R.J. GETTENS, H. KÜHN et W.T. CHASE, "Lead White", dans *Artists' Pigments. A Handbook of Their History and Characteristics*, vol.2, (A. ROY ed.), National Gallery of Art, Washington, 1993, p. 67-81. H. KÜHN, "Zinc White", dans *Artists' Pigments. A Handbook of Their History and Characteristics*, vol.1 (R.L. FELLER ed.), National Gallery of Art, Washington, 1986, p. 169-186. M. LAVER, "Titanium Dioxide Whites", dans *Artists' Pigments. A Handbook of Their History and Characteristics*, vol. 3, (E. WEST FITZHUGH ed.), National Gallery of Art, Washington, 1997, p. 295- 355. J. COUVERT, "Historique, repères chronologiques et méthodes d'identification des blancs de plomb, de zinc et de titane", dans *Revue des archéologues et historiens d'art de Louvain,* XXXI, Louvain-la-Neuve, 1998, p. 159-164.

[14] A. D'ALBIS, *Traité de la porcelaine de Sèvres,* Dijon, 2003. J. LEMAIRE, *La porcelaine de Tournai. Histoire d'une Manufacture,* 1999. T. PREAUD et A. D'ALBIS, *La Porcelaine de Vincennes,* 1991.

[15] S. STEFANOV et S. BATSCHWAROV, *Keramik-glasuren; Ceramic Glazes,* Wiesbaden et Berlin, 1988, p. 66 et 198.

bien réalisés sur des porcelaines tournaisiennes totalement blanches ou sorties de l'atelier avec un décor bleu de grand feu original (œuvres de faussaires ou œuvres d'imitation). Les résultats obtenus permettent de constater que le zinc et parfois le chrome sont présents dans ces pièces. La date de l'apparition du zinc dans ces pièces semble se rapprocher de 1835, date d'introduction du zinc dans la peinture de chevalet.

Quelques exemples présentés ci-après permettront d'illustrer ces résultats. Les éléments chimiques détectés dans la couverte et dans les couleurs de petit feu d'une pièce ancienne seront comparés avec la composition chimique de la couverte et des couleurs

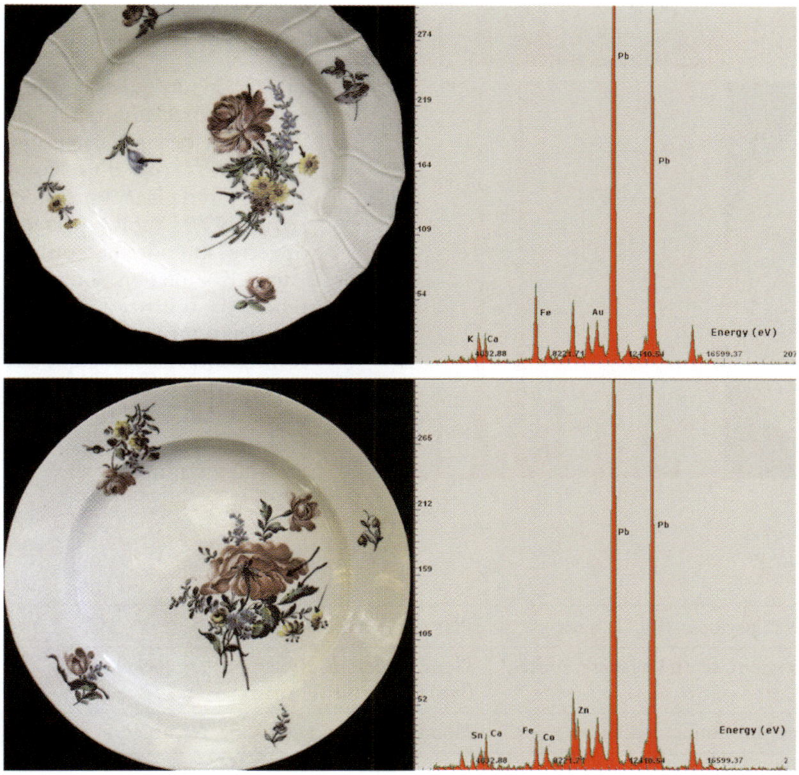

Fig. 1. Haut: assiette en porcelaine tendre de Tournai à décor d'un bouquet de fleurs polychrome décentré. Fin de la 2ème période, 1763-1775 (coll. privée). Spectre XRF de la couleur pourpre. Bas: assiette en porcelaine tendre de Tournai à décor d'un bouquet de fleurs polychrome décentré. Fausse marque aux épées croisées bleues. Sur-décor tardif de la fin du 19ème siècle (coll. privée). Spectre XRF de la couleur pourpre.

	Assiette 2ème période (fig. 1, haut)	**Assiette sur-décorée (fig. 1, bas)**
Couverte	Plomb, potassium, calcium, fer, cuivre	Plomb, potassium, calcium, fer, cuivre
Bleu	Plomb, calcium, fer, cobalt (élevé), nickel, potassium, cuivre	Plomb, potassium, calcium, cuivre, fer, cobalt (élevé), nickel, **zinc**
Pourpre	Plomb, cuivre, potassium, fer, or	Plomb, potassium, étain, calcium, fer, cobalt, cuivre, **zinc**, or
Jaune	Plomb, calcium, antimoine, manganèse, fer, cuivre, nickel	Plomb, calcium, manganèse, fer, cuivre, **zinc**
Vert	Plomb, potassium, étain, calcium, fer, cuivre (élevé)	Plomb, potassium, calcium, fer, cuivre (élevé), **zinc**

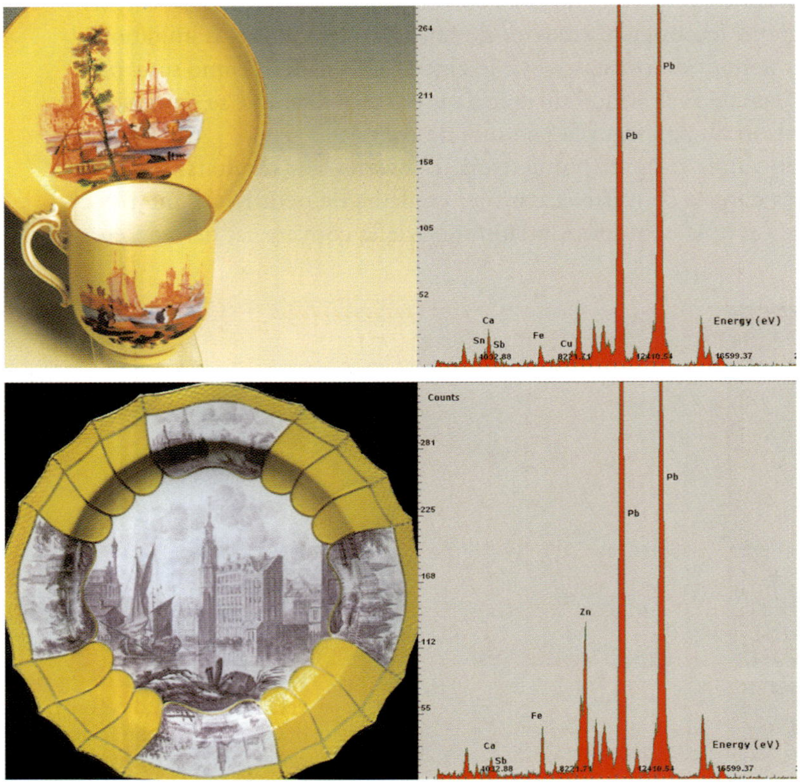

Fig. 2. Haut: tasse et sous-tasse en porcelaine tendre de Tournai à décor polychrome maritime cerné d'un fond jaune. Fin de la 2ème période, *circa* 1770 (coll. privée). Spectre XRF de la couleur jaune. Bas: plat en porcelaine tendre de Tournai à décor en camaïeu pourpre maritime et fond jaune. Fausse marque aux épées croisées en bleu. Sur-décor de la fin du 19ème siècle (coll. privée). Spectre XRF de la couleur jaune.

	Tasse et sous-tasse 2ème période (fig. 2, haut)	**Plat sur-décoré (fig. 2, bas)**
Couverte	Plomb, calcium, potassium, fer, manganèse, nickel, cuivre (bas)	Plomb, calcium, potassium, fer, nickel, cuivre (bas)
Jaune fond	Plomb, étain, calcium, antimoine, fer, cuivre	Plomb, calcium, antimoine, fer, **zinc**
Vert	Plomb, étain, calcium, antimoine, fer, cuivre	
Pourpre	Plomb, potassium, calcium, fer, cuivre, or	Plomb, potassium, calcium, fer, cobalt, nickel, **zinc**, or
Bleu	Plomb, potassium, calcium, manganèse, fer, cobalt, cuivre	
Brun	Plomb, calcium, antimoine, cuivre, nickel, fer	

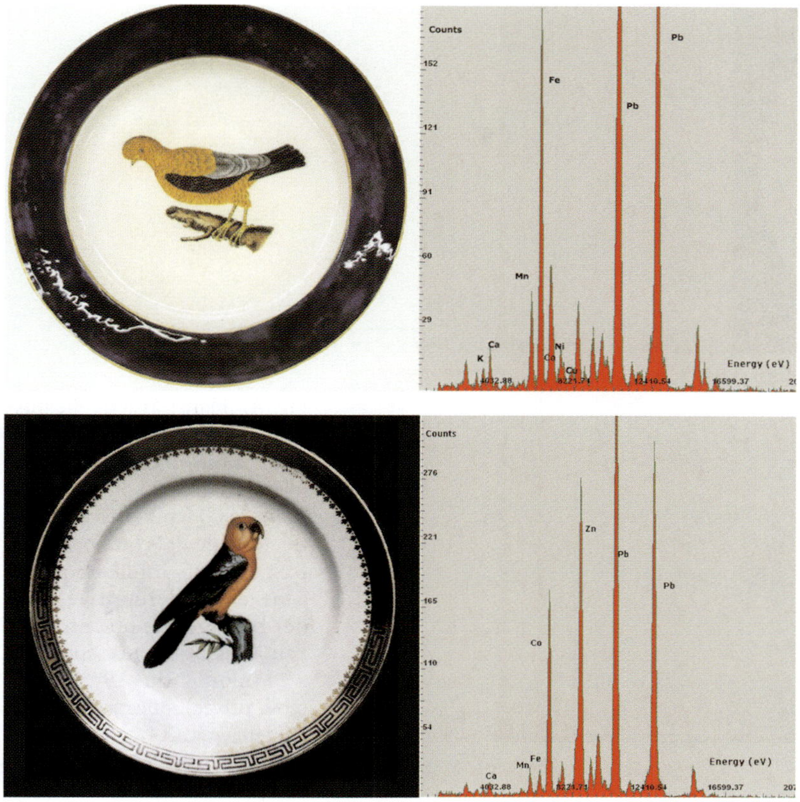

Fig. 3. Haut: assiette en porcelaine tendre de Tournai à décor dit «au grand oiseau», service dit «de Cock», *circa* 1820 (coll. musée de Mariemont). Spectre XRF de la couleur bleue. Bas: assiette en porcelaine tendre de Tournai à décor «au grand oiseau». Représentation de «la perruche tricolore». Fausse marque aux épées or. Sur-décor de la fin du 19ème siècle (coll. privée). Spectre XRF de la couleur bleue.

	Assiette (service aux grands oiseaux) (fig. 3, haut)	**Assiette «La perruche tricolore» (fig. 3, bas)**
Couverte	Plomb, potassium, calcium, fer, cuivre	Plomb, potassium, calcium, fer, manganèse, nickel, cuivre
Brun	Plomb, potassium, calcium, antimoine, chrome, (bas), manganèse, fer (élevé), nickel, cuivre (bas)	Plomb, calcium, étain, fer, **zinc**, baryum
Vert		Plomb, calcium, antimoine, chrome, manganèse, fer, cobalt, **zinc**
Jaune orangé	Plomb, potassium, calcium, antimoine, fer, nickel	Plomb, potassium, étain, calcium, antimoine, fer, nickel, cobalt, **zinc**
Jaune clair	Plomb, potassium, étain, calcium, antimoine, manganèse, fer, nickel	
Bleu de petit feu	Plomb, potassium, cobalt, calcium, manganèse, fer, nickel, cuivre	Plomb, calcium, antimoine, manganèse, fer, cobalt, nickel, **zinc**
Bleu de grand feu	Plomb, potassium, calcium, fer, cobalt, nickel, cuivre	Plomb, calcium, fer, cobalt, nickel, cuivre
Noir	Plomb, potassium, calcium, manganese, fer, nickel, cuivre	

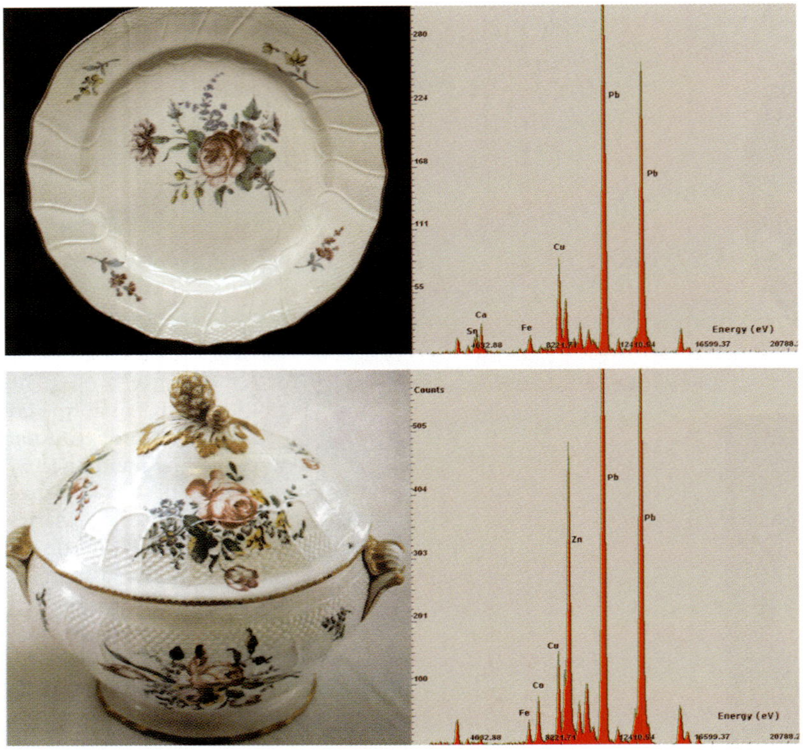

Fig. 4. Haut: assiette en porcelaine de Tournai à décor d'un bouquet polychrome «Rose à l'œillet». 3ème période, 1775-1800 (coll. privée). Spectre XRF de la couleur verte. Bas: soupière ronde en porcelaine tendre de Tournai à décor de bouquets polychromes. Fausse marque aux épées or. 2ème moitié du 19ème siècle (coll. privée). Spectre XRF de la couleur verte.

	Assiette «Rose à l'œillet» (fig. 4, haut)	**Soupière (bouquets polychromes) (fig. 4, bas)**
Couverte	Plomb, potassium, étain, calcium, manganèse, fer, cuivre, nickel	Plomb, fer, calcium, potassium, nickel, cuivre
Vert	Plomb, potassium, étain calcium, antimoine, cuivre (élevé), fer, nickel	Plomb, potassium, calcium, fer, cobalt, nickel, cuivre, **zinc**
Jaune	Plomb, antimoine, calcium, potassium, fer, manganèse, nickel	Plomb, potassium, antimoine, calcium, fer, nickel, **zinc** (élevé)
Pourpre	Plomb, potassium, étain, calcium, manganèse, fer, nickel, cuivre	Plomb, étain, calcium, fer, nickel, cuivre
Bleu	Plomb, potassium, calcium, manganèse, fer, cobalt (élevé), nickel, cuivre	Plomb, calcium, baryum, fer, cobalt, nickel, cuivre, **zinc** (élevé)

d'une pièce fausse, proche de l'autre par le décor ou les couleurs. Par groupe de deux pièces (la vraie et la fausse), deux spectres microfluorescence X (XRF) sont choisis pour illustrer la composition d'une couleur. Un tableau reprend les résultats de la microfluorescence X pour l'ensemble des couleurs, toujours par groupe de deux pièces.

Les résultats sont donc assez concluants à ce stade de nos recherches en ce qui concerne les couleurs de petit feu employées pour le décor de la porcelaine de Tournai. Parfois certaines couleurs d'objets tardifs ou faux ne présentent pas de zinc (c'est le cas du pourpre). Il est donc important d'analyser toutes les couleurs de la décoration d'un objet.

Les essais réalisés sur les bleus de type grand feu, mais dont notre échantillonnage est encore insuffisant, ne semblent pas se conformer à notre méthode de datation. En effet, les décors analysés ne contiennent pas de zinc quelle que soit la période d'exécution. Cela pourrait s'expliquer par le fait que l'oxyde de zinc se liquéfie à haute température.

Enfin, des pièces (porcelaine ou faïence) provenant d'autres manufactures belges (Bruxelles), françaises (Mennecy, Rouen, Samson, Bourg la Reine, Schoelcher), italiennes (Castel Durante, pseudo Urbino), autrichienne (Vienne), allemande (Meissen) ont également été analysées. Les mêmes remarques concernant le zinc peuvent y être appliquées. Toutefois, nous espérons poursuivre nos recherches. L'augmentation du nombre de pièces examinées devrait nous conduire à affiner nos résultats.

Jacqueline COUVERT

Voir Louise DECQ, Jacqueline COUVERT et Hélène VEROUGSTRAETE, *Une tête de vieillard sur parchemin (Jordaens?).*

Laurence LENNE
Chaussée de Bruxelles, 344
B – 7800 Ath
Tel: +32.(0)476.54.26.36
Fax/Tel: +32.(0)68.45.66.35
artetpatrimoine@skynet.be

Antiquaire, spécialisée en céramiques anciennes (porcelaines de Tournai et de Chine aux XVII-XVIIIème siècles). Diplômée du CREPAC et formation à l'institut d'Études Supérieures des Arts (Paris). Membre de l'Association des Spécialistes de la Céramique de Collection à Paris.

Ludovic RECCHIA
Musée royal de Mariemont
B – 7140 Morlanwez
Tel: +32.(0)64.27.37.88
Fax: +32.(0)64.26.29.24
ludovic.recchia@musee-mariemont.be

Historien de l'art (Université libre de Bruxelles). Thèse: «Céramique de studio en Belgique». Assistant de recherche au Musée de Mariemont, en charge des collections d'arts décoratifs industriels.

BIBLIOGRAPHIE DE L'INFRAROUGE ET DU DESSIN SOUS-JACENT

2005-2007

+ ADDENDUM

Anne Dubois

Avertissement

La présente bibliographie porte sur les années 2005-2007 et le reliquat des années précédentes (voir addendum). Nous avons dépouillé les revues parues de 2005 à 2007 et en donnons la liste ci-après. Nous avons limité notre travail, dans la plupart des cas, aux périodiques. Cependant, de plus en plus d'ouvrages importants paraissent sur le sujet. Ils ont été recensés dans la mesure du possible. Il ne s'agit donc pas d'une revue exhaustive des ouvrages parus.

La présente livraison est divisée en deux parties:

1. La bibliographie de l'infrarouge et du dessin sous-jacent. 2005-2007.
2. Le reliquat des années précédentes.

Les choix éditoriaux pris dans les livraisons précédentes sont maintenus, à savoir:

– limiter le numéro d'ordre des documents à un seul chiffre qui pourra, dans la table des illustrations, compléter le numéro d'ordre de l'article; ex. I.R.A.2005.23.1. Cependant, afin d'éviter de répéter les références des œuvres dont il existe plusieurs illustrations dans l'article, le numéro d'ordre signalera le nombre des documents relatifs à la même œuvre; ex. I.R.A.2005.23.8-9. Maître de l'Adoration Von Groote, *Triptyque de l'Adoration des Mages*...
– donner le nom de l'artiste, le titre de l'œuvre, le lieu de conservation et le type de documents.
– limiter le plus possible les références aux illustrations prises par une quelconque méthode à l'infrarouge.

Cependant, les techniques infrarouges évoluent et le montage digital des réflectogrammes est devenu la norme plutôt que l'exception. C'est pourquoi l'abréviation IRR sera utilisée pour de tels documents. Les publications de montages manuels de réflectogrammes ou d'autres techniques spécifiques seront quant à elles précisées (ex. IRR montage). De même, l'utilisation de photographies digitales à l'infrarouge a fait son apparition dans les publications. Elles seront désignées sous le nom générique de IR.

Cette bibliographie peut toujours être complétée; il nous a été impossible d'accéder à toutes les revues ou ouvrages traitant du sujet qui nous occupe. Les conseils et compléments nous sont indispensables pour améliorer cette recension. Ils seront donc bienvenus.

A. Dubois

Abréviations

IR : photographie à l'infrarouge
IRR : réflectographie à l'infrarouge (digitalisée)
IRR montage : assemblage de réflectogrammes à l'infrarouge
IRC : photographie à l'infrarouge en couleur
LO : photographie en lumière ordinaire
LR : photographie en lumière rasante
RX : radiographie
IR scan : scanner digital à infrarouge

LISTE DES REVUES DÉPOUILLÉES

Acta historiae artium. Budapest
XLIV, 2003
XLV, 2004
XLVI, 2005
XLVII, 2006
Annales d'histoire de l'art et d'archéologie. Bruxelles. Université Libre
27, 2005
28, 2006
Anzeiger des Germanischen Nationalmuseum. Nürnberg
2006
2007
Apollo. International Magazine of the Arts. London
2005 (décembre)
2006
2007 (janvier-mars)
Arca Lovaniensis. Artes atque historiae Reserans documenta. Jaarboek. Leuven
33, 2004
35, 2006
Art Bulletin. New York
87, 2005 (4)
88, 2006
89, 2007
Art et Fact. Revue des historiens de l'art, des archéologues, des musicologues et des orientalistes de l'Université de Liège
24, 2005
25, 2006
Art Matters. Netherlandish Technical Studies in Art. Zwolle
3, 2005
4, 2007
Arte Lombarda
2005 (144-145)
2006
2007 (149-150)
Boletín del Museo del Prado. Madrid
38, 2002
39, 2003
40, 2004
41, 2005
Bolletino ICR. Istituto centrale per il restauro. Nouvelle série
10-11, 2005
12-13, 2006
Bulletin de l'Institut royal du Patrimoine artistique. Bruxelles
31, 2004/2005

Bulletin des Musées Hongrois des Beaux-Arts, Budapest
101, 2004
102-103, 2005
104, 2006
Bulletin des Musées royaux d'art et d'histoire. Bruxelles
73, 2002
74, 2003
Bulletin of the National Gallery in Prague/Bulletin Národní Galerie v Praze
XIV-XV, 2004-2005
XVI-XVII, 2006-2007
Bulletin van het Rijksmuseum. Amsterdam
53, 2005 (2-3)
54, 2006
55, 2007 (1-3)
Burlington Magazine. London
147, 2005 (décembre)
148, 2006
149, 2007
Cleveland Studies in the History of Art
9, 2005
Coré. Conservation et restauration du patrimoine culturel. Paris
16-17, 2006
18-19, 2007
Genava
LIII, 2005
LIV, 2006
Jahrbuch des Kunsthistorischen Museums Wien
6/7, 2004-2005
8/9, 2006/2007
Jahrbuch Preussischer Kulturbesitz. Berlin
41, 2004
42, 2005
Journal of the American Institute for Conservation
44, 2005 (2-3)
45, 2006
Journal of Cultural Heritage. A Multidisciplinary Journal of Science and Technology for Conservation and Promotion. Rome
6, 2005 (4)
7, 2006
8, 2007
Journal of the Warburg and Courtauld Institute. London
67, 2004
68, 2005
69, 2006
Kermes. Arte e tecnica del restauro. Firenze
60, 2005
62-64, 2006
Koninklijk Museum voor Schone Kunsten Antwerpen. Jaarboek
2004/2005

Konsthistorisk Tidsckrift. Journal of Art History. Stockholm
- 74, 2005 (3-4)
- 75, 2006
- 76, 2007

Master Drawings
- 43, 2005 (4)
- 44, 2006
- 45, 2007 (1-2)

Metropolitan Museum Journal. New York
- 40, 2005
- 41, 2006
- 42, 2007

Mitteilungen des Kunsthistorischen Institutes in Florenz
- XLVIII, 2004 (3)
- XLIX, 2005
- L, 2006 (1/2)

Münchner Jahrbuch der bildenden Kunst
- 54, 2003
- 55, 2004
- 56, 2005
- 57, 2006

National Gallery Technical Bulletin. London
- 27, 2006
- 28, 2007

Nederlands Kunsthistorisch Jaarboek
- 55, 2004
- 56/57, 2005/2006

Oud Holland
- 118, 2005 (3/4)
- 119, 2006
- 120, 2007 (1/2)

Restauro. München
- 2005 (8)
- 2006
- 2007 (1-4, 6-8)

Revue de l'art. Paris
- 150, 2005
- 151-154, 2006
- 155-158, 2007

Revue belge d'archéologie et d'histoire de l'art. Bruxelles
- LXXIV, 2005
- LXXV, 2006
- LXXVI, 2007

Revue des musées de France. Revue du Louvre, Paris
- 2005 (5)
- 2006
- 2007

RKD. Bulletin. La Haye
- 2005

2006
2007
Simiolus. Kunsthistorisch tijdschrift. Amsterdam
31, 2004-2005 (4)
32, 2006
Studies in Conservation. London
50, 2005 (4)
51, 2006 (1-3)
52, 2007
Technè. Paris, Laboratoire de recherche des Musées de France
23-24, 2006
25, 2007
Technologische Studien Kunsthistorisches Museum. Konservierung-Restaurierung-Forschung-Technologie. Vienne
1, 2004
2, 2005
4, 2007
Umeni. Praha
LIII, 2005 (6)
LIV, 2006
LV, 2007 (1-5)
Wallraf-Richartz-Jahrbuch. Cologne
LXVI, 2005
LXVII, 2006
Zeitschrift für Kunstgeschichte. München-Berlin
69, 2006
70, 2007
Zeitschrift für Kunsttechnologie und Konservierung, Worms am Rhein
19, 2005 (2)
20, 2006
21, 2007 (1)

APPLICATIONS

I.R.A.2005.20

M.W. AINSWORTH, *Intentional Alterations of Early Netherlandish Paintings*, dans *Metropolitan Museum Journal*, 40, 2005, p. 51-65.

1-2. Maître au Feuillage brodé, *Vierge et Enfant*, Williamstown (Mass.), The Sterling and Francine Clark Art Institute. IRR (détail: tête de la Vierge, main de la Vierge)

I.R.A.2005.21

R. BELLUCI, C. FROSININI, B.G. BRUNETTI, L. PEZZATI et P. POGGI, *Diagnostica italiana per un progetto europeo. La Vergine delle Rocce di Leonardo: sorprese nell'underdrawing della versione di Londra*, dans *Kermes*, 60, 2005, p. 39-53.

Rapport de l'étude de la Vierge au Rocher de Léonard de Vinci réalisée en juillet 2005 par une équipe internationale. L'étude du dessin sous-jacent a été publiée par ailleurs dans le *Burlington Magazine* (juillet 2005): Voir I.R.A.2005.18

1. Léonard de Vinci, *Vierge au rocher*, Londres, National Gallery. IRR (ens.)
2-4. Idem. IRR (détails: tête de la Vierge, paysage, Jean-Baptiste)
5-7. Léonard de Vinci, *Annonciation*, Florence, Galleria degli uffizi. IR (détails: tête de la Vierge, Vierge, ange)
8. Léonard de Vinci, *Madonna dei Fusi*, New-York, collection privée. IRR (ens.)
9-10. Idem. IRR (détails: Enfant, coin supérieur gauche)

I.R.A.2005.22

E. BUIJSEN, *Roodkrijttekeningen naar schilderijen van Adriaens van de Venne en hun mogelijke functie*, dans *Oud Holland*, 118, 2005, p. 131-202.

Étude de quinze dessins à la pierre rouge de Adriaens van de Venne (1589-1662). L'auteur pose la question de leur fonction (copies ou dessins préparatoires?).

1. Adriaen van de Venne, *Jammerlijck*, Amsterdam, Amsterdam Historisch Museum. IRR (ens.)
2-3. Idem. IRR (détails: jambes du personnage de droite, têtes)
4-5. Adriaen van de Venne, *Moines dans une taverne*, Cuijk, Kruisherenklooster St. Agatha. IRR (détails: femme, moine)
6. Adriaen van de Venne, *Goed Rond*, Banbury, Upton House. IRR (détail: couple de danseurs à droite)
7-9. Adriaen van de Venne, *Mans-verdriet*, Pays-Bas, collection privée. IRR (détail: femme devant le feu, seau, partie droite)
10-11. Idem. LO (superposition de détails du dessin à la craie rouge (Paris, Institut néerlandais) sur le tableau d'un collection privée néerlandaise: femme, instrument de musique)

I.R.A.2005.23

ExtravagAnt! A forgotten chapter of Antwerp painting 1500-1530. Catalogue (cat. expo.), Anvers, 2005.

Catalogue d'une exposition consacrée à la peinture anversoise au tout début du 16e siècle.

1. Goossen van der Weyden, *Triptyque de l'abbé Antonius TsGrooten*, Anvers, Koninklijke Museum voor Schone Kunsten. IRR (détail: partie inférieure du panneau central)
2. Anonyme, *Triptyque de la sainte Famille*, Lisbonne, Museu de Arte Antiga. IRR (détail-panneau central: anges musiciens à droite)
3. Jan de Beer, *Triptyque de l'Adoration des Bergers*, Cologne, Wallraf-Richartz-Museum. IRR (détail-panneau central: groupe central de la Nativité)
4. Jan de Beer, *Crucifixion*, Cologne, Erzbischöfliches Diözesanmuseum. IRR (détail: Vierge, saint Jean et Madeleine)
5. Maître d'Amiens, «*Pré ministrant pasture salutaire*», Amiens, Musée de Picardie. IRR (détail: têtes)
6. Maître de 1518, *Triptyque de la Crucifixion*, Bruges, Heilige Bloedmuseum. IRR (détail-*Portement de croix*: Christ et bourreaux)
7. Maître de 1518, *Crucifixion*, Londres, National Gallery. IRR (détail: personnages au pied de la croix)

8-9. Maître de l'Adoration Von Groote, *Triptyque de l'Adoration des Mages*, Allemagne, collection privée. IRR (détails: Melchior, personnages centraux)

10. Maître de l'Adoration d'Anvers, *Triptyque de l'Adoration des Mages*, Anvers, Koninklijke Museum voor Schone Kunsten. IRR (détail: Gaspard)
11. Adriaen van Overbeke et atelier, *Triptyque de la Crucifixion*, collection privée. IRR (détail-*Résurrection*: soldat)
12. Atelier du Maître de 1518, *L'ange apparaissant en rêve à Joseph*, Stuttgart, Staatsgalerie. IRR (ens.)
13. Anonyme, *Adoration des Mages*, Cologne, Wallraf-Richartz-Museum. IRR (détail: partie centrale)

I.R.A.2005.24

C. Hale, *Dating Velázquez's The Supper at Emmaus*, dans *Metropolitan Museum Journal*, 40, 2005, p. 67-78.

1. Diego Velásquez, *Repas à Emmaus*, New York, The Metropolitan Museum of Art. IRR (détail: manche de l'homme au bras tendu)

I.R.A.2005.25

I. Hopfner, *Jan Gossaert gen. Mabuse Der Hl. Lukas malt die Madonna. Komposition-Maltechnik-Restaurierung*, dans *Technologische Studien Kunsthistorisches Museum*, 2, 2005, p. 132-161.

1-7. Jan Gossaert, *Saint Luc peignant la Vierge*, Vienne, Kunsthistorisches Museum. IRR (détails: buste de saint Luc, bas du drapé de saint Luc, Enfant, angelot à gauche, architecture, visage de la Vierge, angelot à droite de la Vierge)

I.R.A.2005.26

M. Lawson, *The Belles Heures of Jean, Duc de berry. The Materials and Techniques of the Limbourg Brothers*, dans R. Dückers et P. Roelofs, *The Limbourg Brothers. Nijmegen Masters at the French Court. 1400-1416* (cat. expo.), Nijmegen, 2005, p. 148-163.

1. Frères Limbourg, *Christ cloué sur la croix*, f. 141v des *Belles Heures du Duc de Berry*, The Metropolitan Museum of Art, The Cloisters Collection. IR (détail: partie inférieure de la miniature)

2. Frères Limbourg, *Lamentation*, f. 149v des *Belles Heures du Duc de Berry*, The Metropolitan Museum of Art, The Cloisters Collection. IR (ens. de la miniature)
3. Frères Limbourg, *Le rêve de saint Jérôme*, f. 183v des *Belles Heures du Duc de Berry*, The Metropolitan Museum of Art, The Cloisters Collection. IR (ens.)

I.R.A.2005.27

D. MAHON et S. CENTENO, *A Technical Study of John Singer Sargent's Portrait of Madame Pierre Gautreau*, dans *Metropolitan Museum Journal*, 40, 2005, p. 121-129.

1-2. John Singer Sargent, *Madame X (Madame Pierre Gautreau)*, New York, The Metropolitan Museum of Art. IRR (détails: bras droit, dessus de table)

I.R.A.2005.28

Le Maître au Feuillage brodé. Primitifs flamands. Secrets d'ateliers (cat. expo.), Lille, 2005.

Catalogue d'exposition. Un colloque, dont les actes sont référencés sous le no. I.R.A.2007.10, a été organisé durant cette exposition.

1. Groupe au Feuillage brodé, *Déploration du Christ*, Amiens, Musée de Picardie. IRR (détail: drapé)
2. Groupe au Feuillage brodé (attribué au Maître de la Légende de sainte Barbe), *Saint Pierre*, revers du volet droit du *Triptyque des Miracles du Christ*, Melbourne, National Gallery of Victoria. IRR montage (détail: saint)
3. Groupe au Feuillage brodé, *Vierge à l'Enfant*, Williamstown, Sterling and Francine Clark Art Institute. IRR (détail: visage de la Vierge)
4-6. Groupe au Feuillage brodé, *Vierge et Enfant dans un paysage*, Philadelphia, Philadelphia Museum of Art. IRR (buste de la Vierge et Enfant, drapé, bâtiment dans le paysage)

I.R.A.2005.29

G. MCKIM-SMITH, I. FIELDER, R. MACBETH, R. NEWMAN et F. ZUCCARI, *Velásquez: Painting from Life*, dans *Metropolitan Museum Journal*, 40, 2005, p. 79-91.

Étude de deux tableaux de Velázquez conservés aux États-Unis.

1-2. Diego Velázquez, *Scène de Cuisine*, Chicago, Art institute. IRR (détails: pot et ustensiles à droite, visage de la servante)
3. Diego Velázquez, *Portrait de Góngora*, Boston, Museum of Fine Arts. IRR (détail: tête du personnage)

I.R.A.2005.30

C. PÉRIER-D'IETEREN, *Thierry Bouts. L'œuvre complet*, Bruxelles, 2005.

1. Dirk Bouts, *Retable du Saint Sacrement*, Louvain, Sint-Pieterskerk. IR (détail-panneau central: personnage à l'extrême droite)
2. Idem. IRR (détail-panneau central: Christ)
3. Dirk Bouts, *Justice d'Otton*, Bruxelles, Musées royaux des Beaux-Arts de Belgique. IR (détail-*Épreuve du feu*: visage de deux personnages près de l'empereur)
4-7. Idem. IRR (détail-*Épreuve du feu*: drapé de l'empereur, jambes d'un courtisan, tête du comte; *Exécution de l'innocent*: tête du comte)
8. Dirk Bouts, *Chute des Damnés*, Lille, Palais des Beaux-Arts. IRR (détail: Damnés)

9. Dirk Bouts, *Ascension des Élus*, Lille, Palais des Beaux-Arts. IRR (détail: Élus)
10. Dirk Bouts, *Justice d'Otton*, Bruxelles, Musées royaux des Beaux-Arts de Belgique. IR (détail-*Épreuve du feu*: drapé de la comtesse)
11-13. Dirk Bouts, *Retable du Saint Sacrement*, Louvain, Sint-Pieterskerk. IRR (détail-panneau central: apôtre; *Élie dans le désert*: drapé; *Pâque juive*: drapé)
14. Dirk Bouts, *Ecce Agnus Dei*, Munich, Bayerische Staatsgemäldesammlungen. IRR montage (détail: arbre central)
15. Rogier van der Weyden, *Vierge et Enfant*, Caen, Musée des Beaux-Arts. IRR montage (détail: Enfant et bras de la Vierge)
16. Dirk Bouts, *Retable du Saint Sacrement*, Louvain, Sint Pieterskerk. IR (détail-*Récolte de la manne*: drapé)
17. Dirk Bouts, *Vierge et Enfant*, Francfort, Städelsches Kunstinstitut. IRR (ens.)
18. Dirk Bouts, *Vierge et Enfant*, Paris, Musée du Louvre. IRR (ens.)
19. Dirk Bouts, *Lamentation*, Paris, Musée du Louvre. IRR (détail: saint Jean)
20. Dirk Bouts, *Vierge et Enfant*, Cambridge (Mass.), Fogg Art Museum. IRR (détail: tête de la Vierge)
21. Dirk Bouts, *Lamentation*, Paris, Musée du Louvre. IRR (détail: jambes du Christ)
22. Gerard David, *Nativité*, Budapest, Szépmüvészeti Múzeum. IRR (détail: paysage)
23. Dirk Bouts, *Repas chez Simon le Pharisien*, Berlin, Staatliche Museen. IRR (détail: drapé de saint Jean)
24. Dirk Bouts, *Chute des Damnés*, Lille, Palais des Beaux-Arts. IRR (détail: Damné à droite)
25. Dirk Bouts, *Retable de la Vierge*, Madrid, Museo del Prado. IRR (détails-*Annonciation*: drapé de la Vierge; *Adoration des Mages*: saint Joseph)
26-27. Dirk Bouts, *Moïse au buisson ardent*, Philadelphia, Philadelphia Museum of Art. IRR (détails: buisson, drapé de Moïse)
28-33. Dirk Bouts, *Triptyque du Martyre de saint Hippolyte*, Bruges, Sint-Salvatorskerk. IRR (détails-panneau central: cavalier à droite; volet gauche: ens.; panneau central: cavalier à gauche, saint Hippolyte, ens. avec schéma de construction; volet droit: personnage à genoux)

I.R.A.2005.31

M. SHELLEY, *Pastelists at Work: Two Portraits at the Metropolitan Museum by Maurice Quentin de La Tour and Jean Baptiste Perronneau*, dans *Metropolitan Museum Journal*, 40, 2005, p. 105-119.

1. Jean-Baptiste Perronneau, *Olivier Journu (L'homme aux trois roses)*, New York, The Metropolitan Museum of Art. IRR (détail: buste du personnage)
2. Maurice Quentin de La Tour, *Jean-Charles Garnier d'Îsles*, New York, The Metropolitan Museum of Art. IRR (ens.)

I.R.A.2005.32

I. VERSLYPE, *A Preliminary Study on Paulus Potter's (1625-1654) Painting Technique*, dans *Art Matters*, 3, 2005, p. 97-110.

Étude de *Orphée charmant les animaux* exécutée lors du traitement de conservation de l'œuvre. Comparaison avec six œuvres attribuées au même artiste.

1-4. Paulus Potter, *Orphée charmant les animaux*, Amsterdam, Rijksmuseum. IRR (détails: tête d'un des cerfs, bouc, vache, cerf dans le paysage)

I.R.A.2005.33

M. WITLOX, *Many Hands make Light Work: The seventeenth-century Antwerp «Interior with figures before a picture collection»*, dans *Art Matters*, 3, 2005, p. 82-96.

1-3. Gonzales Coques et autres, *Intérieur avec personnages devant une collection de tableaux*, La Haye, Mauritshuis. IRR (détails: table, enfants, zone située à gauche)

I.R.A.2006.1

M.W. AINSWORTH, *Romanism as a Catalyst for Change in Bernard van Orley's Workshop Practices*, dans *Making and Marketing: Studies of the Painting Process in Fifteenth- and Sixteenth-Century Netherlandish Workshops*, éd. M. FARIES, Turnhout, 2006, p. 99-118.

1. Bernard van Orley, *Retable des saints Thomas et Matthieu*, Vienne, Kunsthistorisches Museum. IRR montage (détail: roi Gondophares)
2-5. Bernard van Orley, *Triptyque de Job*, Bruxelles, Musées royaux des Beaux-Arts de Belgique. IRR (détails-panneau central: servante à gauche, servante à droite; volet extérieur droit: homme nu couché, lit de mort)
6. Bernard van Orley, *Soldat et homme accroupi*, Leipzig, Museum der Bildende Kunste. IRR (détail: pied de l'homme accroupi)
7. Atelier de Bernard van Orley, *Vierge et Enfant*, Bruxelles, Musées royaux des Beaux-Arts de Belgique. IRR montage (détail: tête de l'Enfant)
8-11. Bernard van Orley, *Triptyque du Jugement dernier*, Anvers, Koninklijke Museum voor Schone Kunsten. IRR (détail-panneau central: Élus à gauche, ange à droite, masse de personnages au plan moyen; volet intérieur droit: moitié inférieure)
12. Bernard van Orley, *Crucifixion*, Rotterdam, Museum Boijmans Van Beuningen. IRR (détail: saint Jean)
13. Bernard van Orley, *Vierge et Enfant*, New York, collection privée. IRR montage (ens.)
14. Bernard van Orley, *Circoncision*, Vienne, Kunsthistorisches Museum. IRR (ens.)
15-16. Bernard van Orley, *Soldat et homme accroupi*, Leipzig, Museum der Bildende Kunste. IRR (détails: tête du soldat, tête de l'homme accroupi)
17. Bernard van Orley, *Vierge et Enfant*, collection privée. IRR (ens.)

I.R.A.2006.2

M.W. AINSWORTH et M. FARIES, *Jan Gossaert's Malvagna Triptych: A Study of Two Versions*, dans *Le dessin sous-jacent et la technologie dans la peinture. Colloque XV. La peinture ancienne et ses procédés: copies, répliques, pastiches*, éd. par H. VEROUGSTRAETE et J. COUVERT, avec la collaboration de R. VAN SCHOUTE et A. DUBOIS, Louvain, 2006, p. 138-149.

Comparaison du panneau central du *Triptyque Malvagna* de Jan Gossaert avec un panneau de même attribution et du même sujet conservé dans une collection privée allemande.

1-4. Jan Gossaert (attribué à), *Vierge et Enfant accompagnés d'anges musiciens*, Allemagne, collection privée. IRR (détails: anges à gauche, anges à droite, architecture à gauche, Vierge et Enfant)
5. Jan Gossaert, *Triptyque Malvagna*, Palerme, Galleria Regionale della Sicilia. IRR (ens. du panneau central)
6. Jan Gossaert (attribué à), *Vierge et Enfant accompagnés d'anges musiciens*, Allemagne, collection privée. IRR (détail: architecture en haut au centre)
7. Jan Gossaert, *Tripyque Malvagna*, Palerme, Galleria Regionale della Sicilia. IRR (détail-volet gauche: sainte Catherine)

I.R.A.2006.3

M. Bellavitis, *À propos de deux volets et d'un petit panneau du XVIe siècle au Musée de Bassano del Grappa*, dans *Le dessin sous-jacent et la technologie dans la peinture. Colloque XV. La peinture ancienne et ses procédés: copies, répliques, pastiches*, éd. par H. Verougstraete et J. Couvert, avec la collaboration de R. Van Schoute et A. Dubois, Louvain, 2006, p. 185-188.

1. Peintre anonyme inspiré par l'école d'Anvers, *Fuite en Égypte*, Bassano del Grappa, Museo Civico. IRR (détail: Vierge)
2. Peintre anonyme inspiré par l'école d'Anvers, *Nativité*, Bassano del Grappa, Museo Civico. IRR (détail: Vierge)
3-4. Peintre anonyme inspiré par l'école d'Anvers, *Saint Christophe*, Bassano del Grappa, Museo Civico. IRR (détail: Enfant, ermite)

I.R.A.2006.4

R. Bellucci et C. Frosinini, *The Replicas after the Venus and Cupid cartoon by Michelangelo*, dans *Le dessin sous-jacent et la technologie dans la peinture. Colloque XV. La peinture ancienne et ses procédés: copies, répliques, pastiches*, éd. par H. Verougstraete et J. Couvert, avec la collaboration de R. Van Schoute et A. Dubois, Louvain, 2006, p. 51-58.

Étude de quatre versions tirées d'un carton représentant Vénus et Cupidon réalisé par Michel-Ange au moyen d'un scanner digital infrarouge à haute définition mis au point au Istituto Nazionale di Ottica Applicata de Florence.

1. Jacopo del Pontormo, *Vénus et Cupidon*, Florence, Accademia. IR scan (ens.)
2. Giorgio Vasari (attribué à), *Vénus et Cupidon*, Pise, Scuola Normale. IR scan (ens.)
3. Michele di Ridolfo del Ghirlandaio (attribué à), *Vénus et Cupidon*, Rome, Galleria Colonna. IR scan (détail: main de Vénus)
4. Heindrik van der Broeck, *Vénus et Cupidon*, Naples, Museo di Capodimonte. IR scan (détail: main de Vénus)
5. Michelangelo (attribué à), *Vénus et Cupidon*, carton, Naples, Museo di Capodimonte. IR scan (ens.)

I.R.A.2006.5

D. Bertani, *La reflectografía infrarroja*, dans *El Trazo oculto. Dibujos subyacentes en pinturas de los siglos XV y XVI* (cat. expo.), éd. G. Finaldi et C. Garrido, Madrid, 2006, p. 54-63.

Explicitation de la technique de la réflectographie à l'infrarouge dans le cadre d'une exposition consacrée à l'étude du dessin sous-jacent qui s'est tenue au Museo del Prado de Madrid, ainsi que de l'évolution de cette technique. Voir: I.R.P.2006.5.

I.R.A.2006.6

R. Billinge, *Technical Examination of a Virgin and Child by Luis de Morales in the National Gallery (NG1229)*, dans *Le dessin sous-jacent et la technologie dans la peinture. Colloque XV. La peinture ancienne et ses procédés: copies, répliques, pastiches*, éd. par H. Verougstraete et J. Couvert, avec la collaboration de R. Van Schoute et A. Dubois, Louvain, 2006, p. 1-7.

Étude technique d'une *Vierge et Enfant* de Luis de Morales conservée à la National Gallery de Londres.

1. Luis de Morales, *Vierge et Enfant*, Londres, National Gallery. IRR (ens.)

2-3. Luis de Morales, *Naissance de la Vierge*, Londres, Sotheby's, 10/07/2003, no. 41. IRR (détails: tête de femme à droite, buste de la femme avec le bol de soupe)

I.R.A.2006.7

T.-H. BORCHERT, *The Bruges Master of the Legend of Saint Ursula. Investigating Workshop-Practices of Anonymous Bruges Painters by Looking into Underdrawings*, dans *Le dessin sous-jacent et la technologie dans la peinture. Colloque XV. La peinture ancienne et ses procédés: copies, répliques, pastiches*, éd. par H. VEROUGSTRAETE et J. COUVERT, avec la collaboration de R. VAN SCHOUTE et A. DUBOIS, Louvain, 2006, p. 18-26.

Étude du dessin sous-jacent de l'œuvre éponyme du Maître de la Légende de sainte Ursule conservée au Groeningemuseum de Bruges.

1-8. Maître de la Légende de sainte Ursule, Volets avec scènes de la légende de sainte Ursule, Bruges, Groeningemuseum. IRR (détails-revers: tête de saint Luc, buste de saint Jean l'Évangéliste, drapé de saint Luc, buste de saint Grégoire; *Annonce à sainte Ursule*: bustes de la sainte et de l'ange; *Mariage de sainte Ursule*: drapé de la sainte; *Martyre de sainte Ursule*: sainte Ursule; *Arrivée à Bâle*: architecture)

I.R.A.2006.8

M. BURRELL, *Reynolds's Mona Lisa*, dans *Apollo*, 2006 (sept.), p. 64-71.

Étude d'une copie de la Joconde de Léonardo da Vinci ayant appartenu à Sir Joshua Reynolds.

1-2. Artiste français (?), *Mona Lisa*, Angleterre, collection privée. IRR montage (détails: visage, mains)

I.R.A.2006.9

E. BUZZEGOLI et D. KUNZELMAN, *Re-Use of Cartoons on Paintings by Andrea del Sarto and Pontormo: A Study with High Resolution Digital Scanned IRR*, dans *Le dessin sous-jacent et la technologie dans la peinture. Colloque XV. La peinture ancienne et ses procédés: copies, répliques, pastiches*, éd. par H. VEROUGSTRAETE et J. COUVERT, avec la collaboration de R. VAN SCHOUTE et A. DUBOIS, Louvain, 2006, p. 67-78.

Étude de quelques peintures florentines du 16e siècle présentant des sujets similaires au moyen d'un scanner digital infrarouge à haute définition mis au point au Istituto Nazionale di Ottica Applicata de Florence. Voir I.R.P.2006.3.

I.R.A.2006.10

A. CHÂTELET et N. GOETGHEBEUR, *Le Musée des Beaux-Arts de Lille* (*Corpus de la Peinture du XVe siècle dans les Pays-Bas Méridionaux et la Principauté de Liège*, 21), Bruxelles, 2006.

1-7. Groupe Dirk Bouts, Deux volets d'un triptyque: *Paradis terrestre* (inv. no. 82) et *Enfer* (inv. no. 1808). IRR (ens. *Paradis*; détails-*Paradis*: paysage, groupe d'Élus; ens. *Enfer*; détails-*Enfer*: Damnés à droite, ciel, Damnés à gauche)

8. Groupe Rogier van der Weyden, *Portrait de Philippe le Bon*, inv. no. 824. IRR (ens.)

9-10. Groupe Maître au Feuillage en broderie, *Louis Quarré et Barbe de Croesinck*, inv. no. 980. IRR (détails: visage de Barbe de Croesinck, tour de sainte Barbe)

11-13. Groupe du Maître au Feuillage en broderie, *Vierge à l'Enfant entourée d'anges*, inv. no. 840. IRR (détails-panneau central: Vierge; volet gauche: viole; volet droit: harpe et mains de l'ange)

14-15. Maître de 1499, Fragments d'un panneau de retable: *Épisodes de la vie de saint Gilles* (inv. no. 823) et *Saint Benoît* (inv. no. 846). IRR (détails-*Épisodes de la vie de saint Gilles*: partie gauche; *Saint Benoît*: visage)
16. Maître anonyme, *Crucifixion*, inv. no. 746. IRR (détail: Vierge)
17-18. Maître anonyme, *Deux donateurs présentés par saint Quirin et saint Amand*, inv. no. 860. IRR (détails: saint Quirin, saint Amand)
19. Maître anonyme, *Portrait de Philippe le Hardi*, inv. no. 1514. IRR (détail: visage)
20. Maître anonyme, *Portrait d'Isabelle de Portugal*, inv. no. 1526. IRR (détail: inscription)

I.R.A.2006.11

I. CIULISOVÁ, *The Use of Workshop Patterns in The Seller of Fruits and Vegetables of the Slovak National Gallery in Bratislava (inv. 0249)*, dans *Le dessin sous-jacent et la technologie dans la peinture. Colloque XV. La peinture ancienne et ses procédés: copies, répliques, pastiches*, éd. par H. VEROUGSTRAETE et J. COUVERT, avec la collaboration de R. VAN SCHOUTE et A. DUBOIS, Louvain, 2006, p. 168-172.

Étude de certains procédés techniques basée sur le composition attribuée à Joachim Beuckelaer représentant une vendeuse de fruits et légumes (Bratislava, Galerie Nationale slovaque).

1-3. Joachim Beuckelaer (atelier ou suiveur de), *La vendeuse de fruits et de légumes*, Bratislava, Galerie Nationale slovaque. IR (détails: tête de la vendeuse, panier tenu par la vendeuse, tête de l'homme)

I.R.A.2006.12

C. CURRIE et B. GHYS, *Design Transfer in Pieter Brueghel the Younger's Workshop: a Step-by-step Reconstruction Based on Technical Examination of his Paintings*, dans *Le dessin sous-jacent et la technologie dans la peinture. Colloque XV. La peinture ancienne et ses procédés: copies, répliques, pastiches*, éd. par H. VEROUGSTRAETE et J. COUVERT, avec la collaboration de R. VAN SCHOUTE et A. DUBOIS, Louvain, 2006, p. 196-206.

Essais de reconstruction pratique du support et des couches préparatoires, y compris du dessin sous-jacent, dans l'œuvre de Pieter Brueghel le Jeune.

1. Pieter Brueghel le Jeune, *Combat entre Carnaval et Carême*, Bruxelles, Musées royaux des Beaux-Arts de Belgique. IRR (détail)
2. Photographie montrant l'application du poncif sur un panneau (essai de reconstruction)
3-4. Reconstruction de la technique de Pieter Brueghel le Jeune sur un panneau-test: poncif, dessin sous-jacent à la pierre noire par dessus le poncif et enfin époussetage de la poudre du poncif à l'aide d'une plume de pigeon

I.R.A.2006.13

A. DUBOIS et P. SYFER-D'OLNE, *La copie du Polyptyque de l'Agneau Mystique par Michel Coxcie: Étude des panneaux conservés aux Musées royaux des Beaux-Arts de Belgique*, dans *Le dessin sous-jacent et la technologie dans la peinture. Colloque XV. La peinture ancienne et ses procédés: copies, répliques, pastiches*, éd. par H. VEROUGSTRAETE et J. COUVERT, avec la collaboration de R. VAN SCHOUTE et A. DUBOIS, Louvain, 2006, p. 234-241.

1. Michel Coxcie, copie du *Polyptyque de l'Agneau Mystique*, *Les anges chanteurs*, Bruxelles, Musées royaux des Beaux-Arts de Belgique. IRR montage (détail: visage du deuxième ange depuis la droite)

2. Michel Coxcie, copie du *Polyptyque de l'Agneau Mystique*, *Les Juges intègres*, Bruxelles, Musées royaux des Beaux-Arts de Belgique. IRR montage (détail: visage d'un cavalier)

I.R.A.2006.14

J. Dunkerton, C. Christensen et L. Syson, *The Master of the Story of Griselda and Paintings for Sienese Palace*, dans *National Gallery Technical Bulletin*, 27, 2006, p. 4-71.

Étude du Maître anonyme nommé d'après trois panneaux représentant l'Histoire de Griselda conservés à la National Gallery de Londres (NG 912-914). Analyse scientifique des panneaux de Londres et vue de l'œuvre de l'artiste.

1-2. Maître de l'Histoire de Griselda, *Histoire de Griselda*. I. *Mariage*, Londres, National Gallery. IRR (détails: chien pourchassant un cerf dans le paysage, tête du marié)
3. Maître de l'Histoire de Griselda, *Histoire de Griselda*. II. *Exil*, Londres, National Gallery. IRR (détail: tiers inférieur)
4. Maître de l'Histoire de Griselda, *Histoire de Griselda*. III. *Réunion*, Londres, National Gallery. IRR (détail: personnages féminins à gauche)
5. Maître de l'Histoire de Griselda, *Histoire de Griselda*. I. *Mariage*, Londres, National Gallery. IRR (détail: personnage masculin de dos)
6. Maître de l'Histoire de Griselda, *Histoire de Griselda*. II. *Exil*, Londres, National Gallery. IRR (détail: Griselda)
7. Maître de l'Histoire de Griselda, *Alexandre le Grand*, Birmingham, The Barber Institute of Fine Arts, the University of Birmingham. IRR (détail: inscription au revers)
8. Francesco di Giorgio et Maître de l'Histoire de Griselda, *Scipion l'Africain*, Florence, Museo Nazionale del Bargello. IRR (ens.)
9-10 Neroccio di Landi et Maître de l'Histoire de Griselda, *Claudia Quinta*, Washington, National Gallery of Art. IRR (détails: buste, personnage féminin dans le paysage à droite)
11. Maître de l'Histoire de Griselda, *Histoire de Griselda*. III. *Réunion*, Londres, National Gallery. IRR (détail: couple au plan moyen)
12. Neroccio di Landi et Maître de l'Histoire de Griselda, *Claudia Quinta*, Washington, National Gallery of Art. IRR (détail: bras gauche)
13. Maître de l'Histoire de Griselda, *Joseph d'Égypte, ou Eunostos de Tanagra*, Washington, National Gallery of Art. IRR (détail: personnages dans le paysage à gauche)
14-15. Maître de l'Histoire de Griselda, *Alexandre le Grand*, Birmingham, The Barber Institute of Fine Arts, the University of Birmingham. IRR (détails: armure, tente à droite dans le paysage)
16-19. Maître de l'Histoire de Griselda, *Tiberius Gracchus*, Budapest, Szépmüvészeti Múzeum. IRR (détails: main gauche, cheveux, personnages à droite dans le paysage, toit du bâtiment avec statues à gauche)
20-22. Maître de l'Histoire de Griselda, *Artemisia*, Milan, Museo Poldi Pezzoli. IRR (détails: tête, personnages à gauche, personnages à droite)
23. Pietro Orioli, *Sulpitia*, Baltimore, The Walters Art Museum. IRR (ens.)
24. Idem. IRR (détail: paysage à droite)
25. Matteo di Giovanni, *Judith et la tête d'Holophernes ou Tomyris de Scytie*, Bloomington, Indiana University Art Museum, Samuel H. Kress Foundation Collection. IRR (ens.)
26. Idem. IRR (détail: paysage à gauche)

I.R.A.2006.15

F.J. DUPARC, *Results on the Recent Art-Historical and Technical Research on Carel Fabritius's Early Work*, dans *Oud Holland*, 119, 2006, p. 76-89.

1. Carel Fabritius, *Mercure et Aglauros*, Boston, Museum of Fine Arts. IR (détail: signature)
2-3. Carel Fabritius, *Hagar et l'ange*, Vienne, Collection Schönborn-Buchheim (prêt à long terme à la Residenzgalerie de Salzburg). IRR (détails: partie inférieure droite, signature)
4. Carel Fabritius, *Hera*, Moscou, Musée Poushkine. IRR (détail: partie supérieure gauche)

I.R.A.2006.16

M. FARIES, *Jan van Scorel's Mary Magdalene: Original and Copy*, dans *Le dessin sous-jacent et la technologie dans la peinture. Colloque XV. La peinture ancienne et ses procédés: copies, répliques, pastiches*, éd. par H. VEROUGSTRAETE et J. COUVERT, avec la collaboration de R. VAN SCHOUTE et A. DUBOIS, Louvain, 2006, p. 150-158.

Comparaison d'une copie de la *Sainte Madeleine* de Jan van Scorel (Palerme, Galleria Regionale della Sicilia) avec l'original (Amsterdam, Rijksmuseum)

1. Jan van Scorel, *Marie-Madeleine*, Amsterdam Rijksmuseum. IRR (détail: visage de la sainte)
2. Jan van Scorel (copie d'après), *Marie-Madeleine*, Palerme, Galleria Regionale della Sicilia. IRR (même détail que le précédent)
3. Jan van Scorel, *Marie-Madeleine*, Amsterdam Rijksmuseum. IRR (détail: pot à onguent)
4. Jan van Scorel (copie d'après), *Marie-Madeleine*, Palerme, Galleria Regionale della Sicilia. IRR (même détail que le précédent)
5. Jan van Scorel, *Marie-Madeleine*, Amsterdam Rijksmuseum. IRR (détail: paysage)
6. Jan van Scorel (copie d'après), *Marie-Madeleine*, Palerme, Galleria Regionale della Sicilia. IRR (même détail que le précédent)

I.R.A.2006.17

M. FARIES, *Making and Marketing: Studies of the Painting Process*, dans *Making and Marketing: Studies of the Painting Process in Fifteenth- and Sixteenth-Century Netherlandish Workshops*, éd. M. FARIES, Turnhout, 2006, p. 1-14.

1. Attribué à Jan Joest, *Les Apôtres Jean l'Évangéliste, Thomas et Jacques le Mineur ou Judas*, Ann Arbor, University of Michigan Museum of Art. IRR (ens.)
2. D'après Hieronymus Bosch, *Tentation de saint Antoine*, Indianapolis, Indianapolis Museum of Art. IRR (ens.)
3. Idem. IRR (détail)

I.R.A.2006.18

M.C. GALASSI, *Jan Massys and Artistic Relationships Between Antwerp and Genoa During the Sixteenth Century*, dans *Making and Marketing: Studies of the Painting Process in Fifteenth- and Sixteenth-Century Netherlandish Workshops*, éd. M. FARIES, Turnhout, 2006, p. 179-200.

1-2. Jan Massys, *Vierge et Enfant*, Genoa, Galleria di Palazzo Bianco. IRR (détail: visage de la Vierge, visage de l'Enfant)
3-4. Jan Massys, *Charité*, Genoa, Galleria di Palazzo Bianco. IRR (détail: poitrine de la Charité, enfant endormi)

5. Jan Massys, *Judith*, Anvers, Koninklijke Museum voor Schone Kunsten. IRR (détail: poitrine de Judith)
6. Anonyme (flamand?), *Portrait d'Andrea Doria*, Genoa, Galleria di Palazzo Bianco. IRR (détail: main droite)
7-9. Jan Massys, *Venus Cythereia*, Stockholm, Nationalmuseum. IR (détail: pieds, vase, épaule gauche)

I.R.A.2006.19

M.C. GALASSI, *On the Legacy of Lorenzo di Credi (1458/59-1537). Replicas and Copies (and one pastiche)*, dans *Le dessin sous-jacent et la technologie dans la peinture. Colloque XV. La peinture ancienne et ses procédés: copies, répliques, pastiches*, éd. par H. VEROUGSTRAETE et J. COUVERT, avec la collaboration de R. VAN SCHOUTE et A. DUBOIS, Louvain, 2006, p. 59-66.

Analyse techniques de diverses versions de peintures de dévotions produites par Lorenzo di Credi, son atelier et ses suiveurs.

1-2. Lorenzo di Credi, *Vierge assise tenant l'Enfant nu*, Turin, Galleria Sabauda (inv. 115). IRR (détails: bras de l'Enfant, pieds de l'Enfant)
3. Lorenzo di Credi, *Vierge assise tenant l'Enfant nu*, Turin, Galleria Sabauda (inv. 116). IRR (détail: tête de l'Enfant)
4. Lorenzo di Credi, *Vierge assise tenant l'Enfant nu*, Oxford, Ashmolean Museum. IRR (détail: main de l'Enfant)
5. Atelier de Lorenzo di Credi (Maître de la Conversazione di Santo Spirito?), *Vierge assise tenant l'Enfant nu*, Coldirodi, Museo Rambaldi. IRR (détail: tête de l'Enfant)
6. Angelo Tricca, copie de Lorenzo du Credi, *Vierge assise tenant l'Enfant nu*, Sansepolcro, Palazzo delle Laudi. IRR montage (détail: ventre de l'Enfant)

I.R.A.2006.20

A. GALLONE et O. PICCOLO, *La Predica di San Marco ad Alessandria d'Egitto di Gentile e Giovanni Bellini: nuovi contributi dalle indagini tecniche*, dans *Arte Lombarda*, 2006, p. 73-93.

Étude technique d'un tableau représentant la Prédication de saint Marc en Alexandrie d'Égypte conservé à Milan.

1-3. Gentile et Giovanni Bellini, *Prédication de saint Marc en Alexandrie d'Égypte*, Milan, Pinacoteca Brera. IR (détails: personnages centraux, Dante et un autre poète, enfant)
4. Idem. IRR (détail: second groupe à droite)
5. Idem. IR (détail: premier groupe à droite)
6. Idem. IRR (détail: scribe derrière saint Marc)

I.R.A.2006.21

C. GARRIDO, *El trazo oculto. Dibujos subyacentes en las tablas del Museo del Prado*, dans *El Trazo oculto. Dibujos subyacentes en pinturas de los siglos XV y XVI* (cat. expo.), éd. G. FINALDI et C. GARRIDO, Madrid, 2006, p. 16-53.

Dans le cadre d'une exposition consacrée à l'étude du dessin sous-jacent qui s'est tenue au Museo del Prado de Madrid, l'auteur s'attache à décrire les divers procédés de création ou de reproduction du dessin sous-jacent et les diverses questions posées par cette étude.

1. Maître de Miraflores, *Décollation de saint Jean-Baptiste*, Madrid, Museo del Prado. IRR (détail: tête de salomé)

2. Michel-Ange, *Vierge et Enfant avec saint Jean-Baptiste et des anges*, Londres, National Gallery. LO (peinture inachevée laissant apparaître le dessin sous-jacent)
3. Albrecht Dürer, *Salvator Mundi*, New York, The Metropolitan Museum of Art. LO (ensemble du panneau où le dessin sous-jacent est bien visible dans le visage du Christ)
4. Luis de Morales, *Prière dans le Jardin*, Arroyo de la Luz (Cáceres), Iglesia parroquial de la Asunción. IR (détail)
5. Hieronymus Bosch, *Triptyque de l'Adoration des Mages*, Madrid, Museo del Prado. IRR (détail-volet droit: drapé)
6. Maître de Miraflores, *Naissance de saint Jean-Baptiste*, Madrid, Museo del Prado. IRR (détail: vitrail dans le fond de la scène)
7. Albrecht Dürer, *Adam*, Madrid, Museo del Prado. IRR (détail: main gauche)
8. Joachim Beuckelaer, *Marché*, Madrid, Museo del Prado. IRR (détail: têtes des figures principales)
9. Hans Memling, *Triptyque de l'Adoration des Mages*, Madrid, Museo del Prado. IRR (détail-*Présentation au Temple*: architecture)
10. Rafaël, *Sainte Famille dite «La Perla»*, Madrid, Museo del Prado. IRR (détail: quadrillage pour la transposition de la composition)
11. Alejo Fernández, *Flagellation*, Madrid, Museo del Prado. IRR (détail: utilisation du poncif)
12. Fernando Yáñde la Almedina, *Sainte Anne, la Vierge et l'Enfant avec sainte Isabelle et saint Jean*, Madrid, Museo del Prado. IRR (détail: dessin transposé au moyen d'un calque)
13. Rafaël, *Sainte Famille*, Madrid, Museo del Prado. IRR (ens.)
14. Rogier van der Weyden, *Descente de croix*, Madrid, Museo del Prado. IRR montage (détail: tête de Joseph d'Arimathie)
15. Anonyme flamand, copie de la *Descente de croix* de Rogier van der Weyden, Madrid, Museo del Prado. IRR (détail: tête de Joseph d'Arimathie)
16. Giovanni Gerolamo Savoldo, *Transfiguration*, Madrid, Museo del Prado. IRR (détail: corps du Christ)
17. Suiveur de Pedro Berruguete, *Vierge et Enfant*, Madrid, Museo del Prado. IRR (détail: Enfant)
18. Andrea del Sarto, *Sacrifice d'Isaac*, Madrid, Museo del Prado. IRR (détail: Isaac)
19. Hieronymus Bosch, *Triptyque de l'Adoration des Mages*, Madrid, Museo del Prado. IRR (détail-panneau central: bergers)
20. Maître de Sopetrán, *Mort de la Vierge*, Madrid, Museo del Prado. IRR (détail: ange dessiné mais non peint)
21. Le Titien, *Vierge et Enfant avec saints Jean et Dorothée*, Madrid, Museo del Prado. IRR (détail: tête de l'Enfant)
22. Diego Velázquez, *La forge de Vulcain*, Madrid, Museo del Prado. IRR (détail: Vulcain)
23. Maître de Miraflores, *Décollation de saint Jean-Baptiste*, Madrid, Museo del Prado. IRR (détail: bourreau)
24. Maître de Don Álvaro de Luna, *La septième Douleur de la Vierge*, Madrid, Museo del Prado. IRR montage (détail: Christ)
25. Anonyme espagnol, *Descente de croix*, Madrid, Museo del Prado. IRR (détail: tête du Christ)
26. Maître de Sopetrán, *Mort de la Vierge*, Madrid, Museo del Prado. IRR (détail: Vierge et apôtres)
27-28. Juan de Flandes, *Descente du Saint Esprit*, Madrid, Patrimonio Nacional, Palacio Real. IRR (détail: saint Jean l'Évangéliste, tête de la Vierge)

29. Juan de Flandes, *Retablo Mayor*, Palencia, Cathédrale. IRR (détail-*Mise au tombeau*: tête de la Vierge)
30. Bartolomé Bermejo, *Saint Dominique de Silos intronisé comme abbé*, Madrid, Museo del Prado. IRR (détail: vêtement)
31. Pedro Berruguete, *Retablo Mayor*, Ávila, cathédrale. IRR (détail-*Saint Jean l'Évangéliste*: tête)
32. Santa Cruz, *Retablo Mayor*, Ávila, cathédrale. IRR (détail-*Adoration des Mages*: tête de Gaspard)
33. Juan de Borgoña, *Retablo Mayor*, Ávila, cathédrale. IRR (détail: *Présentation au Temple*: tête de sainte Anne)
34. Entourage de Fernando Gallego, *Martyre de sainte Catherine*, Madrid, Museo del Prado. IRR (détail: jambes du soldat couché à gauche de la composition)
35. Paolo de San Leocadio (?), *Mater Dolorosa*, Madrid, Museo del Prado. IRR montage (détail: visage de la Vierge)
36-37. Rogier van der Weyden, *Descente de croix*, Madrid, Museo del Prado. IRR montage (détails: tête de l'homme à droite, tête de saint Jean)
38. Albrecht Dürer, *Adam*, Madrid, Museo del Prado. IRR (détail: tête d'Adam)
39. Idem. IRR montage (même détail que le précédent)
40. Pedro Berruguete, *Prédelle du Retablo Mayor*, Palencia, Paredes de Nava, Iglésia Santa Eulalia. IRR (détail: tête du roi Osias)

I.R.A.2006.22

J.O. HAND, C.A. METZGER et R. SPRONK, *Prayers and Portraits. Unfolding the Netherlandish Diptych* (cat. expo.), New Haven-Londres, 2006.

1-2. Albrecht Bouts, *Ecce Homo* et *Mater Dolorosa*, Aachen, Suermondt-Ludwig-Museum. IRR (ens. des deux panneaux)
3-4. Albrecht Bouts, *Ecce Homo* et *Mater Dolorosa*, Harvard University Art Museum, Fogg Art Museum. IRR (ens. des deux panneaux)
5. Robert Campin, *Christ bénissant* et *Vierge en prière*, Philadelphia, Philadelphia Museum of Art. IR (détails: doigt du Christ)
6. Idem. IRR (détail: visage de la Vierge)
7. Jan van Eyck, *Diptyque de l'Annonciation*, Madrid, Museo Thissen-Bornemisza. IR (détail: visage de Gabriel)
8. Idem. IRR (détail: buste et main de Gabriel)
9. Idem. IR (détail: base sous Gabriel)
10. Geertgen tot Sint Jans, *Glorification de la Vierge*, Rotterdam, Museum Boymans van Beuningen. IRR (détail: drapé de la Vierge)
11. Quentin Massys, *Vierge en prière* et *Christ Sauveur*, Madrid, Museo del Prado. IR (détail-*Christ Sauveur*: œil droit du Christ)
12. Quentin Massys, *Vierge en prière* et *Christ Sauveur*, Anvers Koninklijke Museum van Schone Kunsten. IR (détail-*Vierge en prière*: cou de la Vierge)
13. Idem. IRR (ens. de *Vierge en prière*)
14-15. Idem. IR (ensemble du *Christ Sauveur*, détail: cheveux du Christ)
16. Quentin Massys, *Sainte Marie-Madeleine* et *Sainte Marie l'Égyptienne*, Philadelphia, Philadelphia Museum of Art. IR (détail-*Sainte Marie-Madeleine*: bras et cheveux de la sainte)
17. Maître des Portraits Benson, *Portraits d'un homme et d'une femme*, Anvers, Museum Mayer van den Bergh. IR (détail: visage de l'homme)

18-19. Maître de l'Adoration de Lille, *Trinité* et *Saint Jérôme*, collection privée (dépôt à long terme au Harvard University Art Museums, Fogg Art Museum). IRR (ensemble des deux panneaux)

20-22. Maître de la Légende de sainte Madeleine et artiste français anonyme, *Vierge et Enfant* et *Portrait de Willem van Bibaut*, collection privée. IRR (détails: Vierge et Enfant, donateur, revers du portrait du donateur)

23. Maître de la Légende de sainte Ursule, *Vierge et Enfant* et *Trois donateurs*, Anvers, Koninklijke Museum voos Schone Kunsten. IRR (détail: Enfant)

24. Idem. IR (détail: yeux de la Vierge)

25. Idem. IRR (détail: donateurs)

26-27. Hans Memling, *Saint Jean-Baptiste* et *Sainte Véronique*, Washington, National Gallery of Art. IRR (détails: calice au revers de Sainte Véronique, drapé de sainte Véronique)

28-29. Hans Memling, *Diptyque de Martin van Nieuwenhove*, Bruges, Hospitaalmuseum Sint-Janshospitaal. IRR (ens. du volet du donateur; détail: vitrail derrière la Vierge)

30. Idem. IR (détail: coin inférieur droit du volet du donateur)

31-32. Jan Mostaert, *Christ apparaissant à sa mère aux Limbes*, Enschede, Collection Rijksmuseum Twenthe; et *Donatrice avec les Élus du Vieux Testament*, Madrid, Museo Thissen-Bornemisza. IRR (ens. des deux panneaux)

33. Idem. IR (détail: donatrice)

34-35. Anonyme, *Vierge et Enfant et deux donateurs*, Anvers, Koninklijke Museum voor Schone Kunsten. IRR (ens. des deux panneaux)

36-38. Jan Provoost, *Christ portant la croix* et *Portrait d'un donateur franciscain*, Bruges, Hospitaalmuseum Sint-Janshospitaal. IRR (ens. des deux panneaux, ens. du revers du panneau du donateur)

39. Bernard de Ryckere, *Adriaan de Santvoort et ses fils Guillaume et Adrien* et *Anna van Hertsbeeke et sa fille Catharina et son fils Baptiste*, collection privée. IR (détail: doigt du fils aîné)

40. Idem. IRR (détail du revers du panneau avec la donatrice: écu à droite)

41. Michel Sittow, *Vierge et Enfant*, Berlin, Staatliche Museen. IRR (détail: Enfant)

42. Jan Cornelisz Vermeyen, *Cardinal Erard de la Marck*, Maastricht, Bonnefantenmuseum (dépôt du Rijksmuseum d'Amsterdam). IRR (détail: visage du cardinal)

43-44. Rogier van der Weyden, *Saint Georges et le dragon*, Washington, National Gallery of Art et *Vierge et Enfant*, Madrid, Museo Thyssen-Bornemisza. IRR (ensemble des deux panneaux)

45. Rogier van der Weyden, *Portrait de Philippe de Croÿ*, Anvers, Koninklijke Museum voor Schone Kunsten. IRR (ens.)

46. Rogier van der Weyden, *Vierge et Enfant*, Caen, Musée des Beaux-Arts. IRR montage (ens.)

47. Suiveur de Rogier van der Weyden, *Vierge et Enfant* et *Joos van der Burch et Saint Simon de Jérusalem*, Harvard University Art Museums, Fogg Art Museum. IR (détail: visage de la Vierge)

48-49. Idem. IRR (détail: visage de saint Simon de Jérusalem, tête du donateur)

I.R.A.2006.23

V. Henderiks, *Le Triptyque de l'Assomption de la Vierge d'Albrecht Bouts: analyse critique*, dans *Annales d'histoire de l'art et d'archéologie*, 28, 2006, p. 1-24.

1-3. Albrecht Bouts, *Triptyque de l'Assomption de la Vierge*, Bruxelles, Musées royaux des Beaux-Arts de Belgique. IRR montages (détails-panneau central: Vierge, Apôtres, volet gauche: donateur)

I.R.A.2006.24

C. HIGGIT, M. SPRING, A. REEVE et L. SYSON, *Working with Perugino: The Technique of an Annunciation attributed to Giannicola di Paolo*, dans *National Gallery Technical Bulletin*, 27, 2006, p. 96-110.

1. Giannicola di Paolo, *Annonciation*, Londres, National Gallery. IRR (détail: poncif utilisé pour placer le lutrin)

I.R.A.2006.25

H. HUDSON, *The Materials and Technique of Two Panel Paintings: The Oxford Annunciation and the Melbourne Saint George*, dans *Le dessin sous-jacent et la technologie dans la peinture. Colloque XV. La peinture ancienne et ses procédés: copies, répliques, pastiches*, éd. par H. VEROUGSTRAETE et J. COUVERT, avec la collaboration de R. VAN SCHOUTE et A. DUBOIS, Louvain, 2006, p. 8-17.

Résultats de l'analyse technique de deux panneaux attribués à Paolo Uccello, à savoir une *Annonciation* (Oxford, Ashmolean Museum) et un *Saint Georges terrassant le dragon* (Melbourne, National Gallery of Victoria).

1-4. Paolo Uccello (attribué à), *Annonciation*, Oxford, Ashmolean Museum, IRR (détails: Vierge, Gabriel dans la partie inférieure, architecture, chapiteau)

I.R.A.2006.26

L. JANSEN, *Serial Products in the Workshop of Pieter Coecke van Aelst: A Working Hypothesis*, dans *Le dessin sous-jacent et la technologie dans la peinture. Colloque XV. La peinture ancienne et ses procédés: copies, répliques, pastiches*, éd. par H. VEROUGSTRAETE et J. COUVERT, avec la collaboration de R. VAN SCHOUTE et A. DUBOIS, Louvain, 2006, p. 173-180.

Étude de la production des compositions représentant l'Adoration des Mages ou des saintes Familles, qui sont l'essentiel de la production de Pieter Coecke van Aelst et de son atelier.

1. Groupe de Pieter Coecke van Aelst, *Crucifixion*, Varsovie, Museum Narodowe. IRR (détail: manteau de la Madeleine)
2. Groupe de Pieter Coecke van Aelst, *Adoration des Mages*, Palerme, Galleria Nazionale della Sicilia. IRR (détail: Vierge et Enfant)
3. Groupe de Pieter Coecke van Aelst, *Adoration des Mages*, Cologne, Walraff-Richartz-Museum. IRR (détail: drapé de la Vierge)
4. Groupe de Pieter Coecke van Aelst, *Sainte famille*, Essen, Banque Gallinat. IRR (détail: Enfant)

I.R.A.2006.27

L. JANSEN, *Shop Collaboration in the Painting of Background Landscapes in the Workshop of Pieter Coecke van Aelst*, dans *Making and Marketing: Studies of the Painting Process in Fifteenth- and Sixteenth-Century Netherlandish Workshops*, éd. M. FARIES, Turnhout, 2006, p. 119-142.

1. Pieter Coecke van Aelst, *Triptyque de la Crucifixion*, Utrecht, Rijksmuseum Catharijneconvent. IRR (détail-panneau central: Marie-Madeleine)
2. Groupe Pieter Coecke van Aelst, *Crucifixion*, Varsovie, Muzeum Narodowe. IRR (détail: drapé de Marie-Madeleine)
3. Pieter Coecke van Aelst, *Saint Jacques et donateurs masculins*, Madrid, Museo del Prado. IRR (détail: saint Jacques)

4. Pieter Coecke van Aelst, *Triptyque de l'Adoration des Mages*, Oldenzaal, Plechelmus-basiliek. IRR (détail-panneau central: Vierge et Enfant)
5. Pieter Coecke van Aelst, *Triptyque de la Crucifixion*, Bonn, Rheinlandisches Museum. IRR (détail-panneau central: perizonium du Christ)
6. Groupe Pieter Coecke van Aelst, *Triptyque de la Crucifixion*, Leeuwarden, Fries Museum. IRR (détail-panneau central: ange à gauche)
7. Pieter Coecke van Aelst, *Triptyque de la Crucifixion*, Utrecht, Rijksmuseum Catharijne-convent. IRR (détail-panneau central: saint Jean)
8. Pieter Coecke van Aelst, *Christ devant Pilate*, Berlin, Jagdschloss Grünewald. IRR (détail: Christ et soldat)
9. Pieter Coecke van Aelst, *Saint Libertus*, Cape Town, Michaëlis Collection. IRR (détail: paysage à gauche)

I.R.A.2006.28

E. KORTHALS ALTES et W. DE RIDDER, *De ondertekening van Adam Willaert's Schepen voor een rotsachtige kust*, dans *Bulletin van het Rijksmuseum*, 54, 2006, p. 382-399.

1-4. Adam Willaerts, *Vaisseaux devant une côte rocheuse*, Amsterdam, Rijksmuseum (dépôt à La Haye, Mauritshuis). IRR (détails: coin inférieur droit, ville du paysage, canots à gauche, enfant)

5-6. Adam Willaerts, *Bataille entre des vaisseaux espagnols et hollandais*, Bruxelles, Musées royaux des Beaux-Arts de Belgique. IRR (détails: personnages centraux, personnage)

7. Adam Willaerts, *Vaisseaux devant une côte rocheuse*, Institut Collectie Nederland (dépôt à Amsterdam, Het Nederlands Scheepvaartmuseum). IRR (détail: poissons)
8. Adam Willaerts, *Vaisseaux devant la côte*, Amsterdam, Rijksmuseum (dépôt à La Haye, Mauritshuis). IRR (détail: femme assise)

I.R.A.2006.29

O. KOTKOVÁ, *A New Investigation of the Venus and Cupid in Vulcanis Forge by Maerten van Heemskerck (National Gallery, Prague)*, dans *Le dessin sous-jacent et la technologie dans la peinture. Colloque XV. La peinture ancienne et ses procédés: copies, répliques, pastiches*, éd. par H. VEROUGSTRAETE et J. COUVERT, avec la collaboration de R. VAN SCHOUTE et A. DUBOIS, Louvain, 2006, p. 189-195.

Étude technologique d'une *Vénus dans la forge de Vulcain* signée par Maerten van Heemskerck et datée de 1536 qui amène l'auteur à déterminer que, contrairement à l'opinion reçue, cette œuvre n'a pas été réalisée lors du séjour du peintre en Italie.

1-2. Maerten van Heemskerck, *Vénus et Cupidon dans la forge de Vulcain*, Prague, National Gallery. IRR (détails: surpeint dans le perizonium du Cyclope, tête du Cyclope)

I.R.A.2006.30

P. LE CHANU, *Tout ce qu'il y a de beau à Rome*, dans *Le dessin sous-jacent et la technologie dans la peinture. Colloque XV. La peinture ancienne et ses procédés: copies, répliques, pastiches*, éd. par H. VEROUGSTRAETE et J. COUVERT, avec la collaboration de R. VAN SCHOUTE et A. DUBOIS, Louvain, 2006, p. 106-114.

Importance de la pratique de la copie en France aux seizième et dix-septième siècles.

1. Martin Noblet, *Cérès*, Paris, Musée du Louvre. IRR (détail: main droite de Cérès)

I.R.A.2006.31

M. LEEFLANG, *The Adoration of the Magi in Naples and Detroit: original and copy?*, dans *Le dessin sous-jacent et la technologie dans la peinture. Colloque XV. La peinture ancienne et ses procédés: copies, répliques, pastiches*, éd. par H. VEROUGSTRAETE et J. COUVERT, avec la collaboration de R. VAN SCHOUTE et A. DUBOIS, Louvain, 2006, p. 131-137.

Comparaion du dessin sous-jacent de deux triptyques représentant l'Adoration des Mages attribués à Joos van Cleve, conservés respectivement au Museo di Capodimonte de Naples et à l'Institute of Arts de Detroit.

1. Joos van Cleve, *Triptyque de l'Adoration des Mages*, Naples, Museo di Capodimonte. IRR (détail: Joseph et Enfant)
2. Joos van Cleve, *Triptyque de l'Adoration des Mages*, Detroit, Detroit Institute of Arts. IRR (même détail que le précédent)
3. Joos van Cleve, *Triptyque de l'Adoration des Mages*, Naples, Museo di Capodimonte. IRR (détail: tête de la Vierge)
4. Joos van Cleve, *Triptyque de l'Adoration des Mages*, Detroit, Detroit Institute of Arts. IRR (détail: buste de la Vierge)
5. Joos van Cleve, *Triptyque de l'Adoration des Mages*, Naples, Museo di Capodimonte. IRR (détail: personnages au second plan)
6. Joos van Cleve, *Triptyque de l'Adoration des Mages*, Detroit, Detroit Institute of Arts. IRR (même détail que le précédent)

I.R.A.2006.32

M. LEEFLANG, *The Saint Reinhold Altarpiece by Joos van Cleve and his Workshop: New insights into the Influence of Albrecht Dürer on the Working Process*, dans *Making and Marketing: Studies of the Painting Process in Fifteenth- and Sixteenth-Century Netherlandish Workshops*, éd. M. FARIES, Turnhout, 2006, p. 15-42.

Étude du dessin sous-jacent d'un retable réalisé pour la Confrérie de Saint Reinhold par Joos van Cleve et de ses relations avec la gravure, et plus particulièrement de l'influence de celles d'Albrecht Dürer.

1-6. Joos van Cleve, *Retable de saint Reinhold*, Varsovie, Muzeum Narodowe. IRR (détails-volet extérieur droit: autoportrait du peintre (?); *Christ devant Pilate*: personnage; *Dernière Cène*: inscription; volet extérieur gauche: tête de saint Jean-Baptiste; *Christ au Jardin de Gethsemane*: saint Pierre; *Baptême du christ*: Christ)
7. Joos van Cleve, *Adoration des Mages*, Dresden, Staatliche Kunstsammlungen. IRR (détail: personnages à l'arrière-plan)
8-9. Joos van Cleve, *Retable de saint Reinhold*, Varsovie, Muzeum Narodowe. IRR (détails-*Présentation au temple*: grand prêtre et personnages derrière lui; *Christ devant Pilate*: groupe de Pilate)
10. Adriaen van Overbeke et atelier, *Retable de sainte Anne*, Kempen, Propsteikirche. IRR (détail-*Mort de sainte Anne*: groupe central)
11. Maître de 1518, *Joachim et Anne donnant leurs biens aux démunis*, Lübeck, Marienkirche. IRR (détail: démunis)
12. Maître de l'Adoration d'Anvers, *Présentation au temple*, Maastricht, Bonnefantenmuseum. IRR (détail: grand-prêtre)
13. Joos van Cleve, *Retable de saint Reinhold*, Varsovie, Muzeum Narodowe. IRR (détail-*Portement de croix*: groupe du Christ)

I.R.A.2006.33

C. Limentani Virdis, *Les bizarreries d'une Crucifixion conservée au Musée de Bassano del Grappa*, dans *Le dessin sous-jacent et la technologie dans la peinture. Colloque XV. La peinture ancienne et ses procédés: copies, répliques, pastiches*, éd. par H. Verougstraete et J. Couvert, avec la collaboration de R. Van Schoute et A. Dubois, Louvain, 2006, p. 181-184.

Étude technique d'une *Crucifixion* conservée au Museo Civico de Bassano del Grappa.

1-4. Peintre anonyme inspiré par l'école de Leyde, *Crucifixion*, Bassano del Grappa, Museo Civico. IRR (détail: ville dans le paysage, personnage à droite, soldat à droite, figure féminine sous-jacente dans le paysage)

I.R.A.2006.34

A.M. Mesquita e Carmo et P.M. Antunes de Sousa, *La Résurrection de Lazare attribuée à Jorge Afonso. Étude du dessin sous-jacent original et non original*, dans *Le dessin sous-jacent et la technologie dans la peinture. Colloque XV. La peinture ancienne et ses procédés: copies, répliques, pastiches*, éd. par H. Verougstraete et J. Couvert, avec la collaboration de R. Van Schoute et A. Dubois, Louvain, 2006, p. 242-248.

1-6. Jorge Afonso, *Résurrection de Lazare*, Tornar, Charola. IRR montage (détails: têtes du Christ et des apôtres derrière lui, tête des apôtres à droite, tours dans le paysage à droite (2), personnage à gauche, groupe de Lazare)

I.R.A.2006.35

D. Meuwissen, *A «Painter in Black and White»: The Symbiotic Relationship Between the Paintings and Woodcuts of Jacob Cornelisz. van Oostsanen*, dans *Making and Marketing: Studies of the Painting Process in Fifteenth- and Sixteenth-Century Netherlandish Workshops*, éd. M. Faries, Turnhout, 2006, p. 55-81.

1-2. Attribué à Jacob Cornelisz. van Oostsanen, *Crucifixion*, New York, collection privée. IRR (détails: groupe de la Vierge, Vierge et saint Jean)

3. Attribué à Jacob Cornelisz. van Oostsanen, *Nativité*, Naples, Museo e Galleria Nazionale di Capodimonte. IRR (détails: Joseph)

4. Attribué à Jacob Cornelisz. van Oostsanen, *Noli me tangere*, Kassel, Staatliche Museen. IRR montage (détail: Christ)

I.R.A.2006.36

B. Natter, *Het Joodse bruidje. De vroegst bekende duiding opnieuw bekeken*, dans *Bulletin van het Rijksmuseum*, 54, 2006, p. 162-181.

Interprétation de *La fiancée juive* de Rembrandt (Amsterdam, Rijksmuseum) reconsidérée.

1. Rembrandt, *La fiancée juive*, Amsterdam, Rijksmuseum. IR (ens.)

I.R.A.2006.37

M. Pietrogiovanna, *Le rôle du dessin sous-jacent dans deux tableaux flamands conservés au musée de Vicence: Le Portrait d'un jeune homme à l'oeillet et la Madone aux cerises*, dans *Le dessin sous-jacent et la technologie dans la peinture. Colloque XV. La peinture ancienne et ses procédés: copies, répliques, pastiches*, éd. par H. Verougstraete et J. Couvert, avec la collaboration de R. Van Schoute et A. Dubois, Louvain, 2006, p. 207-211.

1-2. Atelier de Hans Memling, *Portrait de jeune homme à l'œillet*, Vicence, Pinacoteca Civica. IRR (détail: visage, main)
3. Hans Memling, *Jeune homme en prière*, Londres, National Gallery. IRR (détail: mains)
4. Copie d'après Josse van Cleve, *Madone aux cerises*, Vicence, Pinacoteca Civica. IRR (ens.)

I.R.A.2006.38

C. PLAZZOTTA, M. O'MALLEY, A. ROY, R. WHITE et M. WYLD, *The Madonna di Loreto: An Altarpiece by Perugino for Santa Maria dei Servi, Perugia*, dans *National Gallery Technical Bulletin*, 27, 2006, p. 72-95.

1-2. Perugino, *Madonna di Loretto*, Londres, National Gallery. LO (détails: genoux de l'Énfant, ailes de l'ange de droite, dessin sous-jacent visible à l'œil nu)
3-4. Idem. IRR (détails: ceinture de Enfant, lys tenu par l'ange de droite)
5. Perugino, *Déposition*, Florence, Galleria dell'Accademia. IRR (détail: Madeleine)
6. Perugino, *Madonna di Loretto*, Londres, National Gallery. IRR (détail: pieds et bas du drapé de l'ange de droite)

I.R.A.2006.39

L.B. RØNBERG et J. WADUM, *Two paintings in Copenhagen re-attributed to Rembrandt*, dans *Burlington Magazine*, 148, 2006, p. 82-88.

1. Rembrandt (attribué à), *Le Croisé*, Copenhague, Statens Museum for Kunst. IRR (ens.)
2. Rembrandt, *Chevalier au faucon (Saint Bavon?)*, Göteborgs Konstmuseum. IRR (ens.)

I.R.A.2006.40

P. SCHATBORN, *Drawings attributed to Carel Fabritius*, dans *Oud Holland*, 119, 2006, p. 130-138.

1. Carel Fabritius, *Hagar et l'ange*, Vienne, Collection Schönborn-Buchheim. IRR (détail: ange)

I.R.A.2006.41

L. SHELDON et N. COSTARAS, *Johannes Vermeer's «Young Woman seated at a Virginal»*, dans *Burlington Magazine*, 148, 2006, p. 89-97.

1. Johannes Vermeer, *Jeune fille au virginal*, USA, collection privée. IR (ens.)

I.R.A.2006.42

P. SILVA, *Le dessin sous-jacent de deux peintures eyckiennes du Musée du Prado:* Le Triomphe de l'Église sur la Synagogue, *École de Van Eyck, et* Saint François recevant les Stigmates, *du maître d'Hoogstraten*, dans *Le dessin sous-jacent et la technologie dans la peinture. Colloque XV. La peinture ancienne et ses procédés: copies, répliques, pastiches*, éd. par H. VEROUGSTRAETE et J. COUVERT, avec la collaboration de R. VAN SCHOUTE et A. DUBOIS, Louvain, 2006, p. 42-50.

Étude du dessin sous-jacent de deux peintures conservées au Musée du Prado, à savoir le *Triomphe de l'Église sur la Synagogue* de l'école de van Éyck et *Saint François recevant les stigmates* du Maître d'Hoogstraten.

1-4. École de Van Eyck, *Triomphe de l'Église*, Madrid, Museo del Prado. IRR (détails: coin inférieur droit, groupe de l'Église, coin inférieur gauche, fontaine)

I.R.A.2006.43

R. Spronk et C. Van Daalen, *Two Scenes from the Passion at the Harvard Art Museums: a Tale of Two Antwerp Workshops?*, dans *Making and Marketing: Studies of the Painting Process in Fifteenth- and Sixteenth-Century Netherlandish Workshops*, éd. M. Faries, Turnhout, 2006, p. 43-53.

1. Anonyme, *Christ au jardin de Gethsemane*, Cambridge (MA.), Fogg Art Museum. IRR (ens.)
2. Anonyme, *Arrestation du Christ*, Cambridge (MA.), Busch-Reisinger Museum. IRR (ens.)
3-4. Anonyme, *Christ au jardin de Gethsemane*, Cambridge (MA.), Fogg Art Museum. IRR (détails: barrière, inscription)
5. Anonyme, *Arrestation du Christ*, Cambridge (MA.), Busch-Reisinger Museum. IRR (détail: inscription)
6. Anonyme, *Christ au jardin de Gethsemane*, Cambridge (MA.), Fogg Art Museum. IRR (détail: bras droit du Christ)

I.R.A.2006.44

P. Syfer-d'Olne, R. Slachmuylders, A. Dubois, B. Fransen et F. Peters (avec la collaboration de G. Patigny et N. Toussaint), *The Flemish Primitives. IV. Masters with Provisional Names. Catalogue of Early Netherlandish Painting in the Royal Museums of Fine Arts of Belgium*, Bruxelles, 2006.

1-9. Maître de la Vie de Joseph, *Retable d'Affligem*, inv. 344-345, 347-354. IRR (détails — *Christ parmi les Docteurs*: buste du Christ; *Présentation au Temple*: groupe à gauche; *Mise au tombeau*: Christ; *Christ parmi les Docteurs*: coin inférieur droit; *Portement de croix*: Christ; *Crucifixion*: Marie-Madeleine; *Christ parmi les Docteurs*: Fuite en Égypte; *Portement de croix*: soldat, ciel)
10-12. Maître de la Vie de Joseph, *Triptyque de Zierikzee*, inv. 2405, 4168, 2406. IRR (détails — volet gauche: buste de Philippe le Beau; volet droit: buste de Jeanne de Castille; volet gauche: pied de Philippe le Beau)
13-16. Maître de la Légende de sainte Madeleine, *Triptyque de la Famille Du Quesnoy-Van der Tommen*, inv. 1330. IRR (détails-panneau central: ange; volet droit: donatrice et sainte Marie l'Égyptienne; volet gauche: donateur et saint Simon Zelote; volet droit: filles des donateurs)
17-21. Maître de la Légende de sainte Madeleine (atelier de), *Auguste et la Sibyle de Tibur*, inv. 8619. IRR (détails: Sibyle et deux femmes à gauche, deux hommes à droite, drapé d'Auguste, Vierge et Enfant, paysage à gauche)
22. Anonyme bruxellois, *Vierge et Enfant*, inv. 8535. IRR (ens.)
23. Anonyme (d'après Pieter van Coninxloo?), *Portrait de Philippe le Beau*, inv. 6355. IRR (ens.)
24. Anonyme, *Portrait de Jean de Luxembourg, seigneur de Ville?*, inv. 3827. IRR (ens.)
25-27. Entourage de Colijn de Coter, *Nativité avec donateur religieux et Circoncision*, inv. 364. IRR (détails: visages de personnages dans la partie supérieure, Vierge, donateur)
28-31. Maître de la Légende de sainte Barbe et collaborateur, *Scènes de la Légende de sainte Barbe*, inv. 6149. IRR (détails: Sainte Barbe et l'ange, sainte Barbe devant le juge, paysage à droite, sainte Barbe poursuivie par son père)
32-37. Maître de la Légende de sainte Catherine, *Scènes de la Légende de sainte Catherine*, inv. 12102. IRR (détails: messager au premier plan, visage de sainte Catherine, personnages dans le bâtiment à gauche, drapé de sainte Catherine agenouillée devant l'icône de la Vierge, paysage à gauche, toit du bâtiment à gauche)

38. Maître de la Légende de sainte Catherine, *Nativité/Fragment d'un drapé*, inv. 10513. IRR (détail: Vierge et anges)

39-46. Anonyme brugeois, *Vierge et Enfant parmi des saintes*, inv. 2576. IRR (détails: Vierge, sainte Christine, ange à gauche, sainte Madeleine, robe de la Madeleine, partie supérieure droite, sainte Agnès, sainte avec un turban)

47-48. Anonyme brugeois, *Vierge et Enfant*, inv. 10816. IRR (détails: visage de la Vierge, Enfant)

49-54. Anonyme brugeois, *Sainte Anne, la Vierge et l'Enfant entourés de saints*, inv. 6719. IRR (détails: Vierge et Enfants, saint Jean-Baptiste, livre de sainte Catherine, sainte Barbe, Baptême du Christ, sainte Anne)

55. Maître à la Vue de Sainte-Gudule, *Présentation de la Vierge au temple*, inv. 4348. IRR (détail: Anne et Joachim)

56. Maître à la Vue de Sainte-Gudule, *Instruction Pastorale*, Paris, Musée du Louvre. IR (détail: prêcheur à droite)

57. Maître à la Vue de Sainte-Gudule (atelier de), *Mariage de la Vierge*, inv. 7559. IR (ens.)

58. Idem. IRR (détail: Joseph)

59-62. Maître de 1473, *Triptyque de Jan de Witte*, inv. 7007. IRR (détails: Vierge et Enfant, buste du donateur, buste de la donatrice, drapé de la Vierge)

I.R.A.2006.45

El Trazo oculto. Dibujos subyacentes en pinturas de los siglos XV y XVI (cat. expo.), éd. G. FINALDI et C. GARRIDO, Madrid, 2006.

1. Robert Campin, *Saint Jacques le Majeur et sainte Claire*, Madrid, Museo del Prado. IRR (ens.)

2-11. Idem. IRR (détails: base de la colonne gauche, détails du drapé de saint Jacques (4), pieds de saint Jacques, bas du drapé de sainte Claire, tête de sainte Claire, tête de saint Jacques, drapé de sainte Claire)

12-14. Hans Memling, *Triptyque de l'Adoration des Mages*, Madrid, Museo del Prado. IRR (ens. du panneau central, ens. de la *Nativité*, ens. de la *Présentation au Temple*)

15-24. Idem. IRR (détails-panneau central: Vierge et Enfant, pieds du mage noir à droite, schéma montrant la superposition des deux perspectives du sol, Vierge et roi âgé, buste du roi noir, drapeaux (2); *Présentation au temple*: architecture, groupe central, femme derrière l'Enfant)

25. Hieronymus Bosch, *Table avec les Péchés mortels*, Madrid, Museo del Prado. IRR (ens.)

26-31. Idem. IRR (détails-scène de la Mort: ange; scène de l'Orgueil: ens.; scène de la Luxure: paysage; scène de l'Avarice: personnage avec une pièce de monnaie en main; scène de la Mort: ens.; scène du jugement dernier: apôtres à droite)

32. Hieronymius Bosch, *Adoration des Mages*, Madrid, Museo del Prado. IRR (détail: tête de l'Enfant)

33. Hieronymus Bosch, *Table avec les Péchés mortels*, Madrid, Museo del Prado. IRR (détails-scène de l'Enfer: Damné)

34. Jan Gossaert, *Christ entre la Vierge et saint Jean-Baptiste*, Madrid, Museo del Prado. IRR (ens.)

35-44. Idem. IRR (détails: buste du Christ, tête de l'ange, tête de saint Jean-Baptiste, main droite du Christ, visage de la Vierge, doigts de la main droite du Christ, cou du Christ, tunique de l'ange, mains de la Vierge, main droite de l'ange)

45. Jan van Scorel, *Vierge aux roses*, Utrecht, Centraal Museum. IRR (détail: main de la Vierge)

46-48. Jan Gossaert, *Christ entre la Vierge et saint Jean-Baptiste*, Madrid, Museo del Prado. IRR (détails: main gauche de saint Jean-Baptiste, ombre portée de la main de saint Jean-Baptiste, main gauche du Christ)
49. Jan Provost, *Zacharias*, Madrid, Museo del Prado. IRR (ens.)
50-62. Idem. IRR (détails: drapés (2), architecture, paysage, tête du personnage, drapés (3), inscriptions (2), main gauche, ens. du revers, tête du saint au revers)
63-64. Pieter Coecke van Aelst, Saint Jean l'Évangéliste et donatrices, Madrid, Museo del Prado. IRR (ens.; ens. du revers représentant saint Adrien)
65-68. Idem. IRR (détails-revers: lion et jambes du saint; face: donatrices (2), saint Jean l'Évangéliste)
69. Sandro Botticelli et Bartolomeo di Giovanni, *Histoire de Nastagio degli onesti*, scène III, Madrid, Museo del Prado. IRR (ens.)
70. Sandro Botticelli et collaborateur, *Histoire de Nastagio degli onesti*, scène I, Madrid, Museo del Prado. IRR (détail: coin supérieur gauche)
71. Sandro Botticelli et collaborateur, *Histoire de Nastagio degli onesti*, scène II, Madrid, Museo del Prado. IRR (détail: coin supérieur gauche)
72. Sandro Botticelli et Bartolomeo di Giovanni, *Histoire de Nastagio degli onesti*, scène III, Madrid, Museo del Prado. IRR (détail: figures féminines à gauche)
73. Sandro Botticelli et collaborateur, *Histoire de Nastagio degli onesti*, scène I, Madrid, Museo del Prado. IRR (détail: femme attaquée par le chien)
74. Sandro Botticelli et Bartolomeo di Giovanni, *Histoire de Nastagio degli onesti*, scène III, Madrid, Museo del Prado. IRR (détail: personnages au fond à droite)
75. Rafaël, *Sainte Famille*, Madrid, Museo del Prado. IRR (ens.)
76-77. Idem. IRR (détails: pied de la Vierge, Enfant et saint Jean)
78. Rafaël, *Vierge et Enfant avec Jean-Baptiste enfant*, Londres, National Gallery. IRR (ens.)
79. Rafaël, *Sainte Famille*, Madrid, Museo del Prado. IRR (détail: tête de la Vierge)
80. Andrea del Sarto, *Vierge et Enfant entre saint Matthieu et un ange*, Madrid, Museo del Prado. IRR (ens.)
81-85. Idem. IRR (détails: drapé du saint, manche de l'ange, ange, Vierge et Enfant, paysage à gauche)
86. Fernando Gallego, *Piétà*, Madrid, Museo del Prado. IRR (ens.)
87-94. Idem. IRR (détails: manche de la Vierge, main gauche du Christ, têtes de la Vierge et du Christ, donateurs à gauche, notation de couleur, inscription, bâtiment dans le paysage, partie supérieure droite)
95. Fernando Gallego, *Christ Bénissant*, Madrid, Museo del Prado. IRR (ens.)
96-100. Idem. IRR (détails: pieds du Christ, buste du Christ, drapé du Christ, Église, Synagogue)
101. Maître de la Vierge des Rois Catholiques, *Vierge des Rois Catholiques*, Madrid, Museo del Prado. IRR (ens.)
102. Maître de Miraflores, *Baptême du Christ*, Madrid, Museo del Prado. IRR (détail: tête de l'ange)
103-110. Maître de la Vierge des Rois Catholiques, *Vierge des Rois Catholiques*, Madrid, Museo del Prado. IRR (détails: Vierge et Enfant, Isabelle la Catholique, femme de dos, Ferdinand le Catholique, Don Juan, saint Thomas, saint Dominique, Pedro de Arbués (?))
111. Pedro Berruguete, *Apparition de la Vierge à une communauté de Dominicains*, Madrid, Museo del Prado. IRR (ens.)
112-115. Idem. IRR (détails: laïcs dans la porte, moine au premier plan à droite, moine au premier plan à gauche, moine au centre)
116. Pedro Berruguete, *Mort de saint Pierre Martyr*, Madrid, Museo del Prado. IRR (ens.)

117-123. Idem. IRR (détails: chasseur, paysage (3), frère Dominique, saint, paysage)
124. Taller Castellano, *Christ présenté au peuple*, Madrid, Museo del Prado. IRR (ens.)
125-126. Idem. IRR (détails: cavalier à gauche, cavalier à droite)
127. Alejo Fernández, *Flagellation*, Madrid, Museo del Prado. IRR (ens.)
128-133. Idem. IRR (détails: groupe central, soldat à droite, architecture, personnages sur la galerie, architecture, roi)
134. Fernando Yáñez de la Almedina, *Sainte Anne, La Vierge et Enfant, sainte Isabelle et Jean-Baptiste enfant*, Madrid, Museo del Prado. IRR (ens.)
135-139. Idem. IRR (détails: partie supérieure de la scène, Vierge et Enfant, Jean-Baptiste, iris, personnage dans la scène de la rencontre à la Porte Dorée)

I.R.A. 2006.46

C.M. VAN DAALEN, *Preparations and Materials: Methods of compositional transfer*, dans *Le dessin sous-jacent et la technologie dans la peinture. Colloque XV. La peinture ancienne et ses procédés: copies, répliques, pastiches*, éd. par H. VEROUGSTRAETE et J. COUVERT, avec la collaboration de R. VAN SCHOUTE et A. DUBOIS, Louvain, 2006, p. 212-217.

Présentation d'exemples de deux moyens de transfert mécanique dans la peinture flamande des 16e et 17e siècles, à savoir le calque et le poncif.

I.R.A.2006.47

W. VERBAKEL et A. WALLERT, *Rembrandt lachend?*, dans *Bulletin van het Rijksmuseum*, 54, 2006, p. 130-143.

1. Rembrandt, *Autoportrait*, Nuremberg, Germanisches Nationalmuseum. IRR (ens.)
2. *Portrait d'un jeune homme riant*, Amsterdam, Rijksmuseum. IR (ens.)
3. Jan Lievens, *Portrait de Rembrandt*, Amsterdam, Rijksmuseum. IRR (détail: visage)

I.R.A.2006.48

I. VERSLYPE et G. WUESTMAN, *Een groepsportret met een bewogen geschiedenis. Moderne onderzoekstechnieken leiden tot nieuwe toeschrijving*, dans *Bulletin van het Rijksmuseum*, 54, 2006, p. 400-411.

1. Willem Cornelisz. Duyster et un peintre anonyme, *Portrait de groupe avec un Africain*, Amsterdam, Rijksmuseum. IRR (détail: homme au centre)

I.R.A.2006.49

J. WADUM et L. BOGH RONBERG, *A Unique Rembrandt Rediscovered: «The Crusader». An oil sketch for «The Knight with the Falcon» (St. Bavo, 1659-1665)*, dans *Restauro*, 2006, p. 110-117.

1. Rembrandt, *Le croisé*, Copenhague, Statens Museum for Kunst. IRR (détail: visage)
2. Rembrandt, *Le chevalier au faucon*, Göteborg, Göteborgs Konstmuseum. IRR (détail: visage)

I.R.A.2006.50

A. WALLERT, *Rembrandt geretucee... Lieve?*, dans *Bulletin van het Rijksmuseum*, 54, 2006, p. 122-129.

Grâce à une inscription récemment découverte, il est apparu que l'œuvre étudiée était une peinture réalisée par Jan Lievens et retouchée par Rembrandt.

1. Jan Lievens et Rembrandt, *Tête d'enfant*, Amsterdam, Rijksmuseum. IRR (détail: inscription)
2. Idem. IRR (ens.)

I.R.A.2006.51

M. Wolters, *Creativity and Efficiency: Aspects of Joachim Beuckelaer's Use of Patterns and Models*, dans *Making and Marketing: Studies of the Painting Process in Fifteenth- and Sixteenth-Century Netherlandish Workshops*, éd. M. Faries, Turnhout, 2006, p. 155-178.

1-3. Joachim Beuckelaer, *Scène de marché avec l'Ecce Homo*, Vienne, Museum im Schottenstift. IRR (détail: animal dépecé, tête d'un vendeur de fruit, chien dessiné mais pas réalisé au stade pictural)
4. Joachim Beuckelaer, *Scène de marché avec l'Ecce Homo*, Stockholm, Nationalmuseum. IRR (détail: personnages)
5. Joachim Beuckelaer, *Scène de marché avec l'Ecce Homo*, Florence, Galleria degli Uffizi. IRR (détail: personnages)
6. Pieter Aertsen, *Cuisinière*, Bruxelles, Musées royaux des Beaux-Arts de Belgique. IRR (détail: légumes)
7-15. Joachim Beuckelaer, *Scène de marché avec l'Ecce Homo*, Florence, Galleria degli Uffizi. IRR (détail: canard, indications de couleurs (6), chiffres (2))
16. Joachim Beuckelaer, *Scène de marché avec l'Ecce Homo*, Vienne, Museum im Schottenstift. IRR (détail: architecture)
17. Joachim Beuckelaer, *Scène de marché avec l'Ecce Homo*, Florence, Galleria degli Uffizi. IRR (détail: architecture)
18. Joachim Beuckelaer, *Scène de marché avec l'Ecce Homo*, Stockholm, Nationalmuseum. IRR (détail: architecture)
19. Joachim Beuckelaer, *Scène de marché avec l'Ecce Hom*o, Vienne, Museum im Schottenstift. IRR (détail: personnage féminin)
20-21. Joachim Beuckelaer, *Scène de marché avec l'Ecce Homo*, Stockholm, Nationalmuseum. IRR (détails: couple, point de fuite)

I.R.A.2006.52

M. Wolters, *Two Vegetable Sellers by Joachim Beuckelaer: A Symbiotic Relationship*, dans *Le dessin sous-jacent et la technologie dans la peinture. Colloque XV. La peinture ancienne et ses procédés: copies, répliques, pastiches*, éd. par H. Verougstraete et J. Couvert, avec la collaboration de R. Van Schoute et A. Dubois, Louvain, 2006, p. 159-167.

Comparaison de deux compositions présentant une vendeuse de légumes (Valenciennes et Kassel) réalisées par Joachim Beuckelaer au début de sa carrière. L'étude du dessin sous-jacent a permis à l'auteur de déterminer que les deux œuvres ont été réalisées en même temps et non pas l'une à la suite de l'autre.

1. Joachim Beuckelaer, *Vendeuse de légumes*, Kassel, Staatliche Museen. IRR (détail: buste de la vendeuse)
2. Joachim Beuckelaer, *Vendeuse de légumes*, Valenciennes, Musée des Beaux-Arts. IRR (détail: tablier de la vendeuse)
3. Joachim Beuckelaer, *Vendeuse de légumes*, Kassel, Staatliche Museen. IRR (détail: main droite de la vendeuse)
4. Joachim Beuckelaer, *Vendeuse de légumes*, Valenciennes, Musée des Beaux-Arts. IRR (détail: cercles et demi-cercles à la droite du chou-fleur)

5. Joachim Beuckelaer, *Vendeuse de légumes*, Kassel, Staatliche Museen. IRR (même détail que le précédent)
6. Joachim Beuckelaer, *Vendeuse de légumes*, Valenciennes, Musée des Beaux-Arts. IRR (détail: plat de prunes)
7. Joachim Beuckelaer, *Vendeuse de légumes*, Kassel, Staatliche Museen. IRR (détail: panier de prunes)
8. Joachim Beuckelaer, *Vendeuse de légumes*, Valenciennes, Musée des Beaux-Arts. IRR (détail: plat de mûres)
9. Joachim Beuckelaer, *Vendeuse de légumes*, Kassel, Staatliche Museen. IRR (détail: plat de mûres)

I.R.A.2006-2007.1

M. Bartlová et H. Logan, *«Madonna Protectress» from the National Gallery in Prague*, dans *Bulletin of the National Gallery in Prague*, XVI-XVII, 2006-2007, p. 53-62.

1. Anonyme, *Vierge protecrice*, Prague, Národní Galerie. IRR (détail: pied droit de l'Enfant)

I.R.A.2007.1

M.W. Ainsworth, *The Master of the Embroidered Foliage Group – a Brussels or Bruges Workshop?*, dans *Le Maître au Feuillage brodé. Démarches d'artistes et méthodes d'attribution d'œuvres à un peintre anonyme des anciens Pays-Bas du XVe siècle. Colloque organisé par la Palais des Beaux-Arts de Lille les 23 et 24 juin 2005*, sous la direction de F. Gombert avec la collaboration de C. Heck, D. Martens et J. Satkowski et l'assistance de M.-P. Rose, Lille, 2007, p. 151-163.

1-2. Maître au Feuillage brodé, *Vierge et Enfant*, Williamstown, Sterling and Francine Clark Art Institute. IRR (détails: visage de la Vierge, corps de l'Enfant)

I.R.A.2007.2

Au-delà de l'image. Les techniques du dessin révélées par la science (cat. expo.), Rennes, 2007.

Catalogue d'une exposition réalisée par le Musée des Beaux-Arts de Rennes en association avec le Centre de Recherche et de Restauration des Musées de France. Une partie de la collection de dessins du musée a fait l'objet d'études approfondies au moyen des techniques de laboratoire. Celles-ci sont décrites dans un chapitre introductif.

1. Jacob Toorenvliet, *Femme au Miroir*, inv. 794.1.3410. IR (ens.)
2. Giovanni Francesco Barbieri dit Guerchin, *Étude de femme nue*, inv. 794.1.3138. IR (ens.)
3. Michel Corneille II, *Paysage avec pêcheurs*, inv. 1998.9.19. IR (ens.)
4. Jan van Bisschop, *La Vierge avec Jésus, Elisabeth et Jean-Baptiste*, inv. 794.1.3340. IR (ens.)
5. Idem. IR (détail: bustes de la Vierge et d'Elisabeth)
6. Maître du Carnet de l'Ermitage, *Noces dans un village*, inv. 794.1.3383. IR (ens.)
7. Idem. IR (détail: bâtiment à gauche)
8. Sébastien Bourdon, *Scène de lapidation*, inv. 1998.9.16. IR (ens.)
9-10. Idem. IR (détails: personnages debout au centre, personnage levant une pierre à droite)
11. Raymond de Lafage, *Tobie et l'ange*, inv. 1998.9.44. IR (ens.)
12. Georges Focus ou Faucus, *Paysage*, inv. 794.1.2651. IR (ens.)
13. Pieter With, *Paysage avec pêcheurs*, Paris, Fondation Custodia, Institut Néerlandais, inv. F.C. 7877. IR (ens.)

14. Raffaellino da Reggio, *Martyre de saint Pierre*, inv. 794.1.3054. IR (ens.)
15. Gilles Marie Oppenord, *Le Temps et les médecins*, inv. 794.1.2807. IR (ens.)

I.R.A.2007.3

A. BORN, *Limites de la Kennerschaft de Max Jacob Friedländer: le maniérisme anversois et la notion problématique d'œuvre d'atelier*, dans *Le Maître au Feuillage brodé. Démarches d'artistes et méthodes d'attribution d'œuvres à un peintre anonyme des anciens Pays-Bas du XVe siècle. Colloque organisé par la Palais des Beaux-Arts de Lille les 23 et 24 juin 2005*, sous la direction de F. GOMBERT avec la collaboration de C. HECK, D. MARTENS et J. SATKOWSKI et l'assistance de M.-P. ROSE, Lille, 2007, p. 9-23.

1. Maître de 1418 (?), *Jésus parmi les Docteurs*, Anvers, Museum Mayer van den Bergh. IRR (détail: visage d'un des Docteurs)
2. Goswijn van der Weyden, *Sainte Catherine parmi les Philosophes*, Southampton, City Art Gallery. IRR (détail: groupe des Philosophes)

I.R.A.2007.4

Cranach der Ältere (cat. expo.), éd. B. BRINKMANN, Stuttgart, 2007.

1. Lucas Cranach l'Ancien, *Les quatorze sauveurs*, Torgau, Marienkirche. IRR (détail: visage d'un saint)
2. Lucas Cranach l'Ancien, *Christ couronné d'épines*, collection privée. IRR montage (détail: visage du Christ)

I.R.A.2007.5

S. DI NEPI, A. ROY et R. BILLINGE, *Bernardo Daddi's Coronation of the Virgin: The Reunion of Two Long Separated Panels*, dans *National Gallery Technical Bulletin*, 28, 2007, p. 5-25.

1. Bernardo Daddi, *Couronnement de la Vierge*, Londres, National Gallery. IRR (détail: manteau du Christ)
2. Bernardo Daddi, *Quatre anges musiciens*, Oxford, Christ Church. IRR (détail: anges de gauche)
3. Bernardo Daddi, *Couronnement de la Vierge*, Londres, National Gallery. IRR (détail: main de saint Jean-Baptiste)

I.R.A.2007.6

J. DUNKERTON, *Tintoretto's Underdrawing for Saint George and the Dragon*, dans *National Gallery Technical Bulletin*, 28, 2007, p. 26-35.

1. Jacoppo Tintoretto, *Saint George et le Dragon*, Londres, National Gallery. IRR (ens.)
2-5. Idem. IRR (détails: remparts, tête de la princesse, personnage non réalisé au premier plan à gauche, personnage mort)

I.R.A.2007.7

F. GOMBERT, *Regrouper et exposer les œuvres attribuées à un atelier de maître anonyme*, dans *Le Maître au Feuillage brodé. Démarches d'artistes et méthodes d'attribution d'œuvres à un peintre anonyme des anciens Pays-Bas du XVe siècle. Colloque organisé par la Palais des Beaux-Arts de Lille les 23 et 24 juin 2005*, sous la direction de F. GOMBERT avec la collaboration de C. HECK, D. MARTENS et J. SATKOWSKI et l'assistance de M.-P. ROSE, Lille, 2007, p. 23-31.

1. Groupe au Feuillage brodé, *Triptyque de la Vierge à l'Enfant dans un paysage*, Burgos, Museo Diocesano-Catedralicio. IRR (détail: visages de la Vierge et de l'Enfant)
2. Groupe au Feuillage brodé, *Triptyque de la Vierge à l'Enfant entourée d'anges musiciens*, Lille, Musée des Beaux-Arts. IRR (détail: drapé de la Vierge)
3. Groupe au Feuillage brodé, *Vierge et Enfant dans un paysage*, Philadelphia, Philadelphia Museum of Art. IRR (détail: végétation)
4. Groupe au Feuillage brodé, *Triptyque de la Vierge à l'Enfant dans un paysage*, Burgos, Museo Diocesano-Catedralicio. IRR (détail: bas du drapé de la Vierge)
5. Groupe au Feuillage brodé, *Triptyque de la Vierge à l'Enfant entourée d'anges musiciens*, Lille, Musée des Beaux-Arts. IRR (détail: bas du drapé de la Vierge — la partie inférieure de l'image est photographiée en lumière ordinaire)
6. Groupe au feuillage brodé, *Vierge et Enfant dans un paysage*, Philadelphia, Philadelphia Museum of Art. IRR (détail: paysage)

I.R.A.2007.8

P. Joannides et J. Dunkerton, *A Boy with a Bird in the National Gallery: Two Responses to a Titian Question*, dans *National Gallery Technical Bulletin*, 28, 2007, p. 36-57.

1. Titien ou atelier du Titien?, *Garçon à l'oiseau*, Londres, National Gallery. IRR (ens.)
2. Idem. Diagramme montrant le dessin d'une gravure superposée sur l'IRR.
3. Idem. IRR (détail: tête de l'Enfant)
4. Titien, *Sainte famille avec un berger*, Londres, National Gallery. IRR (détail: Joseph)

I.R.A.2007.9

E. Le Magadure, *Le Saint Christophe du musée Dobrée, fragment d'un triptyque par un élève de Jérôme Bosch*, dans *Revue des Musées de France. Musée du Louvre*, 2007/5, p. 28-39.

Analyse technique d'un fragment de retable conservé au Musée Dobrée de Nantes. Ce fragment est rapproché d'une *Tentation de saint Antoine* (Norfolk, North Carolina, The Chrysler Museum of Art). Ces deux œuvres appartenaient à l'origine à un même retable.

1. Atelier de Hieronymus Bosch, *Saint Christophe*, Nantes, Musée Dobrée. IR (ens.)

I.R.A.2007.10

Le Maître au Feuillage brodé. Démarches d'artistes et méthodes d'attribution d'œuvres à un peintre anonyme des anciens Pays-Bas du XVe siècle. Colloque organisé par la Palais des Beaux-Arts de Lille les 23 et 24 juin 2005, sous la direction de F. Gombert avec la collaboration de C. Heck, D. Martens et J. Satkowski et l'assistance de M.-P. Rose, Lille, 2007, p. 195-210.

Publication des photographies et des réflectographies à l'infrarouge disponibles d'œuvres attribuées au Groupe au Feuillage brodé.

1. Groupe au Feuillage brodé, *Vierge à l'Enfant couronnée par deux anges*, Bruges, Groeningemuseum. IRR (ens.)
2. Groupe au Feuillage brodé, *Triptyque de la Vierge à l'Enfant entourée d'anges musiciens*. IRR (ens. panneau central)
3. Groupe au Feuillage brodé, *Vierge et Enfant dans un jardin*, Minneapolis, Minneapolis Institute of Arts. IRR (ens.)
4. Groupe au Feuillage brodé, *Vierge et Enfant*, Williamstown, Sterling and Francine Clark Art Institute. IR (ens.)

5. Groupe au Feuillage brodé, *Vierge et Enfant dans un paysage*, Philadelphie, Philadelphia Museum of Art. IR (ens.)
6. Groupe au Feuillage brodé, *Triptyque de la Vierge à l'Enfant dans un paysage*, Burgos, Museo Diocesano-Catedralicio. IRR (ens. panneau central)
7. Groupe au Feuillage brodé: anonyme bruxellois, *Vierge entourée d'anges musiciens*, Paris, Musée du Louvre. IR (ens.)
8. Groupe au Feuillage brodé, *Sainte Catherine*, Rotterdam, Museum Boijmans van Beuningen. IRR (ens.)
9. Groupe au Feuillage brodé, *Sainte Barbe*, Rotterdam, Museum Boijmans van Beuningen. IRR (ens.)
10. Anonyme, *Vierge et Enfant*, Munich, collection particulière. IRR (ens.)
11. Albrecht Bouts (attribué à), *Vierge et Enfant*, Louvain, Stedelijk Museum. IRR (ens.)
12-13. Maître des Saints Crépin et Crépinien (Aert van den Bossche?), *Martyre des saints Crépin et Crépinien*, revers des volets gauches: *Saint Crépinien, deux jurés de la guilde des cordonniers en donateurs*, Moscou, Musée d'État des Beaux-Arts Pouchkine. IRR (ens. des deux panneaux)
14-15. Maître des Saints Crépin et Crépinien (Aert van den Bossche?), *Martyre des saints Crépin et Crépinien*, face des volets gauches: *Saints Crépin et Crépinien ébouillantés, Décollation des deux saints*, Bruxelles, Musée de la Ville de Bruxelles — Maison du Roi. IRR (ens. des deux panneaux)
16. Maître des Saints Crépin et Crépinien (Aert van den Bossche?), *Martyre des saints Crépin et Crépinien*, panneau central: *Martyre de saint Crépin et saint Crépinien*, Varsovie, Muzeum Palac w Wilanovie. IRR (ens.)
17. Maître de la Déploration d'Amiens, *Déploration du Christ*, Amiens, Musée de Picardie. IRR (ens.)

I.R.A.2007.11

P. Noble et A. Van Loon, *Renbrandt's Simeon's Song of Praise, 1631 : Pictorial Devices in the Service of Spatial Illusion*, dans *Art Matters*, 4, 2007, p. 19-36.

1. Rembrandt, *Siméon et l'Enfant Jésus dans le Temple*, La Haye, Mauritshuis. IRR (ens.)

I.R.A.2007.12

M. Odlozil et G. Swoboda, *Orazio Gentileschis Büssende Magdalena und die Danae Sauli gemäldetechnologische untersuchungen und vorschlag einer neuen chronologie*, dans *Technologische Studien Kunsthistorisches Museum*, 4, 2007, p. 44-67.

1-2. Orazio Gentileschi, *La Madeleine expiant*, Vienne, Kunsthistorisches Museum. IRR (détails: main droite de la Madeleine, pied)

I.R.A.2007.13

Patinir. Essays and Critical Catalogue (cat. expo), éd. A. Vergara, Madrid, 2007.

1-5. Joachim Patinir, *Charon traversant le Styx*, Madrid, Museo del Prado. IRR (détails: rive à gauche au niveau de la ligne d'hirizon, rochers au premier plan du Paradis, bâtiment du Purgatoire, monstre non peint, partie supérieure du Purgatoire)
6-8. Joachim Patinir, *Paysage avec la destruction de Sodome et Gomorrhe*, Rotterdam, Museum Boymans van Beuningen (dépôt du Netherlands Institute for Cultural Heritage,

Rijswijck, Amsterdam). IRR (détails: rochers et taillis dans la partie centrale, autel, coin supérieur droit)

9-12. Joachim Patinir, *Paysage avec le Repos pendant la Fuite en Égypte*, Berlin, Staatliche Museen. IRR (détails: paysage à gauche, montagne, arbres à droite, Vierge et Enfant)

13-15. Joachim Patinir, *Paysage avec le Repos pendant la Fuite en Égypte*, Madrid, Museo del Prado. IRR (détails: Vierge et Enfant, pannier, Joseph)

16-17. Joachim Patinir et atelier, *Paysage avec le Repos pendant la Fuite en Égypte*, Minneapolis, The Minneapolis Institute of Arts. IRR (détails: rochers à gauche, Vierge et Enfant)

17. Joachim Patinir et atelier, *Paysage avec le Repos pendant la Fuite en Égypte*, Madrid, museo Thissen-Bornemisza. IRR (ens.)

18. Idem. IRR (détail: Vierge et Enfant)

19-21. Joachim Patinir et atelier, Fragment d'un *Paysage avec le Repos pendant la Fuite en Égypte*, Rotterdam, Museum Boijmans Van Beuningen. IRR (détails: arbres, lac et château)

22-25. Joachim Patinir, *Paysage avec le Baptême du Christ*, Vienne, Kunsthistorisches Museum. IRR (détails: rochers à droite, Dieu le Père et colombe, saint Jean-Baptiste, Christ)

26. Cercle de Joachim Patinir, *Paysage avec la Crucifixion*, Madrid, collection privée. IRR (détail: Crucifixion)

27. Joachim Patinir, *Paysage avec l'Assomption de la Vierge*, Philadelphia, Philadelphia Museum of Art. IRR (détail: ange)

28-30. Joachim Patinir et Quentin Massys, *Paysage avec la Tentation de saint Antoine*, Madrid, Museo del Prado. IRR (détails: saint Antoine dans le bois à droite, bâteau, saint Antoine et deux tentatrices)

31. Joachim Patinir, *Paysage avec le Martyre de sainte catherine*, Vienne, Kunsthistorisches Museum. IRR (ens.)

32-35. Idem. IRR (détails: rochers à gauche, partie de la ville avec bâtiment rond, martyre de sainte Catherine, rocher près de la scène de martyre)

36-38. Joachim Patinir, *Paysage avec saint Christophe*, Escorial, Real Monasterio de San Lorenzo, Patrimonio Nacional. IRR montage (détails: ermite, arbre, échelle)

39. Joachim Patinir, *Paysage avec saint Jérôme*, Karlsruhe, Staatliche Kunsthalle. IRR (ens.)

40-47. Joachim Patinir, *Triptyque avec la Pénitence de saint Jérôme, le Baptême du Christ et la Tentation de saint Antoine,* New York, The Metropolitan Museum of Art. IRR (détails-panneau central: saint Jérôme, montagnes; volet gauche: Christ et saint Jean-Baptiste; volet droit: saint Antoine; revers du volet gauche: sainte Anne Trinitaire; revers du volet droit: saint Sebald, revers du volet gauche: Vierge et Enfant; revers du volet droit: main du saint)

48-55. Joachim Patinir, *Paysage avec saint Jérôme*, Madrid, Museo del Prado. IRR (détails: saint Jérôme et le lion, paysage et ciel (3), village au centre, rochers à gauche (2), église à gauche)

56-59. Joachim Patinir et atelier, *Triptyque avec saint Jérôme, saint Jean, saint Antoine et Marie-Madeleine*, Suisse, collection privée. IRR (détails-panneau central: paysage (2), saint Jérôme et le lion, paysage (2))

60. Joachim Patinir, *Paysage avec saint Jérôme*, Paris, Musée du Louvre. IRR (ens.)

61. Idem. IRR (détail: frondaison de l'arbre à gauche)

62-66. Joachim Patinir et atelier, *Paysage avec la Prédication de saint Jean-Baptiste*, Philadelphia, Philadelphia Museum of Art. IRR (détails: paysage (2), saint Jean-Baptiste, rochers (2))

67. Joachim Patinir et atelier, *Paysage avec la Prédication de saint Jean-Baptiste*, Bruxelles, Musées royaux des Beaux-Arts de Belgique. IRR montage (détail: saint Jean-Baptiste)

I.R.A.2007.14

C. Périer-D'Ieteren, *Le Maître au Feuillage brodé et les pratiques d'atelier. Du panneau de La Vierge à l'Enfant de Philadelphie au Retable de la Vierge à l'Enfant de Polizzi Generosa*, dans *Le Maître au Feuillage brodé. Démarches d'artistes et méthodes d'attribution d'œuvres à un peintre anonyme des anciens Pays-Bas du XVe siècle. Colloque organisé par la Palais des Beaux-Arts de Lille les 23 et 24 juin 2005*, sous la direction de F. Gombert avec la collaboration de C. Heck, D. Martens et J. Satkowski et l'assistance de M.-P. Rose, Lille, 2007, p. 121-134.

1-2. Maître au Feuillage brodé, *Vierge et Enfant dans un paysage*, Philadelphia, Philadelphia Museum of Art. IRR (détail: buste de la Vierge et de l'Enfant, livre et mains de la Vierge et de l'Enfant)
3. Maître au Feuillage brodé (?), *Triptyque de la Vierge à l'Enfant entourée d'anges musiciens*, Lille, Musée des Beaux-Arts. IRR (détail: drapé de la Vierge)
4-5. Maître au Feuillage brodé, *Vierge et Enfant dans un paysage*, Philadelphia, Philadelphia Museum of Art. IRR (détail: drapé de la Vierge, pieds de l'Enfant)
6. Idem. IR (détail: paysage)
7. Maître au Feuillage brodé (?), *Triptyque de la Vierge à l'Enfant entourée d'anges musiciens*, Lille, Musée des Beaux-Arts. IRR (détail: paon)
8-9. Maître au Feuillage brodé (?), *Vierge et Enfant*, Minneapolis, Minneapolis Institute of Art. IR (détails du paysage)
10. Groupe au Feuillage brodé: Colijn de Coter (?), *Triptyque de la Vierge à l'Enfant entourée d'anges musiciens, de sainte Barbe et de sainte Catherine*, Polizzi Generoza, Chiesa Madre. IR (détail: bas du drapé de la Vierge)
11. Colijn de Coter, *Trinité*, Paris, Musée du Louvre. IRR (détail: drapé de l'ange en bas à gauche)

I.R.A.2007.15

A. Rodríguez-López, N. Khandekar, G. Gates et R. Newman, *Materials and Techniques of a Spanish Renaissance Panel Painting*, dans *Studies in Conservation*, 52, 2007, p. 81-100.

Étude technique d'un panneau attribué à Bartolomé Bermejo.

1-3. Bartolomé Bermejo (attribué à), *Saint Jérôme* et *Saint Michel*, New York, Hispanic Society of America. IRR (détail: drapé de saint Jérôme, colonne près de saint Jérôme, colonne près de saint Michel)

I.R.A.2007.16

C. Van der Elst et M. Leeflang, *Triptych with The Dancing Shepherds: Unravelling Mysteries of Production and Construction*, dans *Art Matters*, 4, 2007, p. 6-18.

1-3. Anonyme, *Triptyque avec des bergers dansant*, Pays-Bas, collection privée. IRR (détails-panneau central: bergers au premier plan gauche; volet gauche: buste de saint André, têtes des fils du donateur)

I.R.A.2007.17

A. Von Ulmann, *Rembrandt gepaust. «Das gleichnis der Arbeiter im Weinberg» in der St. Petersburger Ermitage und eine weitere neu entdechte Version*, dans *Anzeiger des Germanischen Nationalmuseums*, 2007, p. 85-102.

Analyse de laboratoire de deux versions de la *Parabole des travailleurs dans le vignoble*, l'une conservée dans une cllection privée de Munich, l'autre au Musée de l'Ermitage de Saint Petersbourg.

1. Rembrandt, *Les travailleurs dans le vignoble*, Munich, collection privée. IRR (ens.)
2. Rembrandt, *Les travailleurs dans le vignoble*, Saint-Petersbourg, Musée de l'Ermitage. IRR (ens.)
3. Rembrandt, *Les travailleurs dans le vignoble*, Munich, collection privée. IRR (détail: partie supérieure droite)
4. Rembrandt, *Les travailleurs dans le vignoble*, Saint-Petersbourg, Musée de l'Ermitage. IRR (détail: même détail que le précédent)
5. Rembrandt, *Les travailleurs dans le vignoble*, Munich, collection privée. IRR (déail: personnage à droite du groupe)
6. Rembrandt, *Les travailleurs dans le vignoble*, Saint-Petersbourg, Musée de l'Ermitage. IRR (détail: même détail que le précédent)
7. Rembrandt, *Les travailleurs dans le vignoble*, Munich, collection privée. IRR (détail: groupe des travailleurs)
8. Rembrandt, *Les travailleurs dans le vignoble*, Saint-Petersbourg, Musée de l'Ermitage. IRR (détail: même détail que le précédent)
9. Rembrandt, *Les travailleurs dans le vignoble*, Munich, collection privée. IRR (détail: homme à table)
10. Rembrandt, *Les travailleurs dans le vignoble*, Saint-Petersbourg, Musée de l'Ermitage. IRR (détail: même détail que le précédent)
11. Rembrandt, *Les travailleurs dans le vignoble*, Munich, collection privée. IRR (détail: zone du chat)
12. Rembrandt, *Les travailleurs dans le vignoble*, Saint-Petersbourg, Musée de l'Ermitage. IRR (détail: même détail que le précédent)
13. Rembrandt, *Les travailleurs dans le vignoble*, Munich, collection privée. IRR (détail: personnages à gauche)
14. Rembrandt, *Les travailleurs dans le vignoble*, Saint-Petersbourg, Musée de l'Ermitage. IRR (détail: même détail que le précédent)

MÉTHODOLOGIE

I.R.M.2006.1

C. Garrido, *El trazo oculto. Dibujos subyacentes en las tablas del Museo del Prado*, dans *El Trazo oculto. Dibujos subyacentes en pinturas de los siglos XV y XVI* (cat. expo.), éd. G. Finaldi et C. Garrido, Madrid, 2006, p. 16-53.

Voir: I.R.A.2006.21

I.R.M.2007.1

Au-delà de l'image. Les techniques du dessin révélées par la science (cat. expo.), Rennes, 2007.

Catalogue d'une exposition réalisée par le Musée des Beaux-Arts de Rennes en association avec le Centre de Recherche et de Restauration des Musées de France. Une partie de la collection de dessins du musée a fait l'objet d'études approfondies au moyen des techniques de laboratoire. Celles-ci sont décrites dans un chapitre introductif. Voir: I.R.A.2007.2.

PROCÉDURES

I.R.P.2006.1

R. Bellucci et C. Frosinini, *The Replicas after the Venus and Cupid cartoon by Michelangelo*, dans *Le dessin sous-jacent et la technologie dans la peinture. Colloque XV. La peinture ancienne et ses procédés : copies, répliques, pastiches*, éd. par H. Verougstraete et J. Couvert, avec la collaboration de R. Van Schoute et A. Dubois, Louvain, 2006, p. 51-58.

Étude de quatre versions tirées d'un carton représentant Vénus et Cupidon réalisé par Michel-Ange au moyen d'un scanner digital infrarouge à haute définition mis au point au Istituto Nazionale di Ottica Applicata de Florence. Voir: I.R.A.2006.4.

I.R.P.2006.2

D. Bertani, *La reflectografía infrarroja*, dans *El Trazo oculto. Dibujos subyacentes en pinturas de los siglos XV y XVI* (cat. expo.), éd. G. Finaldi et C. Garrido, Madrid, 2006, p. 54-63.

Explicitation de la technique de la réflectographie à l'infrarouge dans le cadre d'une exposition consacrée à l'étude du dessin sous-jacent qui s'est tenue au Museo del Prado de Madrid, ainsi que de l'évolution de cette technique.

1. Albrecht Dürer, *Eve*, Madrid, Museo del Prado. IRR (détail: main droite)
2. Pedro Berruguete, *Le sépulchre de saint Pierre martyr*, Madrid, Museo del Prado. IRR (détail: mise en perspective du carrelage)
3. Giovanni Bellini, *Vierge et Enfant*, Milan, Pinacoteca di Brera. IR (détail: manteau de la Vierge)
4. Idem. IRR équipé avec un detecteur CCD InGaAs (même détail que le précédent)
5. Fernando Gallego, *Christ bénissant*, Madrid, Museo del Prado. IRR montage (détail: ange)
6. Idem. IRR équipé avec un detecteur CCD InGaAs (même détail que le précédent)

I.R.P.2006.3

E. Buzzegoli et D. Kunzelman, *Re-Use of Cartoons on Paintings by Andrea del Sarto and Pontormo : A Study with High Resolution Digital Scanned IRR*, dans *Le dessin sous-jacent et la technologie dans la peinture. Colloque XV. La peinture ancienne et ses procédés : copies, répliques, pastiches*, éd. par H. Verougstraete et J. Couvert, avec la collaboration de R. Van Schoute et A. Dubois, Louvain, 2006, p. 67-78.

Étude de quelques peintures florentines du 16e siècle présentant des sujets similaires au moyen d'un scanner digital infrarouge à haute définition mis au point au Istituto Nazionale di Ottica Applicata de Florence.

1. Jacopo del Pontormo, *Légende des 10.000 martyrs*, Florence, Galleria Palatina di Palazzo Pitti. IR scan (ens.)
2. Agnolo Bronzino, *Légende des 10.000 martyrs*, Florence, Galleria degli Uffizi. IR scan (ens.)
3. Jacopo del Pontormo, *Légende des 10.000 martyrs*, Florence, Galleria Palatina di Palazzo Pitti. IR scan (détail)
4. Agnolo Bronzino, *Légende des 10.000 martyrs*, Florence, Galleria degli Uffizi. IR scan (même détail que le précédent)
5. Andrea del Sarto, *Vierge et Enfant*, Florence, Galleria Palatina di Palazzo Pitti. IR scan (ens.)
6. Attribué à l'école d'Andrea del Sarto, *Vierge et Enfant avec le jeune saint Jean*, Florence, Appartamenti Reali di Palazzo Pitti. IR scan (ens.)

7. Idem. IR scan (détail de la peinture dont le dessin correspond aux Vierges et Enfants citées en 5 et 6)

I.R.P.2007.1

G. AITKEN et P. LE CHANU, *Le Maître au Feuillage brodé: base de donnée EROS et études en laboratoire*, dans *Le Maître au Feuillage brodé. Démarches d'artistes et méthodes d'attribution d'œuvres à un peintre anonyme des anciens Pays-Bas du XVe siècle. Colloque organisé par la Palais des Beaux-Arts de Lille les 23 et 24 juin 2005*, sous la direction de F. GOMBERT avec la collaboration de C. HECK, D. MARTENS et J. SATKOWSKI et l'assistance de M.-P. ROSE, Lille, 2007, p. 113-121.

Présentation de la base de donnée EROS contenant les documents techniques réalisés par le Centre de Recherche et de Restauration des Musées de France.

1. Groupe au Feuillage brodé, *Vierge et Enfant entourée d'anges musiciens*, Paris, Musée du Louvre. IR (détail: drapé de la Vierge)
2-6. Groupe au Feuillage brodé (Maître de la Déploration d'Amiens), *Déploration du Christ*, Amiens, Musée de Picardie. IRR (détails: arbres, maison et berger, berger, drapé de saint Jean, Christ et saint Jean)

ADDENDUM

I.R.A.1996.37

B. CORLEY, *Conrad von Soest. Painter among Merchant Princes*, Londres, 1996.

1-7. Conrad von Soest, *Retable de Niederwildungen*, Bad Wildungen, Stadtkirche. IR (détails-panneau central-*Christ au mont des Oliviers*: drapé de saint Pierre; *Crucifixion*: bon larron; volet gauche-*Adoration des Mages*: drapé de la Vierge; panneau central-*Ascension*: ensemble; *Crucifixion*: drapé de la sainte femme à gauche, Vierge; *Résurrection*: soldat)
8-10. Conrad von Soest, *Retable de Dortmund*, Dortmund, Evangelische St. Marienkirche. IR (détails-*Mort de la Vierge*: apôtre lisant, rouleau de parchemin, anges à la tête du lit)
11. Conrad von Soest, *Sainte Dorothée*, Münster, Landesmuseum für Kunst und Kulturgeschichte. IR (détail: drapé)
12. Peintre de Fröndenberg, atelier de Conrad von Soest, *Retable de Fröndenberg*, fragment représentant la Vierge et Enfant, Dortmund, Museum für Kunst and Kulturgeschichte. IR (détail: buste de la Vierge et Enfant)
13-14. Peintre de Fröndenberg, atelier de Conrad von Soest, *Retable de Fröndenberg*, Fröndenberg, Stiftskirche. IR (détails-*Nativité*: Joseph, Vierge et Enfant)
15-16. Conrad von Soest, *Panneau de saint Nicolas*, Soest, Chapelle Saint-Nicolas. IR (détails: trois vierges, quatre clercs)
17-18. Maître de Berswordt et atelier, *Retable de Bielefeld*, Bielefeld, Neustäder Marienkirche. IR (détails-*Résurrection*: ensemble; *Déposition*: ensemble)
19. Maître de Berswordt, *Retable de Berswordt*, Dortmund, Evangelische St. Marienkirche. IR (détail-panneau central: Pilate et bas de la croix)
20. Maître de Warendorf, *Retable de Warendorf*, revers des volets: Saint Laurent devant le Juge, Münster, Landesmuseum. IR (détail: drapé du Juge)
21. Maître de Warendorf, *Retable de Warendorf*, Warendorf, Pfarrkirche St. Lawrence. IR (détail-panneau central: drapé de la Madeleine)

22-25. Maître des Panneaux Dorés, *Panneaux dorés de Lüneburg*, Hanovre, Niedersächsisches Landesmuseum. IR (détails-*Crucifixion*: saintes femmes; *Dernière Cène*: Christ; *Nativité*: Joseph; *Déposition*: Vierge)

I.R.A.2001.54

H. WINE, *National Gallery Catalogues. The seventeenth century French Paintings*, Londres, 2001.

1. Lubin Baugin, *Sainte Famille avec saint Jean-Baptiste enfant, sainte Elisabeth et trois personnages*, NG 2293. IR (ens.)
2. Sébastien Bourdon, *Retour de l'Arche*, NG 64. IR (inscription au revers de la toile)
3. Gaspard Dughet, *Paysage de tempête*, NG 36. IR (ens.)
4. Gaspard Dughet et Carla Maratta, *Paysage avec l'union de Didon et Enée*, NG 95. IR (détail: figure allégorique dans le ciel)
5. Nicolas Poussin, *La nourriture de Bacchus*, NG 39. IR (détail: satyre)

I.R.A.2003.50

O. KOTKOVÁ, *Zwischen invention und eklektismus: Hans Brosamers Kreuzigung aus dem Jahr 1548*, dans *Acta Historiae Artium*, XLIV, 2003, p. 183-189.

1-2. Hans Brosamer, *Crucifixion*, Vonderau, Museum Fulda. IRR (détails: buste des cavaliers à droite, jambes du cheval à gauche)

I.R.A.2003.51

R. VAN SCHOUTE et H. VEROUGSTRAETE, *Le jardin des Délices dit «du comte de Pomereu»*, dans *Acta Historiae Artium*, XLIV, 2003, p. 151-159.

1. D'après Hieronymus Bosch, *Jardin des Délices dit «du comte de Pomereu»*, Bruxelles, collection privée. IR (détail: femme nue au centre)
2. Idem. IRR montage (même détail que précédent)

3-6. Idem. IRR montage (détails: construction fantastique à gauche, groupe central au premier plan, sirènes, personnages à droite de la pièce d'eau)

I.R.A.2004.29

B. CORLEY, *Einige Bemerkungen zu Conrad von Soest und seiner Werkstatt*, dans *Conrad von Soest. Neue Forshungen über den Maler und die Kulturgeschichte der Zeit um 1400*, éd. B. BUBERL, Bielefeld, 2004, p. 60-82.

1. Conrad von Soest, *Panneau de saint Nicolas*, Soest, Chapelle Saint-Nicolas. IR (détail: trois vierges)
2. Maître de Berswordt, *Retable de Berswordt*, Dortmund, Evangelische St. Marienkirche. IR (détail-panneau central: Pilates et bas de la croix)

I.R.A.2004.30

L. RUIZ GÓMEZ, *El Retablo de la Anunciacion de Juan Correa de Vivar*, dans *Boletín del Museo del Prado*, 40, 2004, p. 6-21.

1-4. Juan Correa de Vivar, *Triptyque de l'Annonciation*, Madrid, Museo del Prado. IRR (détails-revers du volet droit: main gauche de saint Laurent, coin supérieur droit; revers du volet gauche: tête de saint Etienne, main droite)

I.R.A.2004.31

I. SANDNER, *Unterzeichnungen auf Gemälden des Conrad von Soest*, dans *Conrad von Soest. Neue Forshungen über den Maler und die Kulturgeschichte der Zeit um 1400*, éd. B. BUBERL, Bielefeld, 2004, p. 37-59.

Étude du dessin sous-jacent dans les œuvres principales de Conrad von Soest.

1-7. Conrad von Soest, *Retable de Niederwildungen*, Bad Wildungen, Stadtkirche. IRR (détails-panneau central-*Pentecôte*: apôtre à gauche, ensemble; *Résurrection*: ensemble; *Dernière Cène*: ensemble; *Christ au Mont des Oliviers*: ensemble; *Crucifixion*: groupe de la Vierge; revers du volet gauche: saint Jean-Baptiste)
8. Conrad von Soest, *Sainte Dorothée*, Münster, Landesmuseum für Kunst und Kulturgeschichte. IRR (ens.)
9-10. Conrad von Soest, *Panneau de saint Nicolas*, Soest, Chapelle Saint-Nicolas. IRR (détails: donateur, inscription)
11-14. Conrad von Soest, *Retable de Dortmund*, Dortmund, Evangelische St. Marienkirche. IRR (détails-*Nativité*: Vierge et saint Joseph; *Adoration des Mages*: partie supérieure; *Mort de la Vierge*: Vierge; *Couronnement de la Vierge*: Vierge et Christ)
15-18. Atelier de Conrad von Soest, *Retable de Fröndenberg*, Fröndenberg, Stiftskirche. IRR (détails- *Fuite en Égypte*: personnages; *Visitation*: buste des personnages; *Annonciation*: ensemble; *Présentation de la Vierge au Temple*: ensemble)
19. Conrad von Soest, *Sainte Anne*, Essen, Kunstforum der Gallinat-Bank. IRR (ens.)

I.R.A.2004.32

M. STROLZ, *Zu Maltechnik und Restaurierung des Porträts der Jane Seymour von Hans Holbein d.j.*, dans *Technologische Studien Kunsthistorisches Museum*, 1, 2004, p. 8-31.

1-2. Hans Holbein le Jeune, *Portrait de Jane Seymour*, Vienne Kunsthistorisches Museum. IRR (détails: visage, mains)

I.R.A.2004/2005.1

A. BORN, *Antwerp Mannerism: a fashionable style?*, dans *Koninklijk museum voor schone kunsten Antwerpen. Jaarboek*, 2004/2005, p. 20-45.

1. Maître de 1518, *Retable de la Vie de la Vierge*, Lübeck, Marienkirche. IRR (détails: visage de sainte Anne)

I.R.A.2004/2005.2

M. LEEFLANG, *Workshop practices in early sixteenth-century Antwerp studios*, dans *Koninklijk museum voor schone kunsten Antwerpen. Jaarboek*, 2004/2005, p. 232-273.

1. Maître de 1518, *Retable de sainte Anne*, Lübeck, Marienkirche. IRR (détail-*Adoration des Mages*: Vierge et Enfant)
2-4. Joos van Cleve et atelier, *Retable de Reinhold*, Varsovie, Muzeum Narodowe. IRR (détails-*Dernière Cène*: inscription; *Présentation au temple*: trois femmes à gauche; *Christ devant Pilate*: Pilate et ses serviteurs)
5. Maître de 1518, *Retable de sainte Anne*, Lübeck, Marienkirche. IRR (détail-*Anne et Joachim donnant leurs biens terrestres aux pauvres*: pauvres à droite)
6. Adriaen van Overbeke et atelier, *Retable de sainte Anne*, Kempen, Propdteikirche Sankt Maria Geburt. IRR (détail-*Mort de sainte Anne*: sainte Anne)

7. Maître de l'Adoration d'Anvers, *Raising of Lazarus*, Cologne, Wallraf-Richartz-Museum. IRR (détail: partie gauche)
8. Joos van Cleve et atelier, *Retable de Reinhold*, Varsovie, Muzeum Narodowe. IRR (détail-*Présentation au temple*: prêtre)
9. Maître de 1518, *Retable marial*, Stuttgart, Staatliche Kunstsammlungen. IRR (détail-*Retour de Joseph*: ange et Vierge)
10. Joos van Cleve et atelier, *Retable de Reinhold*, Varsovie, Muzeum Narodowe. IRR (détail-*Christ devant Pilate*: Christ et soldats)
11. Maître de l'Adoration d'Anvers, *Multiplication des pains*, Cologne, Wallraf-Richartz-Museum. IRR (détail: deux personnages dessinés mais non peints)
12. Maître de l'Adoration d'Anvers, *Messe de saint Grégoire*, Cologne, Wallraf-Richartz-Museum. IRR (détail: donateurs)

13-14. Joos van Cleve et atelier, *Retable de Reinhold*, Varsovie, Muzeum Narodowe. IRR (détail-*Baptême du Christ*: saint Jean-Baptiste et le Christ; *Crucifixion*: notation de couleurs)

I.R.A.2004/2005.3

M. POSTEC, *La Sainte Anne Trinitaire de Neerlanden. Traitement de restauration et étude matérielle*, dans *Bulletin de l'Institut royal du Patrimoine artistique*, 31, 2004/2005, p. 69-90.

1. Anonyme préeyckien, *Sainte Anne Trinitaire*, Neerlanden, Heilige Maria Magdalenakerk. IRR (détail: visage de sainte Anne)

I.R.A.2004/2005.4

P. VAN DEN BRINK, *The Artist at work: the crucial role of drawings in early sixteenth-century Antwerp workshops*, dans *Koninklijk museum voor schone kunsten Antwerpen. Jaarboek*, 2004/2005, p. 158-231.

1. Adriaen van Overbeke et atelier, *Retable de sainte Anne*, Kempen, Propdteikirche Sankt Maria Geburt. IRR (détail-volet droit: *Mort de sainte Anne*: drapé de sainte Anne)

I.R.P.2003.5

A. GONZÁLEZ MOZO, *El tratamiento digital de imagen en el estudio de las obras de arte*, dans *Boletín del Museo del Prado*, 39, 2003, p. 88-98.

1. Pedro Berruguete, *Ezechiel*, Paredes de Nava (Palencia), Museo de Santa Eulalia. IRR (détail: main d'Ezechiel-détail pris en 1979 en photographiant un écran)
2. Idem (même détail traité digitalement au moyen de filtres)
3. Hieronymus Bosch, *Le chariot de foin*, Madrid, Museo del Prado. IRR montage (détail: Adam et Eve)
4. idem. Même document traité digitalement.

PRINTED ON PERMANENT PAPER • IMPRIME SUR PAPIER PERMANENT • GEDRUKT OP DUURZAAM PAPIER - ISO 9706
N.V. PEETERS S.A., WAROTSTRAAT 50, B-3020 HERENT